T0331783

Classical Analysis

A conceptually clear induction to fundamental analysis theorems, a tutorial for creative approaches for solving problems, a collection of modern challenging problems, a pathway to undergraduate research—all these desires gave life to the pages here.

This book exposes students to stimulating and enlightening proofs and hard problems of classical analysis mainly published in *The American Mathematical Monthly.*

The author presents proofs as a form of exploration rather than just a manipulation of symbols. Drawing on the papers from the Mathematical Association of America's journals, numerous conceptually clear proofs are offered. Each proof provides either a novel presentation of a familiar theorem or a lively discussion of a single issue, sometimes with multiple derivations.

The book collects and presents problems to promote creative techniques for problem-solving and undergraduate research and offers instructors an opportunity to assign these problems as projects. This book provides a wealth of opportunities for these projects.

Each problem is selected for its natural charm—the connection with an authentic mathematical experience, its origination from the ingenious work of professionals, develops well-shaped results of broader interest.

Hongwei Chen received his Ph.D. from North Carolina State University in 1991. He is currently a professor of mathematics at Christopher Newport University. He has published more than 60 research papers in analysis and partial differential equations. He also authored *Monthly Problem Gems* published by CRC Press and *Excursions in Classical Analysis* published by the Mathematical Association of America.

Textbooks in Mathematics

Series editors:

Al Boggess, Kenneth H. Rosen

Introduction to Linear Algebra
Computation, Application and Theory
Mark J. DeBonis

The Elements of Advanced Mathematics, Fifth Edition
Steven G. Krantz

Differential Equations
Theory, Technique, and Practice, Third Edition
Steven G. Krantz

Real Analysis and Foundations, Fifth Edition
Steven G. Krantz

Geometry and Its Applications, Third Edition
Walter J. Meyer

Transition to Advanced Mathematics
Danilo R. Diedrichs and Stephen Lovett

Modeling Change and Uncertainty
Machine Learning and Other Techniques
William P. Fox and Robert E. Burks

Abstract Algebra
A First Course, Second Edition
Stephen Lovett

Multiplicative Differential Calculus
Svetlin Georgiev, Khaled Zennir

Applied Differential Equations
The Primary Course
Vladimir A. Dobrushkin

Introduction to Computational Mathematics: An Outline
William C. Bauldry

Mathematical Modeling the Life Sciences
Numerical Recipes in Python and MATLAB™
N. G. Cogan

Classical Analysis
An Approach through Problems
Hongwei Chen

https://www.routledge.com/Textbooks-in-Mathematics/book-series/CANDHTEXBOOMTH

Classical Analysis
An Approach through Problems

Hongwei Chen

CRC Press
Taylor & Francis Group
Boca Raton London New York

CRC Press is an imprint of the
Taylor & Francis Group, an **informa** business

A CHAPMAN & HALL BOOK

First edition published 2023
by CRC Press
6000 Broken Sound Parkway NW, Suite 300, Boca Raton, FL 33487-2742

and by CRC Press
4 Park Square, Milton Park, Abingdon, Oxon, OX14 4RN

CRC Press is an imprint of Taylor & Francis Group, LLC

© 2023 Hongwei Chen

Library of Congress Cataloging-in-Publication Data

Names: Chen, Hongwei, author.
Title: Classical analysis : an approach through problems / Hongwei Chen.
Description: First edition. | Boca Raton : CRC Press, 2023. | Includes
   bibliographical references and index.
Identifiers: LCCN 2022025390 | ISBN 9781032302478 (hardback) | ISBN
   9781032302485 (paperback) | ISBN 9781003304135 (ebook)
Subjects: LCSH: Mathematical analysis--Problems, exercises, etc.
Classification: LCC QA301 .C435 2023 | DDC 515.076--dc23/eng20221007
LC record available at https://lccn.loc.gov/2022025390

ISBN: 978-1-032-30247-8 (hbk)
ISBN: 978-1-032-30248-5 (pbk)
ISBN: 978-1-003-30413-5 (ebk)

DOI: 10.1201/9781003304135

Publisher's note: This book has been prepared from camera-ready copy provided by the authors.

To Ying, Alex, and Abby,
for whom my love has no upper bound

Contents

Preface

Mathematics is like looking at a house from different angles.

— Thomas Storer

A conceptually clear induction to fundamental analysis theorems, a tutorial for creative approaches for solving problems, a collection of modern challenging problems, a pathway to undergraduate research—all these desires gave life to the pages that follow.

The Mathematical Association of America (MAA) journals have strongly influenced teaching and research for over a century. They have been an indispensable source that provides new proofs of classical theorems and publishes challenging problems. This is especially true for *The American Mathematical Monthly*, which, according to the records on JSTOR, is the most widely read mathematics journal in the world. Since its inception in 1894, the *Monthly* has printed 50 articles on the gamma function or Stirling's asymptotic formula, including the magisterial 1959 paper by Phillip J. Davis, winner of the 1963 Chauvenet prize. This is also a place where one finds proofs of "period three implies chaos." [67] Meanwhile, its problem section has become the single most challenging and interesting problem section, unrivaled since *SIAM Review* canceled its problem section in 1998. Problems from the *Monthly* regularly lead to further research. For example, the Erdös distance problem that served as the genesis of Euclidean Ramsey theory was posed in 1946 and finally solved by Guth and Katz in 2010. Recently, Lagarias made a surprising observation: the Riemann hypothesis is equivalent to the following elementary statement:

$$\sum_{d \mid n} d \le H_n + \ln(H_n)e^{H_n}, \text{ where } H_n = 1 + \tfrac{1}{2} + \cdots + \tfrac{1}{n}, \text{ for all } n \ge 1.$$

A weaker version of this inequality appeared as Monthly Problem 10949.

Unfortunately, these gems are lost among new papers and new problems, and are relegated to the background. Consequently, from time to time it is rewarding to revisit these marvelous papers and problems. These gems form the foundation of this book. The purpose of this book is to expose students to these stimulating and enlightening proofs and hard problems of classical analysis mainly published in the *Monthly*. The goal is two-fold. First, we present proofs as a form of exploration rather than just a manipulation of symbols. Knowing something is true is far from understanding why it is true and how it is connected to the rest of what we know. The search for a proof is the first step in the search for understanding. Drawing on the papers from the MAA journals, numerous proofs that are conceptually clear have been chosen. I have reexamined the proofs of these many standard theorems with an eye toward understanding them. Each proof provides either a novel presentation of a familiar theorem or a lively discussion of a single issue, sometimes with multiple derivations of many important theorems. Special attention is paid to informal exploration of the essential assumptions, suggestive heuristic considerations, and roots of the basic concepts and theorems. Second, we collect problems that will promote creative techniques for problem-solving and undergraduate research. Each problem is selected for its natural charm—the connection with an authentic mathematical experience, its origination from the ingenious work of professionals, develops well-shaped results of broader interest.

Involving undergraduates in research has been a long-standing practice in the experimental sciences; however, it has only recently been that undergraduates have been involved in mathematical research in significant numbers. In my experience of supervising student research, the selection of appropriate problems and methods for undergraduate research is of fundamental importance to success. To achieve the sophisticated blend of knowledge and creativity required, students need experience in working with abstract ideas at a nontrivial level. However, they receive little training in common math courses, especially in view of the "cookbook" experience of many calculus courses.

Historically, deep and beautiful mathematical ideas have often emerged from simple cases of challenging problems. Examining simple cases allows students to experience working with abstract ideas at a nontrivial level—something they do little of in standard mathematics courses.

I have used the following three criteria to select the problems:

1. The statement is fairly short.

2. The problem is challenging, and not solvable by a straight-forward application of standard methods or by a lengthy calculation.

3. The conclusion is pleasing and its solution requires at least one ingenious idea.

This book is not merely gathering published theorems and problems together. Indeed, each selected proof and problem follows a central theme, contains kernels of sophisticated ideas connected to important current research, and opens new perspectives in the understanding of mathematics. It seeks to shed light on the principles underlying these ideas and immerse readers in the processes of problem solving and research. It also emphasizes historical connections and consolidates results that have been relegated to widely scattered books and journals.

My intention is for this book to be a pathway to advanced problem-solving and undergraduate research, offering a systematic illustration of how to organize the natural transition from problem-solving to exploring, investigating, and discovering new results. It emphasizes the rich and elegant interplay that exists between continuous and discrete mathematics. Experimental mathematics is a newly developed approach to discovering mathematical truths through the use of computers. This book illustrates how these techniques can be applied to help solve problems via Mathematica. I hope this book invites the reader to enjoy the beauty of mathematics, to share their discoveries with others, and to become involved in the process of constructing new proofs that lead to publications.

This book is intended for math students who have completed a one-year introductory sequence in calculus. In Chapter 1, we give a fairly detailed treatment of the completeness theorems. We introduce six theorems that are the fundamental results in analysis and show they are, indeed, logically equivalent. Each theorem asserts that $\mathbb{R}$ is complete, and together enables us to derive many important properties of continuity, differentiability, and integrability. By applying these theorems in conjunction with the Stolz theorem, we establish the convergence of sequences in many different respects.

Equipped with a working knowledge of sequences, in Chapter 2, we study the convergence of infinite numerical series. Our focus is on various convergence tests far beyond those encountered in a first-year calculus sequence. We provide examples and discuss the motivations of the ideas frequently before a new theorem is introduced. We also explore the behavior of a convergent series after rearrangements and prove the famous Riemann series theorem. The importance of continuity can hardly be overemphasized in the context of analysis. Sequential theorems related to continuity and set properties, including the extreme value theorem, the intermediate value theorem, and the uniform continuity theorem, constitute the fundamental core of analysis.

In Chapter 3, we give various proofs of three fundamental theorems based on the completeness theorems of the real numbers. Surprisingly, we will see that the completeness of the real numbers is equivalent to each of these three fundamental theorems. As a consequence of the intermediate value theorem, we present Li-York's beautiful result that "periodic three implies chaos."

In Chapter 4, we give a rigorous treatment of differentiation and related concepts based on limits. First, we focus on two fundamental theorems of derivatives. The first proves that a derivative has an intermediate value property. The second gives the mean value theorem that plays a central role in analysis. As an application of the mean value theorem, we present L'Hôpital's rule, which provides us with a powerful tool to compute limits of some indeterminate forms. Next, we study some basic properties of convex functions including Jensen's inequality, which provides an effective approach to establish inequalities. Finally, we extend the mean value theorem to Taylor's theorem, which is considered to be the pinnacle of the differential calculus.

Chapter 5 studies the Riemann integral on a bounded closed interval. We assume familiarity with this concept from calculus, but try to develop the theory in a more precise manner than typical calculus textbooks. We establish the criteria of integrability and the sense in which differentiation is the inverse of integration.

Chapter 6 extends the study of numerical sequences and series to the study of sequences and series of functions. Given a sequence or series of functions, it is often important to know whether or not certain desirable properties of these functions carry over to the limit function. For example, if each function in a sequence or series of functions is continuous or differentiable or integrable, can we assume then that the limit function will be continuous or differentiable or integrable? In doing so, we introduce uniform convergence and illustrate its nature and significance.

In Chapter 7, with an extra limit process, we develop rules that enable us to integrate certain unbounded functions or functions that are defined on an unbounded interval. When the integrand involves a parameter, a study of uniform convergence of the integral is needed in order to determine whether or not the differentiation and integration on the parameter are allowable under the integral sign. We will establish the required theorems in Section 7.2. As an application, we introduce the basic theory of the gamma function and its relation to the beta function. The gamma function is often the first of the so-called *special functions* that students meet beyond the level of calculus, and frequently appears in formulas of analysis. As a showcase, lots of analysis techniques have been invoked during the process in establishing various properties of the gamma function.

As an integral part of this book, the exercises at the end of each chapter provide three levels of problems (routine, demanding, and challenging) in nature. The challenging problems include many *Putnam* and *The American Mathematical Monthly* problems. These problems do not require more advanced knowledge than what is in this book but often need a bit more persistence and some clever ideas. To help the reader with the transition from elementary problem-solving to challenging problem-solving, in each chapter, we present various examples to give the reader ample opportunity to see where we introduce new ideas and how to put them into action. For more help, the reader can refer to [45] and [26]; the former is focused on *Putnam* and the latter is on *Monthly*. Of course, the fun in such problems is doing it yourself rather than in seeing someone else's solution.

Acknowledgments. A number of my colleagues have made helpful comments and suggestions for turning the manuscript into a book. Among these are Professors Brian Bradie, Neville Fogarty, James Kelly, James Martin, and most of all Professor Christopher Kennedy, who read through the manuscript and provided much valuable feedback which made this book more reader friendly. Thanks also to Robert Ross and the CRC Press's book publication staff for their expertise in preparing this book for publication. Finally, I wish to thank my wife, Ying, and children, Alex and Abby, for their love, patience, support, and encouragement.

Naturally, all possible errors are my own responsibility. Comments, corrections, and suggestions from the reader are always welcome at `hchen@cnu.edu`. Thank you in advance.

1

Sequences

Poets write the words you have heard before but in a new sequence.

— Brian Harris

The infinite! No other question has ever moved so profoundly the spirit of man.

— David Hilbert

In this chapter, we first introduce six fundamental theorems of analysis and show they are logically equivalent. Each theorem will assert that the real number system $\mathbb{R}$ is complete, and together enable us to derive many important properties of continuity, differentiability, and integrability. By applying these theorems and the Stolz-Cesàro theorem, we subsequently establish the convergence of sequences in many different respects.

1.1 Completeness Theorems for the Real Number System

I think it [the calculus] defines more unequivocally than anything else the inception of modern mathematics, and the system of mathematical analysis, which is its logical development, still constitutes the greatest technical advance in exact thinking.

— John von Neumann

The calculus was one of human most significant scientific achievements. John von Neumann said: "It is difficult to overestimate its importance." Although its roots trace back into classical Greek times, the significant contributions were made by Newton when developing his laws of motion and gravitation, and Leibniz, who created the notation we still use today. In the first year calculus course, the techniques of differentiation and integration are taught to transform previously intractable physical problems into routine calculation. Most often, the principal theorems upon which techniques are based are stated without proof, so analysis becomes a course to compensate for the missing proofs. To underpin the theory behind the calculus, Cauchy and Weierstruss placed calculus squarely under limits. Their developments, together with Dedekind and Cantor's work, revealed that limits rested upon properties of the real number system $\mathbb{R}$, foremost among which is what we now call *completeness*. Completeness enriches $\mathbb{R}$ with an abundance of properties not shared by its subsystem the rational numbers $\mathbb{Q}$. Geometrically, completeness implies that the real number line has no holes in it.

There are two popular ways to construct the real numbers from the rationals: Dedekind cuts and Canton's Cauchy sequences. The constructions offer competing views of what the completeness of the real numbers should entail.

Motivated by the nature of the set $\{r \in \mathbb{Q} : r < a\}$ for each $a \in \mathbb{R}$, Dedekind introduced

Definition 1.1 (Dedekind Cut). *A Dedekind cut is a subset S of $\mathbb{Q}$ such that*

1. $S \neq \emptyset, S \neq \mathbb{Q}$,

DOI: 10.1201/9781003304135-1

  2. If $x \in S$ and $y < x$, then $y \in S$, and

  3. S has no largest rational.

With this definition, $\mathbb{R}$ is the set of all Dedekind cuts. Moreover, every subset S of $\mathbb{Q}$ that satisfies (1)–(3) has the form $\{r \in \mathbb{Q} : r < a\}$ for some $a \in \mathbb{R}$.

Definition 1.2. *Let S be a subset of $\mathbb{R}$. S is bounded above if there is some $m \in \mathbb{R}$ such that $x < m$ whenever $x \in S$. In this case, m is called an upper bound of S. M is a least upper bound of S if M is an upper bound for S and $M \leq m$ for all upper bounds m of S. We write $M = \sup S$.*

With this definition, we have $a = \sup\{r \in \mathbb{Q} : r < a\}$. Consequently, the completeness of $\mathbb{R}$ is characterized by

Theorem 1.1 (Least-Upper-Bound Property). *Every nonempty subset S of $\mathbb{R}$ that is bounded above has a least upper bound. In other words, $\sup S$ exists and belongs to $\mathbb{R}$.*

By contrast with the Dedekind cut, to fill in the gaps in rationals, Cantor's construction is through approximation. For example, $\sqrt{2}$ is approximated by the rational sequence

$$1, \frac{14}{10}, \frac{141}{100}, \frac{1414}{1000}, \frac{14142}{10000}, \cdots,$$

which derived from its decimal expansion. Canton defines each real number as a *Cauchy sequence* like this. Recall that a Cauchy sequence a_n means that if, for every $\epsilon > 0$, there exists some $N \in \mathbb{N}$ such that $|a_m - a_n| < \epsilon$ whenever $m, n > N$. Now, $\mathbb{R}$ is complete mainly due to

Theorem 1.2 (Cauchy Completeness). *Every Cauchy sequence in $\mathbb{R}$ converges in $\mathbb{R}$.*

Since every convergent sequence is a Cauchy sequence, Theorem 1.2 is strengthened to

Theorem 1.3 (Cauchy Criterion). *A sequence in $\mathbb{R}$ converges in $\mathbb{R}$ if and only if it is a Cauchy sequence.*

Both constructions provide a foundation for the calculus that did not rely upon geometric intuition. Each guarantees that when we encounter the least upper bound that a set possesses or a number to which a sequence converges, then we can be assured that there is indeed a real number there that we can call it as a limit.

A thorough discussion for constructing the real numbers from the rationals is rather long and a bit esoteric for an elementary analysis course. E. Bloch's recent book [15] offers a careful and rigorous description of how $\mathbb{R}$ can be logically constructed from $\mathbb{Q}$. Interested readers can refer this book for details.

Rather than constructing the set of $\mathbb{R}$ so as to justify its completeness, here we assume that $\mathbb{R}$ exists, postulate the properties of $\mathbb{R}$ that we will need and take them for granted. When we prove assertions about real numbers and functions, we will use exactly the properties here and not rely on our intuition.

Definition 1.3. *There are two operations called addition $+$ and multiplication $\cdot$, respectively, and a relation $<$ on $\mathbb{R}$. For all $a, b, c \in \mathbb{R}$, they satisfy the properties:*

A1. $a + b = b + a$ and $a \cdot b = b \cdot a$. (Commutative laws)

A2. $(a + b) + c = a + (b + c)$ and $(a \cdot b) \cdot c = a \cdot (b \cdot c)$. (Associative laws)

A3. $a \cdot (b + c) = a \cdot b + a \cdot c$. (Distributive law)

A4. There are distinct reals 0 and 1 such that $a + 0 = a$ and $a \cdot 1 = a$ for all a. (Identity laws)

A5. For each a there exists a real number $-a$ such that $a + (-a) = 0$, and if $a \neq 0$, then there is a real number b such that $a \cdot b = 1$. (Inverse laws) Here we write $b = 1/a$.

A6. For each pair of a and b, exactly one of the following is true:

$$a = b, \quad a < b, \quad \text{or} \quad b < a.$$

A7. If $a < b$ and $b < c$, then $a < c$. (Transitive law)

A8. If $a < b$, then $a + c < b + c$ for any c, and $a \cdot c < b \cdot c$ if $0 < c$.

Properties **A1-A5** are the *field axioms* while **A6-A8** are the *order axioms*. Thus, the axioms **A1-A8** assures us that $\mathbb{R}$ is an *ordered field*. Note that $\mathbb{Q}$ is also an ordered field. But there is no rational number in $\mathbb{Q}$ whose square is 2. So $\sqrt{2}$ is missing from $\mathbb{Q}$, which is what we mean by a "hole." Consider

$$S = \{x \mid x \in \mathbb{Q} \text{ and } x^2 < 2\}.$$

This set is bounded above. We can use $\sup S$ to get at the missing $\sqrt{2}$ if $\sup S$ exists. Since this kind of nature is indispensable to the limits, we need one more axiom for the real numbers.

Definition 1.4 (Completeness Axiom). *Every nonempty subset of $\mathbb{R}$ that is bounded above has a least upper bound.*

We now have asserted as axiomatic that $\mathbb{R}$ contains all rational numbers and is complete. It is interesting to see that rather than identify the limit explicitly, completeness provides the assurance that a number is out there somewhere. In mathematics, such results are commonly called *existence theorems*. The existence is often sufficient for theoretical purposes.

The next two existence theorems provide some nice applications of what we can do with completeness.

Theorem 1.4 (Archimedean Property). *If $x, y > 0$, then there is a positive integer n such that $nx > y$.*

Proof. We proceed by contradiction. Let

$$S = \{a \mid a = nx, \ n \text{ is a positive integer}\}.$$

If the statement is false, then y would be an upper bound of S. By the completeness axiom, let $M = \sup S$. Then $nx \leq M$ for all positive integers n. In particular, $(n+1)x \leq M$ and therefore

$$nx \leq M - x$$

for all positive integers n. This implies that $M - x$ is an upper bound, which contradicts the definition of M. $\qquad\square$

Consequently, this theorem yields two facts we will use often:

1. Given a real number x, no matter how large, there exists an integer n such that $n > x$.

2. Given a real number x, no matter how small, there exists an integer n such that $\frac{1}{n} < x$.

Theorem 1.5 (Density of $\mathbb{Q}$). *If $x, y \in \mathbb{R}$ with $x < y$, then there exists some $r \in \mathbb{Q}$ such that $x < r < y$.*

Proof. Since $x < y$, we have $y - x > 0$. By the Archimedean property, there is a positive integer q such that $q(y - x) > 1$. Moreover, there is an integer n such that $n > qx$. Indeed, this statement is obvious when $x \leq 0$, and it follows from the Archimedean property again if $x > 0$. Let p be the smallest integer such that $p > qx$. Then $p - 1 \leq qx$, and so

$$qx < p \leq 1 + qx < qy.$$

Since $q > 0$, it follows that

$$x < \frac{p}{q} < y.$$

This proves the theorem with $r = p/q$. $\qquad\square$

A real number that is not a rational number is called an *irrational number*. This theorem, together with the irrationality of $\sqrt{2}$, enables us to show that the irrational numbers are also dense in the reals. In fact, if $x, y \in \mathbb{R}$ with $x < y$, there are rational numbers r_1 and r_2 such that

$$x < r_1 < r_2 < y.$$

Let

$$z = r_1 + \frac{1}{\sqrt{2}}\left(r_2 - r_1\right).$$

Then z is irrational and $x < z < y$.

The completeness axiom can be put into several equivalent forms. Indeed, there are other versions of completeness of $\mathbb{R}$ cast in terms of monotone sequences, nested intervals and subsequences. Once the real number system is established, the following three theorems, together with the Heine-Borel theorem that we will introduce later, are logically equivalent to the completeness axiom. Each may thereby serve as a way of defining completeness of $\mathbb{R}$. It might be difficult at this point for us to see the extreme importance of these results, but as we penetrate the subject more deeply, we will become more and more aware of their basic characteristics.

Notice that there are bounded sequences which diverge. But, with an additional hypothesis, the following principle enables us to conclude convergence.

Theorem 1.6 (Monotone Convergence Theorem). *Let a_n be a sequence that is monotone increasing and bounded above. Then a_n converges.*

The next theorem is possibly the best mathematical statement equivalent to the statement that $\mathbb{R}$ has no holes.

Theorem 1.7 (Nested Interval Theorem). *Let $I_n = [a_n, b_n]$ be closed intervals such that*

> *1. $I_{n+1} \subset I_n$ for every $n \in \mathbb{N}$, and*
>
> *2. $\lim_{n \to \infty}(b_n - a_n) = 0$.*

Then there is a unique real number that belongs to every I_n.

The third result is one of the most notable theorems in analysis. It asserts that the sequence may diverge with boundedness alone, but some part of it must converge.

Theorem 1.8 (Bolzano-Weierstrass Theorem). *Every bounded sequence has a convergent subsequence.*

We prove the equivalence of these theorems in a simple loop:

$$(1.1) \implies (1.6) \implies (1.7) \implies (1.8) \implies (1.3) \implies (1.1).$$

A clever and surprising approach used in a proof is often called a "trick." But a trick used for a second time becomes a technique. The loop proofs will demonstrate typical techniques that are frequently encountered in analysis. For pedagogical purposes, this will give the reader a foretaste of a refreshing phenomenon that they will meet over and over again as they continue their analysis journey: to a much greater extent than is commonly recognized, completeness turns out to be equivalent to many other theorems in analysis. For example, the extreme value theorem, the mean value theorem, and the Darboux integral theorem are all completeness properties. The process of developing this observation will be reexamined in later chapters.

We now turn to the proof of the list of implications and demonstrate how each theorem asserts the completeness of $\mathbb{R}$ in its own particular language.

Proof of (1.1) $\implies$ (1.6). Let (a_n) be a sequence that is monotone increasing and bounded above. To prove that a_n converges by the definition, we need a candidate for the limit. For this purpose, we define $S = \{a_n : n \in \mathbb{N}\}$. Since S is bounded above, Theorem 1.1 implies that S has a least upper bound, so $a = \sup S$ provides us with the expected candidate. We only need to prove that a_n converges to a. For any $\epsilon > 0$, $a - \epsilon$ is not an upper bound for S, so there exists some $N \in \mathbb{N}$ such that

$$a - \epsilon < a_N \leq a.$$

Since a_n is monotone increasing for any $n > N$, we have

$$a - \epsilon < a_N \leq a_n \leq a < a + \epsilon,$$

and so

$$|a_n - a| < \epsilon \quad \text{whenever } n > N.$$

This proves that a_n converges to a. $\square$

Remark. Here the limit of a bounded increasing sequence is indeed the least upper bound. A similar result holds for a monotone decreasing sequence.

In analysis, Cantor's nested interval theorem plays a central role. It can be viewed as a geometric consequence of the monotone convergence theorem.

Proof of (1.6) $\implies$ (1.7). The assumption $I_{n+1} \subset I_n$ implies that, for every $n \in \mathbb{N}$,

$$a_1 \leq \cdots \leq a_n \leq a_{n+1} < b_{n+1} \leq b_n \leq \cdots \leq b_1.$$

In particular, a_n is monotone increasing and bounded above by b_1, and b_n is monotone decreasing and bounded below by a_1. Thus, by the monotone convergence theorem and the remark above, both a_n and b_n converge. Let

$$\lim_{n \to \infty} a_n = a, \ \lim_{n \to \infty} b_n = b.$$

Then

$$a_n \leq a \leq b \leq b_n, \quad \text{for all } n \in \mathbb{N}.$$

The assumption (2) yields that $a = b$, and so a is the only real number common to every I_n. $\square$

Remark. The nested interval theorem does not tell us what the common point is, but it does say that as n gets larger, both a_n and b_n are approaching a definite fixed number monotonically. In particular, it implies that there exists a unique common point to every nested rational intervals whose lengths shrink to zero. This can be interpreted as a confirmation that there are no holes in the real number system.

Given a sequence a_n, let $n_k \in \mathbb{N}$ satisfy

$$n_1 < n_2 < n_3 < \cdots .$$

Then $b_k = a_{n_k}$ ($k = 1, 2, \ldots$) is called a *subsequence* of a_n. It is often important to study various subsequences of a given sequence in order to learn more about the behavior of the sequence itself. The notion of finding a convergent subsequence from a given sequence occurs frequently in analysis. The Bolzano-Weierstrass theorem is basic in that it establishes the existence of such a convergent subsequence under very simple conditions on the given sequence. To prove the Bolzano-Weierstrass theorem, we use the idea that if some subsequence a_{n_k} converges to a, then there must be infinitely many terms of a_{n_k} (hence a_n) that are arbitrarily close to a. Thus, we need to show the existence such an a with infinitely many a_n arbitrarily close. This is can be done via a trick *bisection method* along with the nested interval theorem.

Proof of (1.7) $\implies$ *(1.8).* Let c_n be bounded below by a and above by b. Then $c_n \in I :=$ $[a, b]$ for all $n \in \mathbb{N}$. Divide I into two equal subintervals by the midpoint $(a + b)/2$. Then at least one of these two intervals must contain infinitely many of the c_n. Denote this interval by $I_1 = \{x : a_1 \leq x \leq b_1\}$. Next divide I_1 into two equal subintervals by its midpoint; again at least one of these subintervals must contain infinitely many of the c_n. Denote this interval by $I_2 = \{x : a_2 \leq x \leq b_2\}$. Repeat the process in this manner to obtain a sequence of closed intervals that satisfy

1. $I_1 \supset I_2 \supset I_3 \supset \cdots$ and each I_n contains c_k for infinitely many of k;

2. The length of I_n is $(b - a)/2^n$, which goes to 0 as $n \to \infty$.

By the nested interval theorem, there is exactly one point c common to all those intervals. It seems reasonable to claim that c is the candidate we were looking for.

We now construct a subsequence of c_n which converges to c. Choose c_{n_1} from I_1. Next choose $c_{n_2} \in I_2$ with $n_2 > n_1$. We can do this because I_2 has infinitely many of c_n. Continuing this process by induction yields the subsequence c_{n_k}. By the approach of selection we have

$$a_k \leq c_{n_k} \leq b_k, \ k = 1, 2, \ldots .$$

Since both a_k and b_k converge to c as $k \to \infty$, it follows that c_{n_k} converges to c as $k \to \infty$. $\qquad\square$

Remark. In the proof, the "bisection method" trick, which has become a popular technique, is often used to setup a collection of bounded and nested closed intervals. We begin with an interval $[a_1, b_1]$, which is then divided in half at its midpoint $(a_1 + b_1)/2$. We next choose one of the resulting intervals called $[a_2, b_2]$. We continue to choose subintervals and halve them. Here, how to pick up $[a_1, b_1]$ and then how to choose between the two subintervals at each step depends on what we are trying to prove. We want the chosen intervals to exhibit a certain nature, and during the bisection we must ensure that this nature is inherited by each subsequent interval, i.e., $[a_1, b_1]$ to $[a_2, b_2]$, $[a_2, b_2]$ to $[a_3, b_3]$, and so on. This forces the nature to be passed onto a neighborhood of $\cap_{n=1}^{\infty} I_n$. For the proof above, we always choose the nature that $[a_n, b_n]$ contains infinitely many terms of the sequence. This process is very similar to the binary search algorithm in computer science.

The convergent subsequence in the Bolzano-Weierstrass theorem may or may not be monotone. But, via the following *peak index* argument, due to Newman [72], we can assert every sequence indeed has a monotone subsequence. In particular, the sequence $\sin n$ has a monotone subsequence. This is quite surprising and not obvious at all!

Theorem 1.8* (Stronger Bolzano-Weierstrass Theorem). *Every sequence a_n has a monotone subsequence. Thus, every bounded sequence has a monotone convergent subsequence.*

Proof. We begin by defining the *peak index*. An index m is called a peak index if

$$a_n \leq a_m \quad \text{for all } n > m.$$

There are two distinct cases:

Case 1. *There are only finitely many peak indices.* Then we can choose an index N such that there are no peak indices greater than N. Define $n_1 = N + 1$. Since n_1 is not a peak index, there is an index $n_2 > n_1$ such that $a_{n_2} > a_{n_1}$. Recursively, we can define a strictly increasing subsequence a_{n_k}. (To see this selection process more clearly, display a_{n_k} on the number line.)

Case 2. *There are infinitely many peak indices.* For each natural number k, let n_k be the k-th peak index. The definition of the peak index implies that $a_{n_{k+1}} < a_{n_k}$ for all k. Thus the sequence a_{n_k} is decreasing. □

We return to prove the Cauchy criterion based on the Bolzano-Weierstrass theorem.

Proof of (1.8) $\Longrightarrow$ (1.3). We first show that a convergent sequence is necessarily Cauchy. Suppose $\lim_{n\to\infty} a_n = A$. Then given any $\epsilon > 0$ there is some $N \in \mathbb{N}$ such that

$$|a_n - A| < \frac{\epsilon}{2} \quad \text{whenever } n > N.$$

Therefore, if $m, n > N$, we have

$$|a_m - a_n| \leq |a_m - A| + |A - a_n| < \frac{\epsilon}{2} + \frac{\epsilon}{2} = \epsilon$$

and so a_n is a Cauchy sequence.

Sufficiency is harder. We need to show that there is a number A to which the sequence converges. Our immediate goal is to use the Bolzano-Weierstrass theorem to find such an A. Suppose that a_n is a Cauchy sequence. First, we claim that a_n is bounded. Letting $\epsilon = 1$ yields that there is a $N \in \mathbb{N}$ such that

$$|a_m - a_n| \leq 1 \quad \text{whenever } m, n > N.$$

Choosing $m_0 > N$, for any $n > N$, we find that

$$|a_n| = |a_n - a_{m_0} + a_{m_0}| \leq |a_n - a_{m_0}| + |a_{m_0}| < 1 + |a_{m_0}|.$$

Let

$$M = \max\{|a_1|, |a_2|, \cdots, |a_N|, 1 + |a_{n_0}|\}.$$

Clearly, $|a_n| \leq M$ for all $n \in \mathbb{N}$ and so a_n is bounded.

Next, by the Bolzano-Weierstrass theorem, the sequence a_n has a convergent subsequence, say a_{n_k}, which converges to A. We show that indeed a_n itself converges to A. To this end, for any $\epsilon > 0$, there exists $N_1 \in \mathbb{N}$ such that

$$|a_m - a_n| \leq \frac{\epsilon}{2} \quad \text{whenever } m, n > N_1.$$

Choose $n_k \in \mathbb{N}$ large enough that

$$|a_{n_k} - A| < \frac{\epsilon}{2} \qquad \text{whenever } n_k > N_1.$$

Thus,

$$|a_n - A| \le |a_n - a_{n_k}| + |a_{n_k} - A| < \frac{\epsilon}{2} + \frac{\epsilon}{2} = \epsilon \qquad \text{whenever } n > N_1,$$

and so the sequence a_n converges to A. $\qquad\qquad\square$

Finally, to close the loop, we prove the Cauchy criterion implies the least-upper-bound property.

Proof of (1.3) $\implies$ (1.1). Let S be bounded above and suppose that b is an upper bound of S. We prove that S has a least upper bound by the bisection method. The proof will consist of three components:

1. Constructing $[a_n, b_n]$ such that b_n is an upper bound of S and a_n is not.

2. Showing the sequence (b_n) is a Cauchy sequence and converges to a limit M.

3. Showing the M is the least upper bound of S.

Choose any number in S and denote it by a_1. Let $b_1 = b$ and define $I_1 = [a_1, b_1]$. Construct nested intervals as follows.

Consider the midpoint $(a_1 + b_1)/2$ of I_1:

(a) If $(a_1 + b_1)/2$ is an upper bound of S, we define $I_2 = [a_2, b_2] = [a_1, (a_1 + b_1)/2]$.

(b) If $(a_1 + b_1)/2$ is not an upper bound of S, then there exists $s \in S$ such that $s > (a_1 + b_1)/2$. We define $I_2 = [a_2, b_2] = [(a_1 + b_1)/2, b_1]$.

In both cases, we have that b_2 is an upper bound of S and $I_2 \cap S \ne \emptyset$. Repeat the process in this manner to obtain a sequence of closed intervals that satisfy

(i) $I_1 \supset I_2 \supset I_3 \supset \cdots$;

(ii) $b_n - a_n = (b_1 - a_1)/2^{n-1}$;

(iii) For every $n \in \mathbb{N}$, b_n is an upper bound of S and $I_n \cap S \ne \emptyset$.

Thus, for $m, n \in \mathbb{N}, m > n$, by (i) and (ii) we have

$$|b_m - b_n| = b_n - b_m < b_n - a_n = \frac{1}{2^{n-1}}(b_1 - a_1),$$

and so b_n is a Cauchy sequence. This implies that b_n converges from the Cauchy criterion.

Let $\lim_{n \to \infty} b_n = M$. We show that $M = \sup S$. Indeed, for every $s \in S$ and every $n \in \mathbb{N}$, we have $s \le b_n$ and so $s \le M$, which means that M is an upper bound of S. On the other hand, for any $\epsilon > 0$, since $b_n - a_n \to 0$ as $n \to \infty$, there is $n_0 \in \mathbb{N}$ such that $b_{n_0} - a_{n_0} < \epsilon$. Since $b_{n_0} \ge M$,

$$a_{n_0} > b_{n_0} - \epsilon \ge M - \epsilon.$$

By (iii) $S \cap I_{n_0} \ne \emptyset$, we deduce that $M - \epsilon$ is not an upper bound of S and so $M = \sup S$. $\quad\square$

Can you figure out where we have used the Archimedean property of $\mathbb{R}$ in the above proof? This implies that Cauchy completeness, together with the the Archimedean property, is equivalent to the completeness axiom.

We now illustrate some applications of these theorems with two examples.

Example 1.1. *The Fibonacci numbers F_n are defined by $F_1 = F_2 = 1$ and $F_{n+2} = F_n + F_{n-1}$ for all $n \in \mathbb{N}$. Show that*

$$\lim_{n \to \infty} \frac{F_{n+1}}{F_n} = \phi,$$

where $\phi = \frac{1+\sqrt{5}}{2}$, which is called the golden ratio.

Proof. Observer the first ten terms of the ratio $r_n := F_{n+1}/F_n$:

$$1, \ 2, \ \frac{3}{2}, \ \frac{5}{3}, \ \frac{8}{5}, \ \frac{13}{8}, \ \frac{21}{13}, \ \frac{34}{21}, \ \frac{55}{34}, \ \frac{89}{55}.$$

In round off decimal form, we have

$$1.0000, \ 2.0000, \ 1.5000, \ 1.6667, \ 1.6000, \ 1.6250, \ 1.6154, \ 1.6190, \ 1.6176, \ 1.6182.$$

This suggests that, although r_n itself is not monotone, the subsequence r_{2n-1} is monotone increasing and bounded above, and r_{2n} is monotone decreasing and bounded below. To confirm the above assertions, we first prove that $1 \leq r_n \leq 2$ for every $n \in \mathbb{N}$. We have $r_1 = 1$. Suppose that $1 \leq r_k \leq 2$. Since

$$r_{k+1} = \frac{F_{n+2}}{F_{n+1}} = \frac{F_{n+1} + F_n}{F_{n+1}} = 1 + \frac{1}{r_k},$$

we have

$$1 < 1 + \frac{1}{2} \leq 1 + \frac{1}{r_k} = r_{k+1} \leq 1 + \frac{1}{1} = 2.$$

By induction, this proves that $1 \leq r_n \leq 2$ for every $n \in \mathbb{N}$. Next, for $n \geq 3$, we have

$$r_n = 1 + \frac{1}{r_{n-1}} = 1 + \frac{1}{1 + 1/r_{n-2}} = 1 + \frac{r_{n-2}}{1 + r_{n-2}}.$$

Hence

$$r_{n+2} - r_n = \frac{r_n}{1 + r_n} - \frac{r_{n-2}}{1 + r_{n-2}} = \frac{r_n - r_{n-2}}{(1 + r_n)(1 + r_{n-2})}.$$

This implies that $r_{n+2} - r_n$ has the same sign as $r_n - r_{n-2}$. Since $r_3 - r_1 = 3/2 - 1 = 1/2 > 0$ and $r_4 - r_2 = 5/3 - 2 = -1/3 < 0$, using induction again yields

$$r_{2n-1} - r_{2n-3} > 0, \qquad r_{2n} - r_{2n-2} < 0.$$

This shows that r_{2n-1} is monotone increasing and r_{2n} is monotone decreasing. By the monotone convergence theorem, r_{2n-1} and r_{2n} both converge. Let

$$L_1 = \lim_{n \to \infty} r_{2n-1} \quad \text{and} \quad L_2 = \lim_{n \to \infty} r_{2n}.$$

Then

$$L_1 = \lim_{n \to \infty} r_{2n-1} = \lim_{n \to \infty} \left(1 + \frac{r_{2n-3}}{1 + r_{2n-3}} \right) = 1 + \frac{L_1}{1 + L_1},$$

$$L_2 = \lim_{n \to \infty} r_{2n} = \lim_{n \to \infty} \left(1 + \frac{r_{2n-2}}{1 + r_{2n-2}} \right) = 1 + \frac{L_2}{1 + L_2}.$$

Thus L_1 and L_2 both satisfy the equation $x^2 - x - 1 = 0$. Consequently, solving the quadratic equation yields

$$L_1 = L_2 = \frac{1 + \sqrt{5}}{2} = \phi.$$

Moreover, for any $\epsilon > 0$, there are N_1 and N_2 such that

$$|r_{2n-1} - \phi| < \epsilon \quad \text{whenever } n > N_1;$$

$$|r_{2n} - \phi| < \epsilon \quad \text{whenever } n > N_2.$$

Let $N = \max\{2N_1, 2N_2\}$. Then

$$|r_n - \phi| < \epsilon \quad \text{whenever } n > N.$$

This proves

$$\lim_{n\to\infty} \frac{F_{n+1}}{F_n} = \lim_{n\to\infty} r_n = \phi$$

as desired.   □

This example indicates that once we know that the limit exists, we can assign it a symbol and find an equation for it. Without knowing the limit exists, such a procedure may lead to a spurious solution. For example, consider the sequence defined by $x_1 = 1, x_{n+1} = 2x_n^2$ for $n \in \mathbb{N}$. Clearly, x_n diverges. But if you let $l = \lim_{n\to\infty} x_n$, then take limit on $x_{n+1} = 2x_n^2$ resulting $l = 2l^2$. Here l must be either 0 or $1/2$.

Recall that an *enumeration* of a set S is a bijection which maps the elements of S into a sequence without repetitions:

$$x_1, x_2, x_3, \ldots.$$

If there exists such an enumeration of S, S is called *countable*. Otherwise, S is *uncountable*.

Example 1.2. *Show that $\mathbb{R}$ is uncountable.*

The most famous proof of the uncountability of $\mathbb{R}$ probably is Cantor's "diagonalization argument." His approach is clever and historically important. We can't help presenting it below. But his proof used the fact that every real number can be represented in decimal notation. As an application of the nested interval theorem, we present a self-contained proof.

Cantor's proof. Notice that $\tan(2x - 1)\pi/2$ is a bijection between $(0, 1)$ and $\mathbb{R}$. So it suffices to show that $(0, 1)$ is uncountable. Cantor did the proof by contradiction. Suppose $\{x_1, x_2, x_3, \ldots\}$ is an enumeration of the numbers in $(0, 1)$. Represent each x_n as a decimal expansion:

$$x_1 = 0.a_{11}a_{12}a_{13}a_{14}a_{15}\cdots,$$
$$x_2 = 0.a_{21}a_{22}a_{23}a_{24}a_{25}\cdots,$$
$$x_3 = 0.a_{31}a_{32}a_{33}a_{34}a_{35}\cdots,$$
$$\cdots$$
$$x_n = 0.a_{n1}a_{n2}a_{n3}a_{n4}a_{n5}\cdots,$$
$$\cdots$$

where each a_{ij} is a digit among $\{0, 1, 2, \cdots, 9\}$ and repeating 9s are chosen as the terminating decimal expansions. For example, use 0.5 instead of $0.49999\ldots$. Let

$$x = 0.b_{11}b_{22}b_{33}b_{44}b_{55}\cdots,$$

where $b_{nn} \neq a_{nn}$ for all $n \in \mathbb{N}$. Since x is different from x_n in the nth decimal place, x is not in the above enumeration list. Thus, there is no one-to-one correspondence between $\mathbb{N}$ and $(0, 1)$. This proves that $(0, 1)$ is uncountable.   □

Proof by the nested interval theorem. We proceed by contradiction as well. Suppose $\mathbb{R}$ is countable. Then there is a bijection between $\mathbb{R}$ and $\mathbb{N}$. Let

$$\mathbb{R} = \{x_1, x_2, \cdots, x_n, \cdots\}.$$

We construct the nested intervals as follows:

1. Let $I_1 = [a_1, b_1]$ such that $I_1 \subset \mathbb{R} \setminus \{x_1\}$.

2. Divide I_1 into two disjoint closed intervals. One of those will not contain x_2; let that interval be I_2.

3. Once I_{n-1} is well-defined, choose I_n such that $I_n \subset I_{n-1} \setminus \{x_n\}$.

It then follows from the nested interval theorem that there is a $y \in I_n$ for all $n \in \mathbb{N}$. On the other hand, since $x_n \notin I_n$ for all $n \in \mathbb{N}$, it implies that $y \neq x_n$ for all $n \in \mathbb{N}$, so the above enumeration is not one-to-one. The contradiction shows that $\mathbb{R}$ is uncountable. $\qquad\square$

We conclude this section with one more completeness theorem of the real numbers. This theorem plays an important role not only in the study of analysis (for example, uniform continuity and the Riemann integral), but also in the study of many different areas of mathematics such as Topology.

We first extend our discussion from a sequence to an arbitrary set S of real numbers. The Bolzano-Weierstrass theorem guarantees that every bounded sequence has a convergent subsequence. As an analogy, we define

Definition 1.5. *A point x_0 is called a limit point (or an accumulation point) of the set S if for any given $\epsilon > 0$, there is a point $x \in S, x \neq x_0$ such that $|x - x_0| < \epsilon$.*

The limit point x_0 may lie in S or not. For example, if $S = (a, b)$, then the points a and b and all real numbers between a and b inclusive are limit points of S. But, neither a nor b belongs to S.

Analogous to the Bolzano-Weierstrass theorem for sequences, we have

Theorem 1.9 (Bolzano-Weierstrass Theorem for Sets). *Every bounded infinite set has at least one limit point.*

Proof. Let S be a bounded infinite set. Then we can select a sequence a_n of distinct numbers from S. Clearly, this sequence is bounded. By the Bolzano-Weierstrass theorem, a_n has a convergent subsequence, which we call a_{n_k}.

Let $x_0 = \lim_{k \to \infty} a_{n_k}$. We claim that x_0 is a limit point of S. Let $\epsilon > 0$. Then there is $K \in \mathbb{N}$ such that

$$|a_{n_k} - x_0| < \epsilon \quad \text{whenever } k > K.$$

In addition, we can require that $a_{n_k} \neq x_0$ since a_n, and hence a_{n_k}, is a sequence of distinct numbers. Thus, for any $\epsilon > 0$, there are $a_{n_k} \in S$ and $a_{n_k} \neq x_0$ such that

$$|a_{n_k} - x_0| < \epsilon \quad \text{whenever } k > K.$$

This proves x_0 is a limit point of S by Definition 1.5. $\qquad\square$

Let

$$U_\epsilon(x_0) = (x_0 - \epsilon, x_0 + \epsilon).$$

$U_\epsilon(x_0)$ is called an ϵ-*neighborhood* of the point x_0. We next define

Definition 1.6. *A set $S \subset \mathbb{R}$ is called open if for every $x \in S$ there is an $\epsilon > 0$ such that $U_\epsilon(x) \subset S$. A set $S \subset \mathbb{R}$ is called closed if its complement $S^c := \mathbb{R} \setminus S$ is open.*

By this definition, both $\mathbb{R}$ and (a,b) are open sets. Since $(-\infty, a) \cup (b, \infty)$ is open, it follows that $[a, b]$ is closed. The following theorem indicates that a closed set can be characterized by the limit points of the set itself.

Theorem 1.10. *A set $S \subset \mathbb{R}$ is closed if and only if S contains all its limit points.*

Proof. Suppose S is closed. Let $x_0 \in S^c$. It suffices to show that x_0 is not a limit point of S. Since S^c is open, there is an $\epsilon > 0$ such that $U_\epsilon(x_0) \subset S^c$. Thus there is no $x \in S$ satisfying $|x - x_0| < \epsilon$, and so x_0 is not a limit point of S.

On the other hand, let S be a set which contains all its limit points. For every point $x \in S^c$, since x is not a limit point of S, there must exist an $\epsilon > 0$ for which no number $y \in S$ satisfies $|x - y| < \epsilon$. Thus, $U_\epsilon(x) \subset S^c$, which implies that S^c is open, so that S itself is closed. $\qquad\square$

Definition 1.7. *A set $S \subset \mathbb{R}$ is called sequentially compact if every sequence $a_n \in S$ has a subsequence that converges to a number in S.*

By this definition, $S_1 = [0, \infty)$ is not sequentially compact. Indeed, let $a_n = n$ for $n \in \mathbb{N}$. Every subsequence of a_n is unbounded and so divergent. The set $S_2 = (0, 1]$ likewise fails to be sequentially compact since the sequence $a_n = 1/n \in S_2$ converges to 0, but $0 \notin S_2$. However, we have the following test.

Theorem 1.11. *If $S \subset \mathbb{R}$ is closed and bounded, then S is sequentially compact.*

Proof. Let a_n be a sequence in S. The boundedness of S implies that a_n is bounded. By the Bolzano-Weierstrass theorem, a_n has a subsequence a_{n_k} that converges to a. Since S is closed, it follows that $a \in S$, and so S is sequentially compact. $\qquad\square$

In particular, the closed interval $[a, b]$ is sequentially compact. Finally, we define

Definition 1.8. *A collection of open sets $\mathcal{G} = \{U_i\}$ is called an open covering of the set S if*

$$S \subseteq \bigcup_{U_i \in \mathcal{G}} U_i.$$

An open covering $\mathcal{G}$ is finite if $|\mathcal{G}| < \infty$.

Example 1.3. *Let $S_1 = (0, 1)$. Define $\mathcal{G} = \{U_i = (1/2^i, 3/2^i)\}$. Clearly, $S_1 \subset \cup_{i=1}^\infty U_i$. Notice that, no matter what index N is selected, there is a positive number less than $1/2^N$ which does not belong to $\cup_{i=1}^N U_i$. Therefore, there is no finite sub-collection of $\mathcal{G}$ that also covers S_1.*

The following theorem asserts the existence of a finite sub-covering for closed intervals.

Theorem 1.12 (Heine-Borel Theorem). *If $\mathcal{G} = \{U_i\}$ is an open covering of $[a, b]$, then there is a finite sub-covering of $\mathcal{G}$ that also covers $[a, b]$.*

Proof. We proceed by contradiction. Assume that no finite subset of $\mathcal{G}$ covers $[a, b]$. We divide $[a, b]$ into two equal subintervals at the midpoint $(a + b)/2$. Then at least one of these two intervals cannot be covered by a finite subset of $\mathcal{G}$. Otherwise, if $\mathcal{G}_1 \subset \mathcal{G}$ is a finite subcovering of $[a, (a + b)/2]$ and $\mathcal{G}_2 \subset \mathcal{G}$ is a finite subcovering of $[(a + b)/2, b]$, then $\mathcal{G}_1 \cup \mathcal{G}_2$ is a finite subcovering of $[a, b]$, contrary to the assumption. Let $I_1 = [a_1, b_1] \subset [a, b]$ be an interval that has no finite subcovering. Next we split I_1 into two equal subintervals at its midpoint; again at least one of these subintervals cannot be covered by a finite subset of $\mathcal{G}$. Denote such an interval by $I_2 = [a_2, b_2]$. Repeating this process, we obtain a sequence of of closed intervals that satisfy

1. $I_1 \supset I_2 \supset I_3 \supset \cdots$;

2. The length of I_n is $(b-a)/2^{n-1}$, which goes to 0 as $n \to \infty$; and

3. Each I_n cannot be covered by a finite subset of $\mathcal{G}$.

By the nested interval theorem, there is a unique point $x_0 \in [a, b]$ common to each I_n.

Because $\mathcal{G}$ is a covering of $[a, b]$, there is a open interval $U_i \subset \mathcal{G}$ such that $x_0 \in U_i$. By (2), taking N large enough, we have

$$I_N \subset U_i.$$

This means that U_i covers I_N, which contradicts (3), and this establishes the theorem. $\square$

Definition 1.9. *A set $S \subset \mathbb{R}$ is called compact if every open covering of S contains a finite sub-covering that also covers S.*

In this terminology, the Heine-Borel theorem says that $[a, b]$ is compact. This property will play an important role in the study of uniform continuity and the Riemann integral.

As an application, we can reprove the previous five completeness theorems by using the Heine-Borel theorem solely. Here we content ourselves with proving just the nested interval theorem.

Proof. By contradiction, we first prove that

$$\bigcap_{n=1}^{\infty} [a_n, b_n] \neq \emptyset.$$

To this end, assume that

$$\bigcap_{n=1}^{\infty} [a_n, b_n] = \emptyset.$$

For every $n \in \mathbb{N}$, let

$$U_n = (-\infty, a_n) \cup (b_n, \infty).$$

Taking complements

$$[a_1, b_1] \setminus \bigcup_{n=1}^{\infty} U_n = \bigcap_{n=1}^{\infty} ([a_n, b_n] \setminus U_n) = \bigcap_{n=1}^{\infty} [a_n, b_n],$$

it then follows that $\cup_{n=1}^{\infty} U_n \supset [a_1, b_1]$. By the Heine-Borel theorem, a finite subset $\{U_{n_k}\}_{k=1}^{M}$ covers $[a_1, b_1]$, i.e.,

$$\bigcup_{k=1}^{M} U_{n_k} \supset [a_1, b_1].$$

This implies that $[a_1, b_1] \setminus \cup_{k=1}^{M} U_{n_k} = \emptyset$, and so $\cap_{k=1}^{M} [a_{n_k}, b_{n_k}] = \emptyset$. But, this contradicts the assumption

$$\bigcap_{k=1}^{M} [a_{n_k}, b_{n_k}] = [a_{n_M}, b_{n_M}].$$

This proves $\cap_{n=1}^{\infty} [a_n, b_n] \neq \emptyset$. Next, we show the uniqueness. Assume that $x, y \in \cap_{n=1}^{\infty} [a_n, b_n]$. Then $x, y \in [a_n, b_n]$ for every $n \in \mathbb{N}$, and so

$$|x - y| \leq |b_n - a_n| \to 0 \text{ as } n \to \infty.$$

Hence $x = y$ and so $\cap_{n=1}^{\infty} [a_n, b_n]$ contains a unique point. $\square$

We now have established the equivalence between the nested interval theorem and the Heine-Borel theorem. Notice that the Heine-Borel theorem actually is a contrapositive-type statement of the nested interval theorem. The nested interval theorem converts an interval property into a point property by constructing suitable nested intervals to single out the required point, while the Heine-Borel theorem extends a local property at points to the entire interval by extracting a finite set from an infinite set. Together these two theorems enable us to interchange arguments between intervals and specific points. The next two examples illustrate this idea.

Example 1.4. *If $f(x)$ is unbounded on $[a, b]$, prove that there exists a point $x_0 \in [a, b]$ such that for any $\delta > 0$, $f(x)$ is unbounded on $(x_0 - \delta, x_0 + \delta) \cap [a, b]$.*

Proof. By assumption, $f(x)$ is unbounded either on $[a, (a + b)/2]$ or on $[(a + b)/2, b]$. Let $I_1 = [a_1, b_1]$ be one of these intervals on which $f(x)$ is unbounded. Next divide I_1 into two equal subintervals by its midpoint; again $f(x)$ is unbounded on at least one of these subintervals. Denote such an interval by $I_2 = [a_2, b_2]$. Repeat the process in this manner to obtain a sequence of closed intervals satisfying

1. $[a_{n+1}, b_{n+1}] \subset [a_n, b_n]$ $(n \in \mathbb{N})$,

2. $b_n - a_n \to 0$ (as $n \to \infty$), and

3. $f(x)$ is unbounded on every $[a_n, b_n]$.

The nested interval theorem guarantees an $x_0 \in [a, b]$ such that $\lim_{n \to \infty} a_n = x_0 = \lim_{n \to \infty} b_n$. Now, for any $\delta > 0$, there is an integer N such that, if $n > N$, $[a_n, b_n] \subset (x_0 - \delta, x_0 + \delta) \cap [a, b]$. Thus, we conclude that $f(x)$ is unbounded on $(x_0 - \delta, x_0 + \delta) \cap [a, b]$ as desired. $\square$

Example 1.5. *Let $f : [a, b] \longrightarrow \mathbb{R}$ be a continuous function. Then f is bounded.*

Proof. For each $x \in [a, b]$, continuity implies that for each x there is a $\delta_x > 0$ such that

$$|f(y) - f(x)| < 1 \text{ whenever } y \in [a, b] \text{ and } |x - y| < \delta_x.$$

Let $I_x = (x - \delta_x, x + \delta_x)$ and

$$\mathcal{G} = \{ I_x \mid x \in [a, b] \}.$$

Then $\mathcal{G}$ is a open covering of $[a, b]$. By the Heine-Borel theorem, there exists a finite sub-covering $\mathcal{G}^*$ of $\mathcal{G}$ that also covers $[a, b]$. Denote the components by

$$\mathcal{G}^* = \{(x_1 - \delta_1, x_1 + \delta_1), (x_2 - \delta_2, x_2 + \delta_2), \cdots, (x_k - \delta_k, x_k + \delta_k)\}.$$

Let

$$M = \max\{ |f(x_1)| + 1, |f(x_2)| + 1, \cdots, |f(x_k)| + 1\}.$$

We now show that $|f(x)| \leq M$ for all $x \in [a, b]$. To this end, for any $x \in [a, b]$, since $\mathcal{G}^*$ covers $[a, b]$, there is an index i such that $x \in I_{x_i}$. Thus,

$$|f(x) - f(x_i)| < 1$$

and so

$$|f(x)| \leq |f(x) - f(x_i)| + |f(x_i)| < 1 + |f(x_i)| \leq M.$$

This proves that f is bounded as desired. $\square$

Here the Heine-Borel theorem is used to extract the finite set $\mathcal{G}^*$ from the infinite set $\mathcal{G}$. This guarantees that we are able to select M. Because of the proprieties of compactness of $[a, b]$, we will see in subsequent chapters why closed intervals are fundamentally different from open intervals.

1.2 Stolz-Cesàro Theorem

To insure the adoration of a theorem for any length of time, faith is not enough, a police force is needed as well. — Albert Camus

The following theorem provides a powerful tool to evaluate limits of sequences, and is viewed as a discrete form of L'Hôpital's rule.

Theorem 1.13 (Stolz-Cesàro Theorem). *Let a_n and b_n be two real sequences.*

1. *Assume that $a_n \to 0, b_n \to 0$ as $n \to \infty$. If b_n is decreasing and*

$$\lim_{n\to\infty} \frac{a_{n+1} - a_n}{b_{n+1} - b_n} = a,$$

where a is finite or $+\infty$, then

$$\lim_{n\to\infty} \frac{a_n}{b_n} = \lim_{n\to\infty} \frac{a_{n+1} - a_n}{b_{n+1} - b_n} = a.$$

2. *Assume that $b_n \to +\infty$ and b_n is increasing. If*

$$\lim_{n\to\infty} \frac{a_{n+1} - a_n}{b_{n+1} - b_n} = a,$$

where a is finite or $+\infty$, then

$$\lim_{n\to\infty} \frac{a_n}{b_n} = \lim_{n\to\infty} \frac{a_{n+1} - a_n}{b_{n+1} - b_n} = a.$$

Proof. Here we prove the case (2) with finite a only. The trick is how to estimate $|a_n/b_n - a|$ by grouping. To this end, for each $\epsilon > 0$, there is an integer N such that $b_n > 0$ and

$$\left| \frac{a_{n+1} - a_n}{b_{n+1} - b_n} - a \right| < \frac{\epsilon}{2} \quad \text{whenever } n > N.$$

Since $b_{n+1} > b_n$ for all $n > N$, we have

$$\left(a - \frac{\epsilon}{2}\right)(b_{N+2} - b_{N+1}) < a_{N+2} - a_{N+1} < \left(a + \frac{\epsilon}{2}\right)(b_{N+2} - b_{N+1}),$$
$$\left(a - \frac{\epsilon}{2}\right)(b_{N+3} - b_{N+2}) < a_{N+3} - a_{N+2} < \left(a + \frac{\epsilon}{2}\right)(b_{N+3} - b_{N+2}),$$
$$\vdots$$
$$\left(a - \frac{\epsilon}{2}\right)(b_{n+1} - b_n) < a_{n+1} - a_n < \left(a + \frac{\epsilon}{2}\right)(b_{n+1} - b_n).$$

Adding these inequalities yields

$$\left(a - \frac{\epsilon}{2}\right)(b_{n+1} - b_{N+1}) < a_{n+1} - a_{N+1} < \left(a + \frac{\epsilon}{2}\right)(b_{n+1} - b_{N+1}),$$

and so

$$\left| \frac{a_{n+1} - a_{N+1}}{b_{n+1} - b_{N+1}} - a \right| < \frac{\epsilon}{2}.$$

Next, appealing to the identity

$$\frac{a_n}{b_n} - a = \frac{a_{N+1} - ab_{N+1}}{b_n} + \left(1 - \frac{b_{N+1}}{b_n}\right)\left(\frac{a_n - a_{N+1}}{b_n - b_{N+1}} - a\right),$$

we have

$$\left|\frac{a_n}{b_n} - a\right| \leq \left|\frac{a_{N+1} - ab_{N+1}}{b_n}\right| + \frac{\epsilon}{2}.$$

Choose an integer $N' > N$ such that

$$\left|\frac{a_{N+1} - ab_{N+1}}{b_n}\right| < \frac{\epsilon}{2} \qquad \text{whenever } n > N'.$$

Therefore, for each $\epsilon > 0$, when $n > N'$, we have

$$\left|\frac{a_n}{b_n} - a\right| < \epsilon.$$

This proves that $\lim_{n\to\infty} a_n/b_n = a$ as needed.   □

The converse of this theorem does not hold, i.e., $\lim_{n\to\infty} a_n/b_n = a$ does not necessarily imply

$$\lim_{n\to\infty} \frac{a_{n+1} - a_n}{b_{n+1} - b_n} = a.$$

Here is a counterexample: $a_n = 3n - (-1)^n, b_n = 3n + (-1)^n$. For further discussion, see Section1.3.6.

1.3   Worked Examples

A good stock of examples, as large as possible, is indispensable for a thorough understanding of any concept, and when I want to learn something new, I make it my first job to build one.

— Paul Halmos

The limits of sequences constitute an important part of the theory of limits. They provide a preliminary step in our understanding of infinite series of numbers and functions and will underlie our future study of the continuity, differentiation, and integration of general functions. They also present wide classes of discrete processes arising in various applications. In this section, we present 33 inspired examples with the expectation of using their solutions to develop common strategies and provide solid preparation for solving challenging problems. Based on the various approaches used, we divide them into nine subsections, which are as follows.

1.3.1 $\epsilon - N$ definition

For example is not proof — Jewish Proverb

To show that $\lim_{n\to\infty} a_n = a$ by using the $\epsilon - N$ definition, we need to show that for any $\epsilon > 0$, there exists a positive integer N that has the desired property: $|a_n - a| < \epsilon$ whenever $n > N$. To find such N, we typically operate on the inequality $|a_n - a| < \epsilon$ algebraically to try to "solve" for N.

Example 1.6. *Let* $\lim_{n\to\infty} a_n = a$, *where a is either finite or infinity. Prove that*

$$\lim_{n\to\infty} \frac{a_1 + a_2 + \cdots + a_n}{n} = a.$$

Proof. First, we assume that a is finite. Since $\lim_{n\to\infty} a_n = a$, for any $\epsilon > 0$, there is an integer $N_1 \in \mathbb{N}$ such that

$$|a_n - a| < \frac{\epsilon}{2} \quad \text{whenever } n > N_1.$$

Notice that, for each $n > N_1$,

$$\frac{a_1 + a_2 + \cdots + a_n}{n} - a = \frac{\sum_{i=1}^{N_1} (a_i - a)}{n} + \frac{\sum_{i=N_1+1}^{n} (a_i - a)}{n}.$$

Thus,

$$\left| \frac{a_1 + a_2 + \cdots + a_n}{n} - a \right| \leq \frac{N_1 \max_{1 \leq i \leq N_1} \{|a_i - a|\}}{n} + \frac{n - N_1}{n} \frac{\epsilon}{2}.$$

Since $N_1 \max_{1 \leq i \leq N_1} \{|a_i - a|\}$ is finite, there is an integer $N_2 \in \mathbb{N}$ such that

$$\frac{N_1 \max_{1 \leq i \leq N_1} \{|a_i - a|\}}{n} < \frac{\epsilon}{2} \quad \text{whenever } n > N_2.$$

Thus, for $n > N = \max\{N_1, N_2\}$, we have

$$\left| \frac{a_1 + a_2 + \cdots + a_n}{n} - a \right| < \frac{\epsilon}{2} + \frac{\epsilon}{2} = \epsilon.$$

This proves the claimed limit when a is finite.

Next, assume that $a = +\infty$. The case of $a = -\infty$ can be proved similarly. Thus, for each $M > 0$, there is an integer $N \in \mathbb{N}$ such that

$$a_n > 3M \quad \text{whenever } n > N.$$

Moreover,

$$\begin{aligned}
\frac{a_1 + a_2 + \cdots + a_n}{n} &= \frac{a_1 + a_2 + \cdots + a_N}{n} + \frac{\sum_{i=N+1}^{n} a_i}{n - N} \left(1 - \frac{N}{n} \right) \\
&> \frac{a_1 + a_2 + \cdots + a_N}{n} + 3M \left(1 - \frac{N}{n} \right).
\end{aligned}$$

Notice that

$$\frac{a_1 + a_2 + \cdots + a_N}{n} \to 0, \quad 1 - \frac{N}{n} \to 1 \text{ as } n \to \infty.$$

There is an integer $N' > N$ such that

$$\left| \frac{a_1 + a_2 + \cdots + a_N}{n} \right| < \frac{M}{2}, 1 - \frac{M}{n} > \frac{1}{2} \quad \text{whenever } n > N'.$$

Thus, when $n > N'$, we have

$$\frac{a_1 + a_2 + \cdots + a_n}{n} > M,$$

and so

$$\lim_{n\to\infty} \frac{a_1 + a_2 + \cdots + a_n}{n} = \infty.$$

$\square$

In this example, by regrouping and enlarging $|a_n - a|$ appropriately, we solved for N from

$$\frac{N_1 \max_{1 \le i \le N_1} \{|a_i - a|\}}{N} < \frac{\epsilon}{2}.$$

Example 1.7 (Putnam Problem 1970-A4). *Let a_n be a sequence such that $\lim_{n \to \infty} (a_n - a_{n-2}) = 0$. Prove that*

$$\lim_{n \to \infty} \frac{a_n - a_{n-1}}{n} = 0.$$

Proof. Let $b_n = |a_n - a_{n-1}|$. Since

$$a_n - a_{n-2} = a_n - a_{n-1} + a_{n-1} - a_{n-2},$$

the triangle inequality yields that $|a_n - a_{n-2}| \ge |b_n - b_{n-1}|$. Thus, $\lim_{n \to \infty} (a_n - a_{n-2}) = 0$ implies that $\lim_{n \to \infty} |b_n - b_{n-1}| = 0$. So for every $\epsilon > 0$, there is an integer N such that

$$|b_n - b_{n-1}| < \frac{\epsilon}{2} \quad \text{whenever } n > N.$$

For this given N, there is an integer $N' > N$ such that

$$\frac{b_N}{n} < \frac{\epsilon}{2} \quad \text{whenever } n > N'.$$

Thus, for $n > N'$,

$$\left| \frac{a_n - a_{n-1}}{n} \right| = \frac{b_n}{n} \le \frac{\sum_{k=N+1}^{n} |b_k - b_{k-1}|}{n} + \frac{b_N}{n}$$

$$\le \frac{n - N}{n} \frac{\epsilon}{2} + \frac{\epsilon}{2} < \epsilon.$$

This completes the proof. $\qquad\qquad\qquad\qquad\qquad\qquad\qquad\qquad\qquad\qquad\qquad\qquad$ □

Applying Example 1.6, we can prove a stronger result: $\lim_{n \to \infty} \frac{a_n}{n} = 0$. Indeed, the assumption implies that

$$\lim_{n \to \infty} (a_{2n} - a_{2(n-1)}) = 0, \quad \lim_{n \to \infty} (a_{2n+1} - a_{2n-1}) = 0.$$

In view of Example 1.6, we have

$$\lim_{n \to \infty} \frac{a_{2n}}{n} = \lim_{n \to \infty} \frac{a_2 + (a_4 - a_2) + \cdots + (a_{2n} - a_{2(n-1)})}{n} = \lim_{n \to \infty} (a_{2n} - a_{2(n-1)}) = 0,$$

and so $\lim_{n \to \infty} \frac{a_{2n}}{2n} = 0$. Similarly, we can show that $\lim_{n \to \infty} \frac{a_{2n+1}}{2n+1} = 0$.

1.3.2 Cauchy criterion

Real mathematics must be justified as art if it can be justified at all.

— Godfrey Harold Hardy

The Cauchy criterion guarantees a sequence converges provided the terms accumulate. This is one of our primary tools for showing that a desired limit actually exists even if we do not explicitly know what it is.

Example 1.8. *For $n \in \mathbb{N}$, let*

$$a_n = 1 - \frac{1}{2} + \frac{1}{3} - \cdots + (-1)^{n+1} \frac{1}{n}.$$

Show that a_n converges.

Proof. For $m, n \in \mathbb{N}$ and $m > n$, if $m - n$ is odd, we have

$$\begin{aligned}
|a_m - a_n| &= \left| \frac{1}{n+1} - \frac{1}{n+2} + \cdots + \frac{1}{m-1} - \frac{1}{m} \right| \\
&= \frac{1}{n+1} - \left(\frac{1}{n+2} - \frac{1}{n+3} \right) - \cdots - \left(\frac{1}{m-2} - \frac{1}{m-1} \right) - \frac{1}{m} \\
&\leq \frac{1}{n+1} < \frac{1}{n}.
\end{aligned}$$

If $m - n$ is even, we have

$$|a_m - a_n| = |a_m - a_{m-1} + a_{m-1} - a_n| \leq |a_{m-1} - a_n| + \frac{1}{m} < \frac{1}{n} + \frac{1}{m} < \frac{2}{n}.$$

Thus, for every $\epsilon > 0$, letting $N > 1/2\epsilon$ yields

$$|a_m - a_n| < \epsilon \qquad \text{whenever } m, n > N.$$

This indicates that a_n is a Cauchy sequence. Therefore, a_n converges. $\square$

We will see that a_n actually converges to $\ln 2$ later.

Example 1.9. *Let $a_1 = 1, a_{n+1} = 1/(1 + a_n), (n = 1, 2, \ldots)$. Prove that a_n converges and find its limit.*

Proof. By induction, it is easy to show that $1/2 \leq a_n \leq 1$ for all $n \in \mathbb{N}$. Thus, for any positive integers n and k, we have

$$\begin{aligned}
|a_{n+k} - a_n| &= \frac{|a_{n-1+k} - a_{n-1}|}{(1 + a_{n-1+k})(1 + a_{n-1})} \\
&\leq \frac{1}{(1 + 1/2)^2} |a_{n-1+k} - a_{n-1}| = \frac{4}{9} |a_{n-1+k} - a_{n-1}|.
\end{aligned}$$

Repeatedly applying this inequality yields

$$|a_{n+k} - a_n| \leq \left(\frac{4}{9} \right)^{n-1} |a_2 - a_1|.$$

Since $(4/9)^n \to 0$ as $n \to \infty$, it follows that a_n is a Cauchy sequence. Hence a_n converges. Denote its limit by a. Taking the limit in $a_{n+1} = 1/(1 + a_n)$ yields $a^2 + a = 1$. Therefore, $a = (\sqrt{5} - 1)/2 = 0.618\ldots$. $\square$

Remarks. The first 20 terms of the sequence are given by Mathematica

```
f[x_]:=1/(1 + x)
NestList[f, 1, 20]
{1, 1/2, 2/3, 3/5, 5/8, 8/13, 13/21, 21/34, 34/55, 55/89,
89/144, 144/233, 233/377, 377/610, 610/987, 987/1597, 1597/2584,
2584/4181, 4181/6765, 6765/10946, 10946/17711}
```

This suggests that $a_n = F_n/F_{n+1}$, where F_n is the nth Fibonacci number. We can confirm this by a simple induction argument on n. In general, the recursion $a_{n+1} = 1/(1+a_n)$ leads to

$$a_n = \frac{a_0 F_{n-1} + F_n}{a_0 F_n + F_{n+1}}, \qquad \text{for } a_0 \neq 0.$$

Example 1.10. *(Example 1.1 Revisited). Show that $r_n = F_{n+1}/F_n$ is a Cauchy sequence, where F_n is the nth Fibonacci number.*

Proof. For any given $\epsilon > 0$, let $N \in \mathbb{N}$ with $N > 1/\epsilon$. Assume that $n, m \in \mathbb{N}$ with $m > n > N$. We have

$$|r_m - r_n| = \left| \frac{F_{m+1}}{F_m} - \frac{F_{n+1}}{F_n} \right|$$

$$= \left| \left(\frac{F_{m+1}}{F_m} - \frac{F_m}{F_{m-1}} \right) + \left(\frac{F_m}{F_{m-1}} - \frac{F_{m-1}}{F_{m-2}} \right) + \cdots + \left(\frac{F_{n+2}}{F_{n+1}} - \frac{F_{n+1}}{F_n} \right) \right|$$

$$= \left| \frac{F_{m+1}F_{m-1} - F_m^2}{F_m F_{m-1}} + \frac{F_m F_{m-2} - F_{m-1}^2}{F_{m-1}F_{m-2}} + \cdots + \frac{F_{n+2}F_n - F_{n+1}^2}{F_{n+1}F_n} \right|.$$

Applying the identity $F_{k+1}F_{k-1} - F_k^2 = (-1)^k$ for $k \geq 2$, we find

$$|r_m - r_n| = \left| \frac{(-1)^m}{F_m F_{m-1}} + \frac{(-1)^{m-1}}{F_{m-1}F_{m-2}} + \cdots + \frac{(-1)^{n+1}}{F_{n+1}F_n} \right|$$

$$= \left| \frac{1}{F_n F_{n+1}} - \frac{1}{F_{n+2}F_{n+1}} + \cdots + \frac{(-1)^{m-n-1}}{F_m F_{m-1}} \right|$$

$$\leq \frac{1}{F_n F_{n+1}} \leq \frac{1}{F_n}.$$

Since $F_n > n$ for $n > 5$, we find that, if $n, m > \max\{5, N\}$ with $m > n$, then

$$|r_m - r_n| \leq \frac{1}{F_n} < \frac{1}{n} < \frac{1}{N} < \epsilon,$$

which proves that r_n is a Cauchy sequence by the definition. $\qquad \square$

In the Cauchy criterion, be aware that the integer m must be arbitrary. Otherwise, for any given m, even though $a_m - a_n \to 0$, a_n itself may still diverge. Here is a counterexample.

Example 1.11. *For any increasing positive integer sequence $\phi(n)$, there exists a divergent sequence a_n such that*

$$\lim_{n \to \infty} (a_{\phi(n)} - a_n) = 0.$$

Proof. By induction, we must have

$$\phi(n) \geq n, \quad \phi(n+k) \geq n + \phi(k), \qquad \text{for all } n, k \in \mathbb{N}.$$

Thus $\phi(n) \to \infty$ as $n \to \infty$. Next we separate into two cases:
(1) $\{\phi(n) - n\}$ is bounded. i.e., for every $n \in \mathbb{N}$, there is a number $M > 0$ such that $\phi(n) - n \leq M$. Let $a_n = H_n := \sum_{k=1}^{n} 1/k$, the nth harmonic number, which is divergent. Then

$$a_{\phi(n)} - a_n = \sum_{k=1}^{\phi(n)} \frac{1}{k} - \sum_{k=1}^{n} \frac{1}{k}$$

$$= \sum_{k=n+1}^{\phi(n)} \frac{1}{k} \leq \frac{M}{n+1} \to 0 \quad \text{as } n \to \infty.$$

(2) $\{\phi(n) - n\}$ is unbounded. Let k_0 be the smallest positive integer such that $\phi(k) > k$. Define

$$a_n = \begin{cases} 1, & n = k_0, \phi(k_0), \phi(\phi(k_0)), \ldots, \\ 0, & \text{otherwise.} \end{cases}$$

Clearly, $\{a_n\}$ has an infinite subsequence with the identical term 1. On the other hand, since $\{\phi(n) - n\}$ is unbounded, $\{a_n\}$ must have an infinite subsequence with the identical term 0. Thus, a_n diverges. However, by the definition of a_n, we have $a_{\phi(n)} = a_n$. $\qquad\square$

1.3.3 The squeeze theorem

Truth is much too complicated to allow anything but approximations.

— John von Neumann

The squeeze theorem states that if sequences a_n, b_n, and c_n satisfy

$$b_n \leq a_n \leq c_n \quad \text{and} \quad \lim_{n\to\infty} b_n = \lim_{n\to\infty} c_n = a,$$

then $\lim_{n\to\infty} a_n = a$. "Trapping" or "squeezing" a_n is based on inequalities. To obtain the identical limits of b_n and c_n, a quite precise estimate is often required.

Example 1.12. *Let $a_n > 0$ for all $n \in \mathbb{N}$ and $\lim_{n\to\infty} a_n = a > 0$. Show that $\lim_{n\to\infty} \sqrt[n]{a_1 a_2 \cdots a_n} = a$.*

Proof. By the assumptions, we have $\lim_{n\to\infty} 1/a_n = 1/a$. Applying the result of Example 1.6 yields

$$\lim_{n\to\infty} \frac{a_1 + a_2 + \cdots + a_n}{n} = a \tag{1.1}$$

and

$$\lim_{n\to\infty} \frac{\frac{1}{a_1} + \frac{1}{a_2} + \cdots + \frac{1}{a_n}}{n} = \frac{1}{a}.$$

The later limit implies

$$\lim_{n\to\infty} \frac{n}{\frac{1}{a_1} + \frac{1}{a_2} + \cdots + \frac{1}{a_n}} = a. \tag{1.2}$$

Recall the *AM-GM-HM inequality*: If $a_1, a_2, \ldots, a_n \in \mathbb{R}$ are positive, then

$$\frac{n}{\frac{1}{a_1} + \frac{1}{a_2} + \cdots + \frac{1}{a_n}} \leq \sqrt[n]{a_1 a_2 \cdots a_n} \leq \frac{a_1 + a_2 + \cdots + a_n}{n}.$$

The desired limit now follows from the facts (1.1), (1.2), and the squeeze theorem. $\qquad\square$

Example 1.13. *Find*

$$\lim_{n\to\infty} \frac{1}{n^2} \sum_{k=1}^{n} \csc\left(\frac{1}{k}\right).$$

Solution. For $x \in (0, \pi/2)$, using the Taylor approximations of $\sin x$ gives

$$x - \frac{x^3}{6} \leq \sin x \leq x.$$

Equivalently,

$$x \leq \csc x \leq \frac{6}{6x - x^3}.$$

Letting $x = 1/k$ gives

$$k \leq \csc\left(\frac{1}{k}\right) \leq \frac{6}{6/k - 1/k^3} = k + \frac{k}{6k^2 - 1} \leq k + \frac{1}{5k}.$$

Summing this inequality for k from 1 to n leads to

$$\frac{n(n+1)}{2} \leq \sum_{k=1}^{n} \csc\left(\frac{1}{k}\right) \leq \frac{n(n+1)}{2} + \frac{1}{5} H_n,$$

where $H_n := \sum_{k=1}^{n} 1/k$ is the nth *harmonic number*. Dividing by n^2 and taking the limit as $n \to \infty$ yields

$$\lim_{n \to \infty} \frac{1}{n^2} \sum_{k=1}^{n} \csc\left(\frac{1}{k}\right) = \frac{1}{2}.$$

□

Example 1.14. *Let*

$$s_n = \sum_{k=1}^{n} \ln\left(1 + \frac{1}{2n + k}\right).$$

Prove that $\lim_{n \to \infty} s_n = \ln(3/2)$.

Proof. We first observe that for all $x \geq 0$,

$$\frac{x}{1 + x} \leq \ln(1 + x) \leq x. \tag{1.3}$$

Setting $x = 1/(2n + k)$ yields

$$\frac{1}{2n + k + 1} \leq \ln\left(1 + \frac{1}{2n + k}\right) \leq \frac{1}{2n + k}.$$

Summing both sides for k from 1 to n gives

$$\sum_{k=1}^{n} \frac{1}{2n + k + 1} \leq s_n \leq \sum_{k=1}^{n} \frac{1}{2n + k}.$$

Rewriting

$$\sum_{k=1}^{n} \frac{1}{2n + k} = \sum_{k=1}^{n} \left(\frac{1}{2 + k/n}\right) \cdot \frac{1}{n}$$

yields the Riemann sum of $f(x) = 1/(2 + x)$ on $[0, 1]$. Thus, we have

$$\lim_{n \to \infty} \sum_{k=1}^{n} \left(\frac{1}{2 + k/n}\right) \cdot \frac{1}{n} = \int_{0}^{1} \frac{dx}{2 + x} = \ln(3/2).$$

Similarly, we have

$$\lim_{n \to \infty} \sum_{k=1}^{n} \frac{1}{2n + k + 1} = \ln(3/2).$$

The desired limit now follows from the squeeze theorem.   □

1.3.4 Monotone convergence theorem

We are here, in fact, a passage to the limit of unexampled audacity. — Felix Klein

Similar to the Cauchy criterion, the monotone convergence theorem also enables us to show the convergence of a large class of sequences without seeing the values of the limits. Applying this theorem requires us to verify two assertions:

1. a_n is monotone.

2. a_n is bounded either below or above.

Example 1.15. *Let $H_n := \sum_{k=1}^{n} 1/k$ be the nth harmonic number. Define*

$$d_n := H_n - \ln n.$$

Prove that $\lim_{n \to \infty} d_n$ exists.

Proof. Applying $x = 1/n$ in (1.3) yields

$$\frac{1}{1+n} < \ln(1 + 1/n) < \frac{1}{n}.$$

Then

$$d_{n+1} - d_n = \frac{1}{n+1} - \ln(1+n) + \ln n = \frac{1}{1+n} - \ln\left(1 + \frac{1}{n}\right) < 0.$$

This shows that d_n is monotone decreasing. On the other hand,

$$d_n = H_n - \ln\left(\frac{n}{n-1} \cdot \frac{n-1}{n-2} \cdots \frac{3}{2} \cdot \frac{2}{1}\right)$$

$$= H_n - \sum_{k=1}^{n-1} \ln\left(1 + \frac{1}{k}\right)$$

$$= \sum_{k=1}^{n-1} \left(\frac{1}{k} - \ln\left(1 + \frac{1}{k}\right)\right) + \frac{1}{n}$$

$$> \frac{1}{n} > 0.$$

Hence d_n is bounded below. It then follows that d_n converges from the monotone convergence theorem. $\square$

Euler introduced the symbol γ as the limit of d_n, namely

$$\lim_{n \to \infty} (H_n - \ln n) = \gamma = 0.5772156649\ldots, \qquad (1.4)$$

which is often called the *Euler–Mascheroni constant*. Similar to π and e, γ appears frequently in many mathematical areas yet maintains a profound air of mystery. For example, although there is compelling evidence that γ is irrational, a proof has not yet been found.

As an application of (1.4), the following elegant problem, due to Euler, presents a surprise link between the *Euler sums* and the *Riemann zeta function* $\zeta(n) := \sum_{k=1}^{\infty} 1/k^n$.

Example 1.16. *Let*

$$E_n = \sum_{k=1}^{n} \frac{H_k}{k(k+1)}, \qquad (n = 1, 2, \ldots).$$

Find $\lim_{n \to \infty} E_n$.

Solution. Using partial fraction decomposition and shifting the summation index, we have

$$E_n = \sum_{k=1}^{n} \left(\frac{H_k}{k} - \frac{H_k}{k+1} \right)$$

$$= \sum_{k=1}^{n} \frac{H_k}{k} - \sum_{k=2}^{n+1} \frac{H_{k-1}}{k}$$

$$= 1 + \sum_{k=2}^{n} \frac{H_k - H_{k-1}}{k} - \frac{1}{n+1} H_n$$

$$= \sum_{k=1}^{n} \frac{1}{k^2} - \frac{H_n - \ln n}{n+1} - \frac{\ln n}{n+1}.$$

Letting $n \to \infty$, in view of (1.4), we find that

$$\lim_{n \to \infty} E_n = \sum_{k=1}^{\infty} \frac{1}{k^2} = \zeta(2) = \frac{\pi^2}{6}.$$

$\square$

Example 1.17. *Let*
$$P_n(x) = x^n + x^{n-1} + \cdots + x - 1.$$

Prove that $P_n(x) = 0$ has a unique positive solution. Denoting this by a_n, find $\lim_{n \to \infty} a_n$.

Proof. Notice that

$$P_n(0) = -1, \ P_n(1) = n - 1 \geq 0, \ P_n'(x) > 0 \ \text{ for } x \in (0, \infty).$$

By the intermediate value theorem, there exists a unique $a_n \in (0, 1]$ such that $P_n(a_n) = 0$. Now we show that a_n is decreasing. To this end, since

$$\begin{aligned} P_n(a_{n+1}) &= a_{n+1}^n + a_{n+1}^{n-1} + \cdots + a_{n+1} - 1 \\ &= (a_{n+1}^{n+1} + a_{n+1}^n + \cdots + a_{n+1}^2 + a_{n+1} - 1) - a_{n+1}^{n+1} \\ &= P_{n+1}(a_{n+1}) - a_{n+1}^{n+1} = -a_{n+1}^{n+1} < 0, \end{aligned}$$

it follows that $P_n(a_n) = 0 > P_n(a_{n+1})$. The fact that $P_n(x)$ is increasing in $(0, \infty)$ implies that $a_n > a_{n+1}$. Thus, by the monotone convergence theorem, $\lim_{n \to \infty} a_n$ exists. Appealing to the identity

$$(x-1)P_n(x) = (x-1)[(x^{n-1} + x^{n-2} + \cdots + x + 1)x - 1] = x^{n+1} - 2x + 1,$$

we have $a_n^{n+1} - 2a_n + 1 = 0$. Thus

$$0 < a_n - \frac{1}{2} = \frac{a_n^{n+1}}{2} < \frac{a_2^{n+1}}{2} \to 0 \quad \text{as } n \to \infty.$$

This implies that $\lim_{n \to \infty} a_n = 1/2$ by the squeeze theorem. $\square$

Example 1.18 (CMJ Problem 965, 2011). *Motivated by Bernoulli's inequality $(1 + x)^n \geq 1 + nx$ for $x \geq -1$, let $G_n(x) = (1+x)^n - (1+nx)$ for positive integers n.*

> 1. *Show that for every odd positive integer $n \geq 3$ there is a unique negative real number x_n such that $G_n(x_n) = 0$.*
>
> 2. *Prove that the sequence $x_3, x_5, x_7, \ldots$ is convergent and find its limit x_∞.*

3. *Verify that for all positive integers n, even and odd, $(1+x)^n \geq 1 + nx$ for $x \geq x_\infty$ and that x_∞ is the smallest number for which this statement is true.*

4. *Show that $x_n - x_\infty = O(\ln(n)/n)$.*

Here, as usual, $a_n = O(b_n)$ means that there exist positive constant c and $N \in \mathbb{N}$ such that $|a_n| \leq c b_n$ for all $n \geq N$.

Solution. (1) Clearly, $G_n(x)$ is continuous. For odd positive integers $n \geq 3$, $G_n(-2) = 2(n-1) > 0$ and $\lim_{x \to -\infty} G_n(x) = -\infty$. Thus, by the intermediate value theorem, there exists $x_n \in (-\infty, -2)$ such that $G_n(x_n) = 0$. Since

$$G_n'(x) = n\left((1+x)^{n-1} - 1\right) = \begin{cases} > 0, & \text{for } x < -2 \\ < 0, & \text{for } -2 < x < 0 \end{cases}$$

$G_n(x)$ is strictly increasing on $(-\infty, -2)$ and strictly decreasing on $(-2, 0)$. It follows that $G_n(x) > G_n(0) = 0$ for $x \in (-2, 0)$, and so $G_n(x) = 0$ has a unique negative solution x_n.

(2) We show that $x_3 < x_5 < x_7 < \cdots \leq -2$ and $x_\infty = -2$. For this part, assume that n is odd. Since $G_n(x)$ is increasing on $(-\infty, -2)$ and

$$G_n\left(-2 - \frac{1}{n}\right) = 2n - \left(1 + \frac{1}{n}\right)^n > 0 = G_n(x_n),$$

we deduce that $x_n < -2 - 1/n$. Thus,

$$\begin{aligned} G_{n+2}(x_n) &= (1+x_n)^{n+2} - (1 + (n+2)x_n) \\ &= (1+x_n)^2(1 + nx_n) - (1 + (n+2)x_n) \\ &= x_n^2(2n + 1 + nx_n) < 0 = G_{n+2}(x_{n+2}). \end{aligned}$$

It follows that $x_n < x_{n+2} < -2$ for all odd positive integers $n \geq 3$. Thus $\lim_{n \to \infty} x_n$ exists. Let $y_n(k) = -2 - k\ln(n)/n$, where k is a positive constant. For sufficiently large odd n and $k > 1$, we have $y_n(k) \in (-3, -2)$ and

$$G_n(y_n(k)) = -\left(1 + \frac{k\ln(n)}{n}\right)^n - 1 + 2n + k\ln(n) \sim -n^k - 1 + 2n + k\ln(n) < 0. \quad (1.5)$$

Thus, $x_n > y_n(k)$ and

$$-2 - \frac{k\ln(n)}{n} < x_n < -2 - \frac{1}{n}.$$

This proves that $x_\infty = -2$ as desired.

(3) We only need to show the inequality holds for $-2 \leq x \leq -1$. Indeed,

$$(1+x)^n \geq -|1+x|^n \geq -|1+x| = 1 + x \geq 1 + nx.$$

If there exists a $t < -2$ such that $(1+t)^n \geq 1 + nt$ for all positive integers n, then $G_n(t) \geq 0$ for all n and so $x_n \leq t$. Thus, $x_\infty \leq t < -2$. This contradicts that $x_\infty = -2$.

(4) In view of (1.5), for large odd n, since

$$G_n(y_n(2)) < 0 = G_n(x_n) < G_n(y_n(1)) > 0,$$

we find that $y_n(2) \leq x_n \leq y_n(1)$. It follows that $x_n - x_\infty = O(\ln(n)/n)$ from the definition of y_n. $\qquad \square$

1.3.5 Upper and lower limits

The essence of mathematics resides in its freedom — George Cantor

Given a bounded sequence a_n, we can always generate two convergent monotone sequences. Indeed, define

$$b_k = \sup\{a_k, a_{k+1}, a_{k+2}, \ldots\}, \quad c_k = \inf\{a_k, a_{k+1}, a_{k+2}, \ldots\}.$$

Clearly,

$$b_1 \geq b_2 \cdots \geq b_k \geq b_{k+1} \geq c_{k+1} \geq c_k \geq \cdots \geq c_2 \geq c_1.$$

Thus, by the monotone convergence theorem, both $\lim_{k \to \infty} b_k = \inf\{b_n\}$ and $\lim_{k \to \infty} c_k = \sup\{c_n\}$ exist. They are called the *upper and lower limits* of a_n, and denoted by $\limsup a_n$ and $\liminf a_n$, respectively. It is easy to see that

$$\inf\{a_n\} \leq \liminf a_n \leq \limsup a_n \leq \sup\{a_n\}.$$

In the applications, the upper and lower limits are often described by the following equivalent forms:

1. The lower limit $\liminf a_n = a$ is the unique number that has the properties

   - For each $\epsilon > 0$, there are *finitely many* a_n's such that $a_n \leq a - \epsilon$.
   - For each $\epsilon > 0$, there are *infinitely many* a_n's such that $a_n < a + \epsilon$.

2. Similarly, the upper limit $\limsup a_n = b$ is the unique number that has the properties

   - For each $\epsilon > 0$, there are *finitely many* a_n's such that $a_n \geq b + \epsilon$.
   - For each $\epsilon > 0$, there are *infinitely many* a_n's such that $a_n > b - \epsilon$.

Moreover, a_n converges if and only if

$$\liminf a_n = \limsup a_n.$$

The following three examples demonstrate how the guaranteed existence of the upper and lower limits offers us an additional bonus to finding the limits.

Example 1.19. *Let x_n be a sequence satisfying $0 \leq x_{m+n} \leq x_m + x_n$ for all $m, n \in \mathbb{N}$. Prove that $\lim_{n \to \infty} x_n/n$ exists.*

Discussion. The proof here is somewhat more involved than those we have encountered so far and we discuss some of the idea before presenting the proof. First, assuming that the limit exists, we want to see the possible values of the limit. For each $k \in \mathbb{N}$, repeatedly using $x_{m+n} \leq x_m + x_n$ yields that $x_{km} \leq k x_m$. Thus,

$$0 \leq \frac{x_{km}}{km} \leq \frac{x_m}{m}, \quad (k, m = 1, 2, \ldots.)$$

and so

$$\lim_{n \to \infty} \frac{x_n}{n} = \lim_{k \to \infty} \frac{x_{km}}{km} \leq \frac{x_m}{m}, \quad (m = 1, 2, \ldots.)$$

This implies that

$$\lim_{n \to \infty} \frac{x_n}{n} \leq \inf\left\{\frac{x_n}{n}\right\}.$$

Hence, if $\lim_{n \to \infty} x_n/n$ exists, then it must be

$$\lim_{n \to \infty} \frac{x_n}{n} = \inf\left\{\frac{x_n}{n}\right\}.$$

Proof. Since $0 \leq x_{m+n} \leq x_m + x_n$, in particular, we have

$$0 \leq x_n \leq nx_1 \quad \Longrightarrow \quad 0 \leq \frac{x_n}{n} \leq x_1.$$

Thus, $\{x_n/n\}$ is bounded below, so $\alpha = \inf\{x_n/n\}$ exists. In the following, we show that $\lim_{n\to\infty} x_n/n = \alpha$. To this end, by the definition of the infimum, for any $\epsilon > 0$, there is an integer N such that

$$\alpha \leq \frac{x_N}{N} < \alpha + \epsilon.$$

When $n > N$, there are unique r and q such that $n = rN + q$ $(0 \leq q < N)$. Thus,

$$x_n \leq x_{rN+q} \leq rx_N + x_q$$

and

$$\alpha \leq \frac{x_n}{n} \leq \frac{x_N}{N} \cdot \frac{rN}{rN+q} + \frac{x_q}{n} \leq \alpha + \epsilon + \frac{x_q}{n}.$$

The claimed limit now follows from the fact that $\lim_{n\to\infty} x_q/n = 0$ and the squeeze theorem. $\square$

As in the Cauchy criterion, here the m and n must be completely arbitrary. Otherwise, assume that

$$0 \leq x_{m+n} \leq x_m + x_n, \quad \text{for all } m, n \in \mathbb{N} \text{ with } |m - n| \leq k. \tag{1.6}$$

Choose $A = \{2^m : m \in \mathbb{N}\}$. Define

$$a_n = \begin{cases} n, & \text{if dist}(n, A) \leq k \\ 0, & \text{otherwise.} \end{cases}$$

Clearly,

$$\liminf \frac{a_n}{n} = 0, \quad \limsup \frac{a_n}{n} = 1.$$

Thus, $\lim_{n\to\infty} a_n$ does not exist. We now verify a_n satisfies the condition (1.6). By contradiction, if there are integers $m, n \geq 1$ such that

$$|m - n| \leq k, \quad a_{m+n} > a_m + a_n, \tag{1.7}$$

this implies that either $a_n = 0$ or $a_m = 0$. Assuming $a_n = 0$, for some positive integer r, we have

$$2^{r-1} + k < n < 2^r - k,$$

which implies

$$2^r + k < n + (n - k) \leq n + m \leq n + (n + m) < 2^{r+1} - k,$$

and so $a_{n+m} = 0$. This contradicts (1.7).

The following example shows the link between the ratio and the nth root of the terms of a positive sequence.

Example 1.20. *Let x_n be a positive sequence. If $\lim_{n\to\infty} x_{n+1}/x_n = l$, prove that*

$$\lim_{n\to\infty} \sqrt[n]{x_n} = l.$$

Proof. By the assumption, for every $\epsilon > 0$, there is a positive integer N such that

$$l - \epsilon \leq \frac{x_{n+1}}{x_n} \leq l + \epsilon \quad \text{whenever } n > N \tag{1.8}$$

For $n > N$, since

$$\frac{x_n}{x_N} = \frac{x_n}{x_{n-1}} \cdot \frac{x_{n-1}}{x_{n-2}} \cdots \frac{x_{N+2}}{x_{N+1}} \cdot \frac{x_{N+1}}{x_N},$$

Applying (1.8) repeatedly yields

$$(l - \epsilon)^{n-N} \leq \frac{x_n}{x_N} \leq (l + \epsilon)^{n-N},$$

which is equivalent to

$$\sqrt[n]{x_N}(l - \epsilon)^{(n-N)/n} \leq \sqrt[n]{x_n} \leq \sqrt[n]{x_N}(l + \epsilon)^{(n-N)/n}.$$

Taking the upper and lower limits, and using the fact that $\lim_{n \to \infty} \sqrt[n]{x_N} = 1$, we find that

$$l - \epsilon \leq \liminf \sqrt[n]{x_n} \leq \limsup \sqrt[n]{x_n} \leq l + \epsilon.$$

Since ϵ is arbitrary, we deduce that

$$\liminf \sqrt[n]{x_n} = \limsup \sqrt[n]{x_n} = l,$$

and so $\lim_{n \to \infty} \sqrt[n]{x_n} = l$. $\qquad\square$

The above proof acutely demonstrates that, for any positive sequence x_n,

$$\liminf \frac{x_{n+1}}{x_n} \leq \liminf \sqrt[n]{x_n} \leq \limsup \sqrt[n]{x_n} \leq \limsup \frac{x_{n+1}}{x_n}.$$

Example 1.21. *Assume the sequence x_n is bounded and $\lim_{n \to \infty} (x_{n+1} - x_n) = 0$. Let*

$$l = \liminf x_n, \quad L = \limsup x_n.$$

Let S be the set of limit points of the sequence $\{x_n\}$. Prove that $S = [l, L]$.

Proof. Since x_n is bounded, the Bolzano–Weierstrass theorem implies that S is not empty. Moreover, $S \subset [l, L]$. If $l = L$, then $\lim_{n \to \infty} x_n = \liminf x_n = \limsup x_n$. The assertion holds clearly. If $l < L$, it suffices to show that every $a \in (l, L)$ is a limit point of the sequence. Let $0 < \epsilon < \frac{1}{2} \min\{a - l, L - a\}$. Then $l < a - \epsilon < a < a + \epsilon < L$. By the assumption that $\lim_{n \to \infty} (x_{n+1} - x_n) = 0$, there exists $N \in \mathbb{N}$ such that

$$|x_{n+1} - x_n| < 2\epsilon, \quad \text{whenever } n > N.$$

On the other hand, by the definition of the upper and lower limits, there are $n_1', n_1'' > N$ such that

$$x_{n_1'} \in (l - \epsilon, l + \epsilon) \quad \text{and} \quad x_{n_1''} \in (L - \epsilon, L + \epsilon).$$

Without loss of generality, assume that $n_1' < n_1''$. Hence,

$$x_{n_1'} < a - \epsilon < a + \epsilon < x_{n_1''}.$$

We now show that there exists $n_1 \in \mathbb{N}$ satisfying $n_1' < n_1 < n_1''$ such that

$$a - \epsilon < x_{n_1} < a + \epsilon.$$

By contradiction, assume that $x_{n_1'+1}, x_{n_1'+2}, \ldots, x_{n_1''-1} \notin (a-\epsilon, a+\epsilon)$. Let $x_{n_1'+p}$ be the first such that $x_{n_1'+p} > a + \epsilon$ among $\{x_{n_1'+1}, x_{n_1'+2}, \ldots, x_{n_1''-1}, x_{n_1''}\}$. Then

$$x_{n_1'+p-1} < a - \epsilon < a + \epsilon < x_{n_1'+p},$$

and so

$$|x_{n_1'+p} - x_{n_1'+p-1}| > 2\epsilon,$$

a contradiction. Thus, there exists $x_{n_1} = x_{n_1'+p} \in (a-\epsilon, a+\epsilon)$, and so a is a limit point of $\{x_n\}$. $\square$

Example 1.22 (Monthly Problem 12143, 2019). *Compute*

$$\lim_{n\to\infty} \sum_{k=1}^n \left(\frac{k}{n}\right)^k.$$

Proof. We show the upper and lower limits both are trapped by $e/(e-1)$. Thus, the desired limit is equal to $e/(e-1)$. We begin with the lower limit. Rewrite

$$\left(\frac{k}{n}\right)^k = \left(\frac{k+(n-k)}{k}\right)^{-k} = \left(1 + \frac{(n-k)}{k}\right)^{-k}.$$

Since $(1+x/k)^{-k}$ is decreasing in k for $x \geq 0$, we have

$$\left(\frac{k}{n}\right)^k \geq e^{-(n-k)}.$$

Thus

$$\sum_{k=1}^n \left(\frac{k}{n}\right)^k \geq \sum_{k=1}^n e^{-(n-k)} = \sum_{m=0}^{n-1} e^{-m} \quad \text{(use } m = n-k\text{)},$$

and so

$$\liminf \sum_{k=1}^n \left(\frac{k}{n}\right)^k \geq \sum_{m=0}^\infty e^{-m} = \frac{e}{e-1}.$$

Next, for $r \in (0,1)$, let

$$J = \{k : 1 \leq k \leq rn\}, \ K = \{k : rn < k < n - n^{1/3}\}, \text{ and } L = \{k : n - n^{1/3} \leq k \leq n\}.$$

For $k \in J$, we have

$$\sum_{k\in J} \left(\frac{k}{n}\right)^k \leq \sum_{k\in J} r^k \leq \sum_{k=1}^\infty r^k = \frac{r}{1-r}.$$

For $k \in K$, since $r < k/n < 1 - n^{-2/3}$, we have

$$\sum_{k\in K} \left(\frac{k}{n}\right)^k \leq \sum_{k\in K} \left(\frac{k}{n}\right)^{rn} \leq n\int_0^{1-n^{-2/3}} x^{rn}\, dx = \frac{n}{1+rn}\left(1 - \frac{1}{n^{2/3}}\right)^{rn+1}.$$

Moreover, using $k/n = 1 - (n-k)/n$, we have

$$\sum_{k \in L}^{n} \left(\frac{k}{n}\right)^k = \sum_{m=0}^{n^{1/3}} {}'\left(1 - \frac{m}{n}\right)^{n-m} \quad \text{(use } m = n - k\text{)}$$

$$= \sum_{m=0}^{n^{1/3}} \exp\left((n-m)\ln\left(1 - \frac{m}{n}\right)\right)$$

$$\leq \sum_{m=0}^{n^{1/3}} \exp\left((n-m)\left(-\frac{m}{n}\right)\right) \quad \text{(use } \ln(1-x) \leq -x \text{ for } x \in (0,1)\text{)}$$

$$= \sum_{m=0}^{n^{1/3}} \exp(-m)\exp(m^2/n) \leq \exp(1/n^{1/3}) \sum_{m=0}^{\infty} e^{-m}$$

$$= \frac{e}{e-1}\exp(1/n^{1/3}).$$

In summary, we find that

$$\sum_{k=1}^{n} \left(\frac{k}{n}\right)^k \leq \frac{r}{1-r} + \frac{n}{1+rn}\left(1 - \frac{1}{n^{2/3}}\right)^{rn+1} + \frac{e}{e-1}\exp(1/n^{1/3}).$$

Therefore,

$$\limsup \sum_{k=1}^{n} \left(\frac{k}{n}\right)^k \leq \frac{r}{1-r} + \frac{e}{e-1}.$$

Since $r \in (0,1)$ is arbitrary, letting $r \to 0$ yields

$$\limsup \sum_{k=1}^{n} \left(\frac{k}{n}\right)^k \leq \frac{e}{e-1}$$

as desired. $\square$

1.3.6 Stolz-Cesàro theorem

In Mathematics, you understand what you build up. — Fan Chung

The Stolz-Cesàro theorem, similar to its continuous analogue L'Hôpital's rule, provides us a systematic method to calculate limits in the form of $0/0$ and ∞/∞. Here we offer four concrete examples to illustrate the power of the Stolz-Cesàro theorem. We begin with revisiting Pólya's classical problem 173 in [77].

Example 1.23. *Define the n times iterated sine function by*

$$\sin_n x = \sin(\sin_{n-1} x), \ \sin_1 x = \sin x.$$

If $\sin x > 0$, *prove that*

$$\lim_{n \to \infty} \sqrt{\frac{n}{3}} \sin_n x = 1.$$

Proof. Since $\sin x < x$ for all $x > 0$, it follows that $0 < \sin_n x < \sin_{n-1} x$ whenever $\sin x > 0$. Thus, by the monotone convergence theorem, $\sin_n x$ converges. Let $\lim_{n\to\infty} \sin_n x = a$. Taking the limit in $\sin_n x = \sin(\sin_{n-1} x)$ yields $a = \sin a$, and so $a = 0$. Define

$$a_n = n, \quad b_n = \frac{1}{\sin_n^2 x}.$$

By the Stolz-Cesàro theorem, we have

$$\lim_{n\to\infty} n\sin_n^2 x = \lim_{n\to\infty} \frac{n}{1/\sin_n^2 x} = \lim_{n\to\infty} \frac{a_{n+1} - a_n}{b_{n+1} - b_n}.$$

Since

$$\lim_{n\to\infty} \frac{a_{n+1} - a_n}{b_{n+1} - b_n} = \lim_{n\to\infty} \frac{1}{\frac{1}{\sin^2(\sin_n x)} - \frac{1}{\sin_n^2 x}}$$

$$= \lim_{t\to 0} \frac{1}{\frac{1}{\sin^2 t} - \frac{1}{t^2}} \quad (t = \sin_n x)$$

$$= \lim_{t\to 0} \frac{t^2 \sin^2 t}{t^2 - \sin^2 t}$$

$$= 3,$$

this proves

$$\lim_{n\to\infty} \sqrt{\frac{n}{3}} \sin_n x = \sqrt{\lim_{n\to\infty} n\sin_n^2 x / 3} = 1$$

as desired. $\qquad\square$

In contrast to Pólya's original approach, this solution is simpler and more straightforward.

The second problem is on the study of asymptotic behavior of a sequence generated by the logistic equation. This equation, which models population growth, was first published by Pierre Verhulst in 1845 [68].

Example 1.24 (Putnam Problem, 1966-A3). *Let* $x_{n+1} = x_n(1-x_n), n = 1, 2, \ldots, x_1 \in (0,1)$. *Show that* $\lim_{n\to\infty} nx_n = 1$.

Proof. Since $x_1 \in (0,1)$, we have $x_2 = x_1(1-x_1) \in (0,1)$. Indeed, by induction, we conclude that $x_n \in (0,1)$ for all $n \in \mathbb{N}$. Notice that

$$0 < \frac{x_{n+1}}{x_n} = 1 - x_n < 1.$$

Then x_n is monotone decreasing and bounded below. It follows from the monotone convergence theorem that $\lim_{n\to\infty} x_n = L$ exists. Letting $n \to \infty$ in $x_{n+1} = x_n(1 - x_n)$ yields that $L = L(1 - L)$, which implies that $L = 0$.

Let $b_n = 1/x_n$. Then $b_n < b_{n+1}$ and $\lim_{n\to\infty} b_n = +\infty$. By the Stolz-Cesàro theorem,

$$\lim_{n\to\infty} nx_n = \lim_{n\to\infty} \frac{n}{b_n} = \lim_{n\to\infty} \frac{1}{b_{n+1} - b_n}.$$

On the other hand,

$$\lim_{n\to\infty} b_{n+1} - b_n = \lim_{n\to\infty} \left(\frac{1}{x_{n+1}} - \frac{1}{x_n} \right) = \lim_{n\to\infty} \frac{1}{1 - x_n} = 1.$$

This shows that $\lim_{n\to\infty} nx_n = 1$ as claimed. $\qquad\square$

The above solution is differ from the one that appeared in Putnam Mathematical Competition Problems and Solutions [3, pp. 51–52]. The application of the Stolz-Cesàro theorem allows us to offer a short alternative solution. Moreover, combining this result with the formula (1.4), we have (see Exercise 89)

$$\lim_{n \to \infty} \frac{n(1 - nx_n)}{\ln n} = 1.$$

In general, the sequence defined by $x_{n+1} = \lambda x_n(1 - x_n)$ will display richer qualitative types of behavior for different values of λ. The change from one form of qualitative behavior to another as λ changes is called a *bifurcation*. To learn more about the bifurcation study for the logistic model, please see Li and York's landmark paper "Period three implies chaos" [67] or [51, pp. 92–101].

The third problem is the **Monthly Problem E1557, 1963**. Both published solutions are based on Stirling's formula for $n!$ and Riemann sums. Here we present an elementary solution by sorely using the Stolz-Cesàro theorem.

Example 1.25. *Let* $S_n = (1/n^2) \sum_{k=0}^{n} \ln C_n^k$ *with* $C_n^k = n!/(k!(n-k)!)$. *Evaluate* $S := \lim_{n \to \infty} S_n$.

Solution. Let $a_n = \sum_{k=0}^{n} \ln C_n^k, b_n = n^2$. By the Stolz-Cesàro theorem,

$$\begin{aligned}
S &= \lim_{n \to \infty} \frac{\sum_{k=0}^{n+1} \ln C_{n+1}^k - \sum_{k=0}^{n} \ln C_n^k}{(n+1)^2 - n^2} \\
&= \lim_{n \to \infty} \frac{\sum_{k=0}^{n} \ln (C_{n+1}^k / C_n^k) + \ln C_{n+1}^{n+1}}{2n+1} \\
&= \lim_{n \to \infty} \frac{\sum_{k=0}^{n} \ln(n+1)/(n+1-k)}{2n+1} \\
&= \lim_{n \to \infty} \frac{(n+1)\ln(n+1) - \sum_{k=1}^{n+1} \ln k}{2n+1}.
\end{aligned}$$

Applying the Stolz-Cesàro theorem again yields

$$\begin{aligned}
S &= \lim_{n \to \infty} \frac{(n+1)\ln(n+1) - n \ln n - \ln(n+1)}{(2n+1) - (2n-1)} \\
&= \lim_{n \to \infty} \frac{1}{2} \ln \left(\frac{n+1}{n} \right)^n \\
&= \frac{1}{2}. \qquad \qquad \square
\end{aligned}$$

Finally, we derive a formula for the sums of the power of integers. It is based on the following well-known fact

$$S_k(n) = \sum_{i=1}^{n} i^k = \sum_{j=1}^{k+1} a_j(k) \, n^j \tag{1.9}$$

for some coefficients $a_j(k)$.

Example 1.26. *Determine the coefficients* $a_j(k)$ *in* (1.9).

Solution. To find the leading coefficient $a_{k+1}(k)$, we divide (1.9) by n^{k+1}, isolate $a_{k+1}(k)$, and then apply the Stolz-Cesàro theorem to get

$$\begin{aligned}
a_{k+1}(k) &= \lim_{n \to \infty} \frac{S_k(n) - \sum_{j=1}^{k} a_j(k)n^j}{n^{k+1}} = \lim_{n \to \infty} \frac{S_k(n)}{n^{k+1}} \\
&= \lim_{n \to \infty} \frac{(n+1)^k}{(n+1)^{k+1} - n^{k+1}} = \frac{1}{k+1}.
\end{aligned}$$

Similarly, for $1 \leq j \leq k$, we have

$$a_j(k) = \lim_{n \to \infty} \frac{S_k(n) - \sum_{i=1, i \neq j}^{k+1} a_i(k)n^i}{n^j}.$$

Applying the Stolz-Cesàro theorem again yields

$$a_j(k) = \lim_{n \to \infty} \frac{S_k(n) - \sum_{i=j+1}^{k+1} a_i(k)n^i}{n^j}$$

$$= \lim_{n \to \infty} \frac{(n+1)^k - a_{k+1}(k)[(n+1)^{k+1} - n^{k+1}] - \cdots - a_{j+1}(k)[(n+1)^{j+1} - n^j]}{(n+1)^j - n^j}.$$

Since

$$(n+1)^j - n^j = jn^{j-1} + C_j^2 n^{j-2} + \cdots + 1$$

and $a_j(k)$ is well-defined, we deduce that all of the terms of order higher than $j-1$ in the numerator must vanish while, all of the terms of order less than $j-1$ have no effect on the limit. Thus, applying the binomial theorem and collecting the coefficients of n^{j-1} in the numerator gives

$$C_k^{j-1} - a_{k+1}(k)C_{k+1}^{j-1} - a_k(k)C_k^{j-1} - \cdots - a_{j+1}(k)C_{j+1}^{j-1}.$$

Therefore,

$$a_j(k) = \frac{C_k^{j-1} - a_{k+1}(k)C_{k+1}^{j-1} - a_k(k)C_k^{j-1} - \cdots - a_{j+1}(k)C_{j+1}^{j-1}}{j}.$$

This, together with $a_{k+1}(k) = 1/(k+1)$, enables us to compute $a_j(k)$ recursively. $\square$

In particular, we have

$$S_2(n) = \sum_{i=1}^{n} i^2 = \frac{1}{3}n^3 + \frac{1}{2}n^2 + \frac{1}{6}n,$$

$$S_3(n) = \sum_{i=1}^{n} i^3 = \frac{1}{4}n^4 + \frac{1}{2}n^3 + \frac{1}{4}n^2,$$

$$S_4(n) = \sum_{i=1}^{n} i^4 = \frac{1}{5}n^5 + \frac{1}{2}n^4 + \frac{1}{3}n^3 - \frac{1}{30}n,$$

$$S_5(n) = \sum_{i=1}^{n} i^5 = \frac{1}{6}n^6 + \frac{1}{2}n^5 + \frac{5}{12}n^4 - \frac{1}{12}n^2.$$

It is interesting to see that, for each $k \in \mathbb{N}$,

$$\sum_{j=1}^{k+1} a_j(k) = 1.$$

1.3.7 Fixed-point theorems

The most practical solution is a good theory. — Albert Einstein

As we have seen in Examples 1.9 and 1.24, sequences are often defined by *recursion*. In this case, once the initial term is given, each further term of the sequence is specified as a function of the preceding terms. More precisely, let $f : \mathbb{R} \longmapsto \mathbb{R}$ be a function. A recursion is defined by

$$x_0 = a \text{ (given)}, \; x_{n+1} = f(x_n), \quad \text{for } n \in \mathbb{N}.$$

For example, the sequences in Examples 1.9 and 1.24 are recursions with $f(x) = 1/(1+x)$ and $f(x) = x(1-x)$, respectively. This kind of feedback mechanism provides an example of *dynamical system* [51], a process that occurs naturally throughout mathematics and the natural sciences, especially in genetics and mathematical biology. The recursive sequence

$$x, f(x), f(f(x)), f(f(f(x))), \ldots$$

is called the *orbit* of x. In dynamical system, the main question we will ask in the sequel: given a function f, as the initial value x changes, what ultimately happens to the fate of orbits? It is usually quite difficult to tell by looking at a function where its orbits will go. The next two problems offer a sufficient condition for the orbits to converge to some limit.

Definition 1.10. *A function* $f : I \to I$ *is called a contraction mapping if there is a constant* k, $0 < k < 1$, *such that*

$$|f(x) - f(y)| \leq k \, |x - y|, \quad \text{for all } x, y \in I. \tag{1.10}$$

The following result, as a direct consequence of the Cauchy criterion, shows that a recursive sequence defined by a contraction mapping always converges.

Example 1.27. *Let* $f : [a, b] \to [a, b]$ *be a contraction mapping. For any* $x_1 \in [a, b]$, *define*

$$x_{n+1} = f(x_n), \quad n = 1, 2, \ldots. \tag{1.11}$$

Then $\lim_{n \to \infty} x_n = x$ *exists and satisfies* $f(x) = x$ *(i.e.,* x *is a* fixed point *of* f*, here* x *is left "fixed" by* f*).*

Proof. We first prove that x_n is a Cauchy sequence. Indeed, for every $p \in \mathbb{N}$, we have

$$|x_{p+1} - x_p| = |f(x_p) - f(x_{p-1})| \leq k \, |x_p - x_{p-1}|.$$

It follows that

$$|x_{p+1} - x_p| \leq k^2 |x_{p-1} - x_{p-2}| \leq \cdots \leq k^{p-1} \, |x_2 - x_1|.$$

Then, for all $m < n$,

$$\begin{aligned}
|x_n - x_m| &\leq |x_n - x_{n-1}| + |x_{n-1} - x_{n-2}| + \cdots + |x_{m+1} - x_m| \\
&\leq (k^{n-2} + k^{n-3} + \cdots + k^{m-1})|x_2 - x_1| \\
&\leq \left(\sum_{i=m-1}^{\infty} k^i \right) |x_2 - x_1| = \frac{k^{m-1}}{1-k} \, |x_2 - x_1|.
\end{aligned}$$

Since $k^{m-1} \to 0$ as $m \to \infty$, it follows that x_n is a Cauchy sequence, so $\lim_{n \to \infty} x_n = x$ exists and $x \in [a, b]$. Note that (1.10) implies that f is continuous. Letting $n \to \infty$ in (1.11) yields that $f(x) = x$. $\qquad \square$

For a contraction mapping f, the fixed point of f is unique. Note that if $f(x) = x$ and $f(y) = y$, then

$$|x - y| \leq |f(x) - f(y)| \leq k\,|x - y|.$$

This is impossible unless $x = y$. It is worth noticing that the result still valid if $[a, b]$ is replaced by $[a, \infty)$ or $\mathbb{R}$. The above result is often referred to as the *contraction mapping principle* or the *Brouwer fixed point theorem* in advanced analysis.

Let $f : I \to I$ be a differentiable function for which

$$|f'(x)| \leq k < 1, \qquad \text{for all } x \in I.$$

The mean value theorem implies that f is a contraction mapping. It is crucial that k is strictly less than 1. Otherwise, let $f(x) = \sqrt{1 + x^2}, x \in \mathbb{R}$. We find that $|f'(x)| < 1$, but $\lim_{|x| \to \infty} |f'(x)| = 1$ and f has no fixed point in $\mathbb{R}$. However, if $I = [a, b]$, there is a remarkable application of the Bolzano-Weierstrass theorem and the monotone convergence theorem that allows us to relax (1.10) a little bit and prove the following fixed point theorem.

Example 1.28. *Let $f : [a, b] \to [a, b]$ for which*

$$|f(x) - f(y)| < |x - y|, \qquad \text{for all } x, y \in [a, b], x \neq y.$$

Then, for any $x_1 \in [a, b]$, the sequence defined by

$$x_{n+1} = f(x_n), \quad (n = 1, 2, \ldots).$$

converges to the unique fixed point of $f(x)$ in $[a, b]$.

Proof. Let

$$D(x) = |f(x) - x|, \qquad x \in [a, b].$$

$D(x)$ is continuous on $[a, b]$, and if $x \neq f(x)$, we have

$$D(f(x)) = |f(f(x)) - f(x)| < |f(x) - x| = D(x). \tag{1.12}$$

Notice that $x_n \in [a, b]$ by the assumption of f. It then follows from the Bolzano-Weierstrass theorem that there exists a subsequence $x_{n_k} \to x_0 \in [a, b]$. Thus,

$$D(x_{n_k}) \to D(x_0), \ D(x_{n_k+1}) = D(f(x_{n_k})) \to D(x_0).$$

On the other hand, by (1.12),

$$D(x_{n+1}) = D(f(x_n)) < D(x_n),$$

so $D(x_n)$ is monotone decreasing and nonnegative, and therefore it converges by the monotone convergence theorem. Since $D(x_{n_k})$ and $D(x_{n_k+1})$ are subsequences of $D(x_n)$, we find that $D(f(x_0)) = D(x_0)$. It follows from (1.12) that $f(x_0) = x_0$. i.e., x_0 is a fixed point of f. Next, if $x_0 = f(x_0), y_0 = f(y_0)$, then

$$|x_0 - y_0| = |f(x_0) - f(y_0)| < |x_0 - y_0|,$$

which is impossible unless $x_0 = y_0$. Therefore, the fixed point is unique. $\qquad \square$

If we replace the recursion $x_{n+1} = f(x_n)$ by

$$x_{n+1} = \alpha x_n + (1 - \alpha) f(x_n), \ (\text{where } 0 < \alpha < 1),$$

for any $x_1 \in [a, b]$, the sequence x_n also converges to the unique fixed point of f. See Exercise 81.

The above two principles can be applied to establish convergence of large classes of recursions.

1. Define $x_1 = 1, x_{n+1} = \sqrt{1 + x_n}$ $(n = 1, 2, \ldots)$.
    Let $f(x) = \sqrt{1 + x}$. For $x \geq 0$, we have

    $$|f'(x)| = \left| \frac{1}{2\sqrt{1 + x}} \right| \leq \frac{1}{2}.$$

    Thus, f is a contraction from $[0, 2]$ to $[0, 2]$, so x_n converges and $\lim_{n \to \infty} x_n = (1 + \sqrt{5})/2$.

2. For $a > 0$, it is well-known that the following recursion

    $$x_{n+1} = \frac{1}{2} \left(x_n + \frac{a}{x_n} \right),$$

    which is based on Newton's method, offers a quick approach to approximate $\sqrt{a}$.
    Here are two more recursions as both converge to $\sqrt{a}$: Let $a > 1$.

    - $x_1 > 0$, $x_{n+1} = a(1 + x_n)/(a + x_n)$. Using

      $$f(x) = \frac{a(1 + x)}{a + x} : [0, \infty) \to [0, \infty), \ |f'(x)| \leq 1 - \frac{1}{a} < 1.$$

    - $x_1 > \sqrt{a}$, $x_{n+1} = (a + x_n)/(1 + x_n)$. Using

      $$f(x) = \frac{a + x}{1 + x} : [\sqrt{a}, \infty) \to [\sqrt{a}, \infty), \ |f'(x)| \leq \frac{a - 1}{a + 1} < 1.$$

It is interesting to notice that x_n defined in the second recursion is not monotone. The confirmation of convergence by the method consists of proving that both x_{2n} and x_{2n+1} are monotone and converge to the same limit.

In general, assume $f : I \to I$ with $f'(x) < 0$ for all $x \in I$. Let x be the only fixed point of f in I. For the recursion sequence $x_{n+1} = f(x_n)$, by the mean value theorem, we have

$$x_{n+1} - x = f(x_n) - f(x) = f'(\xi)(x_n - x), \ \text{(where ξ is between x_n and x)}.$$

This indicates that the sequence x_n is oscillating around the fixed point x. Now, let $F(x) = f(f(x))$. Notice that $F'(x) = f'(f(x)) \cdot f'(x) > 0$. The recursion

$$x_{n+2} = F(x_n)$$

yields two monotone sequences x_{2n} and x_{2n+1}. Both will converge to the fixed point of F provided that F is a contraction.

We end this subsection by presenting an example where the contraction mapping principle is applied to subsequences.

Example 1.29 (Putnam Problem 1953-A6). *Show that the sequence*

$$\sqrt{7}, \sqrt{7 - \sqrt{7}}, \sqrt{7 - \sqrt{7 + \sqrt{7}}}, \sqrt{7 - \sqrt{7 + \sqrt{7 - \sqrt{7}}}}, \ldots$$

converges and find its limit.

Proof. This continued radical has period 2. Thus, let $f(x) = \sqrt{7 - \sqrt{7 + x}}$. Clearly, f maps $[0, 7]$ into $[0, 7]$. Starting with $x_0 = \sqrt{7}$ and $x_1 = \sqrt{7 - \sqrt{7}}$, and appealing to the recursion $x_{n+2} = f(x_n)$, we obtain two sequences x_{2n} and x_{2n+1}. Since

$$|f'(x)| = \left| \frac{-1}{4\sqrt{7 - \sqrt{7 + x}} \cdot \sqrt{7 + x}} \right| \leq \frac{1}{4\sqrt{7 - \sqrt{14}}\sqrt{7}} = k = 0.052347\ldots,$$

f is a contraction mapping. Thus, both x_{2n} and x_{2n+1} converge to the unique fixed point of $f(x)$ in $[0,7]$. Notice that $x = f(x), x \in [0,7]$ implies that $x = 2$. Thus, we find that $x_n \to 2$ as $n \to \infty$. $\qquad\square$

Continued radicals were a favorite topic of Ramanujan. One of his famous results

$$\sqrt{1 + 2\sqrt{1 + 3\sqrt{1 + 4\sqrt{1 + \cdots}}}} = 3$$

appeared as **Putnam Problem 1966-A6**. One general closed form is given in Exercise 98. For two tests for convergence of continued radicals, see Exercises 106 and 107.

1.3.8 Recursions with closed forms

Structures are the weapons of the mathematician. — Nicolas Bourbaki

We now turn our attention to second-order linear and nonlinear recursions. We begin with a simple linear second-order recursion:

$$x_0 = 0, x_1 = 1, x_{n+1} = \frac{x_n + x_{n-1}}{2}. \tag{1.13}$$

Based on

$$x_{n+1} - x_n = -\frac{1}{2}(x_n - x_{n-1}) = \cdots = \left(-\frac{1}{2}\right)^n (x_1 - x_0) = \left(-\frac{1}{2}\right)^n,$$

we can show that x_{2n} is monotone increasing and x_{2n+1} is monotone decreasing. Moreover, both subsequences converge to the same limit. Therefore, x_n converges. When we try to determine the value of the limit as we did before, we find that taking the limit in (1.13) does not gain the desired answer. The following problem presents an alternative approach to solve general second-order linear recursions. It simultaneously provides the convergence and the limit value. The method is based on establishing a closed form for the general terms.

Example 1.30. *Let $a_1, a_2 \in \mathbb{R}$ with $a_1^2 + a_2^2 \neq 0$. Define*

$$a_{n+1} = a\,a_n + b\,a_{n-1}, \quad n \in \mathbb{N}, n \geq 2, \tag{1.14}$$

where $a, b > 0$. Find $\lim_{n\to\infty} a_{n+1}/a_n$.

Solution. The idea is to seek solutions of the form $a_n = r^n$, where r is a constant to be determined. The substitution of $a_{n-1} = r^{n-1}$ and $a_{n+1} = r^{n+1}$ into (1.14) reduces the problem to finding r from the quadratic equation $r^2 - ar - b = 0$. Let these two solutions be

$$\alpha = \frac{a + \sqrt{a^2 + 4b}}{2}, \quad \beta = \frac{a - \sqrt{a^2 + 4b}}{2},$$

respectively. Then

$$\alpha + \beta = a, \quad \alpha \cdot \beta = -b.$$

Replacing a, b in (1.14) in terms of α, β, we have

$$\begin{aligned} a_{n+1} - \alpha\,a_n &= \beta\,(a_n - \alpha\,a_{n-1}), \quad (n = 2, 3,, \cdots), \\ a_{n+1} - \beta\,a_n &= \alpha\,(a_n - \beta\,a_{n-1}), \quad (n = 2, 3,, \cdots). \end{aligned}$$

Repeatedly applying this process yields

$$
\begin{aligned}
a_{n+1} - \alpha\, a_n &= \beta^{n-1}\,(a_2 - \alpha\, a_1), \quad (n = 2, 3,, \cdots), \\
a_{n+1} - \beta\, a_n &= \alpha^{n-1}\,(a_2 - \beta\, a_1), \quad (n = 2, 3,, \cdots).
\end{aligned}
$$

Solving for a_{n+1} gives

$$
a_{n+1} = \frac{\alpha^n\,(a_2 - \beta\, a_1) - \beta^n\,(a_2 - \alpha\, a_1)}{\alpha - \beta}.
$$

If $a_2 - \beta\, a_1 = 0$, we have $a_{n+1}/a_n = \beta$ for all $n \in \mathbb{N}$ and so $\lim_{n\to\infty} a_{n+1}/a_n = \beta$.

On the other hand, if $a_2 - \beta\, a_1 \neq 0$, in view of the fact that $|\alpha/\beta| > 1$, we can choose n large enough such that

$$
\left|\frac{\alpha}{\beta}\right|^{n-1} > \left|\frac{a_2 - \alpha\, a_1}{a_2 - \beta\, a_1}\right|.
$$

We find that $a_n \neq 0$ and

$$
\frac{a_{n+1}}{a_n} = \frac{\alpha^n\,(a_2 - \beta\, a_1) - \beta^n\,(a_2 - \alpha\, a_1)}{\alpha^{n-1}\,(a_2 - \beta\, a_1) - \beta^{n-1}\,(a_2 - \alpha\, a_1)}.
$$

Since $|\beta/\alpha| < 1$, we obtain that $\lim_{n\to\infty} a_{n+1}/a_n = \alpha$. $\qquad\square$

Let $a = b = 1$ in (1.14) with $a_1 = a_2 = 1$. We obtain the famous *Binet's formula* for the Fibonacci numbers:

$$
F_n = \frac{1}{\sqrt{5}}\left(\left(\frac{1 + \sqrt{5}}{2}\right)^n - \left(\frac{1 - \sqrt{5}}{2}\right)^n\right).
$$

Let $a + b = 1$ in (1.14). We have $\alpha = 1, \beta = -b$, and

$$
a_n = \frac{a_2 + b a_1 - (-b)^{n-1}(a_2 - a_1)}{1 + b}.
$$

Therefore,

$$
\lim_{n\to\infty} a_n = \frac{a_2 + b a_1}{1 + b}.
$$

In particular, for the recursion (1.13), we find that $\lim_{n\to\infty} x_n = 2/3$.

Remark. The approach used to find the explicit formula of a_{n+1} can be transformed into computing the powers of a matrix. To see that, we rewrite (1.14) in a matrix equation

$$
\mathbf{x}^{(n)} = A\mathbf{x}^{(n-1)} \qquad n = 3, 4, \ldots,
$$

where

$$
\mathbf{x}^{(n)} = \begin{pmatrix} a_{n+1} \\ a_n \end{pmatrix}, \quad \mathbf{x}^{(n-1)} = \begin{pmatrix} a_n \\ a_{n-1} \end{pmatrix} \text{ and } A = \begin{pmatrix} a & b \\ 1 & 0 \end{pmatrix}.
$$

From the matrix equation above, it follows that

$$
\mathbf{x}^{(n)} = A\mathbf{x}^{(n-1)} = A^2\mathbf{x}^{(n-2)} = \cdots = A^{n-1}\begin{pmatrix} a_2 \\ a_1 \end{pmatrix}.
$$

Consequently, if we can find an explicit expansion for A^{n-1}, we can obtain an explicit formula for $\mathbf{x}^{(n)}$, and so for a_{n+1}. To find an explicit expansion for A^{n-1}, we first diagonalize A. Clearly, α and β are eigenvalues of A; their corresponding eigenvectors are

$$\begin{pmatrix} \alpha \\ 1 \end{pmatrix} \quad \text{and} \quad \begin{pmatrix} \beta \\ 1 \end{pmatrix},$$

respectively. Thus

$$A = \frac{1}{\alpha - \beta} \begin{pmatrix} \alpha & \beta \\ 1 & 1 \end{pmatrix} \begin{pmatrix} \alpha & 0 \\ 0 & \beta \end{pmatrix} \begin{pmatrix} 1 & -\beta \\ -1 & \alpha \end{pmatrix}.$$

We then have

$$A^{n-1} = \frac{1}{\alpha - \beta} \begin{pmatrix} \alpha & \beta \\ 1 & 1 \end{pmatrix} \begin{pmatrix} \alpha^{n-1} & 0 \\ 0 & \beta^{n-1} \end{pmatrix} \begin{pmatrix} 1 & -\beta \\ -1 & \alpha \end{pmatrix}.$$

Therefore,

$$\mathbf{x}^{(n)} = \frac{1}{\alpha - \beta} \begin{pmatrix} \alpha & \beta \\ 1 & 1 \end{pmatrix} \begin{pmatrix} \alpha^{n-1} & 0 \\ 0 & \beta^{n-1} \end{pmatrix} \begin{pmatrix} 1 & -\beta \\ -1 & \alpha \end{pmatrix} \begin{pmatrix} a_2 \\ a_1 \end{pmatrix}$$

$$= \begin{pmatrix} \frac{\alpha^n (a_2 - \beta a_1) - \beta^n (a_2 - \alpha a_1)}{\alpha - \beta} \\ \frac{\alpha^{n-1} (a_2 - \beta a_1) - \beta^{n-1} (a_2 - \alpha a_1)}{\alpha - \beta} \end{pmatrix}$$

The first component of $\mathbf{x}^{(n)}$ reveals the formula of a_{n+1} again. The method of computing the powers of a matrix enables us to handle the multivariable recursions. See Exercise 70.

The following example shows how a nonlinear recursion can be transformed into a linear one.

Example 1.31. *Let $a_n \neq 0$ be a real sequence such that*

$$a_{n+1} = \frac{a_n^2 - 1}{a_{n-1}} \quad (n = 1, 2, \ldots).$$

Prove there exists a real number α such that $a_{n+1} = \alpha\, a_n - a_{n-1}$.

Proof. We work backward by rewriting the desired result as

$$\alpha = \frac{a_{n+1} + a_{n-1}}{a_n}.$$

It suffices to show that, for all $n \in \mathbb{N}$,

$$\frac{a_{n+1} + a_{n-1}}{a_n} = \text{constant}.$$

To this end, for any integer $n \geq 1$, notice the given recursion implies that $a_n^2 - a_{n-1}a_{n+1} = 1$. Thus, we have

$$a_{n+1}^2 - a_n a_{n+2} = a_n^2 - a_{n-1}a_{n+1}.$$

Regrouping yields

$$a_{n+1}(a_{n+1} + a_{n-1}) = a_n(a_{n+2} + a_n).$$

This is equivalent to

$$\frac{a_{n+1} + a_{n-1}}{a_n} = \frac{a_{n+2} + a_n}{a_{n+1}},$$

which proves that $(a_{n+1} + a_{n-1})/a_n$ is the required constant α. $\qquad\square$

In certain cases a sequence defined by a nonlinear recursion may still yield a closed form. Connections with trigonometry and hyperbolic functions may be useful for recovering these forms. For example, for the logistic equation $x_{n+1} = 4x_n(1 - x_n)$, appealing to the identity $\sin 2\alpha = 2 \sin \alpha \cos \alpha$, we have $x_n = \sin^2(2^n \theta)$ for all $n \in \mathbb{N}$, where $\theta = \sin^{-1}(\sqrt{x_1})/2$. We present two more examples as follows.

Example 1.32 (Monthly Problem 11604, 2011). *Given $0 \leq a \leq 2$, let $\{a_n\}$ be the sequence defined by $a_1 = a$ and $a_{n+1} = 2^n - \sqrt{2^n(2^n - a_n)}$ for $n \geq 1$. Find $\sum_{n=1}^{\infty} a_n^2$.*

Solution. First, we show by induction that

$$a_n = 2^n \sin^2(\theta/2^n), \tag{1.15}$$

where $\theta = 2 \sin^{-1}(\sqrt{a/2})$. For $n = 1$, the definition of θ yields that $a_1 = a = 2 \sin^2(\theta/2)$. Assume (1.15) is true for $n = k$; the half angle formula for sine gives

$$a_{k+1} = 2^k - \sqrt{2^k(2^k - 2^k \sin^2(\theta/2^k))} = 2^k(1 - \cos(\theta/2^k)) = 2^{k+1} \sin^2(\theta/2^{k+1}).$$

Thus, by induction, (1.15) holds for all positive integers.

Next, using the fact that

$$\sin^4 x = \sin^2 x(1 - \cos^2 x) = \sin^2 x - \frac{1}{4}\sin^2(2x),$$

we find that

$$a_n^2 = 2^{2n} \sin^4(\theta/2^n) = 2^{2n} \sin^2(\theta/2^n) - 2^{2(n-1)} \sin^2(\theta/2^{n-1}).$$

Thus, the proposed series is telescoping and

$$\begin{aligned}
\sum_{n=1}^{\infty} a_n^2 &= \sum_{n=1}^{\infty} \left(2^{2n} \sin^2(\theta/2^n) - 2^{2(n-1)} \sin^2(\theta/2^{n-1})\right) \\
&= \lim_{N \to \infty} \left(2^{2N} \sin^2(\theta/2^N) - \sin^2(\theta)\right) \\
&= \theta^2 - \sin^2 \theta = \left[4 \sin^{-1}(\sqrt{a/2})\right]^2 - a(2 - a).
\end{aligned}$$

□

Example 1.33. *Let a_1 and a_2 be positive real numbers. Define*

$$\begin{aligned}
a_{2n+1} &= \frac{a_{2n} + a_{2n-1}}{2}, \quad n = 1, 2, \ldots, \\
a_{2n} &= \sqrt{a_{2n-1} \cdot a_{2n-2}}, \quad n = 2, 3, \ldots.
\end{aligned}$$

Find $\lim_{n \to \infty} a_n$.

Solution. We separate into three cases:
(1) If $a_1 = a_2$, the given recursion implies that $a_n = a_1$ for all $n \in \mathbb{N}$. Hence $\lim_{n \to \infty} a_n = a_1$.
(2) If $a_1 < a_2$, construct a right triangle with adjacent leg a_1 and hypotenuse a_2. Let $\theta \in (0, \pi/2)$ be such that $\cos \theta = a_1/a_2$ and a be the opposite leg. Then

$$a_1 = a \cot \theta, \quad a_2 = a \csc \theta;$$

$$\theta = \cos^{-1}(a_1/a_2), \quad a = \sqrt{a_2^2 - a_1^2}.$$

Appealing to the trigonometric identities, for $0 < \alpha < \pi/2$,

$$\frac{\cot(2\alpha) + \csc(2\alpha)}{2} = \frac{1}{2}\cot\alpha, \quad \sqrt{\csc(2\alpha) \cdot \frac{\cot\alpha}{2}} = \frac{1}{2}\csc\alpha,$$

we find that

$$a_{2n+1} = \frac{a}{2^n}\cot\left(\frac{\theta}{2^n}\right), \quad n = 1, 2, \ldots,$$

$$a_{2n} = \frac{a}{2^{n-1}}\csc\left(\frac{\theta}{2^{n-1}}\right), \quad n = 2, 3, \ldots.$$

Since

$$\lim_{n\to\infty}\frac{a}{2^n}\cot\left(\frac{\theta}{2^n}\right) = \lim_{n\to\infty}\frac{a}{2^{n-1}}\csc\left(\frac{\theta}{2^{n-1}}\right) = \frac{a}{\theta},$$

we have

$$\lim_{n\to\infty} a_n = \frac{a}{\theta} = \frac{\sqrt{a_2^2 - a_1^2}}{\cos^{-1}(a_1/a_2)}.$$

(3) If $a_1 > a_2$, using the identities

$$\frac{\coth(2\alpha) + \operatorname{csch}(2\alpha)}{2} = \frac{1}{2}\coth\alpha, \quad \sqrt{\operatorname{csch}(2\alpha) \cdot \frac{\coth\alpha}{2}} = \frac{1}{2}\operatorname{csch}\alpha,$$

similarly, we have

$$\lim_{n\to\infty} a_n = \frac{\sqrt{a_1^2 - a_2^2}}{\cosh^{-1}(a_1/a_2)}.$$

$\square$

Finally, we present an example to study a nonlinear recursion (without a closed form) by determining its asymptotic behavior. Two interesting solutions have been published in the *Amer. Math. Monthly* (October, 2013) and *Math. Mag.* (Vol. 86, No. 1, 2013), respectively. Here we give another one solely based on the Riemann sums.

Example 1.34 (Putnam Problem 2012-B4). *Suppose that $a_0 = 1$ and that $a_{n+1} = a_n + e^{-a_n}$ for $n = 0, 1, 2, \ldots$. Does $a_n - \ln n$ have a finite limit as $n \to \infty$?*

Solution. We show that $a_n - \ln n \to 0$ as $n \to \infty$ indeed. It suffices to show that $e^{a_n}/n \to 1$ as $n \to \infty$. To begin, notice that a_n is strictly increasing. Partition $[a_0, a_n]$ with subintervals $[a_{k-1}, a_k], k = 1, 2, \ldots, n$. Since e^x is increasing, appealing to $a_k - a_{k-1} = e^{-a_{k-1}}$, the Riemann sum with the left endpoints gives

$$e^{a_n} - e = \int_{a_0}^{a_n} e^x\,dx = \sum_{k=1}^{n}\int_{a_{k-1}}^{a_k} e^x\,dx$$

$$\geq \sum_{k=1}^{n} e^{a_{k-1}}(a_k - a_{k-1}) = \sum_{k=1}^{n} 1 = n.$$

It follows that $a_n > \ln(e + n) > \ln n$ for all $n \in \mathbb{N}$ and

$$\frac{e^{a_n}}{n} \geq 1 + \frac{e}{n}. \tag{1.16}$$

On the other hand, the Riemann sum with the right endpoints yields

$$e^{a_n} - e = \int_{a_0}^{a_n} e^x \, dx = \sum_{k=1}^{n} \int_{a_{k-1}}^{a_k} e^x \, dx$$

$$\leq \sum_{k=1}^{n} e^{a_k}(a_k - a_{k-1}) = \sum_{k=1}^{n} e^{a_k - a_{k-1}} = \sum_{k=1}^{n} e^{e^{-a_{k-1}}}.$$

Notice that $e^{a_{k-1}} > (k-1) + e > k$ from above. Thus, $e^{-a_{k-1}} < 1/k$ and

$$e^{a_n} - e \leq \sum_{k=1}^{n} e^{1/k} \leq e + \int_1^n e^{1/x} \, dx.$$

Hence,

$$\frac{e^{a_n}}{n} \leq \frac{2e}{n} + \frac{1}{n} \int_1^n e^{1/x} \, dx. \tag{1.17}$$

Notice that, for example, using the Stolz-Cesàro theorem,

$$\lim_{n \to \infty} \frac{1}{n} \int_1^n e^{1/x} \, dx = 1.$$

Now, $e^{a_n}/n \to 1$ as $n \to \infty$ follows from (1.16), (1.17) and the squeeze theorem. Therefore,

$$\lim_{n \to \infty} (a_n - \ln n) = 0.$$

$\square$

This problem is an example of the general principle that one can often predict the asymptotic behavior of a recursive sequence by studying solutions of a sufficiently similar-looking differential equation. In this case, as Kedlaya and Ng suggested in `http://kskedlaya.org/putnam-archive/2012s.pdf`: We can start with the equation $a_{n+1} - a_n = e^{-a_n}$, then replace a_n with a function $y(x)$ and replace the difference $a_{n+1} - a_n$ with the derivative $y'(x)$ to obtain the differential equation $y' = e^{-y}$, which has the solution $y = \ln x$. As a continuation of this problem, we consider

Example 1.35 (Monthly Problem 12270, 2021). *Let $a_0 = 1$, and let $a_{n+1} = a_n + e^{-a_n}$ for $n \geq 0$. Show that the sequence whose nth term is $e^{a_n} - n - (1/2)\ln n$ converges.*

Proof. Let $b_n = e^{a_n}$ for $n \geq 0$. Then

$$b_{n+1} = e^{a_{n+1}} = e^{a_n + e^{-a_n}} = b_n e^{1/b_n}.$$

Using the power series expansion of e^x and Taylor's theorem, we have

$$b_{n+1} = b_n + 1 + \frac{1}{2b_n} + R_n,$$

where $0 \leq R_n \leq M/b_n^2$ for some positive constant M. Since $b_0 = e^{a_0} = e$, telescoping the equation above from 0 to $n-1$ yields

$$b_n = e + n + \frac{1}{2} \sum_{k=0}^{n-1} \frac{1}{b_k} + \sum_{k=0}^{n-1} R_k.$$

In view of the fact that $e^{a_k}(a_{k+1} - a_k) = 1$, we have $1/b_k = a_{k+1} - a_k$. Telescoping again gives

$$
\begin{aligned}
b_n &= e + n + \frac{1}{2}\sum_{k=0}^{n-1}(a_{k+1} - a_k) + \sum_{k=0}^{n-1} R_k \\
&= e + n + \frac{1}{2}(a_n - 1) + \sum_{k=0}^{n-1} R_k.
\end{aligned}
$$

Hence

$$
e^{a_n} - n - \frac{1}{2}\ln n = e - \frac{1}{2} + \frac{1}{2}(a_n - \ln n) + \sum_{k=0}^{n-1} R_k.
$$

Using the fact that $a_n - \ln n \to 0$ as $n \to \infty$ from Example 1.34, we find that

$$
\lim_{n\to\infty}\left(e^{a_n} - n - \frac{1}{2}\ln n\right) = e - \frac{1}{2} + \sum_{k=0}^{\infty} R_k.
$$

This shows that $e^{a_n} - n - (1/2)\ln n$ converges as $b_k \geq k + e$ implies that $0 \leq R_k \leq M/(k+e)^2$ and then the series $\sum_{k=0}^{\infty} R_k$ converges. $\qquad\square$

1.3.9 Limits involving the harmonic numbers

The profound study of nature is the most fertile source on mathematical discoveries. — Joseph Fourier

Recall that, for each $n \in \mathbb{N}$, the nth *harmonic number* H_n is defined by

$$
H_n = 1 + \frac{1}{2} + \frac{1}{3} + \cdots + \frac{1}{n} = \sum_{k=1}^{n}\frac{1}{k}.
$$

Although there is no closed formula of H_n for general n, by Example 1.15, we have

$$
\lim_{n\to\infty}(H_n - \ln n) = \gamma,
$$

where γ is the Euler-Mascheroni constant. A detailed history on H_n and its development are very well described in Havil's book [55]. Since H_n appears so often in analysis and many consequences can be drawn from it, so it is worthwhile for us to take a closer look at H_n. We begin with an identity that associates the harmonic numbers with the binomial coefficients.

Example 1.36. *Show that*

$$
H_n = \sum_{k=1}^{n}\binom{n}{k}\frac{(-1)^{k-1}}{k}.
$$

Proof. Observe that $1/k = \int_0^1 x^{k-1}dx$ for all $k \in \mathbb{N}$. We have

$$
\begin{aligned}
H_n &= \sum_{k=1}^n \int_0^1 x^{k-1}\, dx = \int_0^1 \frac{1-x^n}{1-x}\, dx \\
&= \int_0^1 \frac{1-(1-u)^n}{u}\, du = \int_0^1 \frac{1}{u}\left(1 - \sum_{k=0}^n \binom{n}{k}(-1)^k u^k\right) du \\
&= \int_0^1 \sum_{k=1}^n \binom{n}{k}(-1)^{k-1} u^{k-1}\, du = \sum_{k=1}^n \binom{n}{k}(-1)^{k-1}\int_0^1 u^{k-1}\, du \\
&= \sum_{k=1}^n \binom{n}{k}\frac{(-1)^{k-1}}{k}.
\end{aligned}
$$

$\square$

Remark. Using integration by parts, we also have

$$
H_n = \int_0^1 \frac{1-x^n}{1-x}\, dx = -n\int_0^1 x^{n-1}\ln(1-x)\, dx.
$$

Let $H_n^{(p)} = \sum_{k=1}^n 1/k^p$. We further find that

$$
\sum_{k=1}^n \binom{n}{k}\frac{(-1)^{k-1}H_k}{k} = H_n^{(2)},
$$

$$
\sum_{k=1}^n \binom{n}{k}\frac{(-1)^{k-1}}{k^2} = \frac{1}{2}(H_n^2 + H_n^{(2)}),
$$

$$
\sum_{k=1}^n \binom{n}{k}\frac{(-1)^{k-1}}{k^3} = \frac{1}{6}\left(H_n^3 + 3H_n H_n^{(2)} + 2H_n^{(3)}\right).
$$

We leave the proofs to the reader.

The next two examples demonstrate how to manipulate the limit problems involving the harmonic numbers. Some interesting identities such as (1.19)–(1.21) will be established along the way.

Example 1.37 (CMJ Problem 1211, 2021). *Evaluate the following limit, where below,* $H_0 = 0$ *and for* $n > 0, H_n$ *denotes the* n*th harmonic number* $\sum_{k=1}^n 1/k$:

$$
\lim_{n\to\infty}\left(H_n^2 - \sum_{k=1}^n \frac{H_{n-k}}{k}\right).
$$

Solution. We show that the limit is $\pi^2/6$. To this end, since $H_0 = 0$, we have

$$
S_m := \sum_{k=1}^m \frac{H_{m-k}}{k} = \sum_{k=1}^{m-1} \frac{H_{m-k}}{k} \qquad \text{(for } m \geq 1\text{)},
$$

$S_0 := 0$, and

$$
\begin{aligned}
S_m - S_{m-1} &= \sum_{k=1}^{m-1} \frac{H_{m-k}}{k} - \sum_{k=1}^{m-2} \frac{H_{m-1-k}}{k} \\
&= \sum_{k=1}^{m-1} \frac{H_{m-k} - H_{m-k-1}}{k} = \sum_{k=1}^{m-1} \frac{1}{k(m-k)} \\
&= \frac{1}{m} \left(\sum_{k=1}^{m-1} \frac{1}{k} + \sum_{k=1}^{m-1} \frac{1}{m-k} \right) \quad \text{(use partial fractions)} \\
&= \frac{2}{m} \sum_{k=1}^{m-1} \frac{1}{k} \quad \text{(use the symmetry)} \\
&= \frac{2H_{m-1}}{m} = \frac{2H_m}{m} - \frac{2}{m^2}.
\end{aligned}
$$

Telescoping the above identity from 1 to n yields

$$
S_n = \sum_{k=1}^{n} \frac{H_{n-k}}{k} = 2 \sum_{m=1}^{n} \frac{H_m}{m} - 2 \sum_{m=1}^{n} \frac{1}{m^2}. \tag{1.18}
$$

On the other hand, we have

$$
\begin{aligned}
H_n^2 &= \sum_{i=1}^{n} \sum_{j=1}^{n} \frac{1}{ij} = \sum_{i=1}^{n} \frac{1}{i} \left(\sum_{j=1}^{i} \frac{1}{j} + \sum_{j=i}^{n} \frac{1}{j} \right) - \sum_{i=1}^{n} \frac{1}{i^2} \\
&= 2 \sum_{i=1}^{n} \frac{1}{i} \sum_{j=1}^{i} \frac{1}{j} - \sum_{i=1}^{n} \frac{1}{i^2} = 2 \sum_{i=1}^{n} \frac{H_i}{i} - \sum_{i=1}^{n} \frac{1}{i^2}.
\end{aligned}
$$

This, together with (1.18), implies that

$$
\sum_{k=1}^{n} \frac{H_{n-k}}{k} = H_n^2 - \sum_{i=1}^{n} \frac{1}{i^2}. \tag{1.19}
$$

Hence,

$$
\lim_{n \to \infty} \left(H_n^2 - \sum_{k=1}^{n} \frac{H_{n-k}}{k} \right) = \lim_{n \to \infty} \sum_{i=1}^{n} \frac{1}{i^2} = \frac{\pi^2}{6}. \qquad \square
$$

Example 1.38 (Math. Magazine Problem 2136, 2022). *Evaluate*

$$
\lim_{n \to \infty} \left(\left(\sum_{k=1}^{n} \frac{H_k^2}{k} \right) - \frac{H_n^3}{3} \right).
$$

Solution. We show that the limit is $\frac{5}{3}\zeta(3)$, where $\zeta(3)$ is *Apéry's constant* defined by $\zeta(3) = \sum_{k=1}^{\infty} 1/k^3$. To this end, we first show that

$$
\sum_{k=1}^{n} \frac{H_k^2}{k} = \frac{H_n^3}{3} + \sum_{k=1}^{n} \frac{H_k}{k^2} - \frac{1}{3} \sum_{k=1}^{n} \frac{1}{k^3}. \tag{1.20}
$$

Upon proving (1.20), letting $n \to \infty$ in (1.20), and using the well-known *Euler sum formula*

$$
\sum_{k=1}^{\infty} \frac{H_k}{k^2} = 2\zeta(3),
$$

we find that

$$\lim_{n\to\infty}\left(\left(\sum_{k=1}^{n}\frac{H_k^2}{k}\right)-\frac{H_n^3}{3}\right)=2\zeta(3)-\frac{1}{3}\zeta(3)=\frac{5}{3}\zeta(3)$$

as claimed.

To prove (1.20), let $a_k=1/k, b_k=H_k^2$. Then $A_k=\sum_{i=1}^{k}a_i=H_k$. Applying Abel's summation formula

$$\sum_{k=1}^{n}a_k b_k=A_n b_n+\sum_{k=1}^{n-1}A_k(b_k-b_{k+1}),$$

we have

$$\sum_{k=1}^{n}\frac{H_k^2}{k}=H_n^3+\sum_{k=1}^{n-1}H_k(H_k^2-H_{k+1}^2)$$

$$=H_n^3+\sum_{k=1}^{n-1}H_k(H_k+H_{k+1})(H_k-H_{k+1})$$

$$=H_n^3-\sum_{k=1}^{n-1}\left(H_{k+1}-\frac{1}{k+1}\right)\left(2H_{k+1}-\frac{1}{k+1}\right)\frac{1}{k+1}$$

$$=H_n^3-2\sum_{k=1}^{n-1}\frac{H_{k+1}^2}{k+1}+3\sum_{k=1}^{n-1}\frac{H_{k+1}}{(k+1)^2}-\sum_{k=1}^{n-1}\frac{1}{(k+1)^3}$$

$$=H_n^3-2\sum_{k=1}^{n}\frac{H_k^2}{k}+3\sum_{k=1}^{n}\frac{H_k}{k^2}-\sum_{k=1}^{n}\frac{1}{k^3}.$$

This leads to

$$3\sum_{k=1}^{n}\frac{H_k^2}{k}=H_n^3+3\sum_{k=1}^{n}\frac{H_k}{k^2}-\sum_{k=1}^{n}\frac{1}{k^3},$$

which yields (1.20) as desired. $\qquad\square$

Remark. Along the same lines, we can prove that

$$\sum_{k=1}^{n}\frac{\left(H_k^{(p)}\right)^2}{k^p}=\frac{1}{3}\left(H_n^{(p)}\right)^3+\sum_{k=1}^{n}\frac{H_k^{(p)}}{k^{2p}}-\frac{1}{3}H_n^{(3p)}. \qquad (1.21)$$

Therefore

$$\lim_{n\to\infty}\left(\sum_{k=1}^{n}\frac{\left(H_k^{(p)}\right)^2}{k^p}-\frac{1}{3}\left(H_n^{(p)}\right)^3\right)=S_{p,2p}-\frac{1}{3}\zeta(3p), \qquad (1.22)$$

where ζ is the Riemann zeta value and

$$S_{p,q}:=\sum_{k=1}^{\infty}\frac{H_k^{(p)}}{n^q},$$

which is often referred as the Euler sums. Since $S_{2,4}=\zeta^2(3)-\zeta(6)/3$, we have

$$\lim_{n\to\infty}\left(\sum_{k=1}^{n}\frac{\left(H_k^{(2)}\right)^2}{k^2}-\frac{1}{3}\left(H_n^{(2)}\right)^3\right)=\zeta^2(3)-\frac{2}{3}\zeta(6).$$

In general, based on [18], if p is odd, $S_{p,2p}$ can be reducible to Riemann zeta values, so we can find the limit in (1.22) in terms of the Riemann zeta values.

1.4 Exercises

The only way to learn mathematics is to do mathematics. — Paul Halmos

1. Prove the Cauchy criterion and the Heine-Borel theorem are equivalent.

2. Use the least-upper-bound property to prove the Heine-Borel theorem. *Hint.* Let $\mathcal{G}$ be an open covering of $[a, b]$. Define

   $$S = \{s \geq a : \text{There is a finite sub-coveing of } \mathcal{G} \text{ that covers } [a, s]\}.$$

   Show that S is nonempty and $b \in S$.

3. Let $\mathcal{G}$ be an open covering of $[a, b]$. Prove that, for any $x_1, x_2 \in [a, b]$, there is a $\delta > 0$ and $U \in \mathcal{G}$ such that $x_1, x_2 \in U$ whenever $|x_1 - x_2| < \delta$. The number δ is called the *Lebesgue number* of $\mathcal{G}$.

4. Let $f : [a, b] \to [a, b]$ be increasing. Use the nested interval theorem to show that there exists $c \in [a, b]$ such that $f(c) = c$.

5. A function $f : I \to \mathbb{R}$ is called *locally bounded* on I, if for each point $x \in I$ there is a $\delta > 0$ such that f is bounded on $(x - \delta, x + \delta) \cap I$.

   (a) Show that $f(x) = 1/x$ is locally bounded on $(0, 1)$, but it is not bounded on $(0, 1)$.

   (b) Show that $f(x) = x$ is locally bounded on $[0, \infty)$, but it is not bounded on $[0, \infty)$.

   (c) Show that if f is locally bounded on $[a, b]$, then f is bounded on $[a, b]$. *Hint:* This can be done by using one of Theorems 1.7, 1.8, and 1.12.

6. A function $f : I \to \mathbb{R}$ is called *locally positive* on I, if for each point $x \in I$ there is a $\epsilon > 0$ such that $f(y) > \epsilon$ for all $y \in (x - \delta, x + \delta) \cap I$. When $I = [a, b]$, show that there exists some constant $c > 0$ such that $f(x) > c$ for all $x \in I$.

7. For $n \in \mathbb{N}$, let

   $$a_n = \frac{1^p + 3^p + \cdots + (2n-1)^p}{(2n+1)^p + (2n+3)^p + \cdots + (4n-1)^p}.$$

   Prove that a_n is increasing for $p > 1$ and decreasing for $0 < p < 1$.

8. Let a_n be a real sequence. Define

   $$\sigma_n = \frac{a_1 + a_2 + \cdots + a_n}{n}.$$

   (a) Show that

   $$\liminf a_n \leq \liminf \sigma_n \leq \limsup \sigma_n \leq \limsup a_n.$$

   (b) If a_n is increasing and $\lim_{n \to \infty} \sigma_n = a$, show that $\lim_{n \to \infty} a_n = a$.

(c) Find a counterexample to show that the increasing assumption in (b) is necessary.

9. (**Putnam Problem 2001-B6**). Assume that a_n is an increasing sequence of positive real numbers such that $\lim_{n\to\infty} \frac{a_n}{n} = 0$. Must there exist infinitely many positive integers n such that

$$a_{n-i} + a_{n+i} < 2a_n \quad \text{for } i = 1, 2, \ldots, n-1 ?$$

10. (**Putnam Problem 2002-A5**). Define a sequence by $a_0 = 1$, together with the rules $a_{2n+1} = a_n$ and $a_{2n+2} = a_n + a_{n+1}$ for each integer $n \geq 0$. Prove that every positive rational number appears in the set

$$\left\{ \frac{a_{n-1}}{a_n} : n \geq 1 \right\} = \left\{ \frac{1}{1}, \frac{1}{2}, \frac{2}{1}, \frac{1}{3}, \frac{3}{2}, \cdots \right\}.$$

11. (**Putnam Problem 2018-B4**). Given a real number a, we define a sequence by $x_0 = 1, x_1 = x_2 = a$, and $x_{n+1} = 2x_n x_{n-1} - x_{n-2}$ for $n \geq 2$. Prove that if $x_n = 0$ for some n, then the sequence is periodic.

12. Given $x_1 \in \mathbb{R}$, define $x_{n+1} = x_n(x_n + 1/n)$ for $n \in \mathbb{N}$. Prove that there exists a unique x_1 such that $0 < x_n < x_{n+1} < 1$ for every n.

13. Given $x_1 \in \mathbb{R}$, define $x_{n+1} = 2^n - 3x_n$ for $n \in \mathbb{N}$. Find all possible value(s) of x_1 such that the sequence x_n is strictly increasing.

14. Consider the sequence x_n defined by $x_1 = x, x_{n+1} = 1/4 + x_n - x_n^2$ for $n \in \mathbb{N}$.

(a) If $x = 0$, show that x_n is monotone increasing and bounded, then find $\lim_{n\to\infty} x_n$.

(b) Find all possible value of x such that $\lim_{n\to\infty} x_n$ exists.

15. Given $x_1 \in \mathbb{R}$, define $x_{n+1} = x_n(2 - x_n)$ for $n \in \mathbb{N}$. Prove that if $x_1 \in (0, 1)$, then the sequence x_n monotonically converges to one of the steady-state solutions (i.e., the fixed point of $f(x) = x(2 - x)$).

16. (**Monthly Problem 12220, 2020**). Let $a_n = \sum_{k=1}^n 1/k^2$ and $b_n = \sum_{k=1}^n 1/(2k-1)^2$. Prove

$$\lim_{n\to\infty} n\left(\frac{b_n}{a_n} - \frac{3}{4} \right) = \frac{3}{\pi^2}.$$

17. Let $a_1 = \alpha$ and

$$a_{n+1} = \frac{\lambda + \beta a_n}{\beta + a_n} \quad \text{for } n \in \mathbb{N}.$$

If $\alpha^2 < \lambda < \beta^2$, show that a_n converges and find its limit.

18. Let $a, a_1 > 0$ and $n \in \mathbb{N}$. Define

$$a_{n+1} = \frac{a_n(a_n^2 + 3a)}{3a_n^2 + a}.$$

Show that a_n converges and find its limit.

19. For $\lambda > 0$, let

$$a_1 = \frac{1}{2}\lambda, \quad a_{n+1} = \frac{1}{2}(\lambda + a_n^2), \quad (n = 1, 2, \cdots).$$

Determine $\lim_{n\to\infty} a_n$.

20. Let $a_1 = 0, a_2 = 1$ and for $n \geq 3$

$$a_n = (n-1)(a_{n-1} + a_{n-2}).$$

Find a closed form for a_n and evaluate $\lim_{n\to\infty} a_n/n!$. Here a_n is the number of *derangements* of a set of n objects.

21. Let $a_1 = a_2 = 1$ and for $n \geq 1$

$$a_{n+2} = n a_{n+1} a_n.$$

Show that

$$a_{n+2} = 1^{F_n} \cdot 2^{F_{n-1}} \cdot 3^{F_{n-2}} \cdots (n-1)^{F_2} \cdot n^{F_1},$$

where F_n is the nth Fibonacci number. Can you find an asymptotical expression of a_n for sufficiently large n?

22. Let $a_1 = \beta$ and $a_{n+1} = a_n^2 + (1 - 2\alpha)a_n + \alpha^2$ for $n \in \mathbb{N}$. Find all possible values of α and β such that the sequence a_n converges, and also determine the limit.

23. Suppose that a_n has the property that for any $\alpha > 1$, the subsequence

$$a_{\lfloor \alpha^n \rfloor} \to 0 \quad \text{as } n \to \infty,$$

where $\lfloor x \rfloor$ is the greatest integer not exceeding x (for example, $\lfloor 3.14 \rfloor = 3$). Does the sequence a_n itself necessarily converge to zero?

24. Determine

$$\lim_{n\to\infty} \frac{1}{n} \sum_{k=1}^{n} \left\{ \frac{n}{k} \right\}^2,$$

where $\{x\} = x - \lfloor x \rfloor$ is the fractional part of x (for example, $\{3.14\} = 0.14$).

25. (**Monthly Problem 12153, 2020**). For a real number x whose fractional part is not $1/2$, let $\langle x \rangle$ denote the nearest integer to x. For a positive integer n, let

$$a_n = \sum_{k=1}^{n} \frac{1}{\langle \sqrt{k} \rangle} - 2\sqrt{n}.$$

(a) Prove that the sequence a_n is convergent, and find its limit L.

(b) Prove that the set $\{\sqrt{n}(a_n - L) : n \geq 1\}$ is a dense subset of $[0, 1/4]$.

26. For $n \in \mathbb{N}$, define

$$a_{n+1} = \frac{1}{2}(1 + b_n^2), \quad b_{n+1} = \frac{1}{2}(2a_n - a_n^2).$$

If $0 \leq b_n \leq 1/2 \leq a_n$, show that both a_n and b_n converge and find their limits.

27. Let $a_n > 0$ for all $n \in \mathbb{N}$. Define

$$s_n = a_1 + a_2 + \cdots + a_n, \quad t_n = \frac{a_1}{s_1} + \frac{a_2}{s_2} + \cdots + \frac{a_n}{s_n}.$$

Show that $\lim_{n\to\infty} s_n = \infty$ implies that $\lim_{n\to\infty} t_n = \infty$

28. Let $\lim_{n\to\infty} a_n = a$. For $|\alpha| < 1$, find

$$\lim_{n\to\infty} (a_n + a_{n-1}\alpha + a_{n-2}\alpha^2 + \cdots + a_0\alpha^n).$$

29. Let $\lim_{n \to \infty} a_n = a$ and $\lim_{n \to \infty} b_n = b$. Show that

$$\lim_{n \to \infty} \frac{a_1 b_{n-1} + a_2 b_{n-2} + \cdots + a_{n-1} b_1}{n} = ab.$$

30. Show that if $p_n > 0 \, (n = 1, 2, \ldots,)$ and

$$\lim_{n \to \infty} \frac{p_n}{p_1 + p_2 + \cdots + p_n} = 0, \quad \lim_{n \to \infty} a_n = a,$$

then

$$\lim_{n \to \infty} \frac{p_1 a_1 + p_2 a_2 + \cdots + p_n a_n}{p_1 + p_2 + \cdots + p_n} = a.$$

31. Let a_n and b_n be positive sequences. Assume that

$$\lim_{n \to \infty} \frac{a_1 + a_2 + \cdots + a_n}{n a_n} = a, \quad \lim_{n \to \infty} \frac{b_1 + b_2 + \cdots + b_n}{n b_n} = b$$

and $a + b > 0$. Find

$$\lim_{n \to \infty} \frac{a_1 b_1 + 2 a_2 b_2 + \cdots + n a_n b_n}{n^2 a_n b_n}.$$

32. Let $\lim_{n \to \infty} a_n = a$, $b_n > 0$, and $\lim_{n \to \infty} (b_1 + b_2 + \cdots + b_n) = B$. Prove that

$$\lim_{n \to \infty} (a_n \, b_1 + a_{n-1} \, b_2 + \cdots + a_1 \, b_n) = aB.$$

33. (**Toeplitz Theorem**). Let $n, k \in \mathbb{N}$. Assume that
     (a) $t_{nk} \geq 0, \sum_{k=1}^{n} t_{nk} = 1, \lim_{n \to \infty} t_{nk} = 0$, and
     (b) $\lim_{n \to \infty} a_n = a$.
     Prove that

$$\lim_{n \to \infty} \sum_{k=1}^{n} t_{nk} a_k = a.$$

34. Let $\lim_{n \to \infty} a_n = a$. Evaluate

$$\lim_{n \to \infty} \frac{1}{2^n} \sum_{k=0}^{n} \binom{n}{k} a_k,$$

where $\binom{n}{k} = \frac{n!}{k! \, (n-k)!}$ is the binomial coefficient.

35. For $n \in \mathbb{N}$, let

$$a_n = \sum_{k=0}^{n} \frac{1}{\binom{n}{k}}.$$

Show that
(a) for $n \geq 2, a_n = 1 + \frac{n+1}{2n} a_{n-1}$,
(b) $a_n = \frac{n+1}{2^{n+1}} \sum_{k=1}^{n+1} \frac{2^k}{k}$,
(c) $\lim_{n \to \infty} a_n = 2$.

36. For $n \in \mathbb{N}$, show that

    (a) $\sum_{k=0}^{n} \frac{1}{\binom{2n}{2k}} = \frac{2n+1}{2^{2n+1}} \sum_{k=1}^{2n+1} \frac{2^k}{k+1}$,

    (b) $\sum_{k=0}^{n} \frac{x^k}{\binom{n}{k}} = \frac{(n+1)x^{n+1}}{(x+1)^{n+2}} \sum_{k=1}^{n+1} \frac{(1+x^k)(x+1)^k}{kx^k}$.

37. For $n \in \mathbb{N}$, show that

    (a) For $n \geq 3$,

    $$\sum_{k=0}^{n} \frac{1}{k!} - \frac{3}{2n} < \left(1 + \frac{1}{n}\right)^n < \sum_{k=0}^{n} \frac{1}{k!}.$$

    (b) Let $S_n = 1 + 2^2 + 3^3 + \cdots + n^n$. Then

    $$n^n \left(1 + \frac{1}{4(n-1)}\right) < S_n < n^n \left(1 + \frac{2}{e(n-1)}\right).$$

38. Let $a_1 = 1, a_2 = 1/2$ and

    $$a_{n+2} = a_{n+1} - \frac{a_n a_{n+1}}{2}, \quad \text{for } n \in \mathbb{N}.$$

    Find $\lim_{n \to \infty} n a_n$.

39. Let $a_0 = 0, a_{n+1} = 1 + \sin(a_n - 1)$ for all $n \in \mathbb{N}$. Find

    $$\lim_{n \to \infty} \frac{a_1 + a_2 + \cdots + a_n}{n}.$$

40. (**Putnam Problem 2016-A2**). Given a positive integer n, let $M(n)$ be the largest integer m such that

    $$\binom{m}{n-1} > \binom{m-1}{n}.$$

    Evaluate $\lim_{n \to \infty} \frac{M(n)}{n}$.

41. (**Monthly Problem 11528, 2010**). Let $p, a,$ and b be positive integers with $a < b$. Consider a sequence a_n defined by $n a_{n+1} = (n + 1/p)a_n$ and an initial condition $a_1 \neq 0$. Evaluate

    $$\lim_{n \to \infty} \frac{a_{an} + a_{an+1} + \cdots + a_{bn}}{n a_{an}}.$$

42. (**Monthly Problem 11786, 2014**). Let $x_1, x_2, \ldots$ be a sequence of positive numbers such that $\lim_{n \to \infty} x_n = 0$ and

    $$\lim_{n \to \infty} \frac{\ln x_n}{x_1 + x_2 + \cdots + x_n} < 0.$$

    Prove that

    $$\lim_{n \to \infty} \frac{\ln x_n}{\ln n} = -1.$$

43. For $k \in \mathbb{N}$, let $d(k)$ be the largest odd divisor of k. Determine

    $$\lim_{n \to \infty} \frac{1}{n^2} \sum_{k=1}^{n} \frac{n-k+1}{k} d(k).$$

44. Find the limit
$$\lim_{n\to\infty} \left(1+\frac{1}{n^2}\right)\left(1+\frac{2}{n^2}\right)\cdots\left(1+\frac{n}{n^2}\right).$$

45. Evaluate
$$\lim_{n\to\infty} \frac{2\ln 2 + 3\ln 3 + \cdots + n\ln n}{n^2\ln n}.$$

46. Find the limit
$$\lim_{n\to\infty} \left(\sqrt[n+1]{(n+1)!} - \sqrt[n]{n!}\right).$$

47. (**Monthly Problem 11935, 2016**). Let f be a function from $\mathbb{Z}^+$ to $\mathbb{R}^+$ such that $\lim_{n\to\infty} f(n)/n = a$, where $a > 0$. Find

$$\lim_{n\to\infty} \left(\sqrt[n+1]{\prod_{k=1}^{n+1} f(k)} - \sqrt[n]{\prod_{k=1}^{n} f(k)}\right).$$

48. (**Monthly Problem 11771, 2014**). Let $n!! = \prod_{i=0}^{[(n-1)/2]} (n-2i)$. Find

$$\lim_{n\to\infty} \left(\sqrt[n]{(2n-1)!!}\left(\tan\frac{\pi\,\sqrt[n+1]{(n+1)!}}{4\,\sqrt[n]{n!}} - 1\right)\right).$$

49. (**Monthly Problem 11875, 2015**). Let $f_n = (1+1/n)^n((2n-1)!!L_n)^{1/n}$. Find $\lim_{n\to\infty}(f_{n+1} - f_n)$. Here $n!! = \prod_{j=0}^{[(n+1)/2]}(n-2j)$, where L_n denotes the nth Lucas number, given by $L_0 = 2, L_1 = 1$, and for other n, by $L_n = L_{n-1} + L_{n-2}$.

50. (**Monthly Problem 12120, 2019**). For positive integers n and k with $n \geq k$, let
$$a(n, k) = \sum_{j=0}^{k-1} \binom{n}{j} 3^j.$$

   (a) Evaluate
   $$\lim_{n\to\infty} \frac{1}{4^n} \sum_{k=1}^{n} \frac{a(n, k)}{k}.$$

   (b) Evaluate
   $$\lim_{n\to\infty} n\left(4^n L - \sum_{k=1}^{n} \frac{a(n, k)}{k}\right),$$

   where L is the limit in part (a).

51. Let a_n be a positive sequence such that
   $$(a_{n+1} - a_n)(a_{n+1}a_n - 1) \leq 0$$

   for all $n \in \mathbb{N}$. If $\lim_{n\to\infty} a_{n+1}/a_n = 1$, prove that a_n converges.

52. Let H_n be the nth harmonic number. Define
   $$a_n = H_n^2 - \sum_{k=1}^{n} \frac{1}{k} H_{\max(k,n-k)}.$$

   Show that $\lim_{n\to\infty} a_n = \pi^2/12$.

53. Let H_n be the nth harmonic number. Show that

    (a) There is no rational function $R(x)$ such that $R(n) = H_n$ for all $n \in \mathbb{N}$.
    (b) $H_n - H_m$ is never an integer for $m < n$.
    (c) H_n has a non-terminating decimal expansion for $n > 7$.

54. Let H_n be the nth harmonic number. Show that

    (a) $\sum_{k=0}^{n} (-1)^{k-1} \binom{n}{k} H_k = \frac{1}{n}$.
    (b) $\sum_{k=0}^{n} \binom{n}{k} H_k = 2^n \left(H_n - \sum_{k=1}^{n} \frac{1}{2^k k} \right)$.
    (c) $\sum_{k=1}^{n} \frac{(-1)^{k-1}}{k+1} \binom{n}{k} H_k = \frac{H_n}{n+1}$.
    (d) $\sum_{k=0}^{n} \frac{(-1)^k H_k}{\binom{n}{k}} = \frac{(-1)^n (n+1) H_n - 1}{n+2} + \frac{1 - (-1)^n}{(n+2)^2}$.

55. Use the Wilf-Zeilberger algorithm to prove that

    (a) $\sum_{k=1}^{n-1} \frac{1}{2k} \binom{2k}{k} \binom{2(n-k)}{n-k} = \binom{2n}{n} (H_{2n-1} - H_n)$.
    (b) $\sum_{k=0}^{2n} (-1)^k \binom{2n}{k} \binom{2n+k}{k} \binom{2k}{k} 4^{2n-k} = \binom{2n}{n}^2$.

56. Let $p_n(x) = a_{n+2} x^2 + a_{n+1} x - a_n$ be a sequence of polynomials, where $a_0 = a_1 = 1$ and $a_{n+2} = a_{n+1} + a_n$ for $n \geq 0$. Let r_n and s_n be the roots of $p_n(x) = 0$ with $r_n \leq s_n$. Find $\lim_{n \to \infty} r_n$ and $\lim_{n \to \infty} s_n$.

57. Let a_n be a real sequence and $|\alpha| < 1$. Prove that if the sequence $a_{n+1} + \alpha a_n$ converges, then a_n converges. What happens for $|\alpha| \geq 1$?

58. Let a_n and b_n be real sequences with

    $$b_n = c_1 a_{n-1} + c_2 a_n,$$

    where $0 < c_1 < c_2$. Show that b_n converges to A if and only if a_n converges to $A/(c_1 + c_2)$.
    When $c_1 = 1, c_2 = 2$, this is **Putnam Problem 1969-A6**.

59. Based on the result in the previous problem, we see that the convergence of a sequence is equivalent to the convergence of some its linear combinations. Let

    (a) $a_n = \frac{1}{2} - \frac{3}{2^2} + \cdots + (-1)^{n-1} \frac{2n-1}{2^n}$,
    (b) $a_n = \cos^3 \theta - \frac{1}{3} \cos^3 3\theta + \cdots + (-1)^{n-1} \frac{1}{3^n} \cos^3 3^n \theta$.

    Find the limit of a_n by constructing a new sequence b_n.

60. Let S_n be the sum of lengths of all the sides and all the diagonals of a regular n-gon inscribed in a unit circle. Find S_n and $\lim_{n \to \infty} S_n/n^2$.

61. Let

    $$S_n = \sum_{i=0}^{n} \left(\sum_{j=0}^{n-i} \frac{x^j}{i! \, j!} \right).$$

    Find $\lim_{n \to \infty} S_n$.

62. Let a_n be a positive sequence such that $a_{m+n} \leq a_m a_n$ for all $m, n \in \mathbb{N}$. Prove that

    $$\lim_{n \to \infty} \frac{\ln a_n}{n} = \inf_{n \geq 1} \left\{ \frac{\ln a_n}{n} \right\}.$$

63. Let a_n be a nonnegative sequence such that $a_{m+n} \leq a_m a_n$ for all $m, n \in \mathbb{N}$. Prove that sequence $\sqrt[n]{a_n}$ converges.

64. (**Monthly Problem E2860, 1980**). Let $a_n \geq 0$ be a nondecreasing sequence. Assume that $a_{mn} \geq ma_n$ for all $m, n \in \mathbb{N}$ and also $\sup\{a_n/n\} = l < \infty$. Must a_n/n have a limit?

65. Assume the sequence a_n satisfies

$$a_m + a_n - 1 < a_{m+n} < a_m + a_n + 1, \quad (m, n = 1, 2, \ldots).$$

Prove that $\lim_{n \to \infty} a_n/n = a$ exists and $an - 1 \leq a_n \leq an + 1$ for all $n \geq 1$.

66. Let a_n be an increasing positive sequence with $\lim_{n \to \infty} a_n = \infty$. If the sequence $a_{n+1} - a_n$ is also increasing, determine

$$\lim_{n \to \infty} \frac{x_1 + x_2 + \cdots + x_n}{n \sqrt{a_n}}.$$

67. Let a_n be a positive sequence. Prove that

   (a) (**Putnam Problem 1949-B5**).

$$\limsup \left(\frac{a_1 + a_{n+1}}{a_n} \right) \geq e.$$

   (b) (**Putnam Problem 1963-A4**).

$$\limsup n \left(\frac{1 + a_{n+1}}{a_n} - 1 \right) \geq 1.$$

68. Let x_1 and x_2 be positive numbers. Define, for $n \geq 2$

$$x_{n+1} = \sum_{k=1}^{n} x_k^{1/n}.$$

   Find the limit

$$\lim_{n \to \infty} \frac{x_n - n}{n + k}.$$

69. Let

$$x_n = \sum_{k=1}^{n} \frac{1}{n+k} e^{1/(n+k)}.$$

   Prove that x_n converges and find its limit.

70. Start with a positive triple (a, b, c). Define

$$(x_0, y_0, z_0) = (a, b, c), \quad (x_{n+1}, y_{n+1}, z_{n+1})$$
$$= (y_n + z_n - x_n, x_n + z_n - y_n, x_n + y_n - z_n) \text{ (for } n \geq 0).$$

   Show that the limits of x_n, y_n and z_n exist if and only if $a + b + c = 1$. *Hint.* Use matrices.

71. Start with a n-tuple of real number $(x_1, x_2, \ldots, x_n)$. Inductively define

$$x_i^{(1)} = \frac{x_i + x_{i+1}}{2}, \quad x_i^{(k)} = \frac{x_i^{(k-1)} + x_{i+1}^{(k-1)}}{2}, \quad i = 1, 2, \ldots n,$$

   where $x_{n+1} = x_1$. For $1 \leq i \leq n$, find

$$\lim_{k \to \infty} x_i^{(k)}.$$

72. (**Putnam Problem 1950-A3**). Define

$$x_0 = a, x_1 = b, x_{n+1} = \frac{x_{n-1} + (2n-1)x_n}{2n} \quad (n = 1, 2, \ldots)$$

Find $\lim_{n\to\infty} x_n$ concisely in terms of a and b.

73. Let f be a monotone decreasing continuous function on $(0, +\infty)$ and $f(x) > 0$. Define

$$a_n = \sum_{k=1}^{n} f(k) - \int_0^n f(x)\,dx.$$

Prove that a_n converges. In particular, if $f(x) = 1/x$, this limit gives the Euler-Mascheroni constant.

74. Given $x_0, x_1 \geq 0$, let

$$x_{n+1} = \sqrt{x_n \cdot x_{n-1}}, \quad (n = 2, 3, \ldots).$$

Find $\lim_{n\to\infty} x_n$ in terms of x_0 and x_1.

75. Let $x_1 = \sqrt{5}, x_{n+1} = x_n^2 - 2$ for all $n \in \mathbb{N}$. Evaluate

$$\lim_{n\to\infty} \frac{x_1 \cdot x_2 \cdots \cdots x_n}{x_{n+1}}.$$

76. Evaluate

$$\lim_{n\to\infty} \frac{\ln^2 n}{n} \sum_{k=2}^{n-2} \frac{1}{\ln k \cdot \ln(n-k)}.$$

77. (**Monthly Problem 11376, 2008**). Given a real number a and a positive integer n, let

$$S_n(a) = \sum_{an \leq k \leq (a+1)n} \frac{1}{\sqrt{kn - an^2}}.$$

For which a does the sequence $S_n(a)$ converge?

78. Let $f(0) = 0$. Assume that $f'(0)$ exists and defines

$$a_n = f\left(\frac{1}{n^2}\right) + f\left(\frac{2}{n^2}\right) + \cdots + f\left(\frac{n}{n^2}\right).$$

Find $\lim_{n\to\infty} a_n$. Use your result to determine

$$\lim_{n\to\infty} \sum_{k=1}^{n} \left(\sqrt{1 + \frac{k}{n^2}} - 1\right).$$

79. Prove that

(a) $\lim_{n\to\infty} \frac{\sum_{k=1}^{n} \sqrt{k}}{n^{3/2}} = \frac{2}{3}$.

(b) $\lim_{n\to\infty} n\left(\frac{\sum_{k=1}^{n} \sqrt{k}}{n^{3/2}} - \frac{2}{3}\right) = \frac{1}{2}$.

80. Let

$$s_n = \sum_{k=1}^{n} \frac{1}{n}\left(\frac{\ln k}{\ln n}\right)^p, \quad (p \geq 1).$$

Prove that $\lim_{n\to\infty} s_n = 1$.

81. Let a_n be a sequence such that $\lim_{n\to\infty} \left(a_n \sum_{k=1}^{n} a_k^2 \right) = 1$. Find

$$\lim_{n\to\infty} \sqrt[3]{3n}\, a_n.$$

82. Let a_n be a sequence such that $e^{a_n} + na_n = 2$ for all $n \in \mathbb{N}$. Find

$$\lim_{n\to\infty} n(1 - na_n).$$

83. Let $a_0, a_1, a_2, \ldots$ be a sequence of nonnegative numbers. The sequence is called *log-concave* (resp. *log-convex*) if for all $k \geq 1, a_{k-1}a_{k+1} \leq a_k^2$ (resp. $a_{k-1}a_{k+1} \geq a_k^2$).

    (1) Show that $\binom{n}{k}$ is log-concave in k for fixed n.

    (2) If x_k is log-concave, show that

    $$y_k = \sum_{j=0}^{k} \binom{k}{j} x_j$$

    is also log-concave.

84. Let a_n be a positive sequence and $\lim_{n\to\infty} a_n = 0$. Find

$$\lim_{n\to\infty} \sum_{k=1}^{n} \frac{1}{n} \ln\left(a_n + \frac{k}{n} \right).$$

85. Let a_n be a sequence with $1/2 < a_n < 1$ for all $n \geq 0$. Define a sequence b_n by

$$b_0 = a_0, \quad b_{n+1} = \frac{a_{n+1} + b_n}{1 + a_{n+1}b_n}.$$

    Does b_n converge? If so, find its limit.

86. Let $f : [0,1] \longmapsto [0,1]$ be continuous. Prove that the sequence $x_{n+1} = f(x_n)$ converges if and only if

$$\lim_{n\to\infty} (x_{n+1} - x_n) = 0.$$

87. Let $f : [a,b] \longmapsto [a,b]$ such that

$$|f(x) - f(y)| < |x - y| \; (x, y \in [a, b]).$$

    For any $0 < \alpha < 1, x_1 \in [a, b]$, define the sequence

$$x_{n+1} = \alpha\, x_n + (1 - \alpha)f(x_n).$$

    Prove that $x_n \to x$ and $x = f(x)$.

88. (**Putnam Problem 2008-B2**). Let $F_0(x) = \ln x$. For $n \geq 0$ and $x > 0$, let $F_{n+1}(x) = \int_0^x F_n(t)\, dt$. Evaluate

$$\lim_{n\to\infty} \frac{n! F_n(1)}{\ln n}.$$

89. Let $x_{n+1} = x_n(1 - x_n), n = 1, 2, \ldots, x_1 \in (0, 1)$. Show that

$$\lim_{n\to\infty} \frac{n(1 - nx_n)}{\ln n} = 1.$$

90. Define the sequence x_n by

$$x_0 = 0, \; x_1 = 1, \; x_{n+1} = \frac{x_n + n x_{n-1}}{n+1} \; \text{ for } n \geq 1.$$

Find $\lim_{n \to \infty} x_n$.

91. (**Monthly Problem 11559, 2011**). For positive p and $x \in (0,1)$, define the sequence x_n by $x_0 = 1, x_1 = x$, and for $n \geq 1$,

$$x_{n+1} = \frac{p x_{n-1} x_n + (1-p) x_n^2}{(1+p) x_{n-1} - p x_n}.$$

Find positive real numbers α and β such that $\lim_{n \to \infty} n^\alpha x_n = \beta$.

92. Let $P(x) = a_m x^m + \cdots + a_1 x + a_0$ with positive coefficients. Define

$$A_n = \frac{P(1) + P(2) + \cdots + P(n)}{n}, \quad G_n = \sqrt[n]{P(1) \cdot P(2) \cdots P(n)}.$$

Prove that

$$\lim_{n \to \infty} \frac{A_n}{G_n} = \frac{e^m}{m+1}.$$

93. (**Monthly Problem 11811, 2014**). Let $\{a_n\}$ and $\{b_n\}$ be infinite sequences of positive numbers. Let $\{x_n\}$ be the infinite sequence given by

$$x_n = \frac{a_1^{b_1} \cdots a_n^{b_n}}{\left(\frac{a_1 b_1 + \cdots + a_n b_n}{b_1 + \cdots + b_n} \right)^{b_1 + \cdots + b_n}}.$$

(a) Prove that $\lim_{n \to \infty} x_n$ exists.

(b) Find the set of all c that can occur as that limit, for suitably chosen $\{a_n\}$ and $\{b_n\}$.

94. (**Monthly Problem 11821, 2015**). Let p be a positive integer. Prove that

$$\lim_{n \to \infty} \frac{1}{2^n n^p} \sum_{k=0}^{n} (n - 2k)^{2p} \binom{n}{k} = \prod_{j=1}^{p} (2j - 1).$$

95. Let

$$h_n = \sum_{k=1}^{n} (-1)^k \left(1 + \frac{1}{3} + \cdots + \frac{1}{2k-1} - \frac{1}{2} \ln n - \frac{1}{2} \gamma - \ln 2 \right),$$

where γ is the Euler-Mascheroni constant. Find $\lim_{n \to \infty} h_n$.

96. (**Monthly Problem 11973, 2002**). Let

$$R_k(n) = \sqrt{2 - \sqrt{2 + \sqrt{2 + \cdots + \sqrt{2 + \sqrt{n}}}}}$$

with k square roots. Prove that $\lim_{k \to \infty} R_k(2)/R_k(3) = 3/2$.

97. (**Monthly Problem 12129, 2019**). Compute

$$\sqrt{2+\sqrt{2+\sqrt{2+\cdots+\sqrt{2-\sqrt{2+\cdots}}}}},$$

where the sequence of signs consists of $n-1$ plus signs followed by a minus sign and repeats with period n.

98. (**Monthly Problem 11367, 2008**). Let $x_1 = \sqrt{1+2}, x_2 = \sqrt{1+2\sqrt{1+3}}$ and in general, let x_{n+1} be the number obtained by replacing the innermost expression $(1+(n+1))$ in the nested square root formula for x_n with $1+(n+1)\sqrt{1+(n+2)}$. Show that

$$\lim_{n\to\infty} \frac{x_n - x_{n-1}}{x_{n+1} - x_n} = 2.$$

99. Let

$$f(x) = \sqrt{1+x\sqrt{1+(x+1)\sqrt{1+(x+2)\sqrt{1+\cdots}}}}$$

Show that $f(x) = x+1$. The particular case $x = 2$ yields Ramanujan's formula

$$\sqrt{1+2\sqrt{1+3\sqrt{1+4\sqrt{1+\cdots}}}} = 3.$$

100. Let F_n be nth Fibonacci number. Define

$$a_n = \sqrt{F_2^2 + \sqrt{F_4^2 + \sqrt{F_8^2 + \sqrt{\cdots + \sqrt{F_{2^n}^2}}}}}.$$

Prove that $\lim_{n\to\infty} a_n = 3$.

101. (**Putnam problem 1947-A1**). If a_n is a sequence of numbers such that for $n \geq 1$

$$(2 - a_n)\, a_{n+1} = 1,$$

prove that $\lim_{n\to\infty} a_n = 1$.

102. Let $a_0 = 1$, and for $n \in \mathbb{N}$, define

$$a_{n+1} = a_n + \frac{1}{a_n}.$$

Find $\lim_{n\to\infty} a_n/\sqrt{2n}$ and $\lim_{n\to\infty} (a_n - \sqrt{2n})$.

103. (**Monthly Problem 12210, 2020**). Let $x_1 = 1$, and let

$$x_{n+1} = \left(\sqrt{x_n} + \frac{1}{\sqrt{x_n}}\right)^2$$

when $n \geq 1$. For $n \in \mathbb{N}$, let $a_n = 2n + (1/2)\log n - x_n$. Show that the sequence $a_1, a_2, \ldots$ converges.

104. (**Putnam problem 2006-B6**). Let k be an integer greater than 1. Suppose $a_0 > 0$, and

$$a_{n+1} = a_n + \frac{1}{\sqrt[k]{a_n}}$$

for $n \geq 1$. Evaluate $\lim_{n \to \infty} a_n^{k+1}/n^k$.

105. Let α_i be either 1 or -1 for $i = 0, 1, \dots, n-1$. Prove that, for every $n \in \mathbb{N}$,

$$\alpha_0 \sqrt{2 + \alpha_1 \sqrt{2 + \alpha_2 \sqrt{2 + \cdots + \alpha_{n-1}\sqrt{2}}}} = 2\sin\left[\left(\alpha_0 + \frac{\alpha_0\alpha_1}{2} + \cdots + \frac{\alpha_0\alpha_1 \cdots \alpha_{n-1}}{2^{n-1}}\right)\frac{\pi}{4}\right].$$

106. Let $a_n > 0$ for all $n \in \mathbb{N}$. Define

$$x_n = \sqrt{a_1 + \sqrt{a_2 + \sqrt{a_3 + \cdots + \sqrt{a_n}}}}; \quad a = \lim_{n \to \infty} \sup \frac{\ln \ln a_n}{n}.$$

Prove that

(a) If $a < \ln 2$, the sequence x_n converges.
(b) If $a > \ln 2$, the sequence x_n diverges.
(c) If $a = \ln 2$, the sequence x_n may converge or diverge.

107. Let $a_n > 0, b_n > 0$ for all $n \in \mathbb{N}$. Define

$$x_n = \sqrt{a_1 + b_1\sqrt{a_2 + b_2\sqrt{a_3 + \cdots + b_{n-1}\sqrt{a_n}}}} \quad (n \in \mathbb{N}).$$

Prove that x_n converges if and only if

$$y_n = \ln\left(\frac{a_n}{2^n}\right) + \sum_{k=1}^{n} \ln\left(\frac{b_k}{2^k}\right) \quad (n \in \mathbb{N})$$

converges.

108. Let $a_0 = 1, a_{n+1} = \exp(-\sum_{k=0}^{n} a_k)$ for $n \geq 0$. Show that

$$a_n = \frac{1}{n} - \frac{\ln n}{2n^2} + o\left(\frac{\ln n}{n^2}\right).$$

109. (**Monthly 11153, 2005**). Let $x_1 = 1$, and for $n \geq 1$ let $x_{n+1} = x_n + 2 + 1/x_n$. Define

$$y_n = 2n + \frac{1}{2}\ln n - x_n.$$

Show that the sequence $\{y_n\}$ is eventually increasing.

110. Let $p, q \in \mathbb{N}$ and $p \geq q$. Define

$$S_{pq}(x, n) = \sum_{k=0}^{n} x^k \ln\left(\frac{pn}{qk}\right).$$

Show that

(a) If $|x| < 1$, then

$$\lim_{n \to \infty} \frac{S_{pq}(x, n)}{\ln n} = \frac{qx}{(1-x)^2}.$$

(b) If $|x| > 1$ and $p = q$, then

$$\lim_{n \to \infty} \frac{S_{pp}(x, n)}{\ln n} = \frac{px}{(1-x)^2}.$$

(c) If $|x| > 1$ and $p > q$, then

$$\lim_{n \to \infty} \frac{S_{pq}(x, n)}{n\,x^n} = \frac{x}{x-1}\left(p \ln p - q \ln q - (p-q)\ln(p-q)\right).$$

111. (**Monthly Problem 11851, 2015**). For real a and $b \geq 0$ and integer $n \geq 1$, let

$$\gamma_n(a, b) = -\ln(n+a) + \sum_{k=1}^{n} \frac{1}{k+b}.$$

(a) Prove that $\gamma(a, b) = \lim_{n \to \infty} \gamma_n(a, b)$ exists and is finite.
(b) Find

$$\lim_{n \to \infty} \left(\ln\left(\frac{e}{n+a}\right) + \sum_{k=1}^{n} \frac{1}{k+b} - \gamma(a, b) \right)^n.$$

112. (**Monthly Problem 11852, 2015**). For $n \in \mathbb{N}$, let $v_n = k$ if 3^k divides n but 3^{k+1} does not. Let $x_1 = 2$, and for $n \geq 2$ let

$$x_n = 4v_n + 2 - \frac{2}{x_{n-1}}.$$

Show that every positive rational number appears exactly once in the sequence $\{x_n\}_{n \geq 1}$.

113. (**Monthly Problem 11976, 2017**). Given a positive real number s, consider the sequence $\{u_n\}$ defined by $u_1 = 1, u_2 = s$, and $u_{n+2} = u_n u_{n+1}/n$ for $n \geq 1$.

(a) Show that there is a constant C such that $\lim_{n \to \infty} u_n = \infty$ when $s > C$ and $\lim_{n \to \infty} u_n = 0$ when $s < C$.
(b) Calculate $\lim_{n \to \infty} u_n$ when $s = C$.

114. (**Monthly Problem 11995, 2017**). Suppose $0 < x_0 < \pi$, and for $n \geq 1$ define

$$x_n = \frac{1}{n} \sum_{k=0}^{n-1} \sin x_k.$$

Find $\lim_{n \to \infty} x_n \sqrt{\ln n}$.

115. (**Math. Magazine Problem 2087, 2020**). Consider the sequence defined by $x_1 = a > 0$ and

$$x_n = \ln\left(1 + \frac{x_1 + x_2 + \cdots + x_{n-1}}{n-1}\right) \quad \text{for } n \geq 2.$$

compute $\lim_{n \to \infty} x_n \ln n$.

116. (**Monthly Problem 12166, 2020**). Let $a_0 = 0$, and define a_k recursively by $a_k = e^{a_{k-1}-1}$ for $k \geq 1$.

(a)Prove $k/(k+2) < a_k < k/(k+1)$ for $k \geq 1$.
(b)Is there a number c such that $a_k < (k+c)/(k+c+2)$ for all k?

Hint: For part (b), use the Stolz-Cesàro theorem to find an asymptotic expansion of a_k.

2

Infinite Numerical Series

The infinite we shall do right away. The finite may take a little longer.

— *Stanislaw Ulam*

How can intuition deceive us at this point?

— *Henri Poincare*

Equipped with a working knowledge of sequences, in this chapter, we study the convergence of infinite numerical series. Our focus is on various convergence tests far beyond those encountered in a first-year calculus sequence. We provide examples and discuss the motivations of the ideas frequently before a new test is introduced. We also explore the behavior of a convergent series after rearrangements and prove the famous Riemann series theorem. This chapter concludes with some intriguing problems.

2.1 Main Definitions and Basic Convergence Tests

"What's one and one and one and one and one and one and one and one and one and one?"
"I don't know," said Alice, "I lost count."
"She can't do Addition," the Red Queen interrupted. — Lewis Carroll

Infinite series have intrigued mathematicians since antiquity. The question of how an infinite sum of positive terms can yield a finite result was viewed both as a deep philosophical challenge and an important gap in the understanding of infinity. Since it is impossible to add up infinitely many terms directly, the time has arrived for a precise definition of what we mean by the sum of an infinite series.

Definition 2.1. *Given a sequence a_n, an* infinite series *is an expression in the form*

$$\sum_{k=1}^{\infty} a_k = a_1 + a_2 + \cdots + a_n + \cdots .$$

In particular, a_n is called the nth term *of the series.*

Let

$$s_n = \sum_{k=1}^{n} a_k = a_1 + a_2 + \cdots + a_n, \qquad (2.1)$$

which is called the *nth partial sum* of the series. As n increases, the partial sum includes more and more terms of the series. Thus, it is reasonable to view the limit of the sequence s_n as the sum of the series. This leads to

DOI: 10.1201/9781003304135-2

Definition 2.2. *If the sequence s_n defined by (2.1) converges to s, we say that the series $\sum_{k=1}^{\infty} a_k$ converges and write*

$$\sum_{n=1}^{\infty} a_n = s.$$

The number s is called the sum of the series. If the sequence s_n diverges, we say the series diverges. *In this case, a divergent series has no sum.*

Notice that here the sum of the series is *the limit of a sequence of finite sums*, instead of adding the successive terms together. By considering the sequence of partial sums, we reduce the series to a more tractable object—a sequence. Moreover, the theorems on sequences can be stated in terms of series. For example, the Cauchy criterion can be restated as:

Theorem 2.1 (Cauchy Criterion). *The series converges if and only if for every $\epsilon > 0$ there is an $N \in \mathbb{N}$ such that*

$$\left| \sum_{k=n}^{m} a_k \right| < \epsilon \qquad \textit{whenever } m \geq n > N.$$

In particular, taking $m = n$ yields

Theorem 2.2 (Necessary Condition for Convergence). *If $\sum_{k=1}^{\infty} a_k$ converges, then* $\lim_{n \to \infty} a_n = 0$.

Given a positive series (i.e., $a_n > 0$ for all $n \in \mathbb{N}$), clearly, the s_n defined by (2.1) is an increasing sequence. Thus, by the monotone convergence theorem, we find that

$$s_n \text{ converges} \quad \Longleftrightarrow \quad s_n \text{ is bounded.}$$

Thus, to test the convergence of a positive series, it suffices to check the boundedness of the series. This observation leads to a powerful technique.

Theorem 2.3 (Comparison Test). *Let a_n and b_n be both positive sequences with $a_n \leq b_n$ for all $n \in \mathbb{N}$.*

>   *1. If $\sum_{n=1}^{\infty} b_n$ converges, then $\sum_{n=1}^{\infty} a_n$ converges.*
>   *2. If $\sum_{n=1}^{\infty} a_n$ diverges, then $\sum_{n=1}^{\infty} b_n$ diverges.*

Based on the Cauchy criterion, the behavior of a series is determined only by its tails. Thus, in the comparison test, the requirement that $a_n \leq b_n$ does not really need to hold for all $n \in \mathbb{N}$ but just needs to be *eventually* true, i.e., $a_n \leq b_n$ holds for all n greater than some integer N.

When the ratio between consecutive terms is simpler than the terms themselves, which is especially true when the terms are in product form, the following test is often more convenient to apply.

Theorem 2.4 (Ratio Comparison Test). *Let a_n and b_n be both positive sequences with $a_{n+1}/a_n \leq b_{n+1}/b_n$ for all $n \in \mathbb{N}$.*

>   *1. If $\sum_{n=1}^{\infty} b_n$ converges, then $\sum_{n=1}^{\infty} a_n$ converges.*
>   *2. If $\sum_{n=1}^{\infty} a_n$ diverges, then $\sum_{n=1}^{\infty} b_n$ diverges.*

Its justification rests on the comparison test. Indeed, by multiplying the inequalities $a_{i+1}/a_i \leq b_{i+1}/b_i$ from $i = 1$ to $i = n - 1$, we obtain

$$\frac{a_n}{a_1} \leq \frac{b_n}{b_1} \quad \text{or} \quad a_n \leq \frac{a_1}{b_1} b_n \quad (n = 1, 2, 3, \cdots).$$

Since the series $\sum_{n=1}^{\infty} \frac{a_1}{b_1} b_n$ and $\sum_{n=1}^{\infty} b_n$ have the same behavior, the theorem follows directly from the comparison test.

In these two comparison tests, the convergence or divergence of one series is based on another series whose convergence or divergence is known. To choose the right series with which to compare, we often need some standard series that serve as measuring sticks. The three most popular standards are the *geometric series*, *p-series* and $\sum_{n=2}^{\infty} 1/(n \ln n)$.

Example 2.1 (Geometric Series). *A series of the form*

$$a + ar + ar^2 + \cdots = \sum_{n=0}^{\infty} ar^n$$

is called a geometric series. *For $r \neq 1$, the n partial sum is given by*

$$s_n = a + ar + \cdots + ar^{n-1} = \frac{a(1 - r^n)}{1 - r}.$$

Thus, it follows that the geometric series converges if and only if $|r| < 1$. In particular, for $|r| < 1$,

$$a + ar + ar^2 + \cdots = \sum_{n=0}^{\infty} ar^n = \frac{a}{1 - r}. \tag{2.2}$$

Example 2.2 (p-series). *A series in the form of $\sum_{n=1}^{\infty} 1/n^p$ is called a p-series, where p is a real constant.*

1. *If $p > 1$, the series converges.*

2. *If $p \leq 1$, the series diverges.*

To this end, if $p > 1$, let s_n be the nth partial sum. For every positive integer n, we have

$$\begin{aligned} s_{2n+1} &= 1 + \left(\frac{1}{2^p} + \frac{1}{4^p} + \cdots + \frac{1}{(2n)^p} \right) + \left(\frac{1}{3^p} + \frac{1}{5^p} + \cdots + \frac{1}{(2n+1)^p} \right) \\ &< 1 + \left(\frac{1}{2^p} + \frac{1}{4^p} + \cdots + \frac{1}{(2n)^p} \right) + \left(\frac{1}{2^p} + \frac{1}{4^p} + \cdots + \frac{1}{(2n)^p} \right) \\ &= 1 + \frac{1}{2^{p-1}} s_n < 1 + \frac{1}{2^{p-1}} s_{2n+1}. \end{aligned}$$

Hence,

$$s_n < s_{2n+1} < \frac{1}{1 - 1/2^{p-1}},$$

from which the convergence of p-series follows by the monotone convergence theorem. One alternative proof is to apply the mean value theorem to $f(x) = 1/x^{p-1}$ on $[n - 1, n]$, which gives that

$$\frac{1}{(n-1)^{p-1}} - \frac{1}{n^{p-1}} = \frac{p-1}{(n-\theta)^p} \quad (0 < \theta < 1).$$

Thus, for $n \geq 2$,

$$\frac{1}{n^p} \leq \frac{1}{(n-\theta)^p} = \frac{1}{p-1} \left(\frac{1}{(n-1)^{p-1}} - \frac{1}{n^{p-1}} \right).$$

But the series $\sum_{n=2}^{\infty} \left(1/(n-1)^{p-1} - 1/n^{p-1}\right)$ converges because its partial sums converge, so the original series $\sum_{n=1}^{\infty} 1/n^p$ converges by the comparison test. In the case where $p \leq 1$, for each $n \in \mathbb{N}, 1/n^p \geq 1/n$. Since the harmonic series diverges, it follows from the comparison test again that the p-series also diverges.

Example 2.3. *The series $\sum_{n=2}^{\infty} 1/(n \ln n)$ is divergent.*

In fact, applying the mean value theorem to $\ln(\ln x)$ on $[n, n+1]$ yields

$$\ln(\ln(n+1)) - \ln(\ln n) = \frac{1}{(n+\theta) \ln(n+\theta)}, \quad (0 < \theta < 1).$$

Thus

$$\frac{1}{n \ln n} > \frac{1}{(n+\theta) \ln(n+\theta)} = \ln(\ln(n+1)) - \ln(\ln n).$$

But the nth partial sum of $\sum_{n=2}^{\infty} \left(\ln(\ln(n+1)) - \ln(\ln n)\right)$ gives

$$\ln(\ln(n+1)) - \ln(\ln 2),$$

which is divergent. By the comparison test, the series diverges as claimed.

Now, using the convergent geometric series and the divergent series $\sum_{n=1}^{\infty} 1$ as the measuring sticks, the ratio comparison test yields two simple and powerful tests that are closely related. Both tests work exclusively with the terms of the given series—they require neither an initial guess about convergence nor the discovery of a series for comparison.

Theorem 2.5 (The Ratio Test). *Let $\sum_{n=1}^{\infty} a_n$ be a positive series and define*

$$r = \lim_{n \to \infty} \frac{a_{n+1}}{a_n}. \tag{2.3}$$

1. *If $r < 1$, the series converges.*

2. *If $r > 1$, the series diverges.*

3. *If $r = 1$, the test is inconclusive.*

Theorem 2.6 (The Root Test). *Let $\sum_{n=1}^{\infty} a_n$ be a positive series and define*

$$\rho = \lim \sqrt[n]{a_n}. \tag{2.4}$$

1. *If $\rho < 1$, the series converges.*

2. *If $\rho > 1$, the series diverges.*

3. *If $\rho = 1$, the test is inconclusive.*

Remark. Recall that if the limit in (2.3) exists, then so does the limit in (2.4). Moreover, $r = \rho$ (see Example 1.20). Thus, whenever r exists, the ratio test and root test always give the same response. However, when r does not exist, the following example indicates that even though both tests are applicable, they may yield different conclusions on convergence and divergence.

Example 2.4. *Let a and b be distinct positive numbers. Determine the convergence of*

$$1 + a + ab + a^2 b + a^2 b^2 + \cdots + a^n b^{n-1} + a^n b^n + \cdots.$$

Since

$$\liminf \frac{a_{n+1}}{a_n} = \min(a, b), \ \limsup \frac{a_{n+1}}{a_n} = \max(a, b)$$

and

$$\rho = \lim_{n \to \infty} \sqrt[n]{a_n} = \sqrt{ab},$$

the ratio test shows that if $a, b < 1$ the series converges and diverges for $a, b > 1$. But, the root test now yields better results: if $ab < 1$ the series converges and diverges for $ab \geq 1$.

The root test is often more complicated to apply than the ratio test, but it will give an answer in some cases where the ratio test is inconclusive.

Example 2.5. *Let*

$$a_n = \begin{cases} (1/2)^{n/2}, & \textit{when } n \textit{ is even}, \\ 2(1/2)^{(n-1)/2}, & \textit{when } n \textit{ is odd}. \end{cases}$$

Since

$$\liminf \frac{a_{n+1}}{a_n} = \frac{1}{4}, \quad \limsup \frac{a_{n+1}}{a_n} = 2,$$

the ratio test is inconclusive. But

$$\rho = \lim_{n \to \infty} \sqrt[n]{a_n} = \frac{\sqrt{2}}{2} < 1,$$

so the root test implies that the series $\sum_{n=1}^{\infty} a_n$ converges.

Moreover, for any positive sequence a_n, appealing to the fact that

$$\liminf \frac{a_{n+1}}{a_n} \leq \liminf \sqrt[n]{a_n} \leq \limsup \sqrt[n]{a_n} \leq \limsup \frac{a_{n+1}}{a_n},$$

we find that whenever the ratio test shows convergence, the root test does too; whenever the root test is inconclusive, so is the ratio test. Therefore, the root test is stronger than the ratio test.

Although all the tests above require the positivity of series, they are often used in conjunction with the following theorem to handle general series.

Theorem 2.7 (Absolute Convergence Test). *If the series $\sum_{n=1}^{\infty} |a_n|$ converges, then $\sum_{n=1}^{\infty} a_n$ itself converges as well.*

The converse of this theorem is false. Consider the *alternating harmonic series*

$$1 - \frac{1}{2} + \frac{1}{3} - \frac{1}{4} + \cdots + (-1)^{n+1} \frac{1}{n} + \cdots.$$

Taking absolute value term by term yields the harmonic series, which diverges. To prove the alternating harmonic series indeed converges, as usual, we consider two partial sums s_{2n+1} and s_{2n}, and find that

$$s_{2n+1} - s_{2n-1} = -\frac{1}{2n} + \frac{1}{2n+1} < 0,$$

$$s_{2n} - s_{2(n-1)} = \frac{1}{2n-1} - \frac{1}{2n} > 0.$$

Moreover,

$$1 - \frac{1}{2} = s_2 < s_{2n} < s_{2n+1} < s_1 = 1.$$

Thus, s_{2n+1} is monotone decreasing and bounded below while s_{2n} is monotone increasing and bounded above. By the monotone convergence theorem, both s_{2n+1} and s_{2n} converge. On the other hand, since

$$s_{2n+1} = s_{2n} + \frac{1}{2n+1},$$

both even and odd partial sums converge to same number. Hence the alternating harmonic series converges. The above argument can be extended to the following handy and easy-to-use test.

Theorem 2.8 (Alternating Series Test). *Let a_n be a sequence satisfying*

1. *$a_n \geq 0$ and $a_n \geq a_{n+1}$ for all $n \in \mathbb{N}$,*

2. *$a_n \to 0$ as $n \to \infty$.*

Then, the alternating series $\sum_{n=1}^{\infty} (-1)^{n+1} a_n$ converges.

Hardy [53] once gave the following example to emphasize the need for a_n to be strictly decreasing to zero as part of the hypothesis for the convergence of an alternating series.

Example 2.6 (Hardy). *The alternating series*

$$\sum_{n=2}^{\infty} \frac{(-1)^n}{\sqrt{n} + (-1)^n} \tag{2.5}$$

diverges.

The divergence of (2.5) can be readily established by rewriting it as

$$\sum_{n=2}^{\infty} \frac{(-1)^n}{\sqrt{n} + (-1)^n} = \sum_{n=2}^{\infty} \left(\frac{(-1)^n}{\sqrt{n}} - \frac{1}{n + (-1)^n \sqrt{n}} \right)$$

$$= \sum_{n=2}^{\infty} \frac{(-1)^n}{\sqrt{n}} - \sum_{n=2}^{\infty} \frac{1}{n + (-1)^n \sqrt{n}}.$$

Here, the first series converges, but the second series diverges because $1/(n + (-1)^n \sqrt{n}) \geq 1/2n$ for every $n \geq 2$.

In general, let $a_n > 0$ and s_n be the nth partial sum of the alternating series $\sum_{n=1}^{\infty} (-1)^{n+1} a_n$. Then

$$s_{2n} = a_1 - a_2 + \cdots + a_{2n-1} - a_{2n} = \sum_{k=1}^{n} a_{2k-1} - \sum_{k=1}^{n} a_{2k}.$$

To obtain a desired divergent series, we choose a_n so that $a_n \to 0$ and only one of the series $\sum_{k=1}^{n} a_{2k-1}$ and $\sum_{k=1}^{n} a_{2k}$ converges. For examples,

$$1 - \frac{1}{2} + \frac{1}{3^2} - \frac{1}{4} + \frac{1}{5^2} - \frac{1}{6} + \cdots + \frac{1}{(2n-1)^2} - \frac{1}{(2n)} + \cdots.$$

Definition 2.3. *A series $\sum_{n=1}^{\infty} a_n$ is said to converge conditionally if it converges but does not converge absolutely.*

In general, the alternating series test is the most accessible test for conditional convergence. To recall our progress to date, we have encountered the tests that often appear in a first-year calculus sequence. However, there are many simple series for which the tests listed above are inconclusive.

Example 2.7. *Determine the convergence of*

$$\sum_{n=1}^{\infty} \frac{1}{2^{\sqrt{n}}}.$$

It resembles a geometric series, but the presence of the square root complicates matters. Applying the ratio test yields

$$\lim_{n\to\infty} \frac{a_{n+1}}{a_n} = \lim_{n\to\infty} \frac{2^{\sqrt{n}}}{2^{\sqrt{n+1}}} = \lim_{n\to\infty} \frac{1}{2^{\sqrt{n+1}-\sqrt{n}}} = 1,$$

providing us no information. The root test is similarly inconclusive. We will see in Section 2.2 that this series is actually convergent.

Historically, the most interesting series mathematicians and scientists were encountering in the early 19th Century all fell into the inconclusive category of the above tests. To overcome the principal drawback of these tests, various sophisticated and finer tests were developed. For example, Gauss's test completely sets the convergence for all hypergeometric series. Kummer's test gives characterizations for convergence or divergence of all positive series. Cauchy's condensation theorem provides an approach to study the convergence or divergence of a class of series via a rather "thin" subsequence. To demonstrate the progression of these sophisticated and finer tests, we introduce in Section 2.2 the Raabe and Logarithmic tests that use p-series as measuring sticks. In Section 2.3, we present the Kummer and Gauss tests by comparing with the standard series $\sum 1/(n \ln^p n)$. For series with monotonically decreasing terms, we formulate in Section 2.4 a few simple and interesting tests based on Cauchy's condensation theorem. In Section 2.5, we show that no single test can determine convergence or divergence for all positive series. Abel's test and Dirichlet's test, Riemann series theorem, and Euler's infinite prime product are the subjects of the three subsequent sections. In closing we present some interesting applications of these tests.

2.2 Raabe and Logarithmic Tests

The purpose of proof is to understand, not to verify. — Arnold Ross

We begin with the p-series and see what we can learn from a_{n+1}/a_n. Since

$$\frac{a_{n+1}}{a_n} = \left(\frac{n}{n+1}\right)^p = \left(1 + \frac{1}{n}\right)^{-p} = 1 - \frac{p}{n} + O\left(\frac{1}{n^2}\right),$$

it follows that

$$n\left(1 - \frac{a_{n+1}}{a_n}\right) = p + O\left(\frac{1}{n}\right).$$

Keeping this in mind, for any positive series, we define *Raabe's sequence* by

$$\mathcal{R}_n = n\left(1 - \frac{a_{n+1}}{a_n}\right). \tag{2.6}$$

For all sufficiently large n, if $\mathcal{R}_n > r > 1$, equivalently,

$$\frac{a_{n+1}}{a_n} < 1 - \frac{r}{n}.$$

Choose a positive number α such that $1 + \alpha < r$. Then, for all sufficiently large n,

$$\left(1 + \frac{1}{n}\right)^{-(1+\alpha)} - \left(1 - \frac{r}{n}\right) = \frac{r - (1+\alpha)}{n} + O\left(\frac{1}{n^2}\right) > 0.$$

Therefore,

$$\frac{a_{n+1}}{a_n} < \left(1 + \frac{1}{n}\right)^{-(1+\alpha)} = \frac{\frac{1}{(n+1)^{1+\alpha}}}{\frac{1}{n^{1+\alpha}}}.$$

Since $\sum_{n=1}^{\infty} 1/n^{1+\alpha}$ converges, by the ratio comparison test, the series $\sum_{n=1}^{\infty} a_n$ converges. On the other hand, if $\mathcal{R}_n \leq 1$, then

$$\frac{a_{n+1}}{a_n} \geq \frac{n-1}{n} = \frac{\frac{1}{n}}{\frac{1}{n-1}}.$$

By the ratio comparison test again, the divergence of the harmonic series implies that $\sum_{n=1}^{\infty} a_n$ diverges.

To summarize, we have established the following test, which is due to Raabe.

Theorem 2.9 (Raabe's Test). *Let $\sum_{n=1}^{\infty} a_n$ be a positive series. Suppose that $\mathcal{R}_n$ is defined by (2.6).*

    1. If $\mathcal{R}_n \geq r > 1$, the series converges.

    2. If $\mathcal{R}_n \leq 1$, the series divergence.

In many cases, $\mathcal{R}_n$ approaches a limit as $n \to \infty$. In this case, there is a simple form of Raabe's test.

Theorem 2.10 (Limit Raabe's Test). *Let $\sum_{n=1}^{\infty} a_n$ be a positive series. Define*

$$\mathcal{R} = \lim_{n \to \infty} \mathcal{R}_n, \quad \text{(finite or infinite)}.$$

1. If $\mathcal{R} > 1$, the series converges.

2. If $\mathcal{R} < 1$, the series diverges.

3. If $\mathcal{R} = 1$, the test is inconclusive.

Clearly, whenever the ratio test shows convergence, Raabe's test does too. The following two examples show that Raabe's test is stronger than the ratio test.

Example 2.8. *The series in Example 2.7 indeed converges.*

We have already seen that the ratio test is inconclusive for this series. But, by the mean value theorem,

$$\mathcal{R}_n = n\left(1 - \frac{2^{\sqrt{n}}}{2^{\sqrt{n+1}}}\right) = \frac{n}{2^{\sqrt{n+1}}}\left(2^{\sqrt{n+1}} - 2^{\sqrt{n}}\right)$$

$$= \frac{n}{2^{\sqrt{n+1}}} 2^{\sqrt{n+\theta}} \ln 2 \frac{1}{2\sqrt{n+\theta}} \quad (0 < \theta < 1)$$

$$= \frac{n}{2\sqrt{n+\theta}} \frac{2^{\sqrt{n+\theta}}}{2^{\sqrt{n+1}}} \ln 2 \to \infty.$$

Raabe's test yields the convergence as claimed.

Example 2.9. *Determine the convergence of*

$$\sum_{n=1}^{\infty} \frac{1}{n!} \left(\frac{n}{e}\right)^n.$$

Again, the ratio test is not delicate enough for this series because

$$\frac{a_{n+1}}{a_n} = \frac{1}{e} \left(1 + \frac{1}{n}\right)^n \to 1.$$

On the other hand,

$$\mathcal{R}_n = n \left(1 - \frac{1}{e} \left(1 + \frac{1}{n}\right)^n\right).$$

To find $\lim_{n\to\infty} \mathcal{R}_n$, we replace $\mathcal{R}_n$ by a more general expression:

$$\frac{1}{x} \left(1 - \frac{(1+x)^{1/x}}{e}\right) \quad (x \to 0^+).$$

By L'Hôpital's rule, the ratio of the derivatives, after some simplification, is

$$\frac{(1+1/x)^x}{e} \cdot \frac{\ln(1+x) - \frac{x}{1+x}}{x^2}.$$

Since

$$\ln(1+x) = x - \frac{1}{2}x^2 + o(x^2), \quad \frac{x}{1+x} = x - x^2 + o(x^2),$$

where $a_n = o(b_n)$ means that $\lim_{n\to\infty} a_n/b_n = 0$, we find that the desired limit is $1/2$. Therefore, the proposed series diverges by the limit Raabe's test.

Rewriting (2.6) as

$$\frac{a_{n+1}}{a_n} = 1 - \frac{\mathcal{R}_n}{n}.$$

We find that

$$\lim_{n\to\infty} \left(\frac{a_{n+1}}{a_n}\right)^n = \lim_{n\to\infty} \left(1 - \frac{\mathcal{R}_n}{n}\right)^n = e^{-\mathcal{R}}$$

as long as $\mathcal{R} = \lim_{n\to\infty} \mathcal{R}_n$ exists. This observation provides a simple test which is equivalent to the limit Raabe's test.

Theorem 2.11. *Let $\sum_{n=1}^{\infty} a_n$ be a positive series, and define*

$$\mathcal{R}^* = \lim_{n\to\infty} \left(\frac{a_{n+1}}{a_n}\right)^n, \quad \text{(finite or infinite)}.$$

1. *If $\mathcal{R}^* < 1/e$, the series converges.*
2. *If $\mathcal{R}^* > 1/e$, the series diverges.*
3. *If $\mathcal{R}^* = 1/e$, the test is inconclusive.*

Returning now to Example 2.7, we see that

$$\left(\frac{a_{n+1}}{a_n}\right)^n = \left(\frac{2^{\sqrt{n}}}{2^{\sqrt{n+1}}}\right)^n = 2^{-\frac{n}{\sqrt{n+1}+\sqrt{n}}} \to 0.$$

This gives an easy proof of the convergence of the series.

Moreover, as with Raabe's test, if $\ln(1/a_n)/\ln n > l > 1$, we seek a "buffer" between 1 and l and so choose a real number $\alpha > 0$ such that $l > 1 + \alpha$. Then

$$\frac{\ln(1/a_n)}{\ln n} > 1 + \alpha.$$

This is equivalent to $a_n \leq 1/n^{1+\alpha}$, so $\sum_{n=1}^{\infty} a_n$ converges. On the other hand, if $\ln(1/a_n)/\ln n \leq 1$, $\sum_{n=1}^{\infty} a_n$ diverges by comparison with the harmonic series. Putting all this together, we obtain a less known convergence test from Cauchy's arsenal.

Theorem 2.12 (Logarithmic Test). *Let $\sum_{n\to\infty} a_n$ be a positive series. Define*

$$\mathcal{L}_n = \frac{\ln(1/a_n)}{\ln n}. \tag{2.7}$$

1. *If $\mathcal{L}_n \geq l > 1$, the series converges.*

2. *If $\mathcal{L}_n \leq 1$, the series diverges.*

Applying the logarithmic test yields the convergence of the series in Example 2.7 in one line. As we will see in the following example, sometimes the logarithmic test turns out to be more convenient than Raabe's test.

Example 2.10. *Let $p > 1$. Determine the convergence of*

$$\sum_{n=1}^{\infty} \frac{\ln n}{n^p}.$$

Notice that
$$\frac{\ln(1/a_n)}{\ln n} = \frac{p \ln n - \ln(\ln n)}{\ln n} = p - \frac{\ln(\ln n)}{\ln n}.$$

Since
$$\lim_{n\to\infty} \frac{\ln(\ln n)}{\ln n} = 0,$$

there is an integer N such that $n > N$ implies that $\frac{\ln(1/a_n)}{\ln n} > l > 1$. Thus, the series converges by the logarithmic test.

In the comparison of the ratio test and the root test, we have seen that if r exists, then ρ exists and $r = \rho$. Using the Stolz-Cesàro theorem, we prove a similar relationship between Raabe's test and the logarithmic test.

Theorem 2.13. *Let a_n be a positive sequence. Suppose that $\mathcal{R}_n$ and $\mathcal{L}_n$ are given by (2.6) and (2.7), respectively. If $\lim_{n\to\infty} \mathcal{R}_n$ exists, finite or infinite, then $\lim_{n\to\infty} \mathcal{L}_n$ exists and*

$$\lim_{n\to\infty} \mathcal{R}_n = \lim_{n\to\infty} \mathcal{L}_n.$$

Proof. Notice that

$$\lim_{n\to\infty} \frac{\ln(1/a_{n+1}) - \ln(1/a_n)}{\ln(n+1) - \ln n} = \lim_{n\to\infty} [-n \ln(a_{n+1}/a_n)].$$

By the Stolz-Cesàro theorem, if $\lim_{n\to\infty} [-n \ln(a_{n+1}/a_n)] = l$, then

$$\lim_{n\to\infty} \frac{\ln(1/a_n)}{\ln n} = l.$$

First, if we assume that $\lim_{n\to\infty} \mathcal{R}_n$ is finite, then

$$\lim_{n\to\infty} (a_{n+1}/a_n - 1) = 0.$$

Since, for $-1 < x < 1$,

$$\ln(1-x) = -x - \frac{1}{2}x^2 + o(x^2),$$

we have

$$-n\ln\left(\frac{a_{n+1}}{a_n}\right) = -n\ln\left[1 - \left(1 - \frac{a_{n+1}}{a_n}\right)\right]$$

$$= n\left(1 - \frac{a_{n+1}}{a_n}\right) - \frac{1}{2}n\left(1 - \frac{a_{n+1}}{a_n}\right)^2 + o((1 - a_{n+1}/a_n)^2).$$

Hence

$$\lim_{n\to\infty}\left[-n\ln\left(\frac{a_{n+1}}{a_n}\right)\right] = \lim_{n\to\infty} \mathcal{R}_n = \lim_{n\to\infty} \frac{\ln(1/a_n)}{\ln n}.$$

Next, if $\lim_{n\to\infty} \mathcal{R}_n = +\infty$, given $M > 0$, there is an integer N such that $n(1 - a_{n+1}/a_n) \geq M$ for all $n \geq N$. Thus,

$$\ln(1 - M/n) > \ln a_{n+1} - \ln a_n$$

and so

$$\frac{M}{n}\left[1 + \frac{1}{2}(M/n)\right] < \ln(1/a_{n+1}) - \ln(1/a_n);$$

$$M < \frac{M}{n\ln(1+1/n)} < \frac{\ln(1/a_{n+1}) - \ln(1/a_n)}{\ln(n+1) - \ln n}.$$

Therefore,

$$\lim_{n\to\infty} \frac{\ln(1/a_{n+1}) - \ln(1/a_n)}{\ln(n+1) - \ln n} = +\infty = \lim_{n\to\infty} \frac{\ln(1/a_n)}{\ln n}.$$

Finally, the case of $\lim_{n\to\infty} \mathcal{R}_n = -\infty$ parallels the case of $+\infty$. $\square$

When $\lim_{n\to\infty} \mathcal{R}_n$ does not exist, we have

$$\liminf n(1 - a_{n+1}/a_n) \leq \liminf \frac{\ln(1/a_n)}{\ln n} \leq \limsup \frac{\ln(1/a_n)}{\ln n} \leq \limsup n(1 - a_{n+1}/a_n).$$

The following example indicates that the logarithmic test is stronger than Raabe's test.

Example 2.11. *Determine the convergence of*

$$\sum_{n=1}^{\infty} \frac{1}{3^{(-1)^n + \ln n}}.$$

Since

$$\lim_{n\to\infty} \frac{\ln(1/a_n)}{\ln n} = \lim_{n\to\infty} \frac{[(-1)^n + \ln n]\ln 3}{\ln n} = \ln 3 > 1,$$

the series converges by the logarithmic test. But Raabe's test is inconclusive since

$$\frac{a_{n+1}}{a_n} = \begin{cases} 3^{-2+\ln(n/n+1)}, & \text{when } n \text{ is odd}; \\ 3^{2+\ln(n/n+1)}, & \text{when } n \text{ is even} \end{cases}$$

and

$$\liminf n(1 - a_{n+1}/a_n) = -\infty, \quad \limsup n(1 - a_{n+1}/a_n) = +\infty.$$

2.3   The Kummer, Bertrand, and Gauss Tests

Problems worthy of attack prove their worth by hitting back. — Piet Hein

Both the limit Raabe's test and the logarithmic test become inconclusive when $\mathcal{R} = \mathcal{L} = 1$. To cope with this case, we introduce the following more general test due to Kummer. This test comes essentially from telescoping.

Theorem 2.14 (Kummer's Test). *Let $\sum_{n=1}^{\infty} a_n$ be a positive series.*

1. $\sum_{n=1}^{\infty} a_n$ *converges if there is a positive sequence c_n and some constant $k > 0$ such that*
$$\mathcal{K}_n := c_n - \frac{a_{n+1}}{a_n} c_{n+1} \geq k.$$

2. $\sum_{n=1}^{\infty} a_n$ *diverges if there is a positive sequence c_n such that $\sum_{n=1}^{\infty} 1/c_n$ diverges and $\mathcal{K}_n \leq 0$.*

Proof.   (1) If
$$\mathcal{K}_n = c_n - \frac{a_{n+1}}{a_n} c_{n+1} \geq k > 0,$$

multiplying this inequality by a_n yields

$$c_n a_n - c_{n+1} a_{n+1} \geq k a_n > 0. \tag{2.8}$$

This implies that $c_n a_n$ is monotone decreasing and bounded below by 0. By the monotone convergence theorem, $\lim_{n \to \infty} c_n a_n$ exists. On the other hand, telescoping gives

$$\sum_{n=1}^{\infty} (c_n a_n - c_{n+1} a_{n+1}) = c_1 a_1 - \lim_{n \to \infty} c_n a_n. \tag{2.9}$$

By (2.8) and (2.9), the comparison test yields that $\sum_{n=1}^{\infty} a_n$ converges.

The convergence of $\sum_{n=1}^{\infty} a_n$ can also be proved by showing that the nth partial sum is bounded:

$$
\begin{aligned}
s_n = \sum_{k=1}^{n} a_k &\leq \frac{1}{k} \sum_{k=1}^{n} (c_k a_k - c_{k+1} a_{k+1}) \quad \text{(use (2.8))} \\
&= \frac{1}{k} \left(\sum_{k=1}^{n} c_k a_k - \sum_{k=2}^{n+1} c_k a_k \right) \quad \text{(use } k+1 \to k \text{ in the second sum)} \\
&= \frac{1}{k} (c_1 a_1 - c_{n+1} a_{n+1}) \leq \frac{c_1 a_1}{k}.
\end{aligned}
$$

(2) If
$$\mathcal{K}_n = c_n - \frac{a_{n+1}}{a_n} c_{n+1} \leq 0,$$

then
$$\frac{a_{n+1}}{a_n} \geq \frac{1/c_{n+1}}{1/c_n}.$$

The divergence of $\sum_{n=1}^{\infty} a_n$ now follows from the ratio comparison test (Theorem 2.4) and the assumption that $\sum_{n=1}^{\infty} 1/c_n$ diverges.   □

In limit version, Kummer's test can be restated as

Theorem 2.15 (Limit Kummer's Test). *Let $\sum_{n=1}^{\infty} a_n$ be a positive series. For a positive sequence c_n, define*

$$\kappa = \lim_{n \to \infty} \left(c_n - \frac{a_{n+1}}{a_n} c_{n+1} \right).$$

1. *If $\kappa > 0$, then the series converges.*
2. *If $\kappa < 0$ and $\sum_{n=1}^{\infty} 1/c_n$ diverges, then the series diverges.*
3. *If $\kappa = 0$, the test is inconclusive.*

In contrast to all tests we have studied, Kummer's test does not merely give very powerful sufficient conditions for convergence or divergence of a positive series. Surprisingly, the conditions (1) and (2) in the test are indeed necessary. This fact is not well publicized.

Theorem 2.16. *Let $\sum_{n=1}^{\infty} a_n$ be a positive series.*

1. *If the series converges, then there exists a positive sequence c_n for which $c_n - \frac{a_{n+1}}{a_n} c_{n+1} \geq 1$.*
2. *If the series diverges, then there exists a positive sequence c_n for which $\sum_{n=1}^{\infty} 1/c_n$ diverges and $c_n - \frac{a_{n+1}}{a_n} c_{n+1} \leq 0$.*

Proof. Suppose that $\sum_{n=1}^{\infty} a_n$ converges to a real number S. We construct a positive sequence c_n for which satisfies the claimed property. To this end, let

$$c_1 = \frac{S}{a_1}, \quad c_n = \frac{S - \sum_{k=1}^{n-1} a_k}{a_n} \quad (n \geq 2).$$

Clearly, $c_n > 0$ for all $n \in \mathbb{N}$ and

$$c_n - \frac{a_{n+1}}{a_n} c_{n+1} = \frac{S - \sum_{k=1}^{n-1} a_k}{a_n} - \frac{S - \sum_{k=1}^{n} a_k}{a_n} = \frac{a_n}{a_n} = 1.$$

On the other hand, suppose that $\sum_{n=1}^{\infty} a_n$ diverges. Let $c_1, c_n = \left(\sum_{k=1}^{n-1} a_k \right)/a_n$ $(n \geq 2)$. We find that $c_n > 0$ for all $n \in \mathbb{N}$ and

$$c_n - \frac{a_{n+1}}{a_n} c_{n+1} = \frac{\sum_{k=1}^{n-1} a_k}{a_n} - \frac{\sum_{k=1}^{n} a_k}{a_n} = -\frac{a_n}{a_n} \leq 0.$$

Now we prove that $\sum_{n=1}^{\infty} 1/c_n$ diverges. We show that for any given positive integer m, there is an $n > m$ such that $\sum_{k=m}^{n} 1/c_n > 1/2$. To this end, since $\sum_{n=1}^{\infty} a_n$ diverges and $a_k > 0$, for any given positive integer m, there exists an $n > m$ such that

$$a_m + a_{m+1} + \cdots + a_n > a_1 + a_2 + \cdots + a_{m-1}.$$

Hence

$$\sum_{k=m}^{n} 1/c_n = \frac{a_m}{a_1 + a_2 + \cdots + a_{m-1}} + \cdots + \frac{a_n}{a_1 + a_2 + \cdots + a_{n-1}}$$

$$> \frac{a_m + \cdots + a_n}{a_1 + \cdots + a_n} = \frac{1}{1 + (a_1 + \cdots + a_{m-1})/(a_m + \cdots + a_n)}$$

$$> 1/(1+1) = \frac{1}{2}.$$

By the Cauchy's criterion, $\sum_{n=1}^{\infty} 1/c_n$ diverges. $\qquad \square$

In summary, we see that Kummer's test gives characterizations for convergence or divergence of *all positive series*. The drawback of this test is the choice of c_n, which has to come from the user, and usually there is no particular motivation. Thus, using Kummer's test is equally as difficult as using the comparison test. However, for a suitable choice of c_n, Kummer's test is a rich source of many other tests. For example,

1. Let $c_n = 1$. Notice that $\sum_{n=1}^{\infty} 1$ diverges and $\mathcal{K}_n = 1 - \frac{a_{n+1}}{a_n}$. Thus, the ratio test is now a special case of Kummer's test.

2. Let $c_n = n - 1$. Appealing to the divergence of $\sum_{n=2}^{\infty} 1/n$ and

$$(n-1) - \frac{a_{n+1}}{a_n} n = (n-1) - \left(1 - \frac{\mathcal{R}_n}{n}\right) n = \mathcal{R}_n - 1,$$

we see that Kummer's test implies Raabe's test.

Moreover, let $c_n = n \ln n$. Recall that $\sum_{n=2}^{\infty} 1/(n \ln n)$ diverges. We rewrite

$$\mathcal{K}_n = (n-1)\ln(n-1) - \frac{a_{n+1}}{a_n} n \ln n$$

$$= \ln n \left[n\left(1 - \frac{a_{n+1}}{a_n}\right) - 1\right] - \ln\left(1 + \frac{1}{n-1}\right)^{n-1}$$

$$= \mathcal{B}_n - \ln\left(1 + \frac{1}{n-1}\right)^{n-1},$$

where

$$\mathcal{B}_n = \ln n \left[n\left(1 - \frac{a_{n+1}}{a_n}\right) - 1\right] = \ln n(\mathcal{R}_n - 1). \tag{2.10}$$

Since

$$\lim_{n\to\infty} \ln\left(1 + \frac{1}{n-1}\right)^{n-1} = 1,$$

Kummer's test yields a logarithmic scale test which was developed by Bertrand.

Theorem 2.17 (Bertrand's Test). *Let $\mathcal{B}_n$ be defined by (2.10). Suppose that*

$$\mathcal{B} = \lim_{n\to\infty} \mathcal{B}_n, \quad \textit{(finite or infinite)}.$$

Then the positive series $\sum_{n=1}^{\infty} a_n$

1. *converges if $\mathcal{B} > 1$, and*

2. *diverges if $\mathcal{B} < 1$.*

Finally, combining the tests above, we have

Theorem 2.18 (Gauss's Test). *Let $\sum_{n=1}^{\infty} a_n$ be a positive series. Suppose*

$$\frac{a_{n+1}}{a_n} = 1 - \frac{\mu}{n} + \frac{\theta_n}{n^k},$$

where θ_n is a bounded sequence and $k > 1$.

1. *If $\mu > 1$, the series converges.*

2. *If $\mu \leq 1$, the series diverges.*

Its justification consists of two cases. In the case $\mu \neq 1$, $\mathcal{R}_n = \mu + \theta_n/n^{k-1} \to \mu$, Raabe's test takes care of the claimed results directly. When $\mu = 1$, in view of (2.10), we have

$$\mathcal{B}_n = \frac{\ln n}{n^{k-1}} \theta_n \to 0, \quad (\text{as } n \to \infty).$$

Now, the divergence case of Bertrand's test applies.

Gauss's test was custom-made to check the convergence of his famous *hypergeometric series*

$$F(\alpha, \beta, \gamma) = 1 + \sum_{n=1}^{\infty} \frac{\alpha(\alpha+1)\cdots(\alpha+n-1)\beta(\beta+1)\cdots(\beta+n-1)}{n!\,\gamma(\gamma+1)\cdots(\gamma+n-1)}$$

$$= 1 + \frac{\alpha\beta}{1!\,\gamma} + \frac{\alpha(\alpha+1)\beta(\beta+1)}{2!\,\gamma(\gamma+1)} + \frac{\alpha(\alpha+1)(\alpha+2)\beta(\beta+1)(\beta+2)}{3!\,\gamma(\gamma+1)(\gamma+2)} + \cdots,$$

where α, β, and γ are real numbers, none of which is a negative integer or 0. Clearly

$$\frac{a_{n+1}}{a_n} = \frac{(\alpha+n)(\beta+n)}{(1+n)(\gamma+n)} = \frac{n^2 + (\alpha+\beta)n + \alpha\beta}{n^2 + (\gamma+1)n + \gamma}.$$

For all large n the terms a_n have the same sign, so Gauss's test is applicable. Since

$$\frac{1}{1+t/n} = 1 - \frac{t}{n} + \frac{t^2}{1+t/n} \cdot \frac{1}{n^2},$$

we have

$$\frac{a_{n+1}}{a_n} = 1 - \frac{\gamma - \alpha - \beta + 1}{n} + \frac{\theta_n}{n^2},$$

where θ_n is bounded. Thus, Gauss's test shows that the series $F(\alpha, \beta, \gamma)$ converges if and only if $\alpha + \beta < \gamma$. This result is sharp and never returns an inconclusive answer.

Since most of the power series encountered in practice are hypergeometric, when both ratio and root tests return inconclusive results, Gauss's test is the next place to turn. When a series is not hypergeometric, the following example indicates that Gauss's test may return an inconclusive answer.

Example 2.12. *Consider the convergence of*

$$\sum_{n=2}^{\infty} \frac{1}{n \ln n}$$

using Gauss's test.

To simplify the algebra, we study the ratio a_n/a_{n-1} rather than a_{n+1}/a_n. Thus,

$$\frac{a_n}{a_{n-1}} = \frac{(n-1)\ln(n-1)}{n \ln n}$$

$$= \frac{n-1}{n} \cdot \frac{\ln\left(n \cdot \frac{n-1}{n}\right)}{\ln n}$$

$$= \left(1 - \frac{1}{n}\right)\left(1 + \frac{\ln(1 - n^{-1})}{\ln n}\right)$$

$$= 1 - \frac{1}{n} + \frac{\ln(1 - n^{-1})}{\ln n} - \frac{\ln(1 - n^{-1})}{n \ln n}$$

$$= 1 - \frac{1}{n} + \frac{E(n)}{n},$$

where

$$E(n) = -\frac{1}{\ln n} + \frac{1}{2n \ln n} + \frac{1}{6n^2 \ln n} + \cdots .$$

Here $\theta_n = nE(n)$ is not bounded. Therefore, Gauss's test is not applicable.

2.4 More Sophisticated Tests Based on Monotonicity

Still round the corner there may wait a new road or a secret gate.

— John Ronald Tolkien

Many series have monotone decreasing terms. For such series, the following striking test indicates that a rather "thin" subsequence of a_n determines the convergence or divergence of $\sum a_n$.

Theorem 2.19 (Cauchy's Condensation Test). *If $a_n > 0$ is decreasing, then $\sum a_n$ converges if and only if $\sum 2^n a_{2^n}$ converges.*

Proof. By the assumption, it suffices to prove the boundedness of the partial sums. Let $t_n = \sum_{k=1}^n 2^k a_{2^k}$. Notice that

$$s_{2^n} - s_{2^{n-1}} = \underbrace{a_{2^{n-1}+1} + \cdots + a_{2^n}}_{2^{n-1}\, terms} .$$

Thus,

$$2^{n-1} a_{2^n} \leq s_{2^n} - s_{2^{n-1}} \leq 2^{n-1} a_{2^{n-1}}.$$

Telescoping gives

$$\frac{1}{2} t_n \leq s_{2^n} \leq t_{n-1} + a_1,$$

and so these two sequences are either both bounded or both unbounded. This completes the proof. $\square$

In the Cauchy's condensation test, clearly, $2^n a_{2^n}$ can be replaced by $m^n a_{m^n}$ where $m \in \mathbb{N}$. Indeed, we can do even better with the following generalization.

Theorem 2.20. *Let $a_n > 0$ be decreasing. If there is a strictly increasing sequence $1 \leq n_1 < n_2 < \cdots$ such that $(n_{k+1} - n_k)/(n_k - n_{k-1})$ is bounded as a function of k, then $\sum_{n=1}^\infty a_n$ converges if and only if $\sum_{k=1}^\infty (n_{k+1} - n_k) a_{n_k}$ converges.*

Proof. Let

$$t_n = \sum_{k=1}^n (n_{k+1} - n_k) a_{n_k},$$

$$u_n = \sum_{k=2}^n (n_k - n_{k-1}) a_{n_k}.$$

We find that

$$\begin{aligned}
s_{n_2} &= s_{n_1} + a_{n_1+1} + a_{n_1+2} + \cdots + a_{n_2} \\
&\geq s_{n_1} + (n_2 - n_1) a_{n_2}, = s_{n_1} + u_2
\end{aligned}$$

and in general

$$s_{n_k} \geq s_{n_1} + u_k. \tag{2.11}$$

On the other hand,

$$
\begin{aligned}
s_{n_2-1} &= s_{n_1-1} + a_{n_1} + a_{n_1+1} + \cdots + a_{n_2-1} \\
&\leq s_{n_1-1} + (n_2 - n_1)a_{n_1}, = s_{n_1} + t_1
\end{aligned}
$$

and in general

$$s_{n_k-1} \leq s_{n_1} + t_{k-1}. \tag{2.12}$$

Thus, if s_n diverges, the inequality (2.12) shows that t_k diverges as well. Similarly, the inequality (2.11) shows that, if u_n diverges, so does s_n. Moreover, since $(n_{k+1} - n_k)/(n_k - n_{k-1})$ is bounded, we deduce that t_n and u_n converge or diverge together. Thus, all three series converge or diverge simultaneously. $\qquad \square$

Cauchy's condensation test is good enough to settle the question of convergence of Example 2.12 that Gauss's test could not handle. In this case, Cauchy's condensation test yields

$$1 + \sum_{k=1}^{\infty} \frac{2^k}{2^k \ln 2^k} = 1 + \frac{1}{\ln 2} \sum_{k=1}^{\infty} \frac{1}{k}.$$

Since the harmonic series diverges, the series $\sum_{n=2}^{\infty} \frac{1}{n \ln n}$ also diverges. In Example 2.7, let $n_k = k^2$. Notice that $(n_{k+1} - n_k)/(n_k - n_{k-1}) = (2k + 1)/(2k - 1) \leq 3$. Thus, the original series converges if and only if

$$\sum_{k=1}^{\infty} (n_{k+1} - n_k)a_{n_k} = \sum_{k=1}^{\infty} \frac{2k + 1}{2^k} \tag{2.13}$$

converges. Now applying the ratio test to (2.13) yields that $r = 1/2 < 1$ and so the original series converges.

The following test is another application of the ratio test along with Cauchy's condensation test.

Theorem 2.21. *Let $a_n > 0$ be decreasing and $\rho = \lim_{n \to \infty} a_{2n}/a_n$. Then the series $\sum a_n$*

    1. converges if $\rho < 1/2$,

    2. diverges if $\rho > 1/2$,

    3. is inconclusive if $\rho = 1/2$.

Proof. If $\rho < 1/2$, then

$$\lim_{n \to \infty} \frac{2a_{2n}}{a_n} = 2\rho < 1.$$

If $n = 2^k$, then

$$\lim_{k \to \infty} \frac{2^{k+1} a_{2^{k+1}}}{2^k a_{2^k}} = 2\rho < 1.$$

By the ratio test, the series $\sum_{k=1}^{\infty} 2^k a_{2^k}$ converges. It follows that the original series $\sum a_n$ converges from the Cauchy's condensation test.

Along the same lines, we can prove that $\sum a_n$ diverges if $\rho > 1/2$. For each of the series $\sum_{n=1}^{\infty} 1/n$ and $\sum_{n=1}^{\infty} 1/(n \ln^2 n)$, we have $\rho = 1/2$. But the first series diverges while the second converges, so it is inconclusive for $\rho = 1/2$. $\qquad \square$

The series in Example 2.7 is handled very efficiently by this test. Moreover, notice that if $\lim_{n\to\infty} a_{n+1}/a_n = r < 1$, then there is an integer N and an $\epsilon > 0$ such that

$$a_{n+1} < (r + \epsilon)a_n \quad \text{whenever } n > N.$$

Thus, $a_{2n} < (r + \epsilon)a_{2n-1} < \cdots < (r + \epsilon)^n a_n$ and so

$$\lim_{n\to\infty} \frac{a_{2n}}{a_n} = 0.$$

Similarly, if $\lim_{n\to\infty} a_{n+1}/a_n = r > 1$, then

$$\lim_{n\to\infty} \frac{a_{2n}}{a_n} = +\infty.$$

Therefore, whenever the ratio test shows convergence or divergence, this new ratio test will always give the same response. On the other hand, if $\lim_{n\to\infty} a_{2n}/a_n = 1/2$, then either $\lim_{n\to\infty} a_{n+1}/a_n$ does not exists or $\lim_{n\to\infty} a_{n+1}/a_n = 1$. Thus, whenever the new test is inconclusive, so is the ratio test. These facts indicate that the new test is actually stronger than the ratio test. The following example illustrates the power of the new test.

Example 2.13. *Study the convergence of*

$$\sum_{n=1}^{\infty} \frac{\ln n}{n^p}, \quad (p > 0).$$

Clearly, $a_n = \ln n/n^p$ is decreasing for $p > 0$. Appealing to

$$\frac{a_{2n}}{a_n} = \frac{\ln 2n}{(2n)^p} \cdot \frac{n^p}{\ln n} = \frac{1}{2^p} \frac{\ln 2n}{\ln n} \to \frac{1}{2^p},$$

by Theorem 2.21, we find that

1. If $p > 1$, then $\rho = \frac{1}{2^p} < 1/2$, and so the series converges.
2. If $p < 1$, then $\rho = \frac{1}{2^p} > 1/2$, and so the series diverges.

When $p = 1$, the test is inconclusive because $\rho = 1/2$. However, in this case, $a_n = \ln n/n \geq 1/n$ for $n \geq 3$; by the comparison theorem, the divergence of the harmonic series implies that the original series diverges.

In closing we study one more test due to Cauchy. In contrast to all the tests above, this test is based on a comparison between series and improper integrals.

Theorem 2.22 (Integral Test). *Let $f(x)$ be a positive, decreasing and integrable function on $[1, \infty)$. The series $\sum_{n=1}^{\infty} f(n)$ converges if and only if the improper integral*

$$\int_1^{\infty} f(x)\, dx$$

converges.

The justification is based on the self evident inequalities

$$\sum_{n=2}^{N+1} f(n) \leq \int_1^{N+1} f(x)\, dx \leq \sum_{n=1}^{N} f(n),$$

in which the partial sums are the right and left Riemann sums of $f(x)$, respectively.

With this test we have a decided advantage over the previous tests. The other tests, powerful as they are, require the user to pick another series that stands in some relationship to the given series. But, this test is *intrinsic* to the series. Most of time, we have only one obvious choice for the function. For example, the p-series and $\sum_{n=1}^{\infty} 1/(n \ln^p n)$ are handled very efficiently by this test.

2.5 On the Universal Test

If something cannot go on forever, it will stop. — Herbert Stein

In the previous three sections, we have displayed a collection of tests far beyond those in a first-year calculus sequence. Each test is slightly more complicated than the traditional one but handles a few more series, and gives a slightly sharper distinction. Thus, it is natural to ask whether there is a universal test which will resolve the convergence or divergence for *all* positive series. Unfortunately, no such test exists. In other words, whatever test is considered, one can construct a series for which the test is inconclusive. To see this more precisely, we need a few definitions to clarify what is going on.

Definition 2.4. *Consider two positive series*

$$\sum_{n=1}^{\infty} a_n \ \ and \ \ \sum_{n=1}^{\infty} a_n^*.$$

1. Suppose both series converge. Let

$$R_n = \sum_{i=n+1}^{\infty} a_i, \ \ and \ R_n^* = \sum_{i=n+1}^{\infty} a_i^*.$$

We say that $\sum_{n=1}^{\infty} a_n^*$ converges more slowly than $\sum_{n=1}^{\infty} a_n$ if

$$\lim_{n\to\infty} \frac{R_n}{R_n^*} = 0.$$

2. Suppose both series diverge. Let

$$s_n = \sum_{i=1}^{n} a_i, \ \ and \ s_n^* = \sum_{i=1}^{n} a_i^*.$$

We say that $\sum_{n=1}^{\infty} a_n^*$ diverges more slowly than $\sum_{n=1}^{\infty} a_n$ if

$$\lim_{n\to\infty} \frac{s_n^*}{s_n} = 0.$$

Observe that

1. If $\sum_{n=1}^{\infty} a_n$ converges, we can construct a new convergent series as

$$\sum_{n=1}^{\infty} a_n^* = \sum_{n=1}^{\infty} (\sqrt{R_{n-1}} - \sqrt{R_n}),$$

where $R_0 = \sum_{n=1}^{\infty} a_n, R_n^* = \sqrt{R_n}$. Moreover,

$$\frac{a_n}{a_n^*} = \frac{R_{n-1} - R_n}{\sqrt{R_{n-1}} - \sqrt{R_n}} = \sqrt{R_{n-1}} + \sqrt{R_n} \to 0.$$

2. If $\sum_{n=1}^{\infty} a_n$ diverges, we can construct a new divergent series

$$\sum_{n=1}^{\infty} a_n^* = \sqrt{s_1} + \sum_{n=2}^{\infty} \left(\sqrt{s_n} - \sqrt{s_{n-1}}\right),$$

where $s_n^* = \sqrt{s_n}$. Moreover,

$$\frac{a_n}{a_n^*} = \frac{s_n - s_{n-1}}{\sqrt{s_n} - \sqrt{s_{n-1}}} = \sqrt{s_n} + \sqrt{s_{n-1}} \to +\infty.$$

In general, we have

Theorem 2.23. *Let $\sum_{n=1}^{\infty} a_n$ be a positive series.*

1. *If $\sum_{n=1}^{\infty} a_n$ converges, then there is a monotone sequence b_n such that $\lim_{n\to\infty} b_n = \infty$ and the series $\sum_{n=1}^{\infty} a_n b_n$ converges.*

2. *If $\sum_{n=1}^{\infty} a_n$ diverges, then there is a monotone sequence b_n such that $\lim_{n\to\infty} b_n = 0$ and the series $\sum_{n=1}^{\infty} a_n b_n$ diverges.*

With all of this machinery in place, we see that with any convergent (divergent) positive series we can associate another convergent (divergent) positive series constructed from the first, that converges (diverges) much more slowly. So there is nonexistence of the "slowest" convergent (divergent) series. Therefore, for every positive series there exists a comparison test which can decide its convergence or divergence (for example, Kummer's test). However, there is no such *universal comparison test* that can decide the convergence or divergence of *all* positive series.

2.6   Tests for General Series

There is no comparison tests for convergence of conditionally convergent series.
— Godfrey Harold Hardy

Under the absolute convergence test (Theorem 2.7), all of the preceding tests for positive series become tests for absolute convergence of general series. But none of the convergence tests that we have examined so far are applicable to a series that is not alternating and does not converge absolutely. We now present two tests that are applicable to the general series of the form

$$\sum_{n=1}^{\infty} a_n b_n = a_1 b_1 + a_2 b_2 + a_3 b_3 + \cdots.$$

Both tests are based on the following algebraic manipulation, which is often called the *Abel's summation formula.*

Theorem 2.24 (Abel's Summation Formula). *Let a_n and b_n be arbitrary sequences. Suppose that*

$$s_n = \sum_{k=1}^{n} a_k.$$

Then

$$\sum_{k=1}^{n} a_k b_k = \sum_{k=1}^{n-1} s_k (b_k - b_{k+1}) + s_n b_n. \tag{2.14}$$

Proof. In view of the fact that $a_k = s_k - s_{k-1}$, we have

$$\sum_{k=1}^{n} a_k b_k = s_1 b_1 + (s_2 - s_1) b_2 + \cdots + (s_n - s_{n-1}) b_n$$

$$= (s_1 b_1 + s_2 b_2 + \cdots + s_n b_n) - (s_1 b_2 + s_2 b_3 + \cdots + s_{n-1} b_n)$$
$$= s_1(b_1 - b_2) + s_2(b_2 - b_3) + \cdots + s_{n-1}(b_{n-1} - b_n) + s_n b_n$$
$$= \sum_{k=1}^{n-1} s_k(b_k - b_{k+1}) + s_n b_n.$$

$\square$

Thus, if $|s_k| \leq L$ for all $1 \leq k \leq n$ and b_k is monotone (increasing or decreasing), then (2.14) implies that

$$\left| \sum_{k=1}^{n} a_k b_k \right| \leq \sum_{k=1}^{n-1} |b_k - b_{k+1}| L + |b_n| L$$
$$= L(|b_1 - b_n| + |b_n|) \leq L(|b_1| + 2|b_n|). \tag{2.15}$$

In particular, if $b_1 \geq b_2 \geq \cdots \geq b_n \geq 0$, we have

$$\left| \sum_{k=1}^{n} a_k b_k \right| \leq L b_1. \tag{2.16}$$

As a corollary of the inequality (2.15), we obtain

Theorem 2.25 (Abel's Test). *Let $\sum_{n=1}^{\infty} a_n$ be a convergent series, and let b_n be a bounded and monotone sequence. Then $\sum_{n=1}^{\infty} a_n b_n$ converges.*

Proof. By the assumption that $\sum_{n=1}^{\infty} a_n$ converges, the Cauchy criterion implies that given $\epsilon > 0$, for any positive integer p, there is an integer N such that

$$|a_{n+1} + a_{n+2} + \cdots + a_{n+p}| < \epsilon \text{ whenever } n > N.$$

Let $|b_n| \leq M$. Taking $L = \epsilon$ in (2.15) yields

$$\left| \sum_{k=n+1}^{n+m} a_k b_k \right| = \left| \sum_{k=1}^{m} a_{n+k} b_{n+k} \right| \leq 3M\epsilon.$$

This proves the convergence of $\sum_{n=1}^{\infty} a_n b_n$ as desired. $\square$

If we replace the convergence of $\sum_{n=1}^{\infty} a_n$ with bounded partial sums, but strengthen the condition on the sequence b_n, we obtain another test from (2.16) as follows.

Theorem 2.26 (Dirichlet's Test). *Let $\sum_{n=1}^{\infty} a_n$ be a series with bounded partial sums, and let b_n be positive, decreasing and $\lim_{n\to\infty} b_n = 0$. Then $\sum_{n=1}^{\infty} a_n b_n$ converges.*

Proof. By the assumption that $\lim_{n\to\infty} b_n = 0$, for any given $\epsilon > 0$, there is an integer N such that

$$|b_n| < \epsilon \text{ whenever } n > N.$$

Let $|s_n| \leq M$ for all $n \in \mathbb{N}$. Then

$$|a_{n+1} + a_{n+2} + \cdots + a_{n+p}| = |s_{n+p} - s_n| \leq 2M.$$

Applying the inequality (2.16) yields

$$\left| \sum_{k=n+1}^{n+m} a_k b_k \right| = \left| \sum_{k=1}^{m} a_{n+k} b_{n+k} \right| \leq 2Mb_{n+1} \leq 2M\epsilon.$$

This proves the convergence of $\sum_{n=1}^{\infty} a_n b_n$.   □

Both Abel's test and Dirichlet's test enable us to test a series that is not necessarily alternating and does not converge absolutely.

Example 2.14. *Determine the convergence of*

$$\sum_{n=1}^{\infty} b_n \sin(nx) \ and \ \sum_{n=1}^{\infty} b_n \cos(nx)$$

where b_n is positive, decreasing and approaching to zero as $n \to \infty$.

By the trigonometric identities

$$\sum_{k=1}^{n} \sin(kx) = \frac{\cos(x/2) - \cos((n+1/2)x)}{2\sin(x/2)},$$

$$\sum_{k=1}^{n} \cos(kx) = \frac{\sin((n+1/2)x) - \sin(x/2)}{2\sin(x/2)},$$

both partial sums are bounded by

$$\frac{1}{|\sin(x/2)|}$$

for all $x \neq 2k\pi$ ($k \in \mathbb{Z}$). By Dirichlet's test, both series converge for $x \neq 2k\pi$ ($k \in \mathbb{Z}$).

It is worth noticing that these two series do not converge absolutely unless $\sum_{n=1}^{\infty} b_n$ converges. We can prove this with an argument similar to that above. First, Dirichlet's test yields that $\sum_{n=1}^{\infty} b_n \cos(2nx)$ converges. Next, since

$$|b_n \sin(nx)| \geq b_n \sin^2(nx), \quad |b_n \cos(nx)| \geq b_n \cos^2(nx),$$

and

$$b_n \sin^2(nx) = \frac{1}{2} b_n (1 - \cos(2nx)),$$
$$b_n \cos^2(nx) = \frac{1}{2} b_n (1 + \cos(2nx)),$$

by the comparison theorem, we find that these two series converge absolutely. Hence both series $\sum_{n=1}^{\infty} b_n \sin^2(nx)$ and $\sum_{n=1}^{\infty} b_n \cos^2(nx)$ converge, and therefore $\sum_{n=1}^{\infty} b_n$ converges.

In particular, the following series

$$\sum_{n=1}^{\infty} \frac{\sin(nx)}{\ln n}, \quad \sum_{n=1}^{\infty} \frac{\sin(nx)}{n}, \quad \sum_{n=1}^{\infty} \frac{H_n}{n} \sin(nx)$$

are all convergent conditionally, where H_n is the nth harmonic number.

Observe that a bound on the partial sums in Dirichlet's test is independent of n. The bound we found in Example 2.14 satisfies this requirement, but it does depend on x. Choosing a specific value for x, we obtain an explicit bound that enables us to apply Dirichlet's

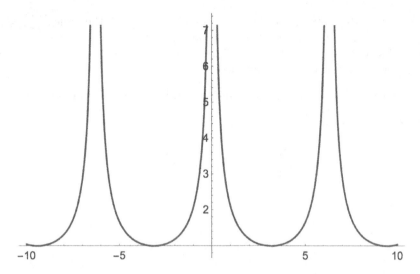

FIGURE 2.1
The graph of $1/|\sin(x/2)|$.

test. However, the graph of this bound as a function of x becomes unbounded as x approaches an even multiple of π (see Figure 2.1).

The strange behavior of the bound is directly related to uniform continuity and the behavior of the Fourier series. Here we do not need a uniform bound in order to get convergence, but we do need a uniform bound in order to have the sum of these continuous functions to again be continuous. We will explore this topic further in Chapter 6.
Next, let

$$a_n = \frac{(-1)^n}{\sqrt{n}}, \; b_n = \frac{(-1)^n}{\sqrt{n} + (-1)^n}.$$

In view of Example 2.6, Hardy showed that $\sum_{n=2}^{\infty} b_n$ diverges even if $\sum_{n=2}^{\infty} \frac{(-1)^n}{\sqrt{n}}$ converges and that $\lim_{n\to\infty} a_n/b_n = 1$. Thus, the comparison test for general series needs additional condition. As an application of Abel's test, we show that monotonicity of a_n/b_n is one of such additional conditions.

Theorem 2.27 (Comparison Test for General Series). *Let $\sum_{n=1}^{\infty} a_n$ and $\sum_{n=1}^{\infty} b_n$ satisfy that*

$$\lim_{n\to\infty} \frac{a_n}{b_n} = c \; and \; \frac{a_n}{b_n} \; is \; monotone.$$

Then

1. *if $\sum_{n=1}^{\infty} b_n$ converges, then $\sum_{n=1}^{\infty} a_n$ converges;*

2. *if $c \neq 0$, then both series are either both divergent, both conditionally convergent, or both absolutely convergent.*

Proof. In view of that $a_n = b_n \cdot \frac{a_n}{b_n}$ and the assumptions, (1) follows from Abel's test directly. To show (2), we separate into two cases.
(a) Since

$$\lim_{n\to\infty} \frac{|a_n|}{|b_n|} = \lim_{n\to\infty} \left| \frac{a_n}{b_n} \right| = |c| \neq 0,$$

by the comparison test on the positive series, it follows that the series $\sum_{n=1}^{\infty} |a_n|$ and $\sum_{n=1}^{\infty} |b_n|$ are either both convergent or both divergent.
(b) Assume that neither $\sum_{n=1}^{\infty} a_n$ nor $\sum_{n=1}^{\infty} b_n$ converges absolutely. By (1), we see that both series converges or diverges simultaneously.   □

Finally, using Abel's summation formula, we establish the following identity:

Example 2.15. *Let $s_n = \sum_{k=1}^{n} a_k$ and $|x| < 1$. Then*

$$\sum_{n=1}^{\infty} a_n x^n = (1 - x) \sum_{n=1}^{\infty} s_n x^n.$$

Proof. By Abel's summation formula, we have

$$\sum_{k=1}^{n} a_n x^n = \sum_{k=1}^{n-1} s_k (x^k - x^{k+1}) + s_n x^n.$$

To establish the expected identity, we must show that $s_n x^n \to 0$ as $n \to \infty$. To this end, let $|x| < r < 1$. Then $|a_k| r^k \le M$ for some positive number M. Therefore,

$$|s_n x^n| \le M \left(1 + \frac{1}{r} + \cdots + \frac{1}{r^n} \right) |x|^n = \frac{M}{r^n} (1 + r + \cdots + r^n) |x|^n$$

$$= \frac{M}{r^n} \frac{1 - r^{n+1}}{1 - r} |x|^n = \frac{M}{1 - r} \left(\frac{|x|}{r} \right)^n - \frac{Mr}{1 - r} |x|^n,$$

which implies that $s_n x^n \to 0$ as $n \to \infty$.   □

In particular, setting $a_n = 1/n$ and using that $\sum_{n=1}^{\infty} x^n/n = -\ln(1 - x)$, we find that

$$\sum_{n=1}^{\infty} H_n x^n = -\frac{\ln(1 - x)}{1 - x}, \quad \text{for } |x| < 1,$$

which is the *generating function* of the harmonic numbers.

2.7 Properties of Convergent Series

All happy families resemble one another, each unhappy family is unhappy in its own way. — Leo Tolstoy

As we have seen, the main difference between finite sums and infinite series lies in the limiting process. The properties of finite sums often fail to extend to infinite series. For example, if we were free to associate terms of the following series at will, then we would have

$$
\begin{aligned}
1 - 1 + 1 - 1 + \cdots &= (1 - 1) + (1 - 1) + \cdots = 0, \\
1 - 1 + 1 - 1 + \cdots &= 1 - (1 - 1) - (1 - 1) - \cdots = 1.
\end{aligned}
$$

Here, the associative law fails. However, we can justify that

$$
\begin{aligned}
\frac{\pi}{4} &= 1 - \frac{1}{3} + \frac{1}{5} - \frac{1}{7} + \cdots \\
&= \left(1 - \frac{1}{3}\right) + \left(\frac{1}{5} - \frac{1}{7}\right) + \cdots \\
&= \frac{2}{1 \cdot 3} + \frac{2}{5 \cdot 7} + \frac{2}{9 \cdot 11} + \cdots \\
&= 1 - \left(\frac{1}{3} - \frac{1}{5}\right) - \left(\frac{1}{7} - \frac{1}{9}\right) + \cdots \\
&= 1 - \frac{2}{3 \cdot 5} - \frac{2}{7 \cdot 9} - \frac{2}{11 \cdot 13} + \cdots .
\end{aligned}
$$

In general, let $\sum_{n=1}^{\infty} a_n$ be a convergent series. If we regroup consecutive summands without changing their order, we get

$$
a_1 + \cdots + a_{n_1}, a_{n_1+1} + \cdots + a_{n_2}, \cdots, a_{n_{k-1}+1} + \cdots + a_{n_k}, \cdots,
$$

where n_k is a monotone subsequence of $\mathbb{N}$. Let $s_n = a_1 + \cdots + a_n$. Then

$$
(a_1 + \cdots + a_{n_1}) + \cdots + (a_{n_{k-1}+1} + \cdots + a_{n_k}) = s_{n_k}.
$$

The existence of $\lim_{n \to \infty} s_n$ implies that $\lim_{k \to \infty} s_{n_k}$ exists and

$$
\lim_{k \to \infty} s_{n_k} = \lim_{n \to \infty} s_n.
$$

This proves

Theorem 2.28. *The associative law holds for the convergent series.*

Remark. Observe that

$$
\left(2 - 1\frac{1}{2}\right) + \left(1\frac{1}{3} - 1\frac{1}{4}\right) + \left(1\frac{1}{5} - 1\frac{1}{6}\right) = \frac{1}{1 \cdot 2} + \frac{1}{3 \cdot 4} + \frac{1}{5 \cdot 6} + \cdots .
$$

Thus, grouping of terms may convert a divergent series into a convergent series. Equivalently, removal of brackets may destroy convergence. However, if the sum of the absolute values of the terms in the brackets tends to 0, in particular, if the terms in each bracket are finite and have same sign, the convergence of the bracketed series implies the convergence of the series obtained by omitting the brackets. Both series converge to the same sum (see Problem 7 for the proof). We illustrate this case using the following example.

Example 2.16. *Determine the convergence of $\sum_{n=1}^{\infty} (-1)^{\lfloor \sqrt{n} \rfloor}/n$, where $\lfloor x \rfloor$ is the greatest integer function.*

Proof. Notice that

$$
\begin{aligned}
\sum_{n=1}^{\infty} \frac{(-1)^{\lfloor \sqrt{n} \rfloor}}{n} &= -1 - \frac{1}{2} - \frac{1}{3} + \frac{1}{4} + \frac{1}{5} + \frac{1}{6} + \frac{1}{7} + \frac{1}{8} \\
&\quad - \frac{1}{9} - \frac{1}{10} - \frac{1}{11} - \frac{1}{12} - \frac{1}{13} - \frac{1}{14} - \frac{1}{15} + \frac{1}{16} + \frac{1}{17} + \cdots .
\end{aligned}
$$

Grouping same-sign terms together by letting $n_k = (k+1)^2 - 1 \, (k = 1, 2, \cdots)$ yields an alternating series

$$
\sum_{k=1}^{\infty} (-1)^k \left(\frac{1}{k^2} + \frac{1}{k^2 + 1} + \cdots + \frac{1}{(k+1)^2 - 1}\right).
$$

Let

$$b_k = \frac{1}{k^2} + \frac{1}{k^2 + 1} + \cdots + \frac{1}{(k+1)^2 - 1}.$$

Notice that the sum of the first k terms is less than $k \cdot \frac{1}{k^2} = 1/k$ and the sum of the rest of $k + 1$ terms is less than $(k+1) \cdot \frac{1}{k^2+k} = 1/k$. We find that

$$\frac{2}{k+1} < b_k < \frac{2}{k}.$$

This shows that $b_k > b_{k+1}$ and $b_k \to 0$ as $k \to \infty$. The convergence of the above alternating series now follows from the alternating series test. Based on Remark of Theorem 2.28, we conclude that the original series converges. □

Next, we turn to the commutative property. The following example shows that different rearrangements can lead to different sums. It is well-known that

$$1 - \frac{1}{2} + \frac{1}{3} - \frac{1}{4} + \frac{1}{5} - \frac{1}{6} + \cdots = \ln 2, \tag{2.17}$$

We rearrange of the terms where one positive term is followed by two negative terms:

$$1 - \frac{1}{2} - \frac{1}{4} + \frac{1}{3} - \frac{1}{6} - \frac{1}{8} + \frac{1}{5} - \frac{1}{10} - \frac{1}{12} + \cdots .$$

By adding each positive term to the negative term following it, we obtain

$$\left(1 - \frac{1}{2}\right) - \frac{1}{4} + \left(\frac{1}{3} - \frac{1}{6}\right) - \frac{1}{8} + \left(\frac{1}{5} - \frac{1}{10}\right) - \frac{1}{12} + \cdots$$
$$= \frac{1}{2} - \frac{1}{4} + \frac{1}{6} - \frac{1}{8} + \frac{1}{10} - \frac{1}{12} + \frac{1}{14} - \frac{1}{16} + \cdots$$
$$= \frac{1}{2}\left(1 - \frac{1}{2} + \frac{1}{3} - \frac{1}{4} + \frac{1}{5} - \frac{1}{6} + \cdots\right) = \frac{1}{2}\ln 2,$$

which is different from the original sum. Thus, the commutative law does not hold for infinite series. As we will see in the following two theorems, this happens because the convergence is *conditional*. First, we see that for absolutely convergent series, rearranging the terms has no impact at all.

Theorem 2.29. *If the series $\sum_{n=1}^{\infty} a_n$ converges absolutely, then any rearrangement of this series converges to the same limit.*

Proof. Assume $\sum_{n=1}^{\infty} a_n$ converges absolutely to S. We begin with the simple case where all $a_n > 0$. Let $\sum_{n=1}^{\infty} b_n$ be a rearrangement of $\sum_{n=1}^{\infty} a_n$ and

$$t_k = b_1 + b_2 + \cdots + b_k.$$

be the partial sums of the rearranged series. Since

$$b_1 = a_{n_1}, b_2 = a_{n_2}, \cdots, b_k = a_{n_k},$$

we have $t_k \leq S$. So $\sum_{n=1}^{\infty} b_n$ converges. Let $\sum_{n=1}^{\infty} b_n$ converge to T. Then $T \leq S$. Turning this argument around, since $\sum_{n=1}^{\infty} a_n$ is a rearrangement of $\sum_{n=1}^{\infty} b_n$, along the same lines, we find that $S \leq T$. Therefore, $S = T$. i.e., both series converge to the same limit.

Next, we consider the case where a_n's are not all positive. As before, we assume that $\sum_{n=1}^{\infty} b_n$ is a rearrangement of $\sum_{n=1}^{\infty} a_n$. Since $\sum_{n=1}^{\infty} |a_n|$ converges, we know from our proof of the first part that $\sum_{n=1}^{\infty} |b_n|$ converges and so $\sum_{n=1}^{\infty} b_n$ converges absolutely. Let

$$p_n = \frac{|a_n| + a_n}{2}, \quad q_n = \frac{|a_n| - a_n}{2}.$$

Since $0 \leq p_n \leq |a_n|, 0 \leq q_n \leq |a_n|$, the comparison theorem implies that both series $\sum_{n=1}^{\infty} p_n$ and $\sum_{n=1}^{\infty} q_n$ converge. Moreover,

$$\sum_{n=1}^{\infty} a_n = \sum_{n=1}^{\infty} p_n - \sum_{n=1}^{\infty} q_n.$$

A rearrangement of $\sum_{n=1}^{\infty} a_n$ causes the rearrangements of $\sum_{n=1}^{\infty} p_n$ and $\sum_{n=1}^{\infty} q_n$, but by what we have proved in the first part, the sums do not change. Therefore, the rearranged series converges to the same limit.

Bypassing p_n and q_n, we present here another proof using an $\epsilon - N$ argument. For any given $\epsilon > 0$, there is an integer N_1 such that

$$|s_n - S| < \frac{\epsilon}{2}, \quad \text{whenever } n \geq N_1.$$

On the other hand, since $\sum_{n=1}^{\infty} a_n$ converges absolutely, there is an integer N_2 such that

$$\sum_{i=k+1}^{n} |a_i| < \frac{\epsilon}{2}, \quad \text{whenever } n > k \geq N_2.$$

Let $N = \max\{N_1, N_2\}$ and

$$A = \{a_1, a_2, \ldots, a_N\}.$$

Now we want to move far enough out in the series $\sum_{n=1}^{\infty} b_n$ so that we have included all terms in A. Thus, choose

$$M = \max\{n_i, 1 \leq i \leq N\}.$$

If $k \geq M$, then $t_k - s_N$ consists of a finite set of terms, the absolute values of which appear in the tail $\sum_{i=N+1}^{\infty} |a_i|$. Our selection of N_2 earlier then guarantees $|t_k - s_N| < \epsilon/2$, so

$$\begin{aligned} |t_k - S| &= |t_k - s_N + s_N - S| \\ &\leq |t_k - s_N| + |s_N - S| \\ &< \frac{\epsilon}{2} + \frac{\epsilon}{2} = \epsilon \end{aligned}$$

whenever $k \geq M$. $\qquad\square$

We have seen that a rearrangement of the conditionally convergent series (2.17) converges to half of the original sum. Could a different rearrangement converge to yet another limit? How many possible values are there? Surprisingly, Riemann found that there is a rearrangement of a conditionally convergent series that converges to any number you wish.

Theorem 2.30 (Riemann Series Theorem). *Let $\sum_{n=1}^{\infty} a_n$ converge conditionally. Then there exists a rearrangement of this series that either converges to any preassigned real number a or diverges to $\pm\infty$.*

We first see how this can be done with an example. We take the alternating harmonic series and rearrange it such that the resulting series converges to $\ln 3$ instead of $\ln 2$.

Notice that $\ln 3 = 1.098612288\ldots$. We add 2 consecutive positive terms such that their sum is above $\ln 3$:

$$1 + \frac{1}{3} = 1.33333\ldots.$$

Then we add one negative term such the sum drops below $\ln 3$:

$$1 + \frac{1}{3} - \frac{1}{2} = 0.83333\ldots.$$

We now put in more positive terms until the sum is just above $\ln 3$:

$$1 + \frac{1}{3} - \frac{1}{2} + \frac{1}{5} + \frac{1}{7} = 1.17619\ldots$$

and then add a negative term until the sum is just below $\ln 3$ again:

$$1 + \frac{1}{3} - \frac{1}{2} + \frac{1}{5} + \frac{1}{7} - \frac{1}{4} = 0.92619\ldots$$

and so on. Thus, no matter how far out we go in the series, there are always enough positive or negative terms remaining to move the sum back to the other side of $\ln 3$. Every term will eventually be selected.

$$\ln 3 = 1 + \frac{1}{3} - \frac{1}{2} + \frac{1}{5} + \frac{1}{7} - \frac{1}{4} + \frac{1}{9} + \frac{1}{11} - \frac{1}{6} + \frac{1}{13} + \frac{1}{15}$$

$$- \frac{1}{8} + \frac{1}{17} + \frac{1}{19} + \frac{1}{21} - \frac{1}{10} + \frac{1}{23} + \frac{1}{25} - \frac{1}{12} + \frac{1}{27} + \frac{1}{29} - \cdots.$$

In general, if $\sum_{n=1}^{\infty} a_n$ converges conditionally, we customarily separate the positive terms from the negative ones, preserving their relative order. Let p_k be the kth positive term, q_m be the absolute value of the mth negative term. Then the series $\sum_{k=1}^{\infty} p_k$ and $\sum_{m=1}^{\infty} q_m$ both diverge. We prove this assertion by contradiction. If both were convergent, then

$$\sum_{n=1}^{\infty} |a_n| = \sum_{k=1}^{\infty} p_k + \sum_{m=1}^{\infty} q_m$$

would converge, contrary to assumption. Since

$$\sum_{n=1}^{N} a_n = \sum_{k=1}^{K} p_k - \sum_{m=1}^{N-K} q_m,$$

if one series on the right-hand side converges while the other diverges the series $\sum_{n=1}^{\infty} a_n$ would diverge, again contrary to assumption. With this in mind, we are ready to prove the Riemann series theorem.

Proof. First, let a be an arbitrary real number. We determine the associated rearrangement as follows: Beginning with the positive terms, we find the smallest k_1 such that

$$p_1 + p_2 + \cdots p_{k_1} > a.$$

This can be done because $\sum_{k=1}^{\infty} p_k$ diverges. Then we add consecutive negative terms and choose the smallest m_1 such that the sum drops below a:

$$p_1 + p_2 + \cdots p_{k_1} - q_1 - q_2 - \cdots - q_{m_1} < a.$$

Notice that
$$p_1 + p_2 + \cdots p_{k_1} - q_1 - q_2 - \cdots - q_{m_1}$$
is a rearrangement of terms of the original series whose sum can be no further from a than q_{m_1}. This process is then repeated. We now put in more consecutive positive terms such that the total partial sum first exceeds a:
$$p_1 + \cdots p_{k_1} - q_1 - \cdots - q_{m_1} + p_{k_1+1} + \cdots + p_{k_2} > a$$
and then add consecutive negative terms until the total partial sum first falls short of a:
$$p_1 + \cdots p_{k_1} - q_1 - \cdots - q_{m_1} + p_{k_1+1} + \cdots + p_{k_2} - q_{m_1+1} - \cdots - q_{m_2} < a.$$

Continuing in this manner, we see that every term in $\sum_{k=1}^{\infty} a_k$ eventually appears in the above rearranged series. Since the original series converges, a_n is approaching to zero. For given $\epsilon > 0$, there is a finite number of terms with absolute value greater than or equal to ϵ. Let t_m be the partial sums of the rearranged series. We move along our rearranged series until we have included all terms with absolute value greater than or equal to ϵ. We continue moving along the series until we come to the next pair of consecutive partial sums that lie on opposite sides of a:
$$t_N < a < t_{N+1} \text{ or } t_N > a > t_{N+1}.$$

We know that all of the terms from here on have absolute value less than ϵ, and so $m \geq N$ implies that $|t_m - a| < \epsilon$. Thus, the series rearranged by this alternating scheme converges to a as claimed.

Next, we show that there exists a rearrangement of the original series that diverges to $+\infty$. To this end, we determine the rearrangement by alternating a group of positive terms followed by a single negative term. Appealing to the divergence of $\sum_{k=1}^{\infty} p_k$, we can choose m_1 so large that
$$p_1 + p_2 + \cdots + p_{m_1} - q_1 > 1,$$
then $m_2 > m_1$ so large that
$$p_1 + \cdots + p_{m_1} + p_{m_1+1} + \cdots + p_{m_2} - q_1 - q_2 > 2$$
and, generally, $m_k > m_{k-1}$ so large that
$$p_1 + \cdots + p_{m_k} - q_1 - q_2 - \cdots - q_k > k.$$

Thus, the series
$$p_1 + \cdots + p_{m_1} - q_1 + p_{m_1+1} + \cdots + p_{m_2} - q_2 + p_{m_2+1} + \cdots$$

diverges to $+\infty$. Analogously, we can find a rearrangement of the original series that tends to $-\infty$. □

The Riemann series theorem demonstrates that summing infinite series is a tricky business. As has been observed before, the study of infinite processes can carry us into deep waters. For example, it is interesting to notice that, for the above rearranged partial sums that converge to $\ln 3$, each time only one negative term is used to bring the partial sum below $\ln 3$. Indeed, that is the generic case for all $a > \ln 2$ (see Exercise 119).

There is no general result about the convergence of rearrangements of conditionally convergent series. Each one has to be handled separately. In view of the idea underlying the proof of the alternating series test (Theorem 2.8), we have the following convergence test based on the subsequences of the partial sum.

Theorem 2.31. *Let* $\lim_{n\to\infty} |a_n| = 0$ *and let* $S_n = \sum_{k=1}^{n} a_k$. *If* S_{kn} *converges to* S *for some* $k \in \mathbb{N}$, *then* $\sum_{n=1}^{\infty} a_n$ *also converges to* S.

Proof. For every $m \in \mathbb{N}$ and $m > k$, there is $n \in \mathbb{N}$ such that $kn < m \leq k(n+1)$. Since

$$S_m = S_{nk} + \sum_{i=kn+1}^{m} a_i,$$

it follows that

$$|S - S_m| < |S - S_{kn}| + \sum_{i=kn+1}^{m} |a_i| \leq |S - S_{kn}| + k \max\{a_{nk+1}, \cdots, a_{(k+1)n}\}.$$

By the assumptions, we see that $|S - S_m|$ can be made as small as desired for large enough m. This proves that $\sum_{n=1}^{\infty} a_n$ converges to S. $\qquad\square$

2.8   Infinite Products

It takes money to make money. — Proverb

In this section we give a brief introduction to the theory of infinite products.

Definition 2.5. *Given a sequence* a_n, *an* infinite product *is an expression of the form*

$$\prod_{k=1}^{\infty} a_k = a_1 \cdot a_2 \cdots a_n \cdots . \tag{2.18}$$

In particular, a_n *is called the* nth factor *of the product.*

Let

$$P_n = \prod_{k=1}^{n} a_k, \tag{2.19}$$

which is called the *nth partial product*. By analogy with the infinite series, it is natural to say the product (2.18) converges if the sequence P_n in (2.19) converges. However, this definition has a drawback: every product having one zero factor will be convergent regardless of the behavior of the remaining factors. To exclude this situation, we define

Definition 2.6. *If the sequence* P_n *defined by* (2.19) *converges to* $P \neq 0$, *we say that the product* $\prod_{k=1}^{\infty} a_k$ *converges and write*

$$\prod_{n=1}^{\infty} a_n = P.$$

The number P *is called the* value *of the product. If the sequence* P_n *converges to zero or diverges, we say the product* diverges.

The following examples demonstrate how to use the definition to find the values of some products.

Example 2.17. *Determine the convergence of*

$$\prod_{n=2}^{\infty} \left(1 - \frac{1}{n^2}\right).$$

This product converges because

$$
\begin{aligned}
P_n &= \left(1 - \frac{1}{2^2}\right)\left(1 - \frac{1}{3^2}\right)\cdots\left(1 - \frac{1}{n^2}\right) \\
&= \frac{1 \cdot 3}{2^2} \cdot \frac{2 \cdot 4}{3^2} \cdot \frac{3 \cdot 5}{4^2} \cdots \frac{(n-1) \cdot (n+1)}{n^2} \\
&= \frac{1}{2} \cdot \frac{n+1}{n} \to 1/2.
\end{aligned}
$$

Example 2.18. *Prove that, for $|x| < 1$,*

$$\prod_{n=1}^{\infty} (1 + x^{2^{n-1}}) = \frac{1}{1-x}.$$

Indeed, we have

$$
\begin{aligned}
(1-x) \cdot P_n &= (1-x)(1+x)(1+x^2)\cdots(1+x^{2^{n-1}}) \\
&= (1-x^2)(1+x^2)(1+x^{2^2})\cdots(1+x^{2^{n-1}}) = \cdots \\
&= (1-x^{2^{n-1}})(1+x^{2^{n-1}}) = 1 - x^{2^n},
\end{aligned}
$$

so

$$P_n = \frac{1 - x^{2^n}}{1-x}.$$

The desired result follows from taking limit $n \to \infty$.

Example 2.19. *Let $\theta \neq 0$. Evaluate*

$$\prod_{n=1}^{\infty} \cos\left(\frac{\theta}{2^n}\right).$$

Repeatedly using the double-angle formula for the sine in the form $\sin\phi = 2\sin(\phi/2)\cos(\phi/2)$ yields

$$
\begin{aligned}
\sin\theta &= 2\cos(\theta/2)\sin(\theta/2) \\
&= 2^2 \cos(\theta/2)\cos(\theta/2^2)\sin(\theta/2^2) = \cdots \\
&= 2^n \cos(\theta/2)\cos(\theta/2^2)\cdots\cos(\theta/2^n)\sin(\theta/2^n)
\end{aligned}
$$

Thus, by the well-known limit $\lim_{t \to 0} \sin t/t = 1$, we obtain

$$P_n = \frac{\sin\theta}{2^n \sin(\theta/2^n)} = \frac{\sin\theta}{\theta}\frac{\theta/2^n}{\sin(\theta/2^n)} \to \frac{\sin\theta}{\theta},$$

and so

$$\prod_{n=1}^{\infty} \cos\left(\frac{\theta}{2^n}\right) = \frac{\sin\theta}{\theta}.$$

In particular, letting $\theta = \pi/2$ gives

$$\frac{2}{\pi} = \prod_{n=1}^{\infty} \cos\left(\frac{\pi}{2^{n+1}}\right). \tag{2.20}$$

Recall that

$$\cos\frac{\pi}{4} = \sqrt{\frac{1}{2}} \quad \text{and} \quad \cos\frac{\alpha}{2} = \sqrt{\frac{1}{2} + \frac{1}{2}\cos\alpha}.$$

These identities along with (2.20) yield *Vieta's formula*

$$\frac{2}{\pi} = \sqrt{\frac{1}{2}} \cdot \sqrt{\frac{1}{2} + \frac{1}{2}\sqrt{\frac{1}{2}}} \cdot \sqrt{\frac{1}{2} + \frac{1}{2}\sqrt{\frac{1}{2} + \frac{1}{2}\sqrt{\frac{1}{2}}}} \cdots.$$

This formula and *Wallis's formula*

$$\frac{2}{\pi} = \frac{1}{2} \cdot \frac{3}{2} \cdot \frac{3}{4} \cdot \frac{5}{4} \cdots \frac{2n-1}{2n} \cdot \frac{2n+1}{2n} \cdots \tag{2.21}$$

were two of the earliest infinite products built in analysis. Apart from being intrinsically interesting, both formulas underlie other important developments in both pure and applied mathematics.

Note that $a_n = P_n/P_{n-1}$. If P_n converges, then $a_n \to 1$ as $n \to \infty$. For this reason, we often write $a_n = 1 + b_n$. Since $\ln(1 + b_n) \sim b_n$, it is easy to show

Theorem 2.32. *Let $b_n > 0$ for all $n \in \mathbb{N}$. Then the product $\prod_{n=1}^{\infty}(1 + b_n)$ converges if and only if $\sum_{n=1}^{\infty} b_n$ converges.*

The following result is analogous to this theorem.

Theorem 2.33. *Let $b_n > 0$ for all $n \in \mathbb{N}$. Then the product $\prod_{n=1}^{\infty}(1 - b_n)$ converges if and only if $\sum_{n=1}^{\infty} b_n$ converges.*

We conclude this section by presenting Euler's remarkable identity, which expresses the Riemann zeta function $\zeta(s) = \sum_{n=1}^{\infty} 1/n^s$ as an infinite product extended over all primes.

Theorem 2.34. *Let p_k be kth prime number and $s > 1$. Then*

$$\zeta(s) = \prod_{k=1}^{\infty} \frac{1}{1 - p_k^{-s}}. \tag{2.22}$$

Proof. Let $P_n = \prod_{k=1}^{n}(1 - p_k^{-s})^{-1}$ be the nth partial product. Applying the geometric series expansion, for each prime number p_k, we have

$$\frac{1}{1 - 1/p_k^s} = 1 + \frac{1}{p_k^s} + \frac{1}{p_k^{2s}} + \cdots,$$

which converges absolutely, so

$$P_n = \prod_{k=1}^{n}\left(1 + \frac{1}{p_k^s} + \frac{1}{p_k^{2s}} + \cdots\right).$$

We now multiply these series together and rearrange the terms according to increasing denominators. This yields an absolutely convergent series, a typical term of which is

$$\frac{1}{p_1^{r_1 s} p_2^{r_2 s} \cdots p_n^{r_n s}} = \frac{1}{m^s}, \quad \text{where } m = p_1^{r_1} p_2^{r_2} \cdots p_n^{r_n}, \; r_i \geq 0.$$

Therefore, we have

$$P_n = \sum_* \frac{1}{m^s},$$

where the summation on the right-hand side is over those m with prime factors $\leq p_n$. By the *unique factorization theorem*, each such m occurs exactly once in the sum $\sum_*$. To illustrate, we have

$$
\begin{aligned}
P_3 &= \left(1 + \frac{1}{2^s} + \frac{1}{2^{2s}} + \cdots\right) \cdot \left(1 + \frac{1}{3^s} + \frac{1}{3^{2s}} + \cdots\right) \cdot \left(1 + \frac{1}{5^s} + \frac{1}{5^{2s}} + \cdots\right) \\
&= 1 + \frac{1}{2^s} + \frac{1}{3^s} + \frac{1}{4^s} + \frac{1}{5^s} + \frac{1}{6^s} + \frac{1}{8^s} + \frac{1}{9^s} + \frac{1}{10^s} + \frac{1}{12^s} + \frac{1}{15^s} + \frac{1}{16^s} + \frac{1}{18^s} \\
&\quad \frac{1}{20^s} + \frac{1}{24^s} + \frac{1}{25^s} + \frac{1}{27^s} + \frac{1}{30^s} + \frac{1}{32^s} + \frac{1}{36^s} + \frac{1}{40^s} + \frac{1}{45^s} + \frac{1}{48^s} + \frac{1}{50^s} + \frac{1}{54^s} \cdots
\end{aligned}
$$

Thus

$$\zeta(s) - P_n = \sum_{*'} \frac{1}{m^s},$$

where the sum $\sum_{*'}$ is over those m having at least one prime factor $> p_n$. Since these m occur among the integers $> p_n$, we have

$$\zeta(s) - P_n \leq \sum_{m > p_n} \frac{1}{m^s}.$$

The convergence of the *p*-series implies that $P_n \to \zeta(s)$ as $n \to \infty$. $\qquad\square$

Based on (2.22), Euler gave two beautiful proofs of the fact that there are infinitely many prime numbers. His first proof goes this way. If there were only a finite number of primes, then the product on the right hand side of (2.22) would be an ordinary finite product and would be well defined for every $s > 0$, even for $s = 1$. But, the value on the left-hand side for $s = 1$ is the harmonic series, which diverges to infinity. This contradiction shows that there must be infinitely many prime numbers. His second proof rests on

$$\prod_{k=1}^{n} \frac{1}{1 - 1/p_k} \geq \sum_{n=1}^{p_k} \frac{1}{n} > \int_{1}^{p_n+1} \frac{dx}{x} = \ln(p_n + 1),$$

and his discovery that

$$\sum \frac{1}{p_n} = \frac{1}{2} + \frac{1}{3} + \frac{1}{5} + \frac{1}{7} + \frac{1}{11} + \cdots$$

diverges.

2.9   Worked Examples

Wit lies in recognizing the resemblance among things which differ and the difference between things which are alike. — Madame Stael

In this section, to expose ourselves to a wide variety of strategies for tacking problems, we select 25 problems that we find intriguing for one reason or another. They may confirm a common intuitive notion with a clarifying proof or may provide a counterexample to intuition, while some may present unexpected techniques.

1. Let $a_n > 0$ be decreasing for all $n \in \mathbb{N}$. If $\sum_{n=1}^{\infty} a_n$ converges, then $\lim_{n \to \infty} na_n = 0$.

Proof. Let $s_n = \sum_{k=1}^{n} a_k$. By assumption, we have

$$na_{2n} \le a_{n+1} + a_{n+2} + \cdots + a_{2n} = s_{2n} - s_n.$$

The convergence of s_n implies that $na_{2n} \to 0$ as $n \to \infty$, and so

$$2na_{2n} \to 0 \text{ as } n \to \infty. \tag{2.23}$$

Along the same lines, we have

$$(n+1)a_{2n+1} \le a_{n+1} + a_{n+2} + \cdots a_{2n+1} = s_{2n+1} - s_n \to 0$$

as $n \to \infty$. Thus, $2(n+1)a_{2n+1} \to 0$ as $n \to \infty$, so

$$(2n+1)a_{2n+1} = 2[(n+1)a_{2n+1}] - a_{2n+1} \to 0 \text{ as } n \to \infty. \tag{2.24}$$

The desired limit now follows from the facts (2.23) and (2.24). □

Remark. Using Abel's summation formula (2.14), we have

$$\sum_{k=1}^{n} a_k = \sum_{k=1}^{n-1} k(a_k - a_{k+1}) + na_n.$$

If $a_n > 0$ is decreasing, we find that $\sum_{n=1}^{\infty} a_n$ converges if and only if $\lim_{n \to \infty} na_n = 0$ and $\sum_{n=1}^{\infty} (a_n - a_{n+1})n$ converges. Here the assumption of monotonicity of a_n is necessary. Consider

$$1 + \frac{1}{2^2} + \frac{1}{3^2} + \frac{1}{4} + \frac{1}{5^2} + \frac{1}{6^2} + \frac{1}{7^2} + \frac{1}{8^2} + \frac{1}{9} + \cdots. \tag{2.25}$$

Clearly, this series is less than $2\sum_{n=1}^{\infty} 1/n^2$, which is a convergent p-series. Thus, the series (2.25) converges. Notice that $na_n = 1$ for every $n = k^2$, so that $\lim_{n \to \infty} na_n \ne 0$. Moreover, the divergence of $\sum_{n=2}^{\infty} 1/(n \ln n)$ indicates that the converse of this problem is false.

2. Let a_n be a positive sequence such that $\lim_{n \to \infty} a_n = 0$. Prove that $\sum_{n=1}^{\infty} |1 - a_{n+1}/a_n|$ diverges.

Proof. We proceed by contradiction. Let $b_n = 1 - a_{n+1}/a_n$, and assume that $\sum_{n=1}^{\infty} |b_n|$ converges. Thus, $\lim_{n \to \infty} b_n = 0$. Without loss of generality, we may assume that $|b_n| < 1/2$ for all $n \in \mathbb{N}$. Then

$$\begin{aligned}
a_{n+1} &= a_1 \cdot \frac{a_2}{a_1} \cdot \frac{a_3}{a_2} \cdots \frac{a_{n+1}}{a_n} \\
&= a_1(1 - b_1)(1 - b_2) \cdots (1 - b_{n+1}).
\end{aligned}$$

Taking the logarithm of both sides yields

$$\ln a_{n+1} = \ln a_1 + \ln(1 - b_1) + \ln(1 - b_2) + \cdots + \ln(1 - b_{n+1}).$$

Since $\lim_{n \to \infty} a_n = 0$, it follows that $\lim_{n \to \infty} \ln a_n = -\infty$, and so $\sum_{n=1}^{\infty} \ln(1 - b_n)$ diverges. In view of the fact that

$$\ln(1 - x) \ge -2|x| \quad \text{for } |x| \le 1/2,$$

the comparison test implies that $\sum_{n=1}^{\infty} |b_n|$ diverges, which contradicts the assumption that $\sum_{n=1}^{\infty} |b_n|$ converges. □

3. Let $\sum_{n=1}^{\infty} a_n$ be a convergent positive series and set $r_n = \sum_{k=n}^{\infty} a_k$. Prove that $\sum_{n=1}^{\infty} a_n/r_n$ diverges.

Proof. By the definition, r_n is monotone decreasing. Hence, for all $n \in \mathbb{N}$,

$$\sum_{k=n}^{\infty} \frac{a_k}{r_k} \geq \frac{1}{r_n} \sum_{k=n}^{\infty} a_k = \frac{r_n}{r_n} = 1.$$

It follows that $\sum_{n=1}^{\infty} a_n/r_n$ diverges from the Cauchy criterion (Theorem 2.1). A more careful proof is given as follows. Notice that $r_n = a_n + r_{n+1}$. We have

$$\frac{r_{n+1}}{r_n} = 1 - \frac{a_n}{r_n}, \quad \frac{r_{n+2}}{r_n} = \left(1 - \frac{a_n}{r_n}\right)\left(1 - \frac{a_{n+1}}{r_{n+1}}\right) > 1 - \frac{a_n}{r_n} - \frac{a_{n+1}}{r_{n+1}}.$$

By induction, for all $n, p \in \mathbb{N}$, we find that

$$\frac{r_{n+p}}{r_n} > 1 - \frac{a_n}{r_n} - \frac{a_{n+1}}{r_{n+1}} - \cdots - \frac{a_{n+p}}{r_{n+p}}.$$

Since $\sum_{n=1}^{\infty} a_n$ converges, it follows that $r_{n+p} \to 0$ as $p \to \infty$. Thus, for any fixed $\epsilon \in (0, 1)$ and every $n \in \mathbb{N}$, there exists a $p \in \mathbb{N}$ such that

$$\frac{a_n}{r_n} + \frac{a_{n+1}}{r_{n+1}} + \cdots + \frac{a_{n+p}}{r_{n+p}} > 1 - \frac{r_{n+p}}{r_n} > \epsilon.$$

This implies that $\sum_{n=1}^{\infty} a_n/r_n$ diverges. $\qquad\square$

4. There exists a divergent series such that its partial sums are bounded and $\lim_{n\to\infty} a_n = 0$.

Proof. Define the sequence a_n by

$$\left\{1, -\frac{1}{2}, -\frac{1}{2}, \frac{1}{3}, \frac{1}{3}, \frac{1}{3}, -\frac{1}{4}, -\frac{1}{4}, -\frac{1}{4}, -\frac{1}{4} \cdots \right\}.$$

Let $s_n = \sum_{k=1}^{n} a_k$ be the nth partial sum. Clearly, $0 \leq s_n \leq 1$ and $\lim_{n\to\infty} a_n = 0$. But, since s_n takes on values 0 and 1 infinitely many times as $n \to \infty$, $\lim_{n\to\infty} s_n$ does not exists, so the corresponding series diverges. $\qquad\square$

5. Let $\lim_{n\to\infty} n\left(\frac{a_n}{a_{n+1}} - 1\right) = \alpha > 0$. Prove that the alternating series $\sum_{n=1}^{\infty} (-1)^{n-1} a_n$ converges.

Proof. We verify the conditions in the alternating series test. First, by the assumption, there is an integer N_1 such that

$$n\left(\frac{a_n}{a_{n+1}} - 1\right) > \frac{\alpha}{2}, \quad \text{whenever } n > N_1.$$

This implies that

$$\frac{a_n}{a_{n+1}} > 1 + \frac{\alpha}{2n} > 1.$$

Therefore, a_n is monotonically decreasing for $n > N_1$. Next, let $0 < \beta < \alpha/2$. Since

$$\lim_{n \to \infty} n \left[\left(1 + \frac{1}{n}\right)^\beta - 1 \right] = \beta < \frac{\alpha}{2},$$

it follows that, there is an integer N_2 such that

$$\left(1 + \frac{1}{n}\right)^\beta < 1 + \frac{\alpha}{2n} < \frac{a_n}{a_{n+1}},$$

and so

$$n^\beta a_n > (n+1)^\beta a_{n+1} > 0$$

whenever $n > N_2$. Thus, the sequence $n^\beta a_n$ converges by the monotone convergence theorem. Therefore, $a_n \to 0$ as $n \to \infty$. In summary, we have proved that, for $n > \max\{N_1, N_2\}$, a_n monotonically decreases to zero. Hence the given series converges by the alternating series test. □

6. Let a_n be a decreasing sequence and $\lim_{n \to 0} a_n = 0$. Prove that

$$\sum_{n=1}^{\infty} (-1)^{n+1} \frac{a_1 + a_2 + \cdots + a_n}{n}$$

converges.

Proof. Notice that

$$\frac{a_1 + a_2 + \cdots + a_n}{n} - \frac{a_1 + a_2 + \cdots + a_n + a_{n+1}}{n+1}$$
$$= \frac{a_1 + a_2 + \cdots + a_n - n a_{n+1}}{n(n+1)}$$
$$= \frac{(a_1 - a_{n+1}) + (a_2 - a_{n+1}) + \cdots + (a_n - a_{n+1})}{n(n+1)} > 0.$$

Thus, the general terms of the given series are decreasing. On the other hand, the Stolz-Cesàro theorem implies that

$$\lim_{n \to \infty} \frac{a_1 + a_2 + \cdots + a_n}{n} = \lim_{n \to \infty} a_{n+1} = 0.$$

The desired result then follows from the alternating series test. □

7. Consider some grouping of the series $\sum_{n=1}^{\infty} a_n$. Further, suppose that the sum of the absolute values of the terms in the groups tends to 0. If the grouping series converges, show that the original series converges to the same sum.

Proof. Let the grouped series be

$$\sum_{k=1}^{\infty} \left(\sum_{n=n_{k-1}+1}^{n_k} a_n \right),$$

where $0 = n_0 < n_1 < n_2 \ldots$ and $\lim_{k \to \infty} n_k = +\infty$. For any $N \in \mathbb{N}$, there exists some $K \in \mathbb{N}$ such that $n_K < N \leq n_{K+1}$. Thus,

$$\left| \sum_{n=1}^{N} a_n - \sum_{k=1}^{K} \left(\sum_{n=n_{k-1}+1}^{n_k} a_n \right) \right| = \left| \sum_{n=n_K+1}^{N} a_n \right|$$

$$\leq \sum_{n=n_K+1}^{N} |a_n| \leq \sum_{n=n_K+1}^{n_{K+1}} |a_n|.$$

By the assumption, we have

$$\lim_{K \to \infty} \sum_{n=n_K+1}^{n_{K+1}} |a_n| = 0.$$

Since the grouped series converges, letting $N \to \infty$ in the estimate above, in this case, so does K, we obtain

$$\sum_{n=1}^{\infty} a_n = \lim_{N \to \infty} \sum_{n=1}^{N} a_n = \sum_{k=1}^{\infty} \left(\sum_{n=n_{k-1}+1}^{n_k} a_n \right). \qquad \square$$

8. (**Monthly Problem 12101, 2019**). Find the least upper bound of

$$\sum_{n=1}^{\infty} \frac{\sqrt{x_{n+1}} - \sqrt{x_n}}{\sqrt{(1 + x_{n+1})(1 + x_n)}}$$

over all increasing sequences $x_1, x_2, \ldots$ of positive real numbers.

Solution. We show that the desired least upper bound is $\pi/2$. First, we reformulate the series by trigonometric substitution. Based on the structure of the summands, appealing to the trigonometric identity

$$\arctan x = \arcsin \left(\frac{x}{\sqrt{1 + x^2}} \right),$$

if $0 < a \leq b$, we introduce

$$\sin \alpha = \frac{\sqrt{a}}{\sqrt{1 + a}}, \quad \sin \beta = \frac{\sqrt{b}}{\sqrt{1 + b}}.$$

Then $\alpha = \arctan \sqrt{a}, \beta = \arctan \sqrt{b}$, and

$$\frac{\sqrt{b} - \sqrt{a}}{\sqrt{(1 + a)(1 + b)}} = \sin \beta \cos \alpha - \cos \beta \sin \alpha = \sin(\beta - \alpha).$$

Since x_n is positive and increasing, the above reformulation implies

$$\sum_{n=1}^{\infty} \frac{\sqrt{x_{n+1}} - \sqrt{x_n}}{\sqrt{(1 + x_{n+1})(1 + x_n)}} = \sum_{n=1}^{\infty} \sin(\arctan \sqrt{x_{n+1}} - \arctan \sqrt{x_n})$$

$$\leq \sum_{n=1}^{\infty} \left(\arctan \sqrt{x_{n+1}} - \arctan \sqrt{x_n} \right)$$

$$(\text{use } \sin x \leq x \text{ for } x \geq 0)$$

$$= \arctan \sqrt{\sup\{x_n\}} - \arctan \sqrt{x_1} \quad (\text{telescoping})$$

$$\leq \frac{\pi}{2} - \arctan \sqrt{x_1} < \frac{\pi}{2}.$$

Thus, $\pi/2$ is the upper bound of the series, which is not attained.

Next, we construct a positive sequence as follows: For any positive integer N, let

$$x_n = \begin{cases} \tan^2\left(\frac{n\pi}{2N}\right), & n = 1, 2, \ldots, N-1, \\ x_{N-1} + n - N + 1, & \text{for } n \geq N. \end{cases}$$

Clearly, x_n is strictly increasing, and

$$\sum_{n=1}^{\infty} \frac{\sqrt{x_{n+1}} - \sqrt{x_n}}{\sqrt{(1+x_{n+1})(1+x_n)}} > \sum_{n=1}^{N-2} \frac{\sqrt{x_{n+1}} - \sqrt{x_n}}{\sqrt{(1+x_{n+1})(1+x_n)}}$$

$$= \sum_{n=1}^{N-2} \sin\left(\frac{(n+1)\pi}{2N} - \frac{n\pi}{2N}\right)$$

$$= \sum_{n=1}^{N-2} \sin\left(\frac{\pi}{2N}\right) = (N-2)\sin\left(\frac{\pi}{2N}\right).$$

Since

$$\lim_{N \to \infty} (N-2)\sin\left(\frac{\pi}{2N}\right) = \frac{\pi}{2},$$

this proves that $\pi/2$ is the least upper bound.   □

9. (**Monthly Problem 12084, 2019**). Let $a_1, a_2, \ldots$ be a sequence of nonnegative numbers. Prove that $(1/n)\sum_{k=1}^{n} a_k$ is unbounded if and only if there exists a decreasing sequence $b_1, b_2, \ldots$ such that $\lim_{n \to \infty} b_n = 0$, $\sum_{n=1}^{\infty} b_n$ is finite, and $\sum_{n=1}^{\infty} a_n b_n$ is infinite. Is the word "decreasing" essential?

Proof. Let $A_n = \sum_{k=1}^{n} a_k$. Assume that $\frac{1}{n}\sum_{k=1}^{n} a_k$ is unbounded. Then there is a strictly increasing sequence $n_k \in \mathbb{N}$ such that $A_{n_k} \geq kn_k$. Let $n_0 = 1$. Define

$$b_i := \sum_{k=j}^{\infty} \frac{1}{n_k k^2}, \qquad \text{for } i = n_{j-1} + 1, \ldots, n_j \text{ with } j \geq 1$$

Clearly the sequence $b_1, b_2, \ldots$ is decreasing and tends to 0. Moreover,

$$\sum_{n=1}^{\infty} b_n = \sum_{j=1}^{\infty} (n_j - n_{j-1}) \sum_{k=j}^{\infty} \frac{1}{n_k k^2}$$

$$= \sum_{k=1}^{\infty} \frac{1}{n_k k^2} \sum_{j=1}^{k} (n_j - n_{j-1}) = \sum_{k=1}^{\infty} \frac{1}{k^2} = \frac{\pi^2}{6}.$$

Thus, $\sum_{n=1}^{\infty} b_n$ is finite. By the Abel's summation formula (2.14), we have

$$\sum_{n=1}^{\infty} a_n b_n \geq \sum_{n=1}^{\infty} A_n(b_n - n_{n+1}) \geq \sum_{k=1}^{\infty} kn_k \cdot \frac{1}{n_k k^2} = \sum_{k=1}^{\infty} \frac{1}{k},$$

which shows that $\sum_{n=1}^{\infty} a_n b_n$ is infinite.

Conversely, we assume that b_n is decreasing such that $\lim_{n \to \infty} b_n = 0$, $\sum_{n=1}^{\infty} b_n$ is finite, and $\sum_{n=1}^{\infty} a_n b_n$ is infinite. We proceed by contradiction. If $(1/n)\sum_{k=1}^{n} a_k$

is bounded, then there is some positive number M such that $A_n \leq Mn$ for all $n \in \mathbb{N}$. Applying Abel's summation formula (2.14) again, we have

$$\sum_{k=1}^{n} a_k b_k = \sum_{k=1}^{n-1} A_k (b_k - b_{k+1}) + A_n b_n$$

$$\leq \sum_{k=1}^{n-1} Mk(b_k - b_{k+1}) + Mnb_n = M \sum_{k=1}^{n} b_k.$$

This implies that $\sum_{n=1}^{\infty} a_n b_n$ is finite since $\sum_{n=1}^{\infty} b_n$ is finite, which contradicting our assumption.

Finally, we show that the word "decreasing" is essential by an example. For $n \in \mathbb{N}$, let

$$a_{2n} = \frac{1}{\sqrt{n}}, \ a_{2n-1} = 0; \ b_n = \frac{(-1)^n}{\sqrt{n}}.$$

Then $\sum_{n=1}^{\infty} b_n$ is finite by the alternating series test, and

$$\sum_{n=1}^{\infty} a_n b_b = \sum_{n=1}^{\infty} a_{2n} b_{2n} = \frac{1}{\sqrt{2}} \sum_{n=1}^{\infty} \frac{1}{n}$$

is unbounded because the harmonic series diverges. However, $(1/n)\sum_{k=1}^{n} a_k$ is bounded, since

$$\frac{1}{n}\sum_{k=1}^{n} a_k \leq \frac{1}{n}\sum_{k=1}^{2n} a_k = \frac{1}{n}\sum_{k=1}^{n} a_{2k} = \frac{1}{n}\sum_{k=1}^{n} \frac{1}{\sqrt{k}} \leq \frac{1}{n}\int_{0}^{n} \frac{dx}{\sqrt{x}} = \frac{2}{\sqrt{n}} \leq 2.$$

$\square$

10. (**Math Magazine Problem 2097, 2020**). For a real number $x \notin 1/2 + \mathbb{Z}$, denote the nearest integer to x by $\langle x \rangle$. For any real number x, denote the largest smaller than or equal to x and the smallest integer larger than or equal to x by $\lfloor x \rfloor$ and $\lceil x \rceil$, respectively. For a positive integer n, let

$$a_n = \frac{2}{\langle \sqrt{n} \rangle} - \frac{1}{\lfloor \sqrt{n} \rfloor} - \frac{1}{\lceil \sqrt{n} \rceil}.$$

(a) Prove that the series $\sum_{n=1}^{\infty} a_n$ is convergent and find its sum L.

(b) Prove that the set

$$\left\{ \sqrt{n} \left(\sum_{k=1}^{n} a_k - L \right) : n \geq 1 \right\}$$

is dense in $[0, 1]$.

Proof. (a) We show that the sum converges to zero. To see this, first, we can easily check the following facts:

$$\langle \sqrt{n} \rangle = k, \quad \text{for } n \in [k(k-1)+1, k(k+1)],$$
$$\lfloor \sqrt{n} \rfloor = k, \quad \text{for } n \in [k^2, (k+1)^2),$$
$$\lceil \sqrt{n} \rceil = k+1, \quad \text{for } n \in (k^2, (k+1)^2].$$

These imply that $a_{k^2} = 0$ and

$$a_n = \frac{2}{k} - \frac{1}{k} - \frac{1}{k+1} = \frac{1}{k(k+1)}, \qquad \text{for } n \in (k^2, k(k+1)],$$

$$a_n = \frac{2}{k+1} - \frac{1}{k} - \frac{1}{k+1} = -\frac{1}{k(k+1)}, \qquad \text{for } n \in (k(k+1), (k+1)^2).$$

Therefore, for $k^2 \le n \le (k+1)^2$, we have $\sum_{m=1}^{k^2} a_m = 0$ and

$$0 \le \sum_{m=1}^{n} a_m \le \frac{1}{k(k+1)} \cdot [k(k+1) - k^2] = \frac{1}{k+1}.$$

As $n \to \infty$, we have $k \to \infty$ and so

$$\sum_{n=1}^{\infty} a_n = \lim_{n \to \infty} \sum_{m=1}^{n} a_m = 0.$$

(b) Let $x \in [0,1]$. We show that there exists a subsequence from the set $\{\sqrt{n} \sum_{m=1}^{n} a_m\}$ that converges to x. Notice that there exist two integer sequences p_k and q_k with $0 \le p_k \le q_k$ such that $p_k/q_k \to x$ as $k \to \infty$. Let $n_k = q_k^2 + p_k$. Then

$$q_k^2 \le n_k \le q_k^2 + q_k < \left(q_k + \frac{1}{2}\right)^2.$$

This implies that

$$\langle \sqrt{n_k} \rangle = q_k, \ \lfloor \sqrt{n_k} \rfloor = q_k, \ \lceil \sqrt{n_k} \rceil = q_k + 1.$$

Therefore, as $k \to \infty$, we have

$$\sqrt{n_k} \sum_{m=1}^{n_k} a_m = \sqrt{n_k} \cdot \frac{n_k - q_k^2}{q_k(q_k + 1)} = \frac{p_k}{q_k} \cdot \frac{\sqrt{n_k}}{q_k + 1} \to x.$$

This proves that the set $\{\sqrt{n} \sum_{m=1}^{n} a_m\}$ is dense in $[0,1]$. $\qquad \square$

11. **(Monthly Problem 11829, 2015).** Let a_n be a monotone decreasing sequence of real numbers that converges to 0. Prove that $\sum_{n=1}^{\infty} a_n/n < \infty$ if and only if $a_n = O(1/\ln n)$ and $\sum_{n=1}^{\infty} (a_n - a_{n+1}) \ln n < \infty$.

Proof. Let

$$S_n = \sum_{k=1}^{n} \frac{a_k}{k}, \quad T_n = \sum_{k=1}^{n} (a_k - a_{k+1}) \ln k.$$

In view of the assumption of a_n, S_n and T_n both are monotone increasing. Let their limits are S and T, respectively.

" $\Longrightarrow$ " Assume that S is finite. For $k \ge 2$, we have

$$\ln k - \ln(k-1) = \int_{k-1}^{k} \frac{dx}{x} \le \frac{1}{k-1}.$$

Thus,

$$a_n \ln n \le a_n \sum_{k=2}^{n} (\ln k - \ln(k-1)) \le \sum_{k=2}^{n} \frac{a_n}{k-1} \le \sum_{k=2}^{n} \frac{a_{k-1}}{k-1} = S_{n-1} \le S,$$

and so $a_n = O(1/\ln n)$. Next, we have

$$
\begin{aligned}
T_n &= \sum_{k=1}^{n} a_k \ln k - \sum_{k=1}^{n} a_{k+1} \ln k \\
&= \sum_{k=2}^{n} a_k (\ln k - \ln(k-1)) - a_{n+1} \ln n \\
&\leq \sum_{k=2}^{n} \frac{a_k}{k-1} \leq \sum_{k=2}^{n} \frac{a_{k-1}}{k-1} \\
&= S_{n-1} \leq S.
\end{aligned}
$$

This implies that T is finite.

" $\Longleftarrow$ " Assume that $a_n \ln n < M$ for some positive constant M and $T < \infty$. Since

$$
\begin{aligned}
S_n - a_1 &= \sum_{k=2}^{n} \frac{a_k}{k} \leq \sum_{k=2}^{n} a_k(\ln k - \ln(k-1)) \\
&= \sum_{k=2}^{n} a_k \ln k - \sum_{k=2}^{n-1} a_{k+1} \ln k \\
&= \sum_{k=2}^{n-1} (a_k - a_{k+1}) \ln k + a_n \ln n \\
&\leq T + M,
\end{aligned}
$$

it follows that S is finite as desired. $\qquad\square$

A similar problem appears in the 2015 Monthly November issue. **Problem 11865** states that if a_n is a decreasing sequence of nonnegative real numbers, then $\sum_{n=1}^{\infty} a_n/n$ is finite if and only if $\lim_{n \to \infty} a_n = 0$ and $\sum_{n=1}^{\infty} (a_n - a_{n+1}) \ln n < \infty$. In this case, we show that $\lim_{n \to \infty} a_n = 0$, together with $T < \infty$, implies that $a_n \ln n \leq T$. Indeed, for $m > n \geq 1$, we have

$$
(a_n - a_{m+1}) \ln n = \sum_{k=n}^{m} (a_k - a_{k+1}) \ln n \leq \sum_{k=n}^{m} (a_k - a_{k+1}) \ln k \leq T.
$$

In view of that $\lim_{n \to \infty} a_n = 0$, letting $m \to \infty$ yields

$$
a_n \ln n \leq T \quad \text{for all } n \geq 1.
$$

12. For $n \geq 2$, show that the integer nearest to $n!/e$ is divisible by $n-1$ but not by n.

Proof. Since

$$
e^{-1} = \sum_{k=0}^{\infty} \frac{(-1)^k}{k!},
$$

it follows that

$$
\frac{n!}{e} = n! \sum_{k=0}^{n} \frac{(-1)^k}{k!} + n! \sum_{k=n+1}^{\infty} \frac{(-1)^k}{k!}.
$$

Clearly, the first term in the sum is an integer. The second term is bounded by

$$\left| n! \sum_{k=n+1}^{\infty} \frac{(-1)^k}{k!} \right| \le n! \cdot \frac{1}{(n+1)!} = \frac{1}{n+1} \le \frac{1}{3}.$$

Therefore, the integer nearest to $n!/e$ is

$$N := n! \sum_{k=0}^{n} \frac{(-1)^k}{k!}.$$

Rewrite N as

$$N = n \left[(n-1)! \sum_{k=0}^{n-1} \frac{(-1)^k}{k!} \right] + (-1)^n.$$

This indicates that N is not divisible by n. On the other hand, since

$$\begin{aligned}
N &= n(n-1) \left[(n-2)! \sum_{k=0}^{n-2} \frac{(-1)^k}{k!} \right] + (-1)^{n-1} n + (-1)^n \\
&= (n-1) \left\{ n \left[(n-2)! \sum_{k=0}^{n-2} \frac{(-1)^k}{k!} \right] + (-1)^{n-1} \right\},
\end{aligned}$$

the number inside the bracket is an integer. Hence N is divisible by $n-1$. $\square$

13. Let $a_n > 0$ for all $n \in \mathbb{N}$. If $\sum_{n=1}^{\infty} a_n$ converges, prove that $\sum_{n=1}^{\infty} a_n^{(n-1)/n}$ converges as well.

Proof. Applying the AM-GM inequality yields

$$a_n^{(n-1)/n} = (a_n^{1/2} \cdot a_n^{1/2} \cdot a_n^{n-2})^{1/n} \le \frac{2a_n^{1/2} + (n-2)a_n}{n}.$$

Moreover, we have

$$\frac{2a_n^{1/2}}{n} \le a_n + \frac{1}{n^2} \quad (\text{using } 2ab \le a^2 + b^2)$$

and $(n-2)a_n/n \le a_n$. Therefore,

$$a_n^{(n-1)/n} \le 2a_n + \frac{1}{n^2}.$$

By the assumption that $\sum_{n=1}^{\infty} a_n$ converges, and $\sum_{n=1}^{\infty} 1/n^2$ converges, the comparison test implies that $\sum_{n=1}^{\infty} a_n^{(n-1)/n}$ converges. $\square$

An alternative solution is based on the following partition:

$$\mathbb{N}_1 = \{n \in \mathbb{N} : a_n \le 1/2^n\}, \quad \mathbb{N}_2 = \mathbb{N} \setminus \mathbb{N}_1.$$

If $n \in \mathbb{N}_1$, we have

$$a_n^{(n-1)/n} \le \left(\frac{1}{2^n} \right)^{(n-1)/n} = \frac{1}{2^{n-1}}.$$

If $n \in \mathbb{N}_2$,

$$a_n^{(n-1)/n} = \frac{a_n}{a_n^{1/n}} \le 2a_n.$$

Therefore,

$$\sum_{n=1}^{\infty} a_n^{(n-1)/n} \le \sum_{n \in \mathbb{N}_1} \frac{1}{2^{n-1}} + \sum_{n \in \mathbb{N}_2} 2a_n \le 2 + 2 \sum_{n=1}^{\infty} a_n < \infty.$$

Under the same assumption, this approach can be used to show that $\sum_{n=1}^{\infty} a_n^{1-b_n}$ converges as well provided $b_n = O(1/\ln n)$.

14. Let H_n be the nth harmonic number, and let n_p be the smallest integer n satisfying $H_n \ge p$ (for example, $n_1 = 1, n_2 = 4, n_3 = 11$). Prove that $\sum_{p=1}^{\infty} 1/n_p$ converges.

    Since H_n grows like $\ln n$, intuitively, we would expect n_p, which is in a sense an "approximate inverse" to H_n, to grow like e^p. We confirm this as follows.

    Proof. By the assumption, we have

    $$H_{n_p} - \frac{1}{n_p} = H_{n_p - 1} < p \le H_{n_p}.$$

    Thus, $H_{n_p} = p - k_p$, where k_p is a nonnegative constant bounded above by $1/n_p$. In particular, $k_p \to 0$ as $p \to \infty$. Recalling that $H_n = \ln n + 1/n + \gamma_n$, where γ_n converges to Euler-Mascheroni constant γ, we have

    $$p = H_{n_p} + k_p = \ln n_p + \frac{1}{n_p} + \gamma_{n_p} + k_p,$$

    and so

    $$\lim_{p \to \infty} (p - \ln n_p) = \lim_{p \to \infty} \left(\frac{1}{n_p} + \gamma_{n_p} + k_p \right) = \gamma.$$

    Similarly, we also have

    $$\lim_{p \to \infty} (p + 1 - \ln n_{p+1}) = \gamma.$$

    The difference yields

    $$\lim_{p \to \infty} \left(1 - \ln \frac{n_{p+1}}{n_p} \right) = 0.$$

    This is equivalent to

    $$\lim_{p \to \infty} \frac{n_{p+1}}{n_p} = e,$$

    from which it follows the given series converges by the ratio test. □

15. Does there exist a bijective map $\sigma : \mathbb{N} \to \mathbb{N}$ such that

    $$\sum_{n=1}^{\infty} \frac{\sigma(n)}{n^2} < \infty ?$$

Solution. We show that the series $\sum_{n=1}^{\infty} \sigma(n)/n^2$ diverges for any bijection from $\mathbb{N}$ to $\mathbb{N}$. To see this, since σ is a permutation of $\mathbb{N}$, it follows that

$$\sigma(1) + \sigma(2) + \cdots + \sigma(n) \geq 1 + 2 + \cdots + n = \frac{n(n+1)}{2}.$$

Thus, by Abel's summation formula (2.14), we have

$$
\begin{aligned}
\sum_{n=1}^{N} \frac{\sigma(n)}{n^2} &= \sum_{n=1}^{N-1} (\sigma(1) + \cdots + \sigma(n)) \left(\frac{1}{n^2} - \frac{1}{(n+1)^2} \right) + \frac{1}{N^2} \sum_{n=1}^{N} \sigma(n) \\
&\geq \sum_{n=1}^{N-1} \frac{n(n-1)}{2} \cdot \frac{2n+1}{n^2(n+1)^2} + \frac{N+1}{2N} \\
&= \sum_{n=1}^{N-1} \frac{(n-1)(2n+1)}{n(n+1)^2} + \frac{N+1}{2N} \\
&\geq \sum_{n=1}^{N-1} \frac{(n-1)}{n(n+1)} + \frac{1}{2}.
\end{aligned}
$$

$\square$

An alternative proof is based on the Cauchy criterion. Observe that, of $\{\sigma(N+1), \sigma(N+2), \cdots, \sigma(3N)\}$, only N of them are possibly at most N, the rest of them must be strictly greater than N. This yields the estimate

$$\sum_{n=N+1}^{3N} \frac{\sigma(n)}{n^2} \geq \frac{1}{(3N)^2} \sum_{n=N+1}^{3N} \sigma(n) > \frac{1}{9N^2} \cdot N \cdot N = \frac{1}{9}.$$

It implies that $\sum_{n=1}^{\infty} \sigma(n)/n^2$ diverges by the Cauchy criterion.

16. (**Monthly Problem 11954, 2017**). Determine the largest constant c and the smallest constant d such that, for all positive integers n,

$$\frac{1}{n-c} \leq \sum_{k=n}^{\infty} \frac{1}{k^2} \leq \frac{1}{n-d}.$$

Solution. First, we rewrite the proposed inequality equivalently as

$$c \leq n - \frac{1}{\sum_{k=n}^{\infty} \frac{1}{k^2}} = n - \frac{1}{\sum_{k=0}^{\infty} \frac{1}{(n+k)^2}} \leq d \quad \text{for all } n \in \mathbb{N}.$$

Let

$$\psi_1(x) = \sum_{k=0}^{\infty} \frac{1}{(k+x)^2}.$$

We next show that

$$F(x) := x - \frac{1}{\psi_1(x)}$$

to be strictly increasing on $(1, \infty)$. Upon showing that $F(x)$ is increasing, an immediate consequence is

$$c = \inf_{n \in \mathbb{N}} F(n) = F(1) = 1 - \frac{1}{\zeta(2)} = 1 - \frac{6}{\pi^2}$$

$$d = \sup_{n \in \mathbb{N}} F(n) = \lim_{n \to \infty} F(n) = \frac{1}{2}.$$

To prove that $F(x)$ is strictly increasing, we show that $F'(x) > 0$ for $x > 1$. Indeed, since

$$F'(x) = 1 + \frac{\psi_1'(x)}{\psi_1^2(x)} = \frac{\psi_1^2(x) + \psi_1'(x)}{\psi_1^2(x)},$$

it suffices to show that $\psi_1^2(x) + \psi_1'(x) > 0$ for $x > 1$. Let

$$G(x) := \psi_1^2(x) + \psi_1'(x) = \left(\sum_{k=0}^{\infty} \frac{1}{(k+x)^2} \right)^2 - 2 \sum_{k=0}^{\infty} \frac{1}{(k+x)^3}.$$

Since $\lim_{x \to \infty} G(x) = 0$, it remains to show that $G(x+1) - G(x) < 0$ for $x > 1$. To this end, using $\psi_1(x+1) = \psi_1(x) - 1/x^2$, we find

$$G(x+1) - G(x) = \left(\psi_1(x) - \frac{1}{x^2} \right)^2 + \left(\psi_1'(x) + \frac{2}{x^3} \right) - \psi_1^2(x) - \psi_1'(x)$$

$$= -\frac{2\psi_1(x)}{x^2} + \frac{1}{x^4} + \frac{2}{x^3} = -\frac{2}{x^2} \left(\psi_1(x) - \frac{1}{x} - \frac{1}{2x^2} \right).$$

Let

$$H(x) := \psi_1(x) - \frac{1}{x} - \frac{1}{2x^2}.$$

Then $\lim_{x \to \infty} H(x) = 0$. Moreover,

$$H(x+1) - H(x) = \left(\psi_1(x) - \frac{1}{x^2} \right) - \psi_1(x) - \frac{2x+3}{2(x+1)^2} + \frac{2x+1}{2x^2}$$

$$= -\frac{1}{2x^2(x+1)^2} < 0.$$

- This implies that $G(x+1) - G(x) < 0$ for $x > 1$, and so $F'(x) > 0$ for $x > 1$. Therefore, $F(x)$ is strictly increasing for $x > 1$. $\qquad\square$

Remark. As a byproduct, the above proof actually establishes the inequality

$$\sum_{k=0}^{\infty} \frac{2}{(k+x)^3} < \left(\sum_{k=0}^{\infty} \frac{1}{(k+x)^2} \right)^2 \qquad \text{for all } x \ge 1.$$

The function $\psi_1(x)$ used in the above proof is often called the *trigamma function*, which is the second of the *polygamma functions*:

$$\psi_1(x) = \frac{d}{dx} \psi(x) = \frac{d^2}{dx^2} \ln \Gamma(x),$$

where $\psi(x)$ is the *digamma function* and $\Gamma(x)$ is the *gamma function* (see Chapter 7). Taking the derivative of the *asymptotic expansion of the digamma function*, we have

$$\psi_1(x) \sim \frac{1}{x} + \frac{1}{2x^2} + \frac{1}{6x^3} - \frac{1}{30x^5} + \frac{1}{42x^7} - \frac{1}{30x^9} + \cdots .$$

The Riemann series theorem challenges our conceptions in an interesting way. Seeing what can go wrong is often an indispensable way to gain insight and intuition into the foundations of analysis. The following problem shows that, for a positive divergent series, there exists a subseries that converges to any positive number as you wish. The proof is similar to the proof of the Riemann series theorem.

17. Let a_n be a positive sequence such that $\lim_{n \to \infty} a_n = 0$. If the series $\sum_{n=1}^{\infty} a_n$ diverges and $a > 0$, prove that there is a subseries of $\sum_{n=1}^{\infty} a_n$ that converges to a.

Proof. Since $\lim_{n \to \infty} a_n = 0$, there is the least integer n_1 such that $a_n < a/2$ for all $n \geq n_1$. Let m_1 be the greatest integer for which

$$s_1 = a_{n_1} + a_{n_1+1} + \cdots + a_{m_1} < a.$$

Then $a/2 < s_1 < a$ since $s_1 \leq a/2$ will against to the selection of m_1. Let n_2 be the least integer such that $n_2 > m_1$ and $a_n < (a - s_1)/2$ whenever $n \geq n_2$. Let m_2 be the greatest integer for which

$$s_2 = s_1 + a_{n_2} + a_{n_2+1} + \cdots + a_{m_2} < a.$$

Then $a - a/2^2 < s_2 < a$. Repeating this process yields the sequences n_k and m_k such that

$$n_1 < m_1 < n_2 < m_2 < n_3 < m_3 < \cdots,$$

$$s_k = s_{k-1} + a_{n_k} + a_{n_k+1} + \cdots + a_{m_k},$$

and $a - a/2^k < s_k < a$. By the squeeze theorem, the following subseries

$$a_{n_1} + a_{n_1+1} + \cdots + a_{m_1} + a_{n_2} + a_{n_2+1} + \cdots + a_{m_2} + \cdots + a_{n_k} + a_{n_k+1} + \cdots + a_{m_k} + \cdots$$

converges to a as desired.   □

Let $a_n = 1/n$ for all $n \in \mathbb{N}$. In particular, for each $N \in \mathbb{N}$, the harmonic series has a subseries that converges to N. On the other hand, let $s_k = \sum_{n=1}^{k} 1/n$. It is well-known that s_k is never an integer for $k \geq 2$. However, surprisingly, we have

18. Prove that every positive rational number is the finite sum of a subseries of the harmonic series.

Proof. Without loss of generality, we assume that $p/q > 1$. Let n_1 be uniquely determined by

$$s_1 = 1 + \frac{1}{2} + \cdots + \frac{1}{n_1} \leq \frac{p}{q} < 1 + \frac{1}{2} + \cdots + \frac{1}{n_1} + \frac{1}{n_1+1}.$$

Let $p_1/q_1 = p/q - s_1$ and assume that $p_1/q_1 > 0$. Let n_2 be uniquely determined by $n_2 \leq q_1/p_1 < n_2 + 1$ and set

$$\frac{p_2}{q_2} = \frac{p_1}{q_1} - \frac{1}{n_2+1}.$$

Let n_3 be uniquely determined by $n_3 \leq q_2/p_2 < n_3 + 1$ and set

$$\frac{p_3}{q_3} = \frac{p_2}{q_2} - \frac{1}{n_3+1}$$

and so on. Since $p_1 > p_2 > p_3 > \cdots$, after finitely many steps we must have $p_k = 1$, that is

$$\frac{1}{q_k} = \frac{p_k}{q_k} = \frac{p_{k-1}}{q_{k-1}} - \frac{1}{n_k+1}.$$

Therefore,

$$\frac{p}{q} = s_1 + \frac{1}{n_2+1} + \frac{1}{n_3+1} + \cdots + \frac{1}{n_k+1}.$$

                                                                                                □

The process above can be done with an example: $p/q = 9/5$, we have $n_1 = 2, n_2 = 3, n_3 = 20$, and

$$\frac{9}{5} = 1 + \frac{1}{2} + \frac{1}{4} + \frac{1}{20}.$$

These unit fractions are often called *Egyptian fractions*. They continue to be an object of study in modern number theory and recreational mathematics, as well as in modern historical studies of ancient mathematics.

19. Suppose a_n and b_n are decreasing sequences of positive numbers with limits L and M, respectively. Prove that

$$\sum_{n=1}^{\infty} (a_n - a_{n+1})(b_n - b_{n+1}) \le (a_1 - L)(b_1 - M).$$

Proof. Telescoping yields

$$\sum_{k=1}^{n} (a_k - a_{k+1}) = a_1 - a_{n+1} \to a_1 - L, \quad (\text{as } n \to \infty).$$

So the series $\sum_{n=1}^{\infty} (a_n - a_{n+1})$ converges. By the assumption, b_n is a monotone and bounded sequence. Thus, by Abel's test, both $\sum_{n=1}^{\infty} (a_n - a_{n+1})b_n$ and $\sum_{n=1}^{\infty} (a_n - a_{n+1})b_{n+1}$ converge. Let

$$s_k = \sum_{i=1}^{k} (a_i - a_{i+1}) = a_1 - a_{k+1}.$$

Then $0 \le s_k \le a_1 - L$ for all $k \in \mathbb{N}$. Moreover, in view of Abel's summation formula (2.14), we have

$$\sum_{k=1}^{n} (a_k - a_{k+1})b_k = \sum_{k=1}^{n-1} s_k(b_k - b_{k+1}) + s_n b_n$$

$$\le \sum_{k=1}^{n-1} (a_1 - L)(b_k - b_{k+1}) + (a_1 - L)b_n$$

$$= (a_1 - L)b_1.$$

Similarly, we have

$$\sum_{k=1}^{n} (a_k - a_{k+1})b_{k+1} = \sum_{k=1}^{n-1} s_k(b_{k+1} - b_{k+2}) + s_n b_{n+1}$$

$$\ge s_n b_{n+1} = (a_1 - L)b_{n+1} + (L - a_{n+1})b_{n+1}.$$

Letting $n \to \infty$ yields

$$\sum_{k=1}^{\infty} (a_k - a_{k+1})b_k \le (a_1 - L)b_1, \quad \sum_{k=1}^{\infty} (a_k - a_{k+1})b_{k+1} \ge (a_1 - L)M.$$

Their difference gives

$$\sum_{k=1}^{\infty} (a_k - a_{k+1})(b_k - b_{k+1}) \le (a_1 - L)b_1 - (a_1 - L)M = (a_1 - L)(b_1 - M).$$

This proves the inequality as claimed. $\qquad\square$

We next provide one example involving rearrangements of the alternating harmonic series.

20. Let

$$\sum_{n=1}^{\infty} a_n = 1 + \frac{1}{3} - \frac{1}{2} + \frac{1}{5} + \frac{1}{7} - \frac{1}{4} + \frac{1}{9} + \frac{1}{11} - \frac{1}{6} + + - \cdots$$

Show the series converges and find the value to which this series converges.

Proof. Notice that

$$a_{3k} = -\frac{1}{2k}, \quad a_{3k-1} = \frac{1}{4k-1}, \quad a_{3k-2} = \frac{1}{4k-3}.$$

Thus,

$$a_{3k-2} + a_{3k-1} + a_{3k} = \frac{8k-3}{(4k-3)(4k-1)(2k)},$$

and so

$$\begin{aligned}
S_{3n} &= \sum_{k=1}^{3n} a_k = \sum_{k=1}^{n} (a_{3k-2} + a_{3k-1} + a_{3k}) \\
&= \sum_{k=1}^{n} \frac{8k-3}{(4k-3)(4k-1)(2k)}.
\end{aligned}$$

By comparison with the series $\sum_{k=1}^{\infty} 1/k^2$, we see that S_{3n} converges. Therefore the original series converges as well by Theorem 2.23.

To find the value to which this series converges, observe that

$$\begin{aligned}
S_{3n} &= \left(1 + \frac{1}{3} - \frac{1}{2}\right) + \left(\frac{1}{5} + \frac{1}{7} - \frac{1}{4}\right) + \left(\frac{1}{9} + \frac{1}{11} - \frac{1}{6}\right) + \cdots \\
&= \left(1 - \frac{1}{2} + \frac{1}{2} + \frac{1}{3} - \frac{1}{4} - \frac{1}{4}\right) + \left(\frac{1}{5} - \frac{1}{6} + \frac{1}{6} + \frac{1}{7} - \frac{1}{8} - \frac{1}{8}\right) + \cdots \\
&= \left(1 - \frac{1}{2} + \frac{1}{3} - \frac{1}{4}\right) + \left(\frac{1}{2} - \frac{1}{4}\right) + \left(\frac{1}{5} - \frac{1}{6} + \frac{1}{7} - \frac{1}{8}\right) + \left(\frac{1}{6} - \frac{1}{8}\right) + \cdots \\
&= \left(1 - \frac{1}{2} + \frac{1}{3} - \frac{1}{4} + \frac{1}{5} - \frac{1}{6} + \cdots - \frac{1}{4n}\right) + \left(\frac{1}{2} - \frac{1}{4} + \frac{1}{6} - \frac{1}{8} + \cdots - \frac{1}{4n}\right).
\end{aligned}$$

In view of the well-known result $\sum_{n=1}^{\infty} (-1)^{n+1}/n = \ln 2$, we find the series converges to $\frac{3}{2} \ln 2$. $\qquad\square$

The following two problems illustrate how to use definite integrals to evaluate the series in closed form.

21. (**Monthly Problem 11400, 2008**). Let ζ be the Riemann zeta function. Evaluate

$$\sum_{n=1}^{\infty} \frac{\zeta(2n)}{n(n+1)}.$$

Solution. For $0 \le x < 1$, we have

$$\sum_{n=1}^{\infty} \frac{\zeta(2n)}{n} x^n = \sum_{n=1}^{\infty} \frac{1}{n} \sum_{k=1}^{\infty} \left(\frac{\sqrt{x}}{k}\right)^{2n} = \sum_{k=1}^{\infty} \left(\sum_{n=1}^{\infty} \frac{1}{n} \left(\frac{\sqrt{x}}{k}\right)^{2n}\right)$$

$$= \sum_{k=1}^{\infty} -\ln\left(1 - \left(\frac{\sqrt{x}}{k}\right)^2\right) = -\ln\left(\prod_{k=1}^{\infty}\left(1 - \left(\frac{\sqrt{x}}{k}\right)^2\right)\right)$$

$$= -\ln\left(\frac{\sin(\pi\sqrt{x})}{\pi\sqrt{x}}\right), \tag{2.26}$$

where the power series of $\ln(1-x)$ and the infinite product of $\sin x / x$ have been used. Since all terms are positive, reversing the order of summation and integration gives

$$\sum_{n=1}^{\infty} \frac{\zeta(2n)}{n(n+1)} = \int_0^1 \sum_{n=1}^{\infty} \frac{\zeta(2n)}{n} x^n = -\int_0^1 \ln\left(\frac{\sin(\pi\sqrt{x})}{\pi\sqrt{x}}\right) dx$$

$$= -\int_0^1 \ln(\sin(\pi\sqrt{x}) dx + \ln\pi - \frac{1}{2}$$

$$= -\frac{2}{\pi^2} \int_0^{\pi} t\,\ln(\sin t)\,dt + \ln\pi - \frac{1}{2}$$

$$= \ln 2 + \ln\pi - \frac{1}{2} = \ln(2\pi) - \frac{1}{2},$$

where we have used

$$\int_0^{\pi} t\,\ln(\sin t)\,dt = -\frac{1}{2}\pi^2 \ln 2. \tag{2.27}$$

To see this, by symmetry, we have

$$\int_0^{\pi} t\,\ln(\sin t)\,dt = \int_0^{\pi} (\pi - t)\,\ln(\sin(\pi - t))\,dt,$$

so

$$\int_0^{\pi} t\,\ln(\sin t)\,dt = \frac{\pi}{2} \int_0^{\pi} \ln(\sin t)\,dt.$$

Appealing to

$$\int_0^{\pi} \ln(\sin t)\,dt = \int_0^{\pi} \ln[2\sin(t/2)\cos(t/2)]\,dt$$

$$= \pi\ln 2 + 2\int_0^{\pi/2} \ln(\sin t)\,dt + 2\int_0^{\pi/2} \ln(\cos t)\,dt$$

$$= \pi\ln 2 + 2\int_0^{\pi} \ln(\sin t)\,dt,$$

we find that

$$\int_0^{\pi} \ln(\sin t)\,dt = -\pi\ln 2,$$

which leads to (2.27). $\qquad\square$

Replacing x by x^2 in (2.26) yields

$$\sum_{n=1}^{\infty} \frac{\zeta(2n)}{n} x^{2n} = -\ln\left(\frac{\sin(\pi x)}{\pi x}\right).$$

Integrating this with respect to x over $[0, 1]$ gives

$$\sum_{n=1}^{\infty} \frac{\zeta(2n)}{n(2n+1)} = \ln(2\pi) - 1.$$

This, together with the result in Problem 21, leads to the following rational sum

$$\sum_{n=1}^{\infty} \frac{\zeta(2n)}{(n+1)(2n+1)} = \frac{1}{2}.$$

22. Let a_n be the sequence defined by

$$a_0 = 1, \quad a_{n+1} = \frac{1}{n+1} \sum_{k=0}^{n} \frac{a_k}{n-k+2}.$$

Find $\sum_{n=0}^{\infty} a_n/2^n$.

Solution. Let $f(x) = \sum_{n=0}^{\infty} a_n x^n$. By induction, we see that the series converges absolutely for $|x| < 1$. In particular, $f(0) = 1$ and $f(x) > 0$ for $x \in [0, 1)$. By the recurrence formula, we have

$$f'(x) = \sum_{n=0}^{\infty} n a_n x^{n-1} = \sum_{n=0}^{\infty} (n+1) a_{n+1} x^n = \sum_{n=0}^{\infty} \sum_{k=0}^{n} \frac{a_k}{n-k+2} x^n$$

$$= \sum_{k=0}^{\infty} a_k x^k \sum_{n=k}^{\infty} \frac{x^{n-k}}{n-k+2} = f(x) \sum_{i=0}^{\infty} \frac{x^i}{i+2}.$$

Reversing the order of summation is justified because all terms are positive. Hence

$$\ln f(x) = \int_0^x \frac{df}{f} = \sum_{i=0}^{\infty} \frac{x^{i+1}}{(i+1)(i+2)}.$$

Using that

$$\frac{1}{(i+1)(i+2)} = \frac{1}{i+1} - \frac{1}{i+2},$$

we find that

$$\sum_{i=0}^{\infty} \frac{x^{i+1}}{(i+1)(i+2)} = \sum_{i=0}^{\infty} \left(\frac{1}{i+1} - \frac{1}{i+2} \right) x^{i+1} = 1 - \left(1 - \frac{1}{x} \right) \ln(1-x).$$

Therefore,

$$\sum_{n=0}^{\infty} a_n/2^n = f(1/2) = e^{\ln f(1/2)} = e^{1-\ln 2} = \frac{e}{2}.$$

$\square$

23. **(Monthly Problem 11409, 2009).** For positive real α and β, let

$$S(\alpha, \beta, N) = \sum_{n=2}^{N} n \ln n (-1)^n \prod_{k=2}^{n} \frac{\alpha + k \ln k}{\beta + (k+1) \ln(k+1)}.$$

Show that if $\beta > \alpha$, then $\lim_{N \to \infty} S(\alpha, \beta, N)$ exists.

Proof. Let $c_k = k \ln k, b_n = (\alpha + c_2)c_n/(\beta + c_{n+1})$ and

$$a_n = c_k \prod_{k=2}^{n} \frac{\alpha + c_k}{\beta + c_{k+1}} = b_n \prod_{k=3}^{n} \left(1 - \frac{\beta - \alpha}{\beta + c_k}\right). \tag{2.28}$$

We now prove that $\sum_{n=2}^{\infty} (-1)^n a_n$ converges, so $\lim_{N \to \infty} S(\alpha, \beta, N)$ exists. By the alternating series test, and noting $a_n > 0$, it suffices to prove

(i). $a_{n+1}/a_n < 1$ for all sufficiently large n, and

(ii). $a_n \to 0$ as $n \to \infty$.

(i). From the definition of a_n in (2.28),

$$\frac{a_{n+1}}{a_n} = \frac{c_{n+1}(\alpha + c_{n+1})}{c_n(\beta + c_{n+2})},$$

so $a_{n+1}/a_n < 1$ is equivalent to $c_{n+1}\alpha + (c_{n+1}^2 - c_n c_{n+2}) < c_n\beta$. Calculation shows $c_n^2 - c_n c_{n+2} = \ln^2 n + \ln n + 1 + o(1)$. Because $\beta > \alpha$ and $c_{n+1} \sim c_n = n \ln n$, the required result follows.

(ii). Because $\lim_{N \to \infty} b_n$ exists, to show $\lim_{N \to \infty} a_n = 0$ it suffices to show that the infinite product

$$\prod_{k=3}^{n} \left(1 - \frac{\beta - \alpha}{\beta + c_k}\right) \tag{2.29}$$

diverges to zero. Recall Theorem 2.31 in present case, the divergence of

$$\sum_{k=3}^{\infty} \frac{1}{c_k} = \sum_{k=3}^{\infty} \frac{1}{k \ln k}$$

shows that the infinite product in (2.29) diverges to 0. This finishes the proof. $\square$

We now end this section with two examples to evaluate series by using Able's summation formula (limit version).

24. (**Monthly Problem 12241, 2021**). Prove

$$\sum_{n=1}^{\infty} (-1)^n n \left(\frac{1}{4n} - \ln 2 + \sum_{k=n+1}^{2n} \frac{1}{k}\right) = \frac{\ln 2 - 1}{8}.$$

Proof. Let the proposed series be S. For $n \geq 1$, let

$$a_n = (-1)^n n, \quad b_n = \frac{1}{4n} - \ln 2 + \sum_{k=n+1}^{2n} \frac{1}{k},$$

and $A_n = \sum_{k=1}^{n} a_k$. Then

$$A_{2n} = \sum_{k=1}^{2n} a_k = -1 + 2 - 3 + 4 + \cdots - (2n-1) + 2n = n;$$

$$A_{2n-1} = \sum_{k=1}^{2n-1} a_k = A_{2(n-1)} + a_{2n-1} = (n-1) - (2n-1) = -n,$$

and

$$b_n - b_{n+1} = \frac{1}{4n} - \frac{1}{4(n+1)} + \frac{1}{n+1} - \frac{1}{2n+1} - \frac{1}{2(n+1)}$$

$$= \frac{1}{4n(n+1)} - \frac{1}{2n+1} + \frac{1}{2(n+1)}$$

$$= \frac{1}{4n(n+1)(2n+1)}.$$

Applying Abel's summation formula:

$$\sum_{n=1}^{\infty} a_n b_n = \lim_{n\to\infty} A_n b_{n+1} + \sum_{n=1}^{\infty} A_n(B_n - b_{n+1}),$$

we have

$$S = \lim_{n\to\infty} A_n b_{n+1} + \sum_{n=1}^{\infty} \frac{A_n}{4n(n+1)(2n+1)}.$$

Let $H_n = \sum_{k=1}^{n} 1/k$ be the nth harmonic number. Recall that

$$H_n = \ln n + \gamma + \frac{1}{2n} - \frac{1}{12n^2} + O(1/n^4),$$

where γ is Euler-Mascheroni constant. We have

$$b_n = \frac{1}{4n} - \ln 2 + H_{2n} - H_n$$

$$= \frac{1}{4n} - \ln 2 + \left(\ln 2 - \frac{1}{4n} + \frac{1}{16n^2} + O(1/n^4)\right)$$

$$= \frac{1}{16n^2} + O(1/n^4).$$

Using the formulas for A_ns above, we find

$$\lim_{n\to\infty} A_n b_{n+1} = 0.$$

Therefore,

$$S = \sum_{n=1}^{\infty} \frac{A_n}{4n(n+1)(2n+1)}$$

$$= \sum_{n=2k}^{\infty} \frac{A_n}{4n(n+1)(2n+1)} + \sum_{n=2k-1}^{\infty} \frac{A_n}{4n(n+1)(2n+1)}$$

$$= \sum_{k=1}^{\infty} \frac{k}{8k(2k+1)(4k+1)} + \sum_{k=1}^{\infty} \frac{-k}{4(2k-1)(2k)(4k-1)}$$

$$= \frac{1}{8}\left(\sum_{k=1}^{\infty} \frac{1}{(2k+1)(4k+1)} - \sum_{k=1}^{\infty} \frac{1}{(2k-1)(4k-1)}\right)$$

$$= \frac{1}{8}\left(\frac{-4+\pi+2\ln 2}{4} - \frac{\pi-2\ln 2}{4}\right)$$

$$= \frac{1}{8}(\ln 2 - 1).$$

$\square$

25. (**Monthly Problem 12287, 2021**). Prove

$$\sum_{n=1}^{\infty} \left(n \left(\sum_{k=n}^{\infty} \frac{1}{k^2} \right)^2 - \frac{1}{n} \right) = \frac{3}{2} - \frac{1}{2}\zeta(2) + \frac{3}{2}\zeta(3).$$

Proof. Let the proposed series be S. For $n \geq 1$, let

$$a_n = n, \quad b_n = \left(\sum_{k=n}^{\infty} \frac{1}{k^2} \right)^2,$$

and $A_n = \sum_{k=1}^{n} a_k$. Then $A_n = n(n+1)/2$ and

$$b_n - b_{n+1} = \left(\frac{1}{n^2} + \sum_{k=n+1}^{\infty} \frac{1}{k^2} \right)^2 - \left(\sum_{k=n+1}^{\infty} \frac{1}{k^2} \right)^2$$

$$= \frac{1}{n^4} + \frac{2}{n^2} \sum_{k=n+1}^{\infty} \frac{1}{k^2}.$$

Applying Abel's summation formula yields

$$S = \lim_{n \to \infty} \frac{1}{2}n(n+1)b_{n+1} + \sum_{n=1}^{\infty} \left(\frac{1}{2}n(n+1) \left(\frac{1}{n^4} + \frac{2}{n^2} \sum_{k=n+1}^{\infty} \frac{1}{k^2} \right) - \frac{1}{n} \right)$$

$$= \lim_{n \to \infty} \frac{1}{2}n(n+1)b_{n+1} + \frac{1}{2}(\zeta(2) + \zeta(3)) + \sum_{n=1}^{\infty} \left(\frac{n+1}{n} \sum_{k=n+1}^{\infty} \frac{1}{k^2} - \frac{1}{n} \right)$$

$$= \lim_{n \to \infty} \frac{1}{2}n(n+1)b_{n+1} + \frac{1}{2}(\zeta(2) + \zeta(3)) + \sum_{n=1}^{\infty} \frac{1}{n} \sum_{k=n+1}^{\infty} \frac{1}{k^2} + \sum_{n=1}^{\infty} \left(\sum_{k=n+1}^{\infty} \frac{1}{k^2} - \frac{1}{n} \right).$$

$$(2.30)$$

By Stolz-Cesáro theorem,

$$\lim_{n \to \infty} \frac{\sum_{k=n}^{\infty} 1/k^2 - 1/n}{1/n^2} = \lim_{n \to \infty} \frac{-1/n^2 - 1/(n+1) + 1/n}{1/(n+1)^2 - 1/n^2} = \frac{1}{2},$$

and therefore

$$\sum_{k=n}^{\infty} \frac{1}{k^2} = \frac{1}{n} + \frac{1}{2n^2} + o(1/n^2). \qquad (2.31)$$

We have

$$b_n = \left(\frac{1}{n} + \frac{1}{2n^2} + o(1/n^2) \right)^2 = \frac{1}{n^2} + \frac{1}{n^3} + o(1/n^3).$$

Hence,

$$\lim_{n \to \infty} \frac{1}{2}n(n+1)b_{n+1} = \frac{1}{2}.$$

Reversing the summation yields

$$\sum_{n=1}^{\infty} \frac{1}{n} \sum_{k=n+1}^{\infty} \frac{1}{k^2} = \sum_{k=2}^{\infty} \frac{1}{k^2} \sum_{n=1}^{k-1} \frac{1}{n} = \sum_{k=2}^{\infty} \frac{1}{k^2} H_{k-1} = \sum_{k=1}^{\infty} \frac{1}{(k+1)^2} H_k = \zeta(3),$$

where the well-known Euler sum formula is used in the last equality.

Applying Abel's summation formula again with $a_n = 1, b_n = \sum_{k=n+1}^{\infty} 1/k^2 - 1/n$, in view of (2.31), we have

$$\sum_{k=n+1}^{\infty} \frac{1}{k^2} = \sum_{k=n}^{\infty} \frac{1}{k^2} - \frac{1}{n^2} = \frac{1}{n} - \frac{1}{2n^2} + o(1/n^2),$$

and

$$\sum_{n=1}^{\infty} \left(\sum_{k=n+1}^{\infty} \frac{1}{k^2} - \frac{1}{n} \right) = \lim_{n\to\infty} nb_{n+1} + \sum_{n=1}^{\infty} n \left(\frac{1}{(n+1)^2} - \frac{1}{n} + \frac{1}{n+1} \right)$$

$$= \lim_{n\to\infty} \left(-\frac{1}{2n} + o(1/n) \right) - \sum_{n=1}^{\infty} \frac{1}{(n+1)^2} = 1 - \zeta(2).$$

In summary, by (2.30), we find

$$S = \frac{1}{2} + \frac{1}{2}(\zeta(2) + \zeta(3)) + \zeta(3) + 1 - \zeta(2) = \frac{3}{2} - \frac{1}{2}\zeta(2) + \frac{3}{2}\zeta(3)$$

as claimed.   □

2.10   Exercises

I have always grown from my problems and challenges, from the things that don't work out. That's when I've really learned. — Carol Burnett

1. Let $p > 0, q > 0$, and $r > 0$. Determine the convergence of the series

$$\sum_{n=3}^{\infty} \frac{1}{n^p \ln^q n \ln^r (\ln n)}.$$

2. Determine the range of a such that

$$\sum_{n=1}^{\infty} a^{(1+1/2+1/3+\cdots+1/n)}$$

converges.

3. Let $a_n > 0$ for all $n \in \mathbb{N}$. If $\sum_{n=1}^{\infty} a_n$ converges, prove that the series

$$\sum_{n=1}^{\infty} \frac{\sqrt{a_n}}{n^p}$$

converges for all $p > 1/2$.

4. Determine the convergence of the following series

(1) $\sum_{n=1}^{\infty} \sin\left(\pi \sqrt{n^2 + 1} \right)$ (2) $\sum_{n=1}^{\infty} \left(1 + \frac{1}{2} + \cdots + \frac{1}{n} \right) \frac{\sin nx}{n}$.

5. Show that a series of the form

$$\sum_{n=1}^{\infty} \frac{1}{n^{1+r_n}}$$

with $r_n > 0$ and $\lim_{n\to\infty} r_n = 0$ could either converge or diverge.

6. Let $a_n > 0$ for all $n \in \mathbb{N}$. Suppose that $\sum_{n=1}^{\infty} a_n$ diverges.

   (a) Prove that $\sum_{n=1}^{\infty} \frac{a_n}{1+a_n}$ diverges, but $\sum_{n=1}^{\infty} \frac{a_n}{1+n^2 a_n}$ converges.
   (b) Determine the convergences of the series

$$\sum_{n=1}^{\infty} \frac{a_n}{1+a_n^2} \quad \text{and} \quad \sum_{n=1}^{\infty} \frac{a_n}{1+na_n}.$$

7. (**Monthly Problem 12004, 2017**). Let a_n be a strictly increasing sequence of real numbers such that $a_n \le n^2 \ln n$ for all $n \ge 1$. Prove that the series

$$\sum_{n=1}^{\infty} \frac{1}{a_{n+1} - a_n}$$

diverges.

8. Consider two standard series of the logarithmic scale

$$\sum_{n=2}^{\infty} \frac{1}{n \ln^p n} \quad \text{and} \quad \sum_{n=2}^{\infty} \frac{1}{n \ln n \ln^p(\ln n)},$$

It is well-known that both series converge if and only if $p > 1$. If

   (1) $\lim_{n\to\infty} (na_n)^{1/\ln(\ln n)} < \dfrac{1}{e}$,   (2) $\lim_{n\to\infty} (n \ln n \, a_n)^{1/\ln(\ln(\ln n))} < \dfrac{1}{e}$,

prove in each case the corresponding series $\sum_{n=1}^{\infty} a_n$ converges.

9. Let a_n be a positive sequence and let

$$\lambda = \lim_{n\to\infty} \left(\frac{a_{n+1}}{a_n} \right)^n.$$

Show that

   (a) if $\lambda < \frac{1}{e}$, then $\sum_{n=1}^{\infty} a_n$ converges,
   (b) if $\lambda > \frac{1}{e}$, then $\sum_{n=1}^{\infty} a_n$ diverges.

10. Assume that $f(x)$ is positive and monotone decreasing on $[0, \infty)$. Let

$$p = \lim_{x\to\infty} \left(\frac{e^x f(e^x)}{f(x)} \right)^n.$$

Show that

   (a) if $p < 1$, then $\sum_{n=1}^{\infty} f(n)$ converges,
   (b) if $p > 1$, then $\sum_{n=1}^{\infty} f(n)$ diverges.

11. Let $x_1 = 1$ and for $n \ge 1$,

$$x_{n+1} x_n + 1 = \sqrt{x_n^2 + 1}.$$

Does the series $\sum_{n=1}^{\infty} x_n$ converge?

12. Assume that $\sum_{n=1}^{\infty} a_n$ converges and $a_n \geq 0$ for all $n \in \mathbb{N}$. Let $\#(m)$ be the number of terms in $\{a_1, a_2, \ldots, a_m\}$ satisfying $a_n > 1/n$. Prove that

$$\lim_{m \to \infty} \frac{\#(m)}{m} = 0.$$

13. Let a_n be a positive sequence. Prove that $\sum_{n=1}^{\infty} \frac{1}{n} \frac{1+a_{n+1}}{a_n}$ diverges.

14. Let $a_1 > 0$. Define $a_{n+1} = \ln(1 + a_n)$ for $n \in \mathbb{N}$. Prove that $\sum_{n=1}^{\infty} a_n$ diverges.

15. (**Putnam Problem 2020-A3**). Let $a_0 = \pi/2$, and let $a_n = \sin(a_{n-1})$ for $n \geq 1$. Determine whether $\sum_{n=1}^{\infty} a_n^2$ converges.

16. (**Putnam Problem 2008-A4**). Define $f : \mathbb{R} \to \mathbb{R}$ by

$$f(x) = \begin{cases} x, & \text{if } x \leq e, \\ xf(\ln x), & \text{if } x > e. \end{cases}$$

Does $\sum_{n=1}^{\infty} 1/f(n)$ converges?

17. Let a_n be a positive sequence and $b_n = \frac{a_1 + a_2 + \cdots + a_n}{n}$. Prove that if $\sum_{n=1}^{\infty} 1/a_n$ converges then $\sum_{n=1}^{\infty} 1/b_n$ also converges.

18. Let a_n be a positive sequence and $b_n = (a_1 a_2 \cdots a_n)^{1/n}$. Prove that if $\sum_{n=1}^{\infty} a_n$ converges then $\sum_{n=1}^{\infty} b_n^2$ also converges.

19. Let $\sum_{n=1}^{\infty} a_n$ be a convergent series and $a_n > 0$ for all $n \in \mathbb{N}$. Let $b_n = 1/na_n^2$. Prove that

$$\sum_{n=1}^{\infty} \frac{n}{b_1 + b_2 + \cdots + b_n}$$

converges.

20. Let a_n be a positive increasing sequence and the series $\sum_{n=1}^{\infty} 1/a_n$ converges. For every $k \in \mathbb{N}$, prove that $\sum_{n=1}^{\infty} (\ln a_n)^k / a_n$ converges if and only if $\sum_{n=1}^{\infty} (\ln n)^k / a_n$ converges.

21. (**Monthly Problem 11649, 2012**). Let a_n be a nonnegative sequence and $p > 1$. Show that

$$\sum_{j=0}^{\infty} \left(\sum_{i=0}^{\infty} \frac{x_i}{i+j+1} \right)^p < \infty \iff \sum_{j=0}^{\infty} \left(\frac{1}{j+1} \sum_{i=0}^{j} x_i \right)^p < \infty.$$

22. Let $p, q, r \in \mathbb{N}$. Prove

$$\sum_{n=1}^{\infty} \frac{1}{(pn+q)(pn+q+pr)} = \frac{1}{pr} \sum_{k=1}^{r} \frac{1}{q+pk}.$$

23. For $x \neq 0$, show that

$$\frac{1}{x+a_1} + \sum_{k=2}^{n} \frac{a_1 a_2 \cdots a_{k-1}}{(x+a_1)(x+a_2) \cdots (x+a_k)} = \frac{1}{x} - \frac{a_1 a_2 \cdots a_{k-1}}{x(x+a_1)(x+a_2) \cdots (x+a_k)}.$$

24. For $k \in \mathbb{N}$, show that

$$\sum_{n=1}^{\infty} \frac{(-1)^{n-1}}{n(n+1) \cdots (n+k)} = \frac{2^k}{k!} \left(\ln 2 - \sum_{i=1}^{k} \frac{1}{i \, 2^i} \right).$$

25. Show that

$$\sum_{n=1}^{\infty} (-1)^n \frac{\ln n}{n} = \gamma \ln 2 - \frac{1}{2}(\ln 2)^2,$$

where γ is the Euler-Mascheroni constant.

26. (**Pi Mu Epsilon Problem 1351, 2018**). Let $\zeta(3)$ be Apéry's constant which is defined by $\sum_{n=1}^{\infty} 1/n^3$. Show that

$$\zeta(3) = 1 + \sum_{n=1}^{\infty} \frac{1}{n^3 + 4n^7}.$$

This series has the convergence rate $O(n^{-7})$ instead of $O(n^{-3})$.

27. Let F_n be the nth Fibonacci number. Prove that

1. $\sum_{n=0}^{\infty} \frac{1}{F_{2n+1}+1} = \frac{1}{2}\sqrt{5}$.
2. $\sum_{n=0}^{\infty} \frac{1}{F_{2^n}} = \frac{1}{2}(7 - \sqrt{5})$.

28. Let F_n be the nth Fibonacci number. Define $S_n = \sum_{k=1}^{n} F_k^2$. Evaluate

$$\sum_{n=1}^{\infty} (-1)^{n+1} \frac{1}{S_n}.$$

29. (**Putnam Problem 2016-B1**). Let $x_0, x_1, x_2 \ldots$ be sequence such that $x_0 = 1$ and for $n \geq 0$

$$x_{n+1} = \ln(e^{x_n} - x_n).$$

Show that the infinite series

$$x_0 + x_1 + x_2 + \cdots$$

converges and find its sum.

30. (**Putnam Problem 2016-B6**). Evaluate

$$\sum_{k=1}^{\infty} \frac{(-1)^{k-1}}{k} \sum_{n=0}^{\infty} \frac{1}{k2^n + 1}.$$

31. (**Putnam Problem 2021-B2**). Determine the maximum value of the sum

$$S = \sum_{n=1}^{\infty} \frac{n}{2^n} (a_1 a_2 \cdots a_n)^{1/n}$$

over all sequences $a_1, a_2, a_3, \cdots$ of nonnegative real numbers satisfying $\sum_{n=1}^{\infty} a_n = 1$.

32. (**Monthly Problem 11910, 2016**). Let F_n be the nth Fibonacci number. Find

$$\sum_{n=1}^{\infty} \left(\arctan \frac{1}{F_{4n-3}} + \arctan \frac{1}{F_{4n-2}} + \arctan \frac{1}{F_{4n-1}} - \arctan \frac{1}{F_{4n}} \right).$$

33. Let $\sum_{n=1}^{\infty} a_n = s$. Prove that

    1. $\lim_{n \to \infty} \frac{1}{n} \sum_{k=1}^{n} k a_k = 0$,
    2. $\sum_{n=1}^{\infty} \frac{a_1 + 2a_2 + \cdots + n a_n}{n(n+1)} = s$.

34. (**Putnam Problem 2002-A6**). Fix an integer $b > 2$. Let $f(1) = 1, f(2) = 2$, and for each $n \geq 3$, define $f(n) = n f(d)$, where d is the number of base-b digits of n. For which values of b does

$$\sum_{n=1}^{\infty} \frac{1}{f(n)}$$

converge?

35. Let d_n be the number of digits of n in its decimal representation. Show that

$$\sum_{n=1}^{\infty} \frac{1}{d_n!} = \frac{9}{10}(e^{10} - 1).$$

36. Let

$$f(x) = \frac{\pi}{16} \frac{\pi x - \sin(\pi x)}{x^3 \cos(\pi x/2)}.$$

For $x \notin \mathbb{Z}$, find

$$\sum_{n=0}^{\infty} \frac{1}{16^n} f\left(\frac{x}{2^n}\right)$$

in closed form.

37. (**Putnam Problem 2011-A2**). Let $a_1, a_2, \ldots$ and $b_1, b_2, \ldots$ be sequence of positive real numbers such that $a_1 = b_1 = 1$ and $b_n = b_{n-1} a_n - 2$ for each $n \geq 2$. Assume that the sequence b_n is bounded. Prove that

$$S = \sum_{n=1}^{\infty} \frac{1}{a_1 a_2 \cdots a_n}$$

converges, and evaluate S.

38. For $n \in \mathbb{N}$, let $a_n = 1/(\ln(n+1) - \ln n)$. Prove that

$$\sum_{n=1}^{\infty} \frac{(a_n - n)^2}{(a_n + n)(a_n + n + 1)} < \frac{\pi}{8} - \frac{1}{3}.$$

39. (**Hardy-Landau's Inequality**). Let a_n be a positive increasing sequence. If $\sum_{n=1}^{\infty} a_n^p$ converges for $p > 1$, prove that

$$\sum_{n=1}^{\infty} \left(\frac{a_1 + a_2 + \cdots + a_n}{n}\right)^p < \left(\frac{p}{p-1}\right)^p \sum_{n=1}^{\infty} a_n^p.$$

40. (**Carleman's Inequality**). Let $a_n > 0$ for all $n \in \mathbb{N}$ and suppose $\sum_{n=1}^{\infty} a_n$ converges. Prove that

$$\sum_{n=1}^{\infty} \sqrt[n]{a_1 \cdot a_2 \cdots a_n} \leq e \sum_{n=1}^{\infty} a_n,$$

where e is the best possible constant.

41. (**Discrete p-Hardy Inequality**). Let a_n be a sequence with $a_0 = 0$. The celebrated discrete p-Hardy inequality claims that

$$\sum_{n=1}^{\infty} \frac{|a_n|^2}{4n^2} \leq \sum_{n=1}^{\infty} |a_n - a_{n-1}|^2.$$

Prove this inequality can be improved to

$$\sum_{n=1}^{\infty} |a_n|^2 b_n \leq \sum_{n=1}^{\infty} |a_n - a_{n-1}|^2,$$

where

$$b_n = 2 - \sqrt{\frac{n+1}{n}} - \sqrt{\frac{n-1}{n}} > \frac{1}{4n^2}.$$

42. Let $a_n \geq 0$ for all $n \in \mathbb{N}$ and $\sum_{n=1}^{\infty} a_n^2 < \infty$. Show that

$$\sum_{m,n=1}^{\infty} \frac{a_m a_n}{m+n} \leq \pi \sum_{n=1}^{\infty} a_n^2.$$

Hint: First show that, for every $m \in \mathbb{N}$,

$$\sum_{n=1}^{\infty} \frac{\sqrt{m}}{\sqrt{n}(m+n)} \leq \pi.$$

43. Is there a sequence $a_1, a_2, \ldots$ of positive real numbers such that $\sum_{k=1}^{\infty} \frac{1}{a_k}$ converges, and $\prod_{k=1}^{n} a_k < n^n$ for all n?

44. Let $p(n) : \mathbb{Z} \to \mathbb{Z}$ be a quadratical polynomial. Find all possible values of

$$\sum_{n=1}^{\infty} \frac{(-1)^{p(n)}}{n^2}.$$

45. (**Monthly Problem 11999, 2017**). Evaluate

$$\sum_{n=1}^{\infty} \frac{(-1)^{\lfloor \sqrt{n}+\sqrt{n+1} \rfloor}}{n(n+1)}.$$

46. (**Monthly Problem 11384, 2008**). Let p_n denote the nth prime. Show that

$$\sum_{n=1}^{\infty} \frac{(-1)^{\lfloor \sqrt{n} \rfloor}}{p_n}$$

converges.
Remark. The published solution has a defect which made an unjustified use of the alternating series test.

47. Let $p, m \in \mathbb{N}$. Prove that the series

$$\sum_{n=1}^{\infty} \frac{(-1)^{\lfloor \sqrt[m]{n} \rfloor}}{n^p}$$

converges if and only if $p + 1/m > 1$.

48. (**Putnam Problem, 2001-B3**). For any positive integer n, let $\langle n \rangle$ denote the closest integer to $\sqrt{n}$. Evaluate

$$\sum_{n=1}^{\infty} \frac{2^{\langle n \rangle} + 2^{-\langle n \rangle}}{2^n}.$$

49. Given the alternating harmonic series

$$\sum_{n=1}^{\infty} \frac{(-1)^{n+1}}{n} = 1 - \frac{1}{2} + \frac{1}{3} - \frac{1}{4} + \cdots,$$

consider the rearrangement in such a way that p positive terms and q negative terms occur alternately. Show the resulting series converges to $\ln(2\sqrt{p/q})$.

50. It is well-known that $L = 1 - \frac{1}{3} + \frac{1}{5} - \frac{1}{7} + \cdots = \frac{\pi}{4}$. For a fixed $p \in \mathbb{N}$, let R be the rearrangement of L by grouping $2p$ positive terms followed by p negative terms. Show that R converges and find its value.

51. Let S be the series obtained from the harmonic series by deleting the terms contains the digital 9. Prove that S converges and its sum is not exceed 80.

52. Start with the harmonic series

$$1 + \frac{1}{2} + \frac{1}{3} + \cdots + \frac{1}{n} + \cdots.$$

Create a new series as follows: p positive terms and q negative terms occur alternately without changing the order of the terms. Show the resulting series converges if and only if $p = q$.

53. Let $a_1 = 2$ and $a_n = \frac{1}{n} \sum_{k=1}^{n-1} k a_k$ for $n \geq 2$. Evaluate $\sum_{n=1}^{\infty} a_n/4^n$.

54. Determine whether the series

$$\sum_{n=1}^{\infty} \left(\frac{1}{4n-3} - \frac{1}{4n-2} + \frac{1}{4n-1} - \frac{1}{4n} \right)$$

converges. If it converges, find its sum.

55. Find a necessary and sufficient condition on (a_1, a_2, a_3, a_4) for the series

$$\sum_{n=0}^{\infty} \left(\frac{a_1}{4n+1} + \frac{a_2}{4n+2} + \frac{a_3}{4n+3} + \frac{a_4}{4n+4} \right)$$

to converge, and determine the sum of the series when that condition is satisfied.

56. (**BBP-Type Formulas**). Show that

$$\ln 2 = \frac{2}{3} \sum_{n=0}^{n} \frac{1}{9^k} \frac{1}{2k+1}, \quad \ln 3 = \sum_{n=0}^{n} \frac{1}{4^k} \frac{1}{2k+1},$$

$$\arctan(1/3) = \frac{1}{32} \sum_{n=0}^{\infty} \frac{1}{64^n} \left(\frac{8}{4n+1} + \frac{4}{4n+2} + \frac{1}{4n+3} \right),$$

$$\pi = \sum_{n=0}^{\infty} \frac{1}{16^n} \left(\frac{4}{8n+1} - \frac{2}{8n+4} - \frac{1}{8n+5} - \frac{1}{8n+6} \right).$$

Remark. Originally, the BBP (named after Bailey-Borwein-Plouffe) formula is the last formula above discovered by Simon Plouffe [11] in 1995. Amazingly, this formula permits one to calculate the nth hexadecimal or binary digit of π, without computing any of the first $n-1$ digits, by means of a simple algorithm that does not require multiple-precision arithmetic. Following the discovery of this formula, similar formulas for other mathematical constants have been investigated. This type of formulas [10] are now called as BBP-type formulas.

57. Show that

$$\arctan x = \sum_{n=1}^{\infty} \frac{1}{n} \left(\frac{x^2}{1+x^2} \right)^{n/2} \sin(n\theta),$$

where $\theta = \arcsin(1/\sqrt{1+x^2})$. Use this result to derive the BBP-type formula

$$\pi = \sum_{n=0}^{\infty} \frac{(-1)^n}{4^n} \left(\frac{2}{4n+1} + \frac{2}{4n+2} + \frac{1}{4n+3} \right).$$

58. (**Monthly Problem 11930, 2016**). Find

$$\sum_{n=1}^{\infty} \sinh^{-1} \left(\frac{1}{\sqrt{2^{n+2}+2} + \sqrt{2^n+2}} \right).$$

59. (**Monthly Problem 11932, 2016**). Let r be an integer. Prove that

$$\sum_{n=-\infty}^{\infty} \arctan \left(\frac{\sinh r}{\cosh n} \right) = \pi r.$$

60. (**Monthly Problem 11952, 2017**). Prove that

$$\sum_{n=1}^{\infty} \frac{2^{2n-1}}{2n+1} \left(\frac{(n-1)!}{(2n-1)!!} \right) = \pi - 2.$$

61. Evaluate

$$\sum_{n=1}^{\infty} \frac{1}{n(2n+1)} \left(\frac{1}{n+1} + \frac{1}{n+2} + \cdots + \frac{1}{2n+1} \right)^2.$$

62. Let h be the alternating harmonic series. Consider the rearrangement series of h:

$$1 - \frac{1}{2} - \frac{1}{4} + \frac{1}{3} + \frac{1}{5} + \frac{1}{7} - \frac{1}{6} - \frac{1}{8} - \frac{1}{10} - \frac{1}{12} + \frac{1}{9} + \cdots .$$

Is this rearrange series convergent? If so, can you find its value?

63. Assume that $\sum_{n=1}^{\infty} a_n$ converges to S conditionally. Let $\sum_{n=1}^{\infty} a_{\sigma(n)} = T$ be a rearrangement of S. If $S \neq T$, show that, for any positive integer N, there exists $n \in \mathbb{N}$ such that $|n - \sigma(n)| > N$.

64. Let $p > 1$. Show that

$$\sum_{n=1}^{\infty} \frac{1}{n^p} \left(1 + \frac{1}{2^p} + \cdots + \frac{1}{n^p} \right) = \frac{1}{2} \left(\zeta^2(p) + \zeta(2p) \right),$$

where $\zeta(p) = \sum_{n=1}^{\infty} 1/n^p$ is the *Reimann zeta function*.

65. For $|x| < 1$, prove that

$$\sum_{n=0}^{\infty} \zeta(2n)x^{2n} = -\frac{1}{2}\pi x \cot(\pi x).$$

66. Show that

(1) $\sum_{n=1}^{\infty} \frac{\zeta(2n)}{n(2n+1)2^{2n}} = \ln \pi - 1$.

(2) $\sum_{n=1}^{\infty} \frac{\zeta(2n+1)}{(2n+1)2^{2n}} = \ln 2 - \gamma$, where γ is the Euler-Mascheroni constant.

67. Let $f(x) = \sum_{n=2}^{\infty} a_n x^n$. Prove that

$$\sum_{n=2}^{\infty} a_n(\zeta(n) - 1) = \sum_{n=2}^{\infty} f(1/n).$$

Use this formula to show that

(1) $\sum_{k=1}^{\infty} (\zeta(2k) - 1) = 3/4$,

(2) $\sum_{k=2}^{\infty} (-1)^k(\zeta(k) - 1) = 1/2$,

(3) $\sum_{k=2}^{\infty} \frac{\zeta(k)-1}{k} = 1 - \gamma$, where γ is the Euler-Mascheroni constant.

68. Let H_n be the nth harmonic number. Prove that

(1) $\sum_{n=1}^{\infty} \frac{1}{n} \left(\zeta(2) - \sum_{k=1}^{n} \frac{1}{k^2} \right) = \zeta(3)$.

(2) $\sum_{n=1}^{\infty} \frac{1}{n} H_n \left(\zeta(2) - \sum_{k=1}^{n} \frac{1}{k^2} \right) = \frac{7}{4}\zeta(4)$.

(3) $\sum_{n=1}^{\infty} H_n \left(\zeta(3) - \sum_{k=1}^{n} \frac{1}{k^3} \right) = 2\zeta(3) - \zeta(2)$. (**Monthly Problem 11810, 2015**)

(4) $\sum_{n=1}^{\infty} \frac{[1+2(-1)^n]}{n^4} H_n = -\frac{11}{16}\zeta(5)$.

69. (**Monthly Problem 12134, 2019**). Evaluate the series

$$\sum_{n=1}^{\infty} \left(n \sum_{k=n}^{\infty} \frac{1}{k^2} - 1 - \frac{1}{2n} \right)$$

70. (**Monthly Problem 12194, 2020**). Evaluate

$$\sum_{n=1}^{\infty} \left(H_n - \ln n - \gamma - \frac{1}{2n} \right)$$

where $H_n = \sum_{k=1}^{n} 1/k$ and γ is the Euler-Mascheroni constant.

71. (**Monthly Problem 12102, 2019**). Prove

$$\sum_{n=1}^{\infty} H_n^2 \left(\zeta(2) - \sum_{k=1}^{n} \frac{1}{k^2} - \frac{1}{n} \right) = 2 - \zeta(2) - 2\zeta(3)$$

where $H_n = \sum_{k=1}^{n} 1/k$ is the nth harmonic number.

72. (**Monthly Problem 12206, 2020**). Prove

$$\sum_{n=1}^{\infty} \frac{\overline{H}_{2n}}{n^2} = \frac{3}{4}\zeta(3)$$

where $\overline{H}_n = \sum_{k=1}^{n}(-1)^{k+1}/k$ is the nth skew-harmonic number.

73. (**Monthly Problem 12215, 2020**). Calculate

$$\sum_{n=1}^{\infty}\left(\left(\frac{1}{n^2} + \frac{1}{(n+2)^2} + \frac{1}{(n+4)^2} + \cdots\right) - \frac{1}{2n}\right).$$

74. (**Monthly Problem 12262, 2021**). For a nonnegative integer m, let

$$A_m = \sum_{k=0}^{\infty}\left(\frac{1}{(6k+1)^{2m+1}} - \frac{1}{(6k+5)^{2m+1}}\right).$$

Prove $A_0 = \pi\sqrt{3}/6$ and, for $m \geq 1$,

$$2A_m + \sum_{n=1}^{m} \frac{(-1)^n \pi^{2n}}{(2n)!} A_{m-n} = \frac{(-1)^m(4^m+1)\sqrt{3}}{2(2m)!}\left(\frac{\pi}{3}\right)^{2m+1}.$$

75. (**Monthly Problem 11068, 2004**). For a rational number x that equals a/b in lowest terms, let $f(x) = ab$.

   (a) Show that
   $$\sum_{x\in\mathbb{Q}} \frac{1}{f(x)^2} = \frac{5}{2},$$
   where the sum extends over all positive rationals.

   (b) More general, exhibit an infinite sequence of distinct rational exponents s such that
   $$\sum_{x\in\mathbb{Q}} \frac{1}{f(x)^s}$$
   is rational.

76. Show that

   (a) $\sum_{k=0}^{\infty} \frac{1}{(3k+1)^3} + \sum_{k=0}^{\infty} \frac{1}{(3k+2)^3} = \frac{26}{27}\zeta(3)$,

   (b) $\sum_{k=0}^{\infty} \frac{1}{(3k+1)^3} - \sum_{k=0}^{\infty} \frac{1}{(3k+2)^3} = \frac{4\pi^3\sqrt{3}}{243}$,

   (c) $\sum_{k=0}^{\infty} \frac{(-1)^k}{(3k+1)^3} + \sum_{k=0}^{\infty} \frac{(-1)^k}{(3k+2)^3} = \frac{5\pi^3\sqrt{3}}{243}$,

   (d) $\sum_{k=0}^{\infty} \frac{(-1)^k}{(3k+1)^3} - \sum_{k=0}^{\infty} \frac{(-1)^k}{(3k+2)^3} = \frac{13}{18}\zeta(3)$.

77. Let $k \in \mathbb{N}$. Define

$$S_k = \left(1 + \frac{1}{2} + \cdots + \frac{1}{k}\right) - \left(\frac{1}{k+1} + \frac{1}{k+2} + \cdots + \frac{1}{2k}\right)$$

$$+ \left(\frac{1}{2k+1} + \frac{1}{2k+2} + \cdots + \frac{1}{3k}\right) - \left(\frac{1}{3k+1} + \frac{1}{3k+2} + \cdots + \frac{1}{4k}\right) + \cdots.$$

(a) Show that $H_k > S_k > cH_k$ for some $c \in (0, 1)$.

(b) Prove that $\lim_{k \to \infty} (H_k - S_k) = \ln(\pi/2)$.

(c) Find S_k in closed form. Check your solution with $S_2 = (\pi + 2\ln 2)/4, S_3 = (2\pi\sqrt{3} + 3\ln 3)/9$.

(d) Show that

$$\lim_{k \to \infty} \left(\sum_{n=1}^{k-1} \frac{\pi}{2k \sin(n\pi/k)} - \ln \frac{2k}{\pi} \right) = \gamma,$$

where γ is the Euler-Mascheroni constant.

78. (**Monthly Problem 12026, 2018**). For $n \in \mathbb{N}$, let $H_n = \sum_{k=1}^{n} 1/k$ and

$$S_n = \sum_{k=1}^{n} \frac{(-1)^{n-k}}{k} \sum_{j=1}^{k} H_j.$$

Find $\lim_{n \to \infty} S_n / \ln n$ and $\lim_{n \to \infty} (S_{2n} - S_{2n-1})$.

79. (**Mathieu's inequality**). Show that

$$\frac{1}{x^2 + 1/2} < S(x) := \sum_{n=1}^{\infty} \frac{2n}{(n^2 + x^2)^2} < \frac{1}{x^2} \quad \text{for } x \neq 0.$$

Remark. The series $S(x)$, often referred as *Mathieu series* , was introduced by Emile Mathieu in the study of elasticity of solid bodies. Bounds for this series are needed for solving biharmonic equations in a rectangular domain. Various refinement inequalities have been established. For example, Alzer et al. [5] proved that

$$\frac{1}{x^2 + 1/(2\zeta(3))} < S(x) < \frac{1}{x^2 + 1/6}.$$

Although this estimate is asymptotically sharp, it leaves a big gap for small x. Can you find a best possible estimate of $S(x)$ applicable on the entire $\mathbb{R}$?

80. Let $S(x)$ be the Mathieu series defined above. For $|x| < 1$, show that

$$S(x) = 2 \sum_{n=1}^{\infty} (-1)^{n+1} n\zeta(2n+1)x^{2(n-1)}.$$

81. Let $a_n > 0$ for all $n \in \mathbb{N}$ and $s_n = \sum_{k=1}^{n} a_k$. Show that

(1) If $\sum_{n=1}^{\infty} a_n$ converges, then $\sum_{n=1}^{\infty} a_n/s_n^\alpha$ converges for any real number α.

(2) If $\sum_{n=1}^{\infty} a_n$ diverges, then $\sum_{n=1}^{\infty} a_n/s_n^\alpha$ converges for $\alpha > 1$ and diverges for $\alpha \leq 1$.

82. (**Putnam Problem, 1948-A3**). Let $a_n > 0$ be a decreasing sequence with limit 0 such that

$$b_k = a_k - 2a_{k+1} + a_{k+2} \geq 0 \quad \text{(for all } k = 1, 2, \ldots\text{)}.$$

Prove that $\sum_{n=1}^{\infty} nb_n = a_1$.

83. (**Putnam Problem, 1949-B2**). Show that

$$\sum_{n=2}^{\infty} \frac{\cos(\ln(\ln n))}{\ln n}$$

diverges.

84. (**Putnam Problem, 1966-B3**). Let $a_n > 0$ for all $n \in \mathbb{N}$. If $\sum_{n=1}^{\infty} 1/a_n$ converges, prove that

$$\sum_{n=1}^{\infty} \frac{n^2}{(a_1 + a_2 + \cdots + a_n)^2} a_n$$

also converges.

85. (**Putnam Problem, 1994-A1**). Suppose the sequence a_n satisfies $0 < a_n < a_{2n} + a_{2n+1}$ for all $n \geq 1$. Prove that series $\sum_{n=1}^{\infty} a_n$ diverges.
Remark. The inequality can be replaced by $a_n < a_{n+1} + a_{n^2}$.

86. Prove that the series

$$\sum_{n=1}^{\infty} (2 - e^{\alpha})(2 - e^{\alpha/2}) \cdots (2 - e^{\alpha/n})$$

converges when $\alpha = \ln 2$ or $\alpha > 1$, and diverges when $\alpha \neq \ln 2$ and $\alpha \leq 1$.

87. Let na_n be decreasing and converge to zero. If $\sum_{n=1}^{\infty} a_n$ converges, prove that $\lim_{n \to 0} (n \ln n) a_n = 0$.

88. If $\sum_{n=1}^{\infty} a_n$ converges conditionally, prove that the terms can be grouped so as to form an absolutely convergent series.

89. Evaluate

$$\sum_{n=1}^{\infty} \ln\left(1 + \frac{1}{n}\right) \cdot \ln\left(1 + \frac{1}{2n}\right) \cdot \ln\left(1 + \frac{1}{2n+1}\right).$$

90. (**Monthly Problem 11809, 2015**). Let $\{a_n\}$ be a sequence of real numbers.

   (a) Suppose that $\{a_n\}$ consists of nonnegative numbers and is non-increasing, and $\sum_{n=1}^{\infty} a_n/\sqrt{n}$ converges. Prove that $\sum_{n=1}^{\infty} (-1)^{\lfloor \sqrt{n} \rfloor} a_n$ converges.
   (b) Find a non-increasing sequence $\{a_n\}$ of positive numbers such that $\lim_{n \to \infty} \sqrt{n} a_n = 0$ and $\sum_{n=1}^{\infty} (-1)^{\lfloor \sqrt{n} \rfloor} a_n$ diverges.

91. (**Monthly Problem 11260, 2006**). Find those nonnegative values of α and β for which

$$\sum_{n=1}^{\infty} \prod_{k=1}^{n} \frac{\alpha + k \ln k}{\beta + (k+1) \ln(k+1)}$$

converges. For those values of α and β, evaluate the sum.

92. (**Monthly Problem 11982, 2017**). Calculate

$$\lim_{x \to \infty} \left(\sum_{n=1}^{\infty} \left(\frac{x}{n}\right)^n\right)^{1/x}.$$

93. For any $x > 0$, show that

$$\left(1 + \frac{1}{x}\right)^x = e\left(1 - \sum_{n=1}^{\infty} \frac{a_n}{(1+x)^n}\right),$$

where

$$a_1 = \frac{1}{2}, \quad a_{n+1} = \frac{1}{(n+1)(n+2)} - \frac{1}{n+1} \sum_{i=1}^{n} \frac{a_i}{n-i+2}.$$

Remark. As an application, a refinement of Carleman's inequality can be achieved.

94. Let a_n be a decreasing positive sequence. Define

$$b_n = 1 - \frac{a_{n+1}}{a_n}.$$

Then a_n converges to zero if and only if $\sum_{n=1}^{\infty} b_n$ diverges.
Remark. This may use to show $a_n \to 0$ as $n \to \infty$. For example, here we can show both $(2n)!/4^n(n!)^2$ and $n^n/n!e^n$ converge to zero without using Stirling's formula.

95. Let a_n be a monotone increasing positive sequence. Define

$$b_n = 1 - \frac{a_n}{a_{n+1}}.$$

Prove that $\sum_{n=1}^{\infty} b_n$ converges if a_n is bounded.

96. Evaluate

   (a) $\sum_{n=1}^{\infty} 4^n \sin^4(2^{-n}\theta)$. (**Monthly Problem 11515, 2010**)
   (b) $\sum_{n=2}^{\infty} n \coth^{-1} n(4n^2 - 3)$.

97. (**Monthly Problem 11853, 2015**). Find

$$\sum_{n=1}^{\infty} \frac{1}{\sinh 2^n}.$$

98. Let $a_0 = 0, a_1 = 3/2$, and $a_n = \frac{5}{2}a_{n-1} - a_{n-2}$ for $n \geq 2$. Find a closed form for a_n and determine $\sum_{n=0}^{\infty} a_n$.

99. Prove that

$$n! = \left(\frac{n}{e}\right)^n \sqrt{2\pi n} \prod_{k=n}^{\infty} \left\{ (1 + 1/k)^{k+1/2}/e \right\}.$$

100. Let $a_1 = \sqrt{2}, a_{n+1} = \sqrt{\frac{2a_n}{a_n+1}}$. Determine

$$\prod_{n=1}^{\infty} a_n.$$

101. (**Putnam Problem 2004-B5**). Evaluate

$$\lim_{x \to 1^-} \prod_{n=0}^{\infty} \left(\frac{1 + x^{n+1}}{1 + x^n} \right)^{x^n}.$$

102. (**Putnam Problem 2014-A3**). Let $a_0 = 5/2$ and $a_k = a_{k-1}^2 - 2$ for $k \geq 1$. Compute

$$\prod_{k=0}^{\infty} \left(1 - \frac{1}{a_k} \right)$$

in closed form.

103. Let $a \in (0, 1/2)$. Define $a_1 = a$ and for $k \geq 1$,

$$a_{k+1} = \frac{1}{2}\left(\frac{a_k}{1 - a_k}\right)^2.$$

Compute

$$\prod_{k=1}^{\infty}(1 - a_k)$$

in closed form.

104. Let $a_1 = 2$ and for $k \geq 1$,

$$a_{k+1} = \frac{2}{1 + a_k}.$$

Evaluate $\prod_{k=1}^{\infty} a_k$.

105. (**Cantor**). Prove that

$$\sqrt{\frac{x+1}{x-1}} = \prod_{n=1}^{\infty}\left(1 + \frac{1}{q_n}\right),$$

where $q_1 = x > 1$, $q_{n+1} = 2q_n^2 - 1$.

106. Evaluate

$$\prod_{n=1}^{\infty}(1 + e^{-n\pi}).$$

107. Prove that

$$\prod_{n=1}^{\infty}\left(1 + \frac{x^2}{n^2 + n - 1}\right) = \frac{\cos\left(\frac{\sqrt{5 - 4x^2}}{2}\pi\right)}{(1 - x^2)\cos(\sqrt{5}\pi/2)}.$$

108. Show that

$$\frac{13}{8} + \frac{1}{64}\sum_{n=0}^{\infty}\frac{(-1)^{n+1}(2n+1)!}{(n+2)!\,n!\,4^{2n}} = \phi,$$

where ϕ is the golden ratio.

109. (**Monthly Problem 11333, 2007**). Show that

$$\prod_{n=2}^{\infty}\left(\left(\frac{n^2 - 1}{n^2}\right)^{2(n^2-1)}\left(\frac{n+1}{n-1}\right)^n\right) = \pi.$$

110. (**Monthly Problem 11299, 2007**). Show that

$$\prod_{n=2}^{\infty}\left(\frac{1}{e}\left(\frac{n^2}{n^2 - 1}\right)^{n^2-1}\right) = \frac{e\sqrt{e}}{2\pi}.$$

111. (**Monthly Problem 12029, 2018**). For $a > 0$, evaluate

$$\lim_{n\to\infty}\prod_{k=1}^{n}\left(a + \frac{k}{n}\right).$$

112. (**Monthly Problem 12110, 2019**). Let $\alpha_k = (k + \sqrt{k^2 + 4})/2$. Evaluate

$$\lim_{k \to \infty} \prod_{n=1}^{\infty} \left(1 - \frac{k}{\alpha_k^n + \alpha_k} \right).$$

113. Evaluate

$$\prod_{n=1}^{\infty} \left(e^{-2n} \left(1 + \frac{1}{n} \right)^{2n^2 + n - 1/6} \right).$$

114. Show that

$$\prod_{n=1}^{\infty} \left(\frac{n!}{\sqrt{2\pi n}(n/e)^n} \right)^{(-1)^{n-1}} = \frac{A^3}{2^{7/12} \pi^{1/4}},$$

where A is the *Glaisher-Kinkelin constant*.

115. (**Monthly Problem 11739, 2013**). Let $B(x) = \begin{pmatrix} 1 & x \\ x & 1 \end{pmatrix}$. Consider the infinite matrix product

$$M(t) = \prod_p B(p^{-t}) = B(2^{-t})B(3^{-t})B(5^{-t})\cdots,$$

where the product runs over all primes. Find $M(t)$ in closed form and evaluate $M(2)$.

116. (**Monthly Problem 11685, 2013**). Prove that

$$\prod_{k=0}^{\infty} \left(1 + \frac{1}{2^{2^k} - 1} \right) = \frac{1}{2} + \sum_{k=0}^{\infty} \frac{1}{\prod_{i=0}^{k-1}(2^{2^i} - 1)}.$$

117. (**Monthly Problem 11438, 2009**). Let

$$P(x) = \sum_{k=1}^{\infty} \arctan \left(\frac{x - 1}{(k + x + 1)\sqrt{k + 1} + (k + 2)\sqrt{k + x}} \right).$$

(a) Find a closed form expression for $P(n)$ when n is a nonnegative integer.

(b) Show that $\lim_{x \to -1+} P(x)$ exists, and find a closed-form expression for it.

118. (**Monthly Problem 11499, 2010**). Let H_n be the nth harmonic number, and let

$$S_k = \sum_{n=1}^{\infty} (-1)^{n-1}(\ln n - (H_{kn} - H_n)).$$

Prove that for $k \geq 2$,

$$S_k = \frac{k-1}{2k} \ln 2 + \frac{1}{2} \ln k - \frac{\pi}{2k^2} \sum_{l=1}^{[k/2]} (k + 1 - 2l) \cot \left(\frac{(2l - 1)\pi}{2k} \right).$$

119. Let p_n be the number of consecutive positive summands and let q_n be the number of consecutive negative summands that appear in the rearrangement of the alternating harmonic series to sum a prescribed real number a. If $a > \ln 2$, prove that

(a) $q_n = 1$,

(b) $p_n = \lfloor b \rfloor$ or $\lfloor b \rfloor + 1$ eventually, where $b = e^{4a}/4$,

(c) b is rational if and only if p_n is eventually periodic.

120. Let $|q| < 1$. Define $(q)_k = \prod_{j=1}^{k}(1 - q^j)$ for $k \in \mathbb{N}$. Prove that

(a) $\prod_{k=1}^{\infty}(1 + xq^{k-1}) = \sum_{k=0}^{\infty} \frac{q^{k(k-1)/2}}{(q)_k} x^k$.   (Euler)

(b) $\prod_{k=1}^{\infty} \frac{1}{1+xq^{k-1}} = \sum_{k=0}^{\infty} \frac{(-1)^k}{(q)_k} x^k$.   (Euler)

Hint: Let $P_n(x) = \prod_{k=1}^{n}(1 + xq^{k-1})$. Find the coefficient of x^k from $(1 + x)P_n(xq) = (1 + xq^n)P_n(x)$.

121. (**Jacobi's Triple Product Formula**). Let $|q| < 1$ and $z \in \mathbb{C}$. Prove that

$$\prod_{k=1}^{\infty}(1 + zq^{2k-1})(1 + z^{-1}q^{2k-1})(1 - q^{2k}) = \sum_{k=-\infty}^{\infty} q^{k^2} z^k.$$

122. (**Pi Mu Epsilon Problem 1401, 2023**). Let m be a positive integer. For $|q| < 1$, prove

$$\sum_{k=0}^{\infty} \cos\left(\frac{(2k+1)\pi}{m}\right) q^{k(k+1)/2} = \cos(\pi/m) \prod_{k=1}^{\infty}(1 + 2\cos(2\pi/m)q^k + q^{2k})(1 - q^k).$$

Use this result to deduce the recent **Monthly Problem 12289, 2021**:

$$\sum_{k=0}^{\infty} 2\cos\left(\frac{(2k+1)\pi}{3}\right) q^{k(k+1)/2} = \prod_{k=1}^{\infty}(1 - q^{6k-1})(1 - q^{6k-5})(1 - q^k).$$

3

Continuity

The whole of mathematics is nothing more than a refinement of everyday thinking.

— Albert Einstein

In most sciences one generation tears down what another has built, and what one has established another undoes. In mathematics alone each generation adds a new story to the old structure.

— Hermann Hankel

The importance of continuity can hardly be overemphasized in the context of analysis. Sequential theorems related to continuity and set properties, including the extreme value theorem, the intermediate value theorem, and the uniform continuity theorem, constitute the fundamental core of analysis. In this chapter, we give various proofs of these three fundamental theorems based on the completeness theorems of the real numbers. Surprisingly, we see that each of these three fundamental theorems is logically equivalent to the completeness of the real numbers. As a consequence of the intermediate value theorem, we present Li-York's beautiful result "periodic three implies chaos."

3.1 Definition of Continuity

Mathematics is an experimental science, and definitions do not come first, but later on.

— Oliver Heaviside

In Chapter 1, we studied the sequences of real numbers, which can be viewed as real-valued functions defined on $\mathbb{N}$. We now begin the study of real-valued functions defined on a general subset D of $\mathbb{R}$, especially those defined on an interval I in $\mathbb{R}$. *Continuity* and *differentiability* are two essential concepts that display the nature of functions. Before Cauchy and Weierstrass, the continuity of a function is built on geometrically intuitive notions such as "Its graph can be drawn without lifting the pen from the paper" or "Its graph has no jumps or gaps." Why did the formulation of a rigorous definition of continuity take so long? The basic reason for this two-hundred-year waiting period lies in the fact that, for most of that time, functions were thought of as being composed of elementary functions such as polynomials and trigonometric functions, which are always continuous on their relevant domains. In the early 19th century, mainly under the influence of Fourier's heat equation and Dirichlet's study of Fourier series, the gradual liberation of the concept of function led to its modern definition—a rule associating a unique output to a given input. A typical question was whether the continuity of the limiting polynomials (power series) or trigonometric functions (Fourier series) necessarily implied that the limit f would also be continuous. The intuitive notion of continuity no longer met this kind of challenge.

DOI: 10.1201/9781003304135-3

Cauchy (1821) introduced the concept of continuous functions:

The function $f(x)$ will remain continuous with respect to x between the given limits, if between these limits an infinitely small increasing of variable always produces an infinitely small increasing of the function itself.

Weierstrass was more precise: the difference $f(x) - f(x_0)$ must be *arbitrarily* small, if the difference $x - x_0$ is *sufficiently* small.

Definition 3.1. *Let D be a subset of $\mathbb{R}$. The function $f(x) : D \to \mathbb{R}$ is continuous at $x_0 \in D$ if for every $\epsilon > 0$, there exists a $\delta > 0$ such that*

$$|f(x) - f(x_0)| < \epsilon \quad \text{whenever } x \in D \text{ and } |x - x_0| < \delta. \tag{3.1}$$

The function $f(x)$ is called continuous on D, if it is continuous at every point $x_0 \in D$.

This kind of mathematical style of Weierstrass installed rigor in analysis: the cutting up of "epsilons," the systematic use of inequalities and majorization. All of which becomes the daily bread of modern analysis. Historically, Klein referred to these movements as "the arithmetization of analysis."

With a mathematically unambiguous definition for the limit of a sequence in hand, we are well on an alternative way toward a rigorous understanding of continuity—the sequential characterization which is equivalent to the above $\epsilon - \delta$ definition.

Theorem 3.1. *A function $f(x) : D \to \mathbb{R}$ is continuous at $x_0 \in D$ if and only if for every sequence $x_n \in D$ we have*

$$f(x_n) \to f(x_0) \quad \text{whenever } x_n \to x_0. \tag{3.2}$$

Proof. Suppose f is continuous at x_0. For every $\epsilon > 0$, choose $\delta > 0$ as in (3.1). Since $x_n \to x_0$, there exists a positive integer N such that $|x_n - x_0| < \delta$ for all $n > N$. It follows that $|f(x_n) - f(x_0)| < \epsilon$ for all $n > N$ and so (3.2) holds.

Suppose now that (3.2) holds, but that $f(x)$ is not continuous at x_0. Then there exists $\epsilon > 0$ such that for every $\delta > 0$ there exists $x \in D$ with $|x - x_0| < \delta$ but $|f(x) - f(x_0)| \geq \epsilon$. In particular, for every n there exists $x_n \in D$ with $|x_n - x_0| < 1/n$, hence $x_n \to x_0$. But at same time $|f(x_n) - f(x_0)| \geq \epsilon$, which contradicts (3.2). □

As the following examples illustrate, working with the sequential criterion (3.2) on continuity is sometimes more convenient than using $\epsilon - \delta$ definition (3.1).

Example 3.1 (Dirichlet's Function). *Define*

$$D(x) = \begin{cases} 1, & \text{if } x \text{ is rational,} \\ 0, & \text{if } x \text{ is irrational.} \end{cases}$$

Since $\mathbb{Q}$ is dense in $\mathbb{R}$, technically, there is no way to graph $D(x)$ accurately. Moreover, $D(x)$ is nowhere-continuous on $\mathbb{R}$. Indeed, for any $x_0 \in \mathbb{R}$, there is a rational sequence r_n that converges to x_0, and also a irrational sequence q_n that converges to x_0. Since

$$\lim_{n \to \infty} D(r_n) = 1 \neq 0 = \lim_{n \to \infty} D(q_n),$$

this implies that $D(x)$ is not continuous at x_0.

The following example indicates that there exists a function that is continuous at one and only one point.

Example 3.2. *Define*

$$f(x) = \begin{cases} x, & \text{if } x \text{ is rational,} \\ 0, & \text{if } x \text{ is irrational.} \end{cases}$$

For any $x_0 \neq 0$, just as we did in Example 3.1, we can construct rational sequences $r_n \to x_0$ and irrational sequences $q_n \to x_0$ such that

$$\lim_{n \to \infty} f(r_n) = x_0 \quad \text{and} \quad \lim_{n \to \infty} f(q_n) = 0.$$

Hence $f(x)$ is not continuous at every point $x_0 \neq 0$.

We now show that $f(x)$ is continuous at $x_0 = 0$. For any $\epsilon > 0$, let $\delta = \epsilon$. Then

$$|f(\dot{x}) - f(0)| = \begin{cases} |x|, & \text{if } x \text{ is rational,} \\ 0, & \text{if } x \text{ is irrational.} \end{cases}$$

Thus,

$$|f(x) - f(0)| < \epsilon \quad \text{whenever} \quad |x - 0| < \delta.$$

So $f(x)$ is continuous at $x = 0$ only.

This example also forces us to emphasize the continuity at a particular point rather than over an interval. This is a significant departure from thinking of continuous functions as curves that can be drawn without lifting the pen from paper. The following example is even more surprising. It was discovered by Thomae at 1875.

Example 3.3 (Thomae's Function). *Define*

$$T(x) = \begin{cases} 1, & \text{if } x = 0, \\ 1/q, & \text{if } x = p/q \text{ in lowest terms with } q > 0, \\ 0, & \text{if } x \text{ is irrational.} \end{cases}$$

If $x_0 \in \mathbb{Q}$, then $T(x_0) > 0$. On the other hand, there is an irrational sequence q_n converging to x_0, which yields

$$\lim_{n \to \infty} T(q_n) = 0 \neq T(x_0).$$

Thus, $T(x)$ is not continuous at every rational point. The twist comes when x_0 is irrational. We have seen that all irrational sequences converge to zero by $T(x)$. Thus, if x_0 is irrational and $T(x)$ is continuous at x_0, then $T(r_n)$ must converge to zero for all rational sequences r_n with $r_n \to x_0$.

We first see how this can be done with an example. Let $x_0 = e = 2.718281828\dots$. A particular rational approximate sequence r_n is

$$\{2, \frac{27}{10}, \frac{271}{100}, \frac{2718}{1000}, \frac{27182}{10000}, \frac{271828}{100000}, \frac{2718281}{1000000} \cdots \}.$$

In this case, we have $T(r_n)$ as

$$\{1, \frac{1}{10}, \frac{1}{100}, \frac{1}{500}, \frac{1}{5000}, \frac{1}{25000}, \frac{1}{1000000} \cdots \},$$

which is fast approaching $0 = T(e)$. We prove that this always happens. To this end, for any irrational number α, let k be the closest integer to α. Then $\alpha \in (k-1, k+1)$. For any given $\epsilon > 0$, choose $N \in \mathbb{N}$ such that $1/N < \epsilon$. Notice that, on $(k-1, k+1)$, the number of fractions in the form of p/q with $0 < q \leq N$ is finite (for example, the only such fractions with denominators 5 or smaller in $(-1, 1)$ are $\pm 1/2, \pm 1/3, \pm 2/3, \pm 1/4, \pm 3/4, \pm 1/5, \pm 2/5, \pm 3/5,$

and $\pm 4/5$). Hence we can find some δ small enough that interval $(\alpha - \delta, \alpha + \delta) \subset (k-1, k+1)$ and it contains none of these fractions. For any x with $|x - \alpha| < \delta$, if $x = p/q$ in lowest terms, then

$$|T(x) - T(\alpha)| = |T(x)| = \frac{1}{q} < \frac{1}{N} < \epsilon$$

since $q > N$. Alternately, if x is irrational, then

$$|T(x) - T(\alpha)| = |T(x)| = 0 < \epsilon$$

as well. In either case, we have found a $\delta > 0$ such that

$$|T(x) - T(\alpha)| < \epsilon \quad \text{whenever } |x - \alpha| < \delta.$$

This proves that $\lim_{x \to \alpha} T(x) = 0 = T(\alpha)$ as desired. In summary, we demonstrade that $T(x)$ is continuous at every irrational number and discontinuous at every rational number.

In general, to prove discontinuity of a function at a given point x_0, we can phrase the discontinuity in terms of the sequential characterization.

Theorem 3.2 (Criterion for Discontinuity). *Let $f(x) : D \to \mathbb{R}$ and $x_0 \in D$. If there exists a sequence $x_n \in D$ with $x_n \to x_0$ but such that $f(x_n)$ does not converge to $f(x_0)$, then f is discontinuous at x_0.*

Example 3.4. *Define*

$$f(x) = \begin{cases} \sin(1/x), & \text{if } x \neq 0, \\ 0, & \text{if } x = 0. \end{cases}$$

Clearly, the sequence $x_n = 1/(2n\pi + \pi/2) \to 0$ as $n \to \infty$, but

$$f(x_n) = \sin\left(2n\pi + \frac{\pi}{2}\right) = \sin\left(\frac{\pi}{2}\right) = 1 \neq f(0),$$

so f is not continuous at $x = 0$ (see Figure 3.1).

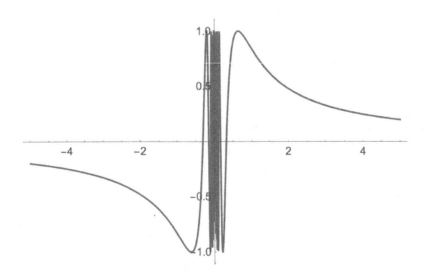

FIGURE 3.1
$f(x) = \sin(1/x), \ x \neq 0.$

3.2 Limits of Functions

Mathematical knowledge adds a manly vigor to the mind, frees it from prejudice, credulity, and superstition. — John Arbuthnot

Assume that $f(x)$ is not continuous at x_0 or not even defined there; in such a situation it is interesting to know what happens to $f(x)$ as x approaches x_0. We begin with a general definition of a functional limit $\lim_{x \to x_0} f(x) = L$, which is similar to the definition of limit of a sequence.

Definition 3.2. *Let* $f : D \to \mathbb{R}$ *and let* x_0 *be a limit point of* D. *We say that* $\lim_{x \to x_0} f(x) = L$ *if for any* $\epsilon > 0$, *there exists a* $\delta > 0$ *such that*

$$|f(x) - L| < \epsilon \qquad \text{whenever} \ \ 0 < |x - x_0| < \delta.$$

In contrast to the definition of continuity, here the additional inequality $0 < |x - x_0|$ is just a short-cut for saying $x \neq x_0$. We see from this definition that the issue of what happens when $x = x_0$ is irrelevant from the point of view of functional limits. Indeed, x_0 may not even be in the domain of f.

Equivalently, in terms of sequences, we have

Definition 3.3. $\lim_{x \to x_0} f(x) = L$ *if for every sequence* $x_n \in D$ *satisfying* $x_n \neq x_0$ *and* $x_n \to x_0$, *we have*

$$\lim_{n \to \infty} f(x_n) = L.$$

Since the limit of a sequences is unique, we see that the limit of a function, when it exists, is unique too. We now introduce two standard but weaker limits—right and left-sided limits, which are of particular interest in connection with monotone functions, as we will study in Section 4.

Definition 3.4. *Let* $f : D \to \mathbb{R}$ *and let* x_0 *be a limit point of* D. *We say that the right-hand limit* $\lim_{x \to x_0^+} f(x) = L$ *if for any* $\epsilon > 0$, *there exists a* $\delta > 0$ *such that*

$$|f(x) - L| < \epsilon \ \ \text{whenever} \ \ x_0 < x < x_0 + \delta.$$

Similarly, we define the left-hand limit $\lim_{x \to x_0^-} f(x) = L$ *if for any* $\epsilon > 0$, *there exists a* $\delta > 0$ *such that*

$$|f(x) - L| < \epsilon \ \ \text{whenever} \ \ x_0 - \delta < x < x_0.$$

As an application of one-sided limits, we can classify the discontinuity of a function into one of the following three categories.

Definition 3.5. *Let* f *be discontinuous at* x_0.

(1) *If* $\lim_{x \to x_0} f(x)$ *exists but differs from* $f(x_0)$, *then* f *has a removable discontinuity at* x_0.

(2) *If both sided limits exist but are distinct, then* f *has a jump discontinuity at* x_0.

(3) *If* $\lim_{x \to x_0} f(x)$ *does not exist for some other reason, then* f *has an essential discontinuity at* x_0.

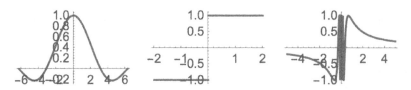

FIGURE 3.2
Examples of three types of discontinuity.

The following three examples illustrate each discontinuity at $x = 0$, respectively (see Figure 3.2):

$$(1)\ f(x) = \frac{\sin x}{x}, \quad (2)\ f(x) = \frac{|x|}{x}, \quad (3)\ f(x) = \sin\left(\frac{1}{x}\right).$$

It is also interesting to see that Dirichlet's function in Example 3.1 has an essential discontinuity everywhere, while Thomae's function in Example 3.3 has a jump discontinuity at every rational point.

It is easy to prove algebraic limit theorems for function limits using established properties of sequences. In particular, we have the following economical way to show the nonexistence of a limit. All proofs are left to the reader.

Theorem 3.3. *Let $f : D \to \mathbb{R}$ and let x_0 be a limit point of D. If there are two sequences $x_n, y_n \in D$ satisfying $x_n \neq x_0, y_n \neq x_0$ and*

$$\lim_{n\to\infty} x_n = \lim_{n\to\infty} y_n = x_0 \quad but \quad \lim_{n\to\infty} f(x_n) \neq \lim_{n\to\infty} f(y_n),$$

then $\lim_{x\to x_0} f(x)$ does not exist.

3.3   Three Fundamental Theorems

All the truths of mathematics are linked to each other, and all means of discovering them are equally admissible. — Adrien-Marie Legendre

The behavior of a continuous function defined on an interval is at the root of the analysis. If $f(x)$ is continuous at x_0, we can rewrite the criterion as

$$\lim_{x\to x_0} f(x) = f\left(\lim_{x\to x_0} x\right).$$

In other words, f commutes with the $\lim_{x\to x_0}$ if f is continuous. An interesting question is to sort out what happens when x_0 is replaced by a particular set $D \subset \mathbb{R}$. In this section, using solely the completeness theorems of the real numbers in Section 1.1, we rigorously establish the basic properties of such functions in three fundamental theorems when $D = [a, b]$, a finite closed interval in $\mathbb{R}$. The requirement of the finite closed interval is not incidental. As we will see, these fundamental theorems depend not only on the continuity of f but also on the topological properties of $[a, b]$. For example, the intermediate value theorem depends on *connectedness*; the extreme value theorem and the uniform continuity theorem rely on *compactness*. Moreover, during the process, we will see how to put the completeness theorems into action.

We begin with the extreme value theorem, which is used (at least implicitly) all the time in solving maximum-minimum problems in calculus because it is taken for granted that a continuous function on a closed interval indeed attains a maximum and a minimum.

Theorem 3.4 (Extreme Value Theorem). *Let f be continuous on $[a, b]$. Then f is bounded. Moreover, f attains its maximum and minimum on $[a, b]$, i.e., there are $x_1, x_2 \in [a, b]$ such that*

$$\min_{x \in [a,b]} \{f(x)\} = f(x_2) \leq f(x) \leq f(x_1) = \max_{x \in [a,b]} \{f(x)\}.$$

Proof. We show that f is bounded by contradiction. Assume that f is not bounded on $[a, b]$. Then for each $n \in \mathbb{N}$ there is an $x_n \in [a, b]$ such that $|f(x_n)| > n$. By the Bolzano-Weierstrass theorem, x_n has a subsequence x_{n_k} that converges to some real number $x_0 \in [a, b]$. Since f is continuous at x_0, we have $\lim_{k \to \infty} f(x_{n_k}) = f(x_0)$, which contradicts $\lim_{k \to \infty} |f(x_{n_k})| = \infty$. This proves that f is bounded.

Let $M = \sup_{x \in [a,b]} \{f(x)\}$. By the preceding argument, we already know that M is finite. We now show by contradiction that f attains M at some real number $x_1 \in [a, b]$. If there is no number $x_1 \in [a, b]$ such that $f(x_1) = M$, then $1/(M - f(x))$ is continuous and bounded on $[a, b]$. Let

$$\frac{1}{M - f(x)} \leq m,$$

where m is some positive constant. Thus,

$$f(x) \leq M - \frac{1}{m}.$$

This contradicts the definition of M. Thus f assumes its maximum at some $x_1 \in [a, b]$. Applying the above argument to $-f$, we conclude that $-f$ assumes its maximum at some $x_2 \in [a, b]$. It follows that f assumes its minimum at x_2. □

An alternative proof of the boundedness based on the nested interval theorem is as follows.

Proof. Assume that f is unbounded. We show that f must be discontinuous at least one point $x_0 \in [a, b]$. For this purpose, we use the nested interval theorem to find such a point x_0. By induction, we define

1. $a_1 = a, b_1 = b, m_1 = (a_1 + b_1)/2$. The function f must be unbounded on at least one of $[a_1, m_1]$ and $[m_1, b_1]$. Choose one of them on which f is unbounded and define the interval as $[a_2, b_2]$.

2. Suppose that $[a_n, b_n]$ is an interval on which f is unbounded. Let $m_n = (a_n + b_n)/2$. Choose one of $[a_n, m_n]$ and $[m_n, b_n]$ on which f is still unbounded as $[a_{n+1}, b_{n+1}]$.

Thus we see that f is unbounded on every $[a_n, b_n]$ and

$$a \leq a_n \leq a_{n+1} < b_{n+1} \leq b_n \leq b,$$

and

$$b_{n+1} - a_{n+1} = \frac{1}{2}(b_n - a_n) = \frac{1}{2^n}(b - a).$$

By the nested interval theorem, both a_n and b_n converge to $x_0 \in [a, b]$. We now show that f is not continuous at x_0. To see this, let ϵ be any positive number and suppose there exists a $\delta > 0$ such that $|x - x_0| < \delta$ implies

$$|f(x) - f(x_0)| < \epsilon. \tag{3.3}$$

On the other hand, there exists a sufficiently large n such that $b_n - a_n < \delta$, so every $x \in [a_n, b_n]$ satisfies $|x - x_0| < \delta$. Recall that f is unbounded on $[a_n, b_n]$; it follows that there is at least one $x \in [a_n, b_n]$ for which $f(x) > f(x_0) + \epsilon$ or $f(x) < f(x_0) - \epsilon$. This contradicts (3.3). □

As an example of the "existence theorems," the above proof only shows us the maximum exists, but does not reveal where the maximum occurs. Based on the Bolzano-Weierstrass theorem, we now give another proof of the extreme value theorem which also demonstrates how to find the maximum.

Proof. Let $M = \sup_{x \in [a,b]} \{f(x)\}$. The definition of supremum implies that $f(x) \leq M$ for all $x \in [a, b]$. We now show how to use the Bolzano-Weierstrass theorem to find the point $x_0 \in [a, b]$ such that $f(x_0) = M$. To this end, let $n \geq 1$ be any integer. Since $M - 1/n$ is not an upper bound, there exists a $x \in [a, b]$ such that $M - 1/n < f(x)$.

We now construct a sequence x_n by choosing

$$\text{for each } n, \ x_n \in [a, b] \text{ such that } M - \frac{1}{n} < f(x_n).$$

By the Bolzano-Weierstrass theorem, there exists a subsequence x_{n_k} which converges to a number $x_0 \in [a, b]$. We now show that $f(x_0) = M$. Since f is continuous, as usual, we have

$$\lim_{k \to \infty} f(x_{n_k}) = f(x_0).$$

On the other hand, in view of the construction of the sequence, we have

$$f(x_{n_k}) > M - \frac{1}{n_k} \geq M - \frac{1}{k}$$

and so by taking limits of both sides we see that

$$f(x_0) = \lim_{k \to \infty} f(x_{n_k}) \geq \lim_{k \to \infty} \left(M - \frac{1}{k} \right) = M.$$

Combining the inequality $f(x) \leq M$ we have $f(x_0) = M$ as desired. □

Let $f(x) = x(1 - x)$ on $[0, 1]$. The sequence $x_n = 1/2 - 1/n$ for $n \geq 3$ is one of the desired sequences in the above proof to approach the maximum value $1/4$. The following examples show that Theorem 3.4 is false if either f is discontinuous or $[a, b]$ is replaced by an open interval.

Example 3.5. *The function $f(x) = 1/x$ is continuous but unbounded on $(0, 1)$. The function $g(x) = x^2$ is continuous and bounded on $(0, 1)$, but it does not have a maximum value on $(0, 1)$.*

Example 3.6. *Define*

$$f(x) = \begin{cases} \frac{1}{x} \sin \left(\frac{1}{x} \right), & x \in (0, 1], \\ 0, & x = 0. \end{cases}$$

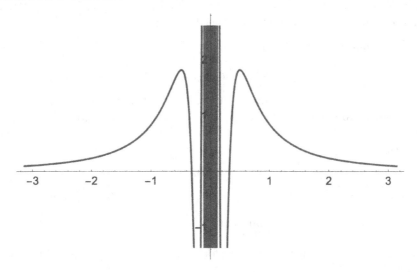

FIGURE 3.3
$f(x) = \sin(1/x)/x, \ x \neq 0.$

This function f is not bounded on $[0, 1]$ even though f is continuous on $(0, 1]$ (see Figure 3.3).

The extreme value theorem demonstrates that a continuous function f attains both maximum and minimum. The following theorem shows that f indeed achieves every value in between. This result is pictorially obvious. A curve without break points cannot jump from one height to another. It must past through every intermediate height.

Theorem 3.5 (Intermediate Value Theorem). *Let f be continuous on $[a, b]$. If c lies strictly between $f(a)$ and $f(b)$, then there is a point $x_0 \in (a, b)$ such that $f(x_0) = c$.*

Proof. Without loss of generality, we assume that $f(a) < c < f(b)$. By induction, we define

1. $a_1 = a, b_1 = b$.
2. Suppose that $[a_n, b_n]$ has been defined such that $f(a_n) < c < f(b_n)$. Let $m_n = (a_n + b_n)/2$. If $f(m_n) = c$, then we are done. Otherwise,

    (i) If $f(m_n) < c$, let $a_{n+1} = m_n, b_{n+1} = b_n$.
    (ii) If $f(m_n) > c$, let $a_{n+1} = a_n, b_{n+1} = m_n$.

Thus we see that, for each n,

$$a \leq a_n \leq a_{n+1} < b_{n+1} \leq b_n \leq b,$$

and

$$b_{n+1} - a_{n+1} = \frac{1}{2}(b_n - a_n) = \frac{1}{2^n}(b - a).$$

By the nested interval theorem, both a_n and b_n converge to $x_0 \in (a, b)$. Since f is continuous on $[a, b]$, it follows that

$$f(x_0) = \lim_{n \to \infty} f(a_n) \leq c$$

and

$$f(x_0) = \lim_{n \to \infty} f(b_n) \geq c.$$

Consequently, $f(x_0) = c$. $\qquad\square$

Here is another proof based on the least-upper-bound property.

Proof. Assume that $f(a) < c < f(b)$. Let

$$S = \{x \in [a, b] : f(x) < c\}.$$

Since $a \in S$ and $S \subset [a, b]$, S is nonempty and bounded. So $x_0 = \sup S$ exists by the least-upper-bound property, and $x_0 \in [a, b]$. We now show that $f(x_0) = c$.

For $n \in \mathbb{N}$, $x_0 - 1/n$ is not an upper bound of S, so there exists $x_n \in S$ such that $x_n \geq x_0 - 1/n$. Thus,

$$x_0 - \frac{1}{n} \leq x_n \leq x_0.$$

By the squeeze theorem, we have $\lim_{n \to \infty} x_n = x_0$. The continuity of f implies that

$$\lim_{n \to \infty} f(x_n) = f(x_0).$$

Since $x_n \in S$ for every n, we have $f(x_n) < c$ for every n, and so $f(x_0) \leq c$. Moreover, $x_0 < b$ since $c < f(b)$.

On the other hand, since $x_0 < b$, there exists an N such that $x_0 + 1/n < b$ for all $n > N$. Since $x_0 + 1/n > x_0$, we have $x_0 + 1/n \notin S$ for all $n > N$, and so

$$f\left(x_0 + \frac{1}{n}\right) \geq c.$$

The continuity of f implies that

$$\lim_{n \to \infty} f\left(x_0 + \frac{1}{n}\right) = f(x_0) \geq c.$$

In particular, $x_0 \neq a$ since $f(a) < c$.

In summary, we obtain $f(x_0) = c$ and $x_0 \in (a, b)$, as desired. $\square$

Corollary 3.1. *If $f(x)$ is continuous on $[a, b]$ satisfying $f(a) \cdot f(b) < 0$, then there is some $c \in (a, b)$ such that $f(c) = 0$.*

Recall that a sign chart is often used in calculus to identify where a derivative is positive or negative in order to determine where a function is increasing and decreasing. The construction of the sign chart itself is simply an application of this Corollary. The bisection method used in the construction of the nested intervals indeed provides an approach to approximate the value c.

Example 3.7. *Approximate the solution of $x^3 - 2x - 5 = 0$.*

Let $f(x) = x^3 - 2x - 5$. Since

$$f(2) = -1 < 0 < f(3) = 16,$$

the given equation must have a solution in $(2, 3)$. Using the bisection method, we obtain

Test $f(a_n) < 0 < f(b_n)$	The root range
$f(2) < 0 < f(3)$	$(2, 3)$
$f(2) < 0 < f(2.5)$	$(2, 2.5)$
$f(2) < 0 < f(2.25)$	$(2, 2.25)$
$f(2) < 0 < f(2.215)$	$(2, 2.125)$
$f(2.0625) < 0 < f(2.215)$	$(2.0625, 2.125)$
$f(2.09375) < 0 < f(2.2125)$	$(2.09375, 2.125)$
$f(2.09375) < 0 < f(2.109375)$	$(2.09375, 2.109375)$

Thus, one approximate solution is given by $(2.09375 + 2.109375)/2 = 2.1015625$ with error less than $1/2^7 = 0.007515$.

Corollary 3.2. *If $f : [a,b] \to [a,b]$ is continuous, then there is an $x \in [a,b]$ such that $f(x) = x$, a fixed point of f.*

Corollary 3.3. *If f is a continuous function, then $f([a,b])$, the range of f on $[a,b]$, is also an interval or a single point. More precisely, $f([a,b]) = [\min_{x\in[a,b]} f(x), \max_{x\in[a,b]} f(x)]$.*

Definition 3.6. *A function f has the intermediate value property on $[a,b]$ if for all $x_1, x_2 \in [a,b]$ with $x_1 < x_2$ and every c strictly between $f(x_1)$ and $f(x_2)$, then there is a $\xi \in (x_1, x_2)$ with $f(\xi) = c$.*

Thus, every continuous function on $[a,b]$ has the intermediate value property. But, the converse of the intermediate value theorem is false. Let

$$f(x) = \begin{cases} \sin(1/x), & \text{if } x \neq 0, \\ 0, & \text{if } x = 0. \end{cases}$$

By Example 3.4, f is not continuous at 0, but it does exhibit the intermediate value property on $[0,1]$. Moreover, Lebesgue proved that there exists a function $f : [0,1] \to [0,1]$ that has the intermediate value property and is nowhere continuous.

By the $\epsilon - \delta$ definition 3.1, if f is continuous at x_0, then around x_0 there is an "island" $(x_0 - \delta, x_0 + \delta)$ such that $f(x)$ does not stray by more than ϵ from $f(x_0)$. However, the δ measuring the size of the island usually depends on both ϵ and x_0, In the other words, the δ is not necessarily the same for all $x_0 \in D$.

Example 3.8. *Consider $f(x) = x^2$ on $D = \mathbb{R}$.*

For any $x_0 \in \mathbb{R}$, we have

$$|f(x) - f(x_0)| = |x^2 - x_0^2| = |x + x_0|\,|x - x_0|.$$

If $|x - x_0| < 1$, we have $x \in (x_0 - 1, x_0 + 1)$ and

$$|x + x_0| \leq |x| + |x_0| < 2|x_0| + 1.$$

Now, let $\epsilon > 0$. If we choose $\delta = \min\{1, \epsilon/(2|x_0| + 1)\}$, then $|x - x_0| < \delta$ implies

$$|f(x) - f(x_0)| < (2|x_0| + 1)\,|x - x_0| < (2|x_0| + 1)\left(\frac{\epsilon}{2|x_0| + 1}\right) = \epsilon.$$

It is clear that the response δ depends on the value of x_0. For example, let $\epsilon = 1$; a response of $\delta = 1/3$ is good for $x_0 = 1$ but it requires $\delta = 1/21$ for $x_0 = 10$. The function f now becomes "less and less" continuous at every point as x gets bigger. Moreover, for any $\epsilon > 0$, no matter how small δ is, if we let $x_0 = 2\epsilon/\delta$ and $x_1 = 2\epsilon/\delta + \delta/2$, we have $|x_1 - x_0| < \delta$ but

$$|f(x_1) - f(x_0)| > \left(\frac{2\epsilon}{\delta} + \frac{2\epsilon}{\delta}\right)\frac{\delta}{2} = 2\epsilon.$$

Thus no response δ fits all x_0 values.

If a *single response* δ, i.e., a δ that depends only on ϵ and not on $x \in D$, can be chosen, this indicates the function behaves in the same way throughout the entire D. Such a strengthened version of continuity is called *uniformly continuous* on D.

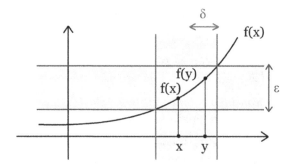

FIGURE 3.4
Uniform continuous.

Definition 3.7. *A function $f : D \to \mathbb{R}$ is uniformly continuous on D if for every $\epsilon > 0$ there exists $\delta > 0$ such that*

$$|f(x) - f(y)| < \epsilon \quad \text{whenever} \quad x, y \in D, |x - y| < \delta.$$

Remark. In contrast to continuity at a single point x_0, uniform continuity is a global property concerning a function and a set on which it is defined. It does not pre-specify a domain point x_0 as is the case for continuity. In the uniformly continuous case, x and x_0 play a symmetric role, so we call them x and y in the above definition (see Figure 3.4).

To better recognize the difference of δ in the definitions of continuity and uniform continuity, we turn to their formulations in terms of quantifiers and logical symbols:

Continuity : $(\forall x_0 \in D)(\forall \epsilon > 0)(\exists \delta > 0)[(x \in D \wedge |x - x_0| < \delta) \to |f(x) - f(x_0)| < \epsilon]$,

Uniform continuity : $(\forall \epsilon > 0)(\exists \delta > 0)(\forall x \in D)(\forall y \in D)[|x - y| < \delta \to |f(x) - f(y)| < \epsilon]$.

In the second formulation, the order of quantifiers shows where the δ depends only on ϵ. As always, a good understanding of the use of quantifiers is a benefit to us for rigorous proofs.

Resembling Theorem 3.1, we can characterize uniform continuity in terms of sequences.

Theorem 3.6. *A function $f : D \to \mathbb{R}$ is uniformly continuous on D if and only if for every sequence $x_n, y_n \in D$,*

$$f(x_n) - f(y_n) \to 0 \quad \text{whenever} \quad x_n - y_n \to 0 \text{ as } n \to \infty.$$

Example 3.9. *Show that $f(x) = \sin(1/x)$ is not uniformly continuous on $(0, 1)$.*

Clearly, f is continuous on $(0, 1)$. Let

$$x_n = \frac{1}{2n\pi + \pi/2}, \quad y_n = \frac{1}{2n\pi}.$$

We have $x_n - y_n \to 0$ as $n \to \infty$, but

$$f(x_n) - f(y_n) = 1,$$

for all $n \in \mathbb{N}$.

By Examples 3.8 and 3.9, we see that there are continuous functions that are not uniformly continuous on $\mathbb{R}$ or (a, b). However, when we turn our attention to a bounded closed interval, the following astonishing theorem asserts that there is a single response δ that works with every point in the closed interval.

Theorem 3.7 (Uniform Continuity Theorem). *If f is continuous on $[a, b]$, then f is uniformly continuous on $[a, b]$.*

Proof. We use the sequential characterization and a proof by contradiction. Assume that f is not uniformly continuous on $[a, b]$. Then there is some $\epsilon > 0$ such that for all $\delta > 0$, there exist $x, y \in [a, b]$ with $0 < |x - y| < \delta$ and yet $|f(x) - f(y)| \geq \epsilon$. In particular, taking $\delta = 1, 1/2, 1/3, \ldots$, we see that for every $n \in \mathbb{N}$ there exist $x_n, y_n \in [a, b]$ such that

$$|x_n - y_n| < \frac{1}{n} \quad \text{but} \quad |f(x_n) - f(y_n)| \geq \epsilon.$$

By the Bolzano-Weierstrass theorem, x_n has a subsequence x_{n_k} such that

$$x_{n_k} \to x_0 \in [a, b].$$

Since

$$|y_{n_k} - x_0| \leq |y_{n_k} - x_{n_k}| + |x_{n_k} - x_0| < \frac{1}{n_k} + |x_{n_k} - x_0|,$$

it follows that $y_{n_k} \to x_0$ as well. The continuity of f now implies that $|f(x_{n_k}) - f(y_{n_k})| \to 0$. Thus, there is $K \in \mathbb{N}$ such that

$$|f(x_{n_k}) - f(y_{n_k})| < \epsilon$$

for $k > K$. This contradicts the hypothesis above. Thus, f must be uniformly continuous on $[a, b]$. $\square$

To reinforce the importance of the completeness theorems in Chapter 1, we offer two more alternative proofs of Theorem 3.7. They are based on the nested interval theorem and the Heine-Borel theorem, respectively.

Proof by the nested interval theorem. Assume that f is not uniformly continuous on $[a, b]$. Then there is no uniform response δ to ϵ on $[a, b]$. Since

$$[a, b] = [a, (a + b)/2] \cup [(a + b)/2, b],$$

then on at least one of these half intervals, there is no uniform response (if δ_1 works for the first half interval and δ_2 works for the second half interval, then $\min\{\delta_1, \delta_2\}$ works for $[a, b]$). By repeating this process, we obtain nested intervals of arbitrarily short length on which there is no uniform response to ϵ.

Let x_0 be the point common to all of these nested intervals. We now show that there can be no response to $\epsilon/2$ at this point. Thus, f is not continuous at x_0, which contradicts the assumption. To this end, there is a nested interval of length less than $\delta/2$. Since there is no uniform response to ϵ in this nested interval, it must contain an x for which $\delta/2$ is not a response to ϵ, this means that there is a y with $|x - y| < \delta/2$ for which

$$|f(x) - f(y)| \geq \epsilon.$$

But, on the other hand, both x and y now lie within δ of x_0. If δ is a response to $\epsilon/2$ at x_0, then we have

$$|f(x) - f(y)| \leq |f(x) - f(x_0)| + |f(x_0) - f(y)| < \frac{\epsilon}{2} + \frac{\epsilon}{2} = \epsilon.$$

This leads to a contradiction. $\square$

Proof by the Heine-Borel theorem. Assume that f is continuous on $[a, b]$. By the definition of continuity, for $\epsilon > 0$ and $x_0 \in [a, b]$, there is some $\delta(x_0)$ such that

$$|f(x) - f(x_0)| < \frac{\epsilon}{2} \quad \text{whenever } x \in [a, b] \text{ and } |x - x_0| < \delta(x_0).$$

Then the family

$$\{(x_0 - \delta(x_0)/2, x_0 + \delta(x_0)/2)\}_{x_0 \in [a,b]}$$

forms a open covering of $[a, b]$. Applying the Heine-Borel theorem, we find that there are $n \in \mathbb{N}$ and $x_1, x_2, \ldots, x_n \in [a, b]$ such that $\cup_{k=1}^{n} (x_k - \delta(x_k)/2, x_k + \delta(x_k)/2)$ covers $[a, b]$. Let

$$\delta = \min \left\{ \frac{\delta(x_1)}{2}, \frac{\delta(x_2)}{2}, \ldots, \frac{\delta(x_n)}{2} \right\}.$$

For any $x, y \in [a, b]$ with $|x - y| < \delta$, there is some $m, 1 \le m \le n$ such that $y \in (x_m - \delta(x_m)/2, x_m + \delta(x_m)/2)$. Hence $|y - x_m| < \delta(x_m)/2$. By the definition of δ, we also have $|x - y| < \delta(x_m)/2$. This implies that

$$|x - x_m| \le |x - y| + |y - x_m| < \frac{\delta(x_m)}{2} + \frac{\delta(x_m)}{2} = \delta(x_m).$$

In view of the choice of $\delta(x_m)$, we have

$$|f(x) - f(x_m)| < \frac{\epsilon}{2} \quad \text{and} \quad |f(y) - f(x_m)| < \frac{\epsilon}{2}.$$

Therefore

$$|f(x) - f(y)| \le |f(x) - f(x_m)| + |f(y) - f(x_m)| < \frac{\epsilon}{2} + \frac{\epsilon}{2} = \epsilon.$$

This shows that f is uniformly continuous on $[a, b]$. $\square$

In general, a continuous function does not necessarily map a Cauchy sequence into a Cauchy sequence. For example, let

$$f(x) = \frac{1}{x} \sin \left(\frac{1}{x} \right), \quad x \in (0, 1).$$

Clearly, the sequence $x_n = \frac{2}{n\pi}$ is Cauchy, but $f(x_n) = \frac{n\pi}{2} \sin(n\pi/2)$ is not. However, the next theorem show that we have a different story if f is uniformly continuous.

Theorem 3.8. *Let f be uniformly continuous on (a, b). If $x_n \in (a, b)$ is a Cauchy sequence, then so is $f(x_n)$. Moreover, both $f(a^+) = \lim_{x \to a^+} f(x)$ and $f(b^-) = \lim_{x \to b^-} f(x)$ exist.*

Proof. Let $\epsilon > 0$. By assumption, there exists a $\delta > 0$ such that

$$|f(x) - f(y)| < \epsilon \quad \text{whenever } x, y \in (a, b), |x - y| < \delta.$$

Since x_n is a Cauchy sequence, there exists a positive integer N such that

$$|x_n - x_m| < \delta \quad \text{whenever } m, n > N.$$

Thus, we see that

$$|f(x_n) - f(x_m)| < \epsilon \quad \text{whenever } m, n > N.$$

This proves that $f(x_n)$ is also a Cauchy sequence.

We now define

$$f(a^+) = \lim_{n \to \infty} f(x_n) \text{ for any sequence } x_n \in (a, b) \text{ with } x_n \to a. \tag{3.4}$$

To guarantee (3.4) is well-defined, we prove that

1. If $x_n \in (a, b)$ and $x_n \to a$, then $f(x_n)$ converges.

2. If $x_n, y_n \in (a, b)$ and both converge to a, then both $f(x_n)$ and $f(y_n)$ converge to the same limit.

Here (1) implies the existence of (3.4) and (2) guarantees its uniqueness, i.e., the limit is independent of the choice of sequences. To prove (1), note that x_n is Cauchy, so $f(x_n)$ is also Cauchy by the preceding argument. Hence $f(x_n)$ converges. To prove (2), based on the choice of sequences x_n and y_n, we construct a third sequence z_n by

$$\{z_n\} = \{x_1, y_1, x_2, y_2, \cdots, x_n, y_n, \cdots\}.$$

It is clear that $z_n \in (a, b)$ with $z_n \to a$, and so $f(z_n)$ converges by (1). Since both $f(x_n)$ and $f(y_n)$ are subsequences of $f(z_n)$, we find that

$$\lim_{n \to \infty} f(x_n) = \lim_{n \to \infty} f(z_n) = \lim_{n \to \infty} f(y_n).$$

Similarly, we can show that $f(b^-)$ exists. $\qquad\square$

Consider the following extended function of $f(x)$:

$$g(x) = \begin{cases} f(a^+), & \text{if } x = a, \\ f(x), & \text{if } a < x < b. \\ f(b^-), & \text{if } x = b. \end{cases}$$

If $f(x)$ is uniformly continuous on (a, b), then $g(x)$ is uniformly continuous on $[a, b]$ by Theorem 3.7.

The proofs of the above three fundamental theorems all rely upon the completeness theorems of the real numbers. In closing this section, we point out that these three fundamental theorems are logically equivalent to the completeness theorems of the real numbers. In particular, we show that all three theorems are equivalent to the least-upper-bound property. To see this, we partition $\mathbb{R}$ in sprit of *Dedekind cuts*, which divides the number line into a left-set and a right-set. We first set the stage by converting the least-upper-bound property into a more intuitively convincing version.

Definition 3.8 (The Cut Principle). *Suppose A and B are nonempty disjoint subsets of $\mathbb{R}$ whose union is all of $\mathbb{R}$, and every element of A is less than every element of B. Then there exists a number $c \in \mathbb{R}$, which is called a cut point for A and B, such that every $x \in A, x \leq c$, and every $x \in B, x \geq c$.*

We then show that

Theorem 3.9. *The cut principle is equivalent to the least-upper-bound property.*

Proof. " $\Longrightarrow$ " Let the set $S \subset \mathbb{R}$ be nonempty and bounded above. Define

$$B = \{b \in \mathbb{R} : b \text{ is an upper bound of } S\} \neq \emptyset, \quad A = \mathbb{R} \setminus B \neq \emptyset.$$

Let $c \in \mathbb{R}$ be the cut point for A and B. We show that the c indeed is the least-upper-bound of S. In fact, assume first that $c \in A$, i.e., c is not an upper bound for S. Then there exists an $s \in S$ such that $c < s$. Let $a = (c + s)/2$. The fact that $a < s$ implies that a is not an upper bound for S, thus $a \in A$. But $a > c$, together with the fact that c is a cut point, implies that $a \in B$. That's impossible. Thus, $c \in B$ and c is an upper bound for S. Moreover, for any $\epsilon > 0$, $c - \epsilon < c$ yields that $c - \epsilon \in A$, which means that $c - \epsilon$ is not an upper bound for S. This proves that c is the least upper bound for S.

" $\Longleftarrow$ " Let A, B be a cut. The least-upper-bound property implies that $c = \sup(A)$ exists. We show that c is the cut point for A, B. For any $x \in A$, we immediately get $x \le c = \sup(A)$. For any $y \in B$, if $y < c$, appealing to the definition of supremum, there is $a' \in A$ such that $y < a' < c$. This contradicts the definition of the cut. Thus $y \ge c$, which completes the proof. $\qquad \square$

The following corollary gives us one implication on the cut when the Cut principle fails.

Corollary 3.4. *If the cut principle is not satisfied, then there exists a cut of open sets.*

Proof. Suppose there exists a cut for A and B that does not possess a cut point. We now show that both sets A and B are open. Here, we show that A is open only, the argument for B is similar. Assume that $c \in A$, then $(-\infty, c] \subset A$. If there is no $x \in A$ such that $x > c$, then $x \le c$ for all $x \in A$ and $c < b$ for all $b \in B$ since $c \in A$. This implies that c is a cut point. The contradiction implies that there must be an $a \in A$ such that $a > c$. Thus, $[c, a] \subset A$ and so $c \in (-\infty, a) \subset A$, which shows that A is open. $\qquad \square$

Thus, if the cut principle does not hold, let A, B be a cut of open set. Clearly, if $f, g : \mathbb{R} \to \mathbb{R}$ are continuous, then the function

$$F(x) = \begin{cases} f(x), & x \in A, \\ g(x), & x \in B \end{cases}$$

is also continuous on $\mathbb{R}$.

Theorem 3.9 indicates that the cut principle is a completeness property. Thus, to prove that each of the three fundamental theorems indeed is a completeness property, it suffices to show that it implies the cut principle. We proceed by contrapositive. Based on Corollary 4, the proof will be accomplished by a suitable choice of functions f and g such that F provides a counterexample for each theorem. Here we give the proofs for the extreme value theorem and the intermediate value theorem. We leave the proof of the uniform continuous theorem to the reader.

The extreme value theorem does not hold, if the cut principle is not satisfied. Let

$$F(x) = \begin{cases} x, & x \in A \cap [a, b], \\ a - 1, & x \in B \cap [a, b]. \end{cases}$$

By Corollary 4, F is continuous on $[a, b]$. We claim that F does not achieve its maximum on $[a, b]$. To see this, assume there is $c \in [a, b]$ such that $F(c) = \max_{x \in [a,b]} f(x)$. Since $F(x) = x \ge a > a - 1 = F(y)$ for all $x \in A \cap [a, b]$ and $y \in B \cap [a, b]$, it follows that $c \notin B \cap [a, b]$. Thus, $c \in A \cap [a, b]$. However, since A is open, there exists a $c' \in A \cap [a, b]$ such that $c' > c$. This implies that $F(c') = c' > c = F(c)$, a contradiction.

The intermediate value theorem does not hold, if the cut principle is not satisfied. Let

$$F(x) = \begin{cases} 0, & x \in A \cap [a, b], \\ 1, & x \in B \cap [a, b]. \end{cases}$$

By Corollary 4, F is continuous on $[a, b]$. Clearly, F violates the conclusion of the intermediate value property. Thus, in addition to the six theorems in Chapter 1, each of these three theorems could well be used to "axiomatize" the completeness of the real numbers.

3.4   From the Intermediate Value Theorem to Chaos

Life is the twofold internal movement of composition and decomposition at once general and continuous. — Henri de Blainville

Let $f : [a, b] \to [a, b]$ be a continuous function. In Section 1.3.7, we studied additional conditions that ensure the orbit of $x \in [a, b]$

$$x, f(x), f(f(x)), f(f(f(x))), \cdots$$

converges to a fixed point of f. A natural generalization of the fixed point is the *periodic point*.

Definition 3.9. *Let $n \in \mathbb{N}$. A point $x \in [a, b]$ is called a periodic point of f with period n if*

$$f^n(x) = x, f^k(x) \neq x, \quad for \; k = 1, 2, \ldots, n - 1,$$

where f^n is the nth iteration of f.

Observe that, if x is a periodic point of f with period n, then the orbit of x is an infinite repetition of the pattern

$$x, f(x), f^2(x), \ldots, f^{n-1}(x).$$

Geometrically, the point of period 1 is a fixed point of f and so lies on the diagonal line $y = x$. Assume that a has period 2. Then $f(a) = b, f(b) = a$, and $a \neq b$. This implies that both points (a, b) and (b, a) are on the graph of $y = f(x)$, they are symmetric with respect to $y = x$. However, in contrast to points of period 1 and 2, the points of period 3 or higher do not have an obvious geometric interpretation in terms of the graph. For example, let (see Figure 3.5)

$$f(x) = \left\{ \begin{array}{ll} x + 1/2, & x \in [0, 1/2], \\ 2(1 - x), & x \in [1/2, 1]. \end{array} \right.$$

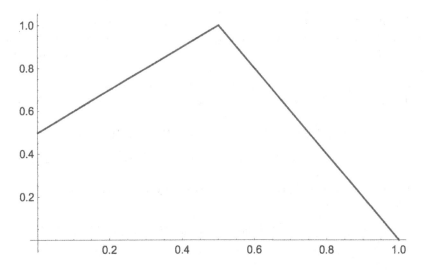

FIGURE 3.5
Periodic points of period 3.

We have $f(0) = 1/2, f^2(0) = f(1/2) = 1, f^3(0) = f(1) = 0$. So f has a point of period 3. Does f have a point of period 5? A point of period 7? Clearly, by just looking at Figure 3.5, there is no hint of the fact that every positive integer n is the period of some point! In 1975, Li and Yorke published an article with a provocative titled "Period three implies chaos" in the *American Mathematical Monthly* [67]. In this paper, solely based on the intermediate value theorem, they asserted the following theorem.

Theorem 3.10. *Let $f : [a,b] \to [a,b]$ be continuous. If f has a periodic point of period three, then f has periodic points of all other periods.*

The proof will be based on the following three elementary facts. The first fact is similar to Corollary 2, in which $f([a,b]) \subset [a,b]$.

F_1 *Let f be continuous on $[a,b]$. If $[a,b] \subset f([a,b])$, then f has at least one fixed point in $[a,b]$.*

Proof. By the assumption, there are $x_1, x_2 \in [a,b]$ such that

$$f(x_1) = a \leq x_1, \quad f(x_2) = b \geq x_2.$$

Let $g(x) = f(x) - x$. Then $g(x_1) \geq 0$ and $g(x_2) \leq 0$. The desired result follows from the intermediate value theorem. □

F_2 *Let f be continuous on $[a,b]$ and let $[c,d] \subset [a,b]$ such that $[c,d] \subset f([a,b])$. Then there is a closed interval $K \subset [a,b]$ satisfying $f(K) = [c,d]$.*

Proof. The intermediate value theorem implies that there are $x_1, x_2 \in [a,b]$ such that $f(x_1) = c$ and $f(x_2) = d$. First, consider the case $x_1 < x_2$. Define

$$\alpha = \sup\{s : f(s) = c, x_1 \leq s < x_2\}, \quad \beta = \inf\{t, : f(t) = d, \alpha \leq s \leq x_2\}.$$

The continuity implies that $f(\alpha) = c, f(\beta) = d$. Geometrically, α and β are the closest two such points with $f(\alpha) = c$ and $f(\beta) = d$. Let $K = [\alpha, \beta]$. Notice that $f(x)$ is not equal to c or d when $x \in (\alpha, \beta)$. Thus, for any $x \in K$, $f(x) \in [c,d]$, and so $f(K) = [c,d]$. The case $x_1 > x_2$ can be proved by applying the above argument to $g(x) = f(a + b - x)$. □

The technique of studying how certain sequences of sets are mapped into or onto each other is often used in studying dynamical systems. For instance, Smale uses this method in his famous "horseshoe example" [51] in which he shows how a homeomorphism on the plane can have infinitely many periodic points.

The last fact is a generalized version of the intermediate value theorem.

F_3 *Let $f : [a,b] \to [a,b]$ be continuous, and let I_0, I_1, I_{n-1} be closed subintervals of $[a,b]$. If*

$$I_{k+1} \subset f(I_k), (k = 0, 1, \ldots, n-2); \quad I_0 \subset f(I_{n-1}),$$

then there exists at least one point $x_0 \in I_0$ such that $f^n(x_0) = x_0$ and $f^k(x_0) \in I_k$ for $k = 0, 1, \ldots, n-1$.

In geometry, this means that, mapped successively by f, x_0 visits $I_0, I_1, \ldots, I_{n-1}$ in order and eventually returns to where it was.

Proof. Since $I_0 \subset f(I_{n-1})$, by F_2 there is a closed interval $K_{n-1} \subset I_{n-1}$ such that $f(K_{n-1}) = I_0$. Similarly, from

$$K_{n-1} \subset I_{n-1} \subset f(I_{n-2}),$$

there is a closed subinterval $K_{n-2} \subset I_{n-2}$ such that $f(K_{n-2}) = K_{n-1}$. Proceeding in this way, we can find $K_1 \subset I_1$ such that $f(K_1) = K_2$ where $K_2 \subset I_2$. Finally there is a subinterval $K_0 \subset I_0$ such that $f(K_0) = K_1$. In summary, we find that

$$f(K_0) = K_1, f^2(K_0) = f(K_1) = K_2, \cdots,$$
$$f^{n-1}(K_0) = K_{n-1}, f^n(K_0) = f(K_{n-1}) = I_0 \supset K_0.$$

Now, by F_1, there is a point $x_0 \in K_0 \subset I_0$ such that $f^n(x_0) = x_0$. It is clear that $f^k(x_0) \in K_k \subset I_k$ for $k = 1, 2, \ldots, n-1$. $\qquad\square$

Equipped with Facts $F_1 - F_3$, now we are ready to prove Li and Yorke's Theorem 3.10.

Proof. Let $\alpha < \beta < \gamma \in [a, b]$, which form an orbit of period three. There are only two possibilities: $f(\alpha) = \beta$ and $f(\alpha) = \gamma$. We take care of the first case:

$$f(\alpha) = \beta, \ f(\beta) = \gamma, \ f(\gamma) = \alpha.$$

The second case can be handled similarly.

To begin, let $I = [\alpha, \beta]$ and $K = [\beta, \gamma]$. Since $\beta, \gamma \in f(I)$, it follows that $K \subset f(I)$. Similarly, in view of the fact of that $f(\beta) = \gamma, f(\gamma) = \alpha$, we have $[\alpha, \gamma] = I \cup K \subset f(K)$.

For any given positive integer n, we now show that there exists a point of period n. For $n = 1$, from $K \subset f(K)$ and Fact 1, there is a fixed point of f in K. Next let $n = 2$. Consider two closed intervals $I_0 = K$ and $I_1 = I$, which meet all the requirements of Fact 2. Thus, there is a point $x_0 \in [\beta, \gamma]$ such that $f(x_0) \in [\alpha, \beta]$ and $f^2(x_0) = x_0$. We claim that x_0 is a point of period 2. For otherwise, since $[\alpha, \beta] \cap [\beta, \gamma] = \{\beta\}$, we must have $x_0 = \beta$. This implies that $\beta = f(\beta) = \gamma > \beta$, a contradiction.

Now, assume that $n > 3$. Let

$$I_0 = I_1 = I_2 = \cdots = I_{n-2} = K, I_{n-1} = I.$$

All the assumptions of Fact 3 are fulfilled. Thus there is a point $x_0 \in K = [\beta, \gamma]$ such that $f^n(x_0) = x_0$. We show that x_0 is a point of period n. By contradiction, if there is a $k \in \mathbb{N}$ and $k < n$ such that $f^k(x_0) = x_0$, then $f^{n-1}(x_0)$ will be one of the following numbers:

$$x_0, f(x_0), f^2(x_0), \ldots, f^{n-2}(x_0)$$

that are all in K. Since $f^{n-1}(x_0) \in I$ as well, it follows that $f^{n-1}(x_0) \in K \cap I$ and so $f^{n-1}(x_0) = \beta$. Thus, $x_0 = f^n(x_0) = f(\beta) = \gamma$. Fact 3 implies that

$$f(x_0) = f(\gamma) = \alpha \in I_1 = K = [\beta, \gamma].$$

That's impossible. Thus, Li and Yoke's theorem is proved. $\qquad\square$

Li and Yorke's theorem is just the beginning of the story. Soon afterward, it was found that Li and Yorke's theorem is only a special case of a remarkable theorem published in 1964 by the Russian mathematician Sharkovsky. He gave a complete account of which periods imply other periods for a continuous function. His theorem can be best stated in terms of the following ordering of the positive integers, which is now known as the *Sharkovsky ordering*:

$$3 \lhd 5 \lhd 7 \lhd \cdots \lhd 2 \cdot 3 \lhd 2 \cdot 5 \lhd 2 \cdot 7 \cdots \lhd$$
$$2^2 \cdot 3 \lhd 2^2 \cdot 5 \lhd 2^2 \cdot 7 \lhd \cdots \lhd 2^4 \lhd 2^3 \lhd 2^2 \lhd 2 \lhd 1.$$

Theorem 3.11 (Sharkovsky Theorem [84]). *Let $f : [a, b] \to [a, b]$ be continuous and let f have a point of period m. If $m \lhd n$, then f also has a point of period n. Moreover, there are functions which have points of period n but no point of period m.*

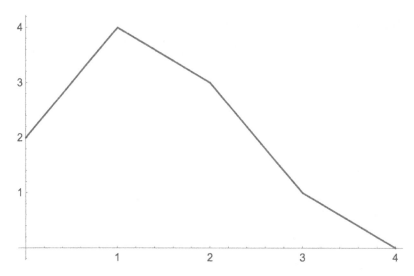

FIGURE 3.6
Period 5 does not imply period 3.

Despite its strangeness, the Sharkovsky ordering does include all positive integers. The number 3 is the "smallest" in this list, so period 3 implies every possible period! Moreover, this theorem also says the following:

- Periodic 5 implies every period except for possibly 3.

- Period 6 implies all of the even periods.

- Any function with finitely many periodic points must only have points with periods equal to powers of 2.

The following example (see Figure 3.6) shows that a function can have a point of period 5 but no point of period 3. Hence the ordering of periods in the Sharkovsky theorem is sharp.

The proof of the Sharkovsky theorem is a more elaborate application of the intermediate value theorem. The original proof is quite difficult. A simplified version can be found in [37, Section 1.11]. We will show a special case in Example 17 of Section 3.6.

Even through the Sharkovsky theorem anticipated Li and Yorke's theorem by 11 years, but, it was Li and Yorke's paper that popularized the concept of *chaos*. Indeed, in this paper, they also showed that if f has a periodic point of period 3, then there exists an uncountable set S such that no point in S is periodic and

(1) For any $x, y \in S, x \neq y$,

$$\limsup |f^n(x) - f^n(y)| > 0.$$

(2) For any $x, y \in S, x \neq y$,

$$\liminf |f^n(x) - f^n(y)| = 0.$$

(3) For any $x \in S$ and any periodic point p,

$$\limsup |f^n(x) - f^n(p)| > 0.$$

In view of the definitions of limit superior and limit inferior in Section 1.3.5, these results indicate that, for any two distinct points x and y in S, there are infinitely many n such that $f^n(x)$ and $f^n(y)$ are as close as you like, but, there are also infinitely many n such that $|f^n(x) - f^n(y)|$ are always positive. In other words, under the successive iteration of f, different points of S are sometimes close, sometimes separated, and none of them is periodic. Thus, the orbits of points of S behave in a unpredictable fashion. Beginning with Li and Yorke, these properties of S above are used to characterize the *chaotic* phenomena in mathematics.

Definition 3.10. *A function f is called chaotic if for every positive integer n, it has a point of period n.*

Examples of chaos are common in our everyday world such as the weather and the rise and fall of the stock market. Even the quadratic maps (see Example 1.24) can behave chaotically when iterated (see [51, Section 3.5]. The mathematical formulation of what constitutes chaotic behavior is one of the most important recent advances in mathematics.

The Sharkovsky theorem reveals a beautiful and deep relation between periodic points of a continuous function. It created a new direction for studying the problem. Functions have been studied for hundreds of years. It is surprising that Sharkovsky's relatively simple theorem was not found earlier. Part of the reason is that the classical analysis was more concentrated on the local properties of functions. For example, continuity, differentiability, and integrability are determined by the local behaviors of functions. The uniform continuity is one exception, but it can be derived simply from the local properties. Since a function can be iterated on $[a, b]$ but may not be iterated on any subinterval of $[a, b]$, it follows that the iteration is a global property. Studying the global properties of functions or mappings now forms a new branch of mathematics called global analysis, which includes dynamical systems, global differential geometry, and the qualitative theory of differential equations. Note that the proof of the Sharkovsky theorem is far from advanced mathematics. This suggests that people need not have advanced knowledge to make important contributions in mathematics. If opportunity doesn't knock, build a door.

3.5   Monotone Functions

The value of a principle is the number of things it will explain. — Ralph Emerson

We now discuss a class of functions which is distinct from the class of continuous functions, but shares some similar properties. We begin with the needed terms.

Definition 3.11. *A function $f : D \to \mathbb{R}$ is increasing if $f(x_1) \leq f(x_2)$ whenever $x_1 \leq x_2$ on D and decreasing if $f(x_1) \geq f(x_2)$ whenever $x_1 \leq x_2$ on D. A monotone function on D is one that is either increasing or decreasing on D.*

The monotone convergence theorem guarantees that every monotone sequence converges if and only if it is bounded. For monotone functions, we now study some special properties not possessed by general functions.

First, we see that a monotone function can have only jump discontinuities.

Theorem 3.12. *Let f be an increasing and bounded function on (a, b). Then for every $t \in (a, b)$ the one sided limits $f(t^-)$ and $f(t^+)$ exist, and $f(t^-) \leq f(t) \leq f(t^+)$. If $f(t^-) = f(t^+)$, then f is continuous at t. Furthermore, $f(a^+)$ and $f(b^-)$ exist, with $f(a) \leq f(a^+)$ and $f(b^-) \leq f(b)$.*

Proof. Let $A(t) = \{f(x) : x < t\}$. Then $A(t)$ is nonempty and bounded above by $f(t)$. The least-upper-bound property guarantees that $\sup\{A(t)\}$ exists. It follows that for any $\epsilon > 0$, $\sup A(t) - \epsilon$ is not an upper bound for this set, so that there is some $x_0 < t$ with $f(x_0) > \sup A - \epsilon$. If $x_0 < x < t$, then

$$\sup A(t) - \epsilon < f(x_0) \le f(x) \le \sup A(t) < \sup A(t) + \epsilon,$$

so that $|f(x) - \sup A(t)| < \epsilon$. This means that

$$f(t^-) = \lim_{x \to t^-} f(x) = \sup A(t).$$

Similarly, define $B(t) = \{f(x) : x > t\}$. We have $f(t^+) = \inf B$. $\qquad\square$

Equipped with this theorem, we next show that the set of discontinuity points of a monotone function is countable.

Theorem 3.13. *Let f be an increasing and bounded function on (a, b). Then*

$$D = \{x : x \in (a, b), \ f \text{ is discontinuous at } x\}$$

is countable.

Proof. By Theorem 3.12, the fact $x \in D$ implies that $f(x^-) < f(x^+)$. Let

$$D_n = \left\{ x : f(x^+) - f(x^-) > \frac{1}{n} \right\}.$$

Thus, $D = \cup_{n=1}^{\infty} D_n$. We show that D_n is finite for all $n \in \mathbb{N}$. Let

$$\{x_1, x_2, \ldots, x_k\} \subset D_n$$

satisfying $a < x_1 < x_2 < \cdots < x_k < b$. Choose $y_1, y_2, \ldots, y_k$ such that

$$a < y_1 < x_1, \ x_i < y_i < x_{i+1} \ (i = 1, 2, \ldots k - 1), \ x_k < y_k < b.$$

Now for each i, $f(y_i) \le f(x_i^-), f(x_i^+) < f(y_{i+1})$ and

$$f(y_{i+1}) - f(y_i) \ge f(x_i^+) - f(x_i^-).$$

Thus

$$
\begin{aligned}
f(b^-) - f(a^+) &= f(b^-) - f(y_{k+1}) \\
&\quad + \sum_{j=2}^{k+1} (f(y_j) - f(y_{j-1})) + f(y_1) - f(a^+) \ge k\frac{1}{n}.
\end{aligned}
$$

Since $f(b^-) - f(a^+) > 0$ is finite, it is necessary that $k \le n(f(b^-) - f(a^+))$. Therefore, D_n is finite for each n. This proves that D is countable as desired. $\qquad\square$

The following floor function provides a nice illustration for this theorem.

Example 3.10. *Consider the greatest integer function $f(x) = \lfloor x \rfloor$.*

f is increasing and discontinuous at every integer point (see Figure 3.7).
In general, the intermediate value property is not enough to guarantee continuity.

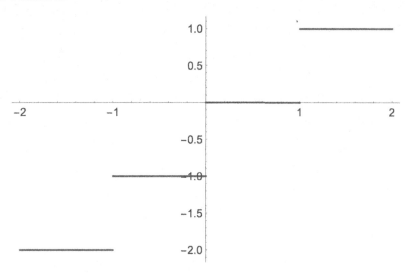

FIGURE 3.7
$f(x) = \lfloor x \rfloor$.

Example 3.11. *Let*

$$f(x) = \begin{cases} x, & \text{if } x \leq 0, \\ x - 1, & \text{if } x > 0. \end{cases}$$

Clearly, f satisfies the intermediate value property, but is discontinuous at $x = 0$ (see Figure 3.8).

However, the following theorem shows that a monotone function with the intermediate value property is continuous, which provides a partial converse to the the intermediate value theorem.

Theorem 3.14. *Let* $f : [a, b] \to \mathbb{R}$ *be increasing. If* f *satisfies the intermediate value property on* $[a, b]$, *then* f *is continuous in* $[a, b]$.

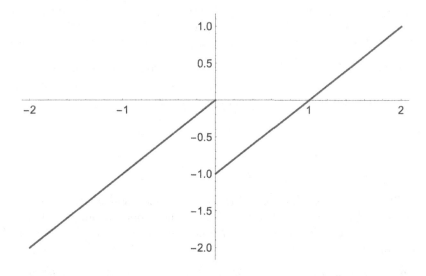

FIGURE 3.8
A function has the IVP, but not continuous.

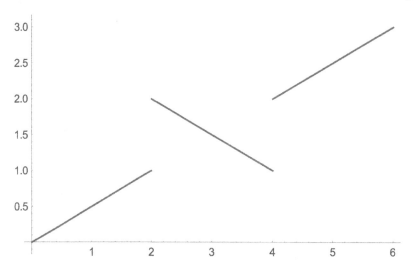

FIGURE 3.9
A function is one-to-not, but not monotone.

Proof. We proceed by contradiction. Assume that f is not continuous at $x_0 \in (a, b)$, a tiny change in the proof can treat the end point case. By Theorem 3.9, there is a jump at x_0. Without loss of generality, we assume that $f(x_0^-) < f(x_0)$. In this case, if $x < x_0$, then $f(x) \le f(x_0^-)$, while if $x > x_0$, then $f(x) \ge f(x_0)$. Thus, $f(x)$ cannot attain the values in $(f(x_0^-), f(x_0))$. This contradicts the intermediate value property.

If you are unsatisfied with the above proof by contradiction, here is one by the $\epsilon - \delta$ definition. For any sequence $x_n \to x_0 \in D$, we show that $f(x_n) \to f(x_0)$, i.e, for any $\epsilon > 0$, we must find a positive integer N such that

$$f(x_0) - \epsilon < f(x_n) < f(x_0) + \epsilon.$$

To prove the right-hand-side inequality, if for n, $f(x_n) < f(x_0)$, we just take $N = 1$. If there is x_m with $f(x_m) > f(x_0)$, we set $y_\epsilon = \min\{f(x_0) + \epsilon/2, x_m\}$. By the intermediate value property, there is x_ϵ such that $f(x_\epsilon) = y_\epsilon$ and $x_\epsilon > x_0$. Hence, there is an index N such that $x_n < x_\epsilon$ for all $n \ge N$ and $f(x_n) \le f(x_\epsilon) < f(x_0) + \epsilon$.   □

We saw that $f([a, b]) = [c, d]$ earlier if f is continuous on $[a, b]$. In addition, if f is strictly increasing on $[a, b]$, then f is one-to-one and so $f^{-1}(x) : [c, d] \to [a, b]$ exists. However, a one-to-one function need not be monotone. For example,

Example 3.12. *Let*

$$f(x) = \begin{cases} x/2, & \text{if } 0 \le x \le 2, \\ -x/2 + 3, & \text{if } 2 < x < 4, \\ x/2, & \text{if } 4 \le x \le 6. \end{cases}$$

Here $f : [0, 6] \to [0, 3]$ is one-to-one, but f is not monotone (see Figure 3.9).

Notice that f also satisfies the intermediate value property, but it is not continuous everywhere on $[0, 6]$. Thus, the continuity of f is necessary to guarantee a one-to-one function is monotone. Moreover,

Theorem 3.15. *If f is a continuous, strictly increasing function on $[a, b]$, then f^{-1} is a continuous, strictly increasing function on $[c, d] = f([a, b])$.*

Proof. We assume that $y_0 \in (c, d)$. The end points can be treated similarly. Then $f^{-1}(y_0) \in (a, b)$ and so there exists $\epsilon_0 > 0$ such that $(f^{-1}(y_0) - \epsilon_0, f^{-1}(y_0) + \epsilon_0) \subset (a, b)$.

For any $0 < \epsilon < \epsilon_0$, there exist $y_1, y_2 \in (c, d)$ such that

$$f^{-1}(y_1) = f^{-1}(y_0) - \epsilon, \quad f^{-1}(y_2) = f^{-1}(y_0) + \epsilon.$$

Clearly we have $y_1 < y_0 < y_2$. Moreover, if $y_1 < y < y_2$, then

$$f^{-1}(y_1) < f^{-1}(y) < f^{-1}(y_2).$$

Thus,

$$f^{-1}(y_0) - \epsilon < f^{-1}(y) < f^{-1}(y_0) + \epsilon,$$

and so $|f^{-1}(y) - f^{-1}(y_0)| < \epsilon$. Now, if we choose $\delta = \min\{y_2 - y_0, y_0 - y_1\}$, then $|y - y_0| < \delta$ implies $y_1 < y < y_2$ and hence $|f^{-1}(y) - f^{-1}(y_0)| < \epsilon$. Hence, $f^{-1}(y)$ is continuous at y_0.

We now show that $f^{-1}(y)$ is strictly increasing. Assume that

$$y_1 < y_2, \quad x_1 = f^{-1}(y_1), \quad x_2 = f^{-1}(y_2).$$

Then

$$y_1 = f(x_1) \quad \text{and} \quad y_2 = f(x_2).$$

If $x_1 > x_2$, then $y_1 > y_2$, a violation of our assumption. If $x_1 = x_2$, then it follows that $y_1 = y_2$, again contrary to our assumption. Thus, we must have $x_1 < x_2$, i.e., $f^{-1}(y)$ is strictly increasing. $\qquad\square$

3.6 Worked Examples

Very often in mathematics the crucial problem is to recognize and to discover what are the relevant concepts; once this is accomplished the job may be more than half done.
— Israel Nathan Herstein

In this section, we continue with some concrete examples of theoretical flavor to illustrate the applications of the theorems we just covered and to reinforce the importance of these theorems.

1. Let $f : (a, +\infty) \to \mathbb{R}$ be bounded on any finite interval. If

$$\lim_{x \to +\infty} (f(x+1) - f(x)) = A,$$

prove that $\lim_{x \to +\infty} f(x)/x = A$.

Proof. We use the typical "$\epsilon/3$ argument." Since $\lim_{x \to +\infty} (f(x+1) - f(x)) = A$, for any $\epsilon > 0$, there exists a positive integer $N_1 > \max\{0, a\}$ such that

$$|f(x+1) - f(x) - A| < \frac{\epsilon}{3} \quad \text{whenever} \quad x > N_1.$$

For this selected N_1, by assumption, if $x \in (a, N_1 + 1)$, then $|f(x)| < M$, where M is a positive integer. Now choose $N > N_1$ such that

$$\frac{M}{N} < \frac{\epsilon}{3}, \quad \left|\frac{(N_1 + 1)A}{N}\right| < \frac{\epsilon}{3}.$$

If $x > N$, there is an integer n such that $x - n \in (N_1, N_1 + 1]$. Thus,

$$\left| \frac{f(x)}{x} - A \right| = \left| \frac{\sum_{i=0}^{n-1} [f(x-i) - f(x-i-1)] + f(x-n)}{x} - A \right|$$

$$\leq \sum_{i=0}^{n-1} \left| \frac{f(x-i) - f(x-i-1) - A}{x} \right| + \left| \frac{(x-n)A}{x} \right| + \left| \frac{f(x-n)}{x} \right|$$

$$< \frac{\epsilon}{3} \cdot \frac{n}{x} + \frac{(N_1 + 1)|A|}{N} + \frac{M}{N}$$

$$< \frac{\epsilon}{3} + \frac{\epsilon}{3} + \frac{\epsilon}{3} = \epsilon.$$

This proves the desired limit. $\qquad\qquad\qquad\qquad\qquad\qquad\qquad\qquad\qquad\quad \square$

2. Let $f, g : [a, \infty)$ satisfy

   (a) $g(x + T) > g(x)$ for $x \geq a$, where T is a positive constant.
   (b) f and g are bounded on any finite interval in $[a, \infty)$.
   (c) $\lim_{x \to \infty} g(x) = \infty$.

   Prove that if

   $$\lim_{x \to \infty} \frac{f(x + T) - f(x)}{g(x + T) - g(x)} = A,$$

   then $\lim_{x \to \infty} f(x)/g(x) = A$.

Proof. By the assumption, for any $\epsilon > 0$, there exists a positive integer $N > a$ such that

$$\left| \frac{f(x + T) - f(x)}{g(x + T) - g(x)} - A \right| < \epsilon \text{ whenever } x > N.$$

For any $x > N + T$, there is $k \in \mathbb{N}$ such that $x = N + kT + r$, where $0 \leq r < T$. Notice that $x \to \infty \Leftrightarrow k \to \infty$. Using the inequality above successively yields

$$A - \epsilon < \frac{f(x) - f(x - T)}{g(x) - g(x - T)} < A + \epsilon,$$

$$A - \epsilon < \frac{f(x - T) - f(x - 2T)}{g(x - T) - g(x - 2T)} < A + \epsilon,$$

$$\vdots$$

$$A - \epsilon < \frac{f(x - (k-1)T) - f(x - kT)}{g(x - (k-1)T) - g(x - kT)} < A + \epsilon.$$

Combining these inequalities gives

$$A - \epsilon$$
$$< \frac{f(x) - f(x-T) + f(x-T) - f(x-2T) + \cdots + f(x-(k-1)T) - f(x-kT)}{g(x) - g(x-T) + g(x-T) - g(x-2T) + \cdots + g(x-(k-1)T) - g(x-kT)}$$
$$< A + \epsilon.$$

Telescoping the numerator and denominator, respectively, leads to

$$A - \epsilon < \frac{f(x) - f(x - kT)}{g(x) - g(x - kT)} = \frac{f(x) - f(N + r)}{g(x) - g(N + r)} < A + \epsilon. \qquad (3.5)$$

By (a), we have $g(x) = g(N + r + kT) > g(N + r + (k-1)T) > \cdots > g(N + r)$. Multiplying (3.5) by $g(x) - g(N + r)$ and then dividing by $g(x)$ yields

$$\frac{f(N + r)}{g(x)} + (A - \epsilon)\left(1 - \frac{g(N + r)}{g(x)}\right) \leq \frac{f(x)}{g(x)} \leq \frac{f(N + r)}{g(x)} + (A + \epsilon)\left(1 - \frac{g(N + r)}{g(x)}\right).$$
$$(3.6)$$

By applying (b) and (c), (3.6) implies that

$$\limsup \frac{f(x)}{g(x)} \leq A + \epsilon \text{ and } \liminf \frac{f(x)}{g(x)} \geq A - \epsilon.$$

Now, the desired limit follows from the arbitrariness of ϵ. $\qquad\square$

Remark. This problem generalizes the Stolz-Cesàro theorem from sequences to functions. Let $g(x) = x, T = 1$. We see that Problem 1 indeed is a special case of this result. Later, we will use this result to prove the well-known L'Hôpital's rule.

Recall that, in Example 1.5, we have proved that a continuous function on $[a, b]$ is bounded by using the Heine-Borel theorem. Subsequently, we reprove the intermediate value theorem and the uniform continuity theorem via the Heine-Borel theorem. Once again, when a local property must be extended to $[a, b]$, the Heine-Borel theorem meets the challenge!

3. Use the Heine-Borel theorem to prove that if f is continuous on $[a, b]$ with $f(a) \cdot f(b) < 0$, then there is a point $c \in (a, b)$ such that $f(c) = 0$.

Proof. Assume the contrary: suppose that f has no zero in $[a, b]$. Thus for every point $x \in [a, b]$, there exists $(x - \delta_x, x + \delta_x)$ in which $f(t)$ has the same sign as $f(x)$'s. The family of open intervals $\{(x - \delta_x, x + \delta_x)\}_{x \in [a,b]}$ cover $[a, b]$. Hence by the Heine-Borel theorem, there are finitely many these intervals, say $\{x_i - \delta_{x_i}, x_i + \delta_{x_i}\}_{i=1}^{N}$, which cover $[a, b]$. Without loss of generality, assume $a \in (x_1 - \delta_{x_1}, x_1 + \delta_{x_1})$. Here $x_1 + \delta_{x_1} \in (x_2 - \delta_{x_2}, x_2 + \delta_{x_2})$, in which $f(t)$ has same sign as $f(a)$. Continuing in this manner, we will reach $b \in (x_N - \delta_{x_N}, x_N + \delta_{x_N})$. By the selection of the intervals, it follows that $f(b)$ has same sign as $f(a)$. This is a contradiction! $\qquad\square$

4. Use the Heine-Borel theorem to prove the uniform continuity theorem.

Proof. For any $\epsilon > 0$ and each $x_0 \in [a, b]$, by the definition of continuity, there is some $\delta_{x_0} > 0$ such that $x \in (x_0 - \delta_{x_0}, x_0 + \delta_{x_0}) \cap [a, b]$ implies

$$|f(x) - f(x_0)| < \frac{\epsilon}{2}.$$

Since the family of open intervals $\{(x_0 - \delta_{x_0}/2, x_0 + \delta_{x_0}/2)\}_{x_0 \in [a,b]}$ cover $[a, b]$, by the Heine-Borel theorem, there are $N \in \mathbb{N}$ and $x_i \in [a, b], i = 1, 2, \ldots, N$ such that

$$[a, b] \subset U_{i=1}^{N}(x_i - \delta_{x_i}/2, x_i + \delta_{x_i}/2)$$

Let $\delta = \min_{1 \leq i \leq N}\{\delta_{x_i}/2\}$. Suppose that $x, y \in [a, b]$ and $|x - y| < \delta$. There exists some $k \in \{1, 2, \ldots, N\}$ such that $y \in (x_k - \delta_{x_k}/2, x_k + \delta_{x_k}/2)$. Hence $|y - x_k| < \delta_{x_k}/2$. The definition of δ implies that $|x - y| < \delta_{x_k}/2$. Hence

$$|x - x_k| \leq |x - y| + |y - x_k| < \frac{\delta_{x_k}}{2} + \frac{\delta_{x_k}}{2} = \delta_{x_k}.$$

It follows that

$$|f(x) - f(y)| \leq |f(x) - f(x_k)| + |f(x_k) - f(y)| < \frac{\epsilon}{2} + \frac{\epsilon}{2} = \epsilon. \qquad \square$$

5. Let f be continuous on $[a, b]$. Prove that functions

$$M(x) = \max_{a \leq t \leq x} f(t) \text{ and } m(x) = \min_{a \leq t \leq x} f(t)$$

are continuous on $[a, b]$ as well.

Proof. We only show that $M(x)$ is continuous. The proof for the case of $m(x)$ is similar. Since f is continuous on $[a, b]$, the extreme value theorem guarantees that $M(x)$ is well-defined and bounded. Clearly, $M(x)$ is also increasing. Thus, by Theorem 3.9, its one-sided limits exist. We assume that $x_0 \in (a, b)$. (A few tiny changes in the proof are needed for endpoints.) We now show that

$$M(x_0^-) = M(x_0) = M(x_0^+).$$

Given any $\epsilon > 0$, since $f(x)$ is continuous at x_0, there exists a $\delta > 0$ such that $|x - x_0| < \delta$ implies that $|f(x) - f(x_0)| < \epsilon$. In particular, if $x_0 < x < x_0 + \delta$, we have

$$f(x) < f(x_0) + \epsilon \leq M(x_0) + \epsilon.$$

On the other hand, if $a \leq x \leq x_0$, we have $f(x) \leq M(x) \leq M(x_0) + \epsilon$. Thus, as $x_0 < x < x_0 + \delta$, we have $M(x) \leq M(x_0) + \epsilon$, and so

$$M(x_0) \leq M(x) \leq M(x_0) + \epsilon.$$

It follows that $M(x_0) = M(x_0^+)$.

To prove that $M(x_0^-) = M(x_0)$, we assume that $M(x_0) = f(x_0)$. Otherwise, if $M(x_0) = f(x_1)$ for some $a \leq x_1 < x_0$, in this case, we have $M(x) \equiv M(x_0)$ for all $x_1 \leq x \leq x_0$. Therefore, $M(x)$ is continuous from the left hand side. Similar to the preceding argument, for any given $\epsilon > 0$, there exists a $\delta > 0$ such that $x_0 - \delta < x < x_0$ implies that $f(x) > f(x_0) - \epsilon = M(x_0) - \epsilon$. Thus, $M(x) \geq M(x_0) - \epsilon$ and

$$M(x_0) - \delta \leq M(x) \leq M(x_0).$$

This proves $M(x_0^-) = M(x_0)$ as desired. $\qquad \square$

6. Let $f : \mathbb{R} \to \mathbb{R}$ be a continuous function such that

$$f\left(r + \frac{1}{n}\right) = f(r), \quad \text{for any } r \in \mathbb{Q} \text{ and } n \in \mathbb{N}.$$

Prove that f is constant.

Proof. Since

$$f\left(r + \frac{2}{n}\right) = f\left(r + \frac{1}{n}\right) = f(r),$$

by induction, it follows that

$$f\left(r + \frac{m}{n}\right) = f(r), \quad \text{for any } r \in \mathbb{Q} \text{ and } m, n \in \mathbb{N}.$$

Let $r = 0, -m/n$, respectively. Then

$$f\left(\frac{m}{n}\right) = f(0) = f\left(-\frac{m}{n}\right).$$

This implies that $f(r) = f(0)$ for all $r \in \mathbb{Q}$. When $x \in \mathbb{R} \setminus \mathbb{Q}$, there exists $r_n \in \mathbb{Q}$ such that $r_n \to x$ as $n \to \infty$. By the continuity of f, we find that

$$f(x) = \lim_{n \to \infty} f(r_n) = f(0).$$

This concludes the proof. $\qquad\square$

7. Let $f : \mathbb{R} \to \mathbb{R}$ be a function satisfying the intermediate value property. If $E_r = \{x : f(x) = r, r \in \mathbb{Q}\}$ is a closed set, prove that f is continuous.

Proof. We proceed by contradiction. Assume that f is not continuous at $x_0 \in \mathbb{R}$. Then there exist an $\epsilon > 0$ and a sequence $x_n \to x_0$ such that

$$|f(x_n) - f(x_0)| \geq \epsilon.$$

For the inequalities

$$f(x_n) \geq f(x_0) + \epsilon \quad \text{and} \quad f(x_n) \leq f(x_0) - \epsilon,$$

at least one of them should be hold for the infinitely many x_n. Without loss of generality, we assume that

$$f(x_{n_k}) \leq f(x_0) - \epsilon, \quad (k = 1, 2, 3, \ldots).$$

Then there is an rational number r such that

$$f(x_{n_k}) < r < f(x_0), \quad (k = 1, 2, 3, \ldots). \tag{3.7}$$

By the intermediate value property, there are y_k between x_{n_k} and x_0 such that $f(y_k) = r$. Since $x_{n_k} \to x_0$, the squeeze theorem yields that $y_k \to x_0$. Moreover, $x_0 \in E_r$ because E_r is closed. It follows that $f(x_0) = r$, which contradicts (3.7). $\qquad\square$

8. Prove that f is uniformly continuous on an interval I if and only if for any $\epsilon > 0, x, y \in I, x \neq y$, there exists a positive integer N such that

$$|f(x) - f(y)| < \epsilon \quad \text{whenever} \quad \left|\frac{f(x) - f(y)}{x - y}\right| > N. \tag{3.8}$$

Before proceeding to the proof, we restate the definition of the uniformly continuous in an equivalent form:

For any $\epsilon > 0$ and $x, y \in I$, there exists a $\delta > 0$, such that

$$|f(x) - f(y)| \geq \epsilon \implies |x - y| \geq \delta. \tag{3.9}$$

Similarly, the equivalent form of (3.8) becomes:

For any $\epsilon > 0$ and $x, y \in I$, there exists an integer $N > 0$, such that

$$|f(x) - f(y)| \geq \epsilon \implies \left|\frac{f(x) - f(y)}{x - y}\right| \leq N. \tag{3.10}$$

Proof. First, we show that (3.8) is sufficient. By its equivalent form (3.10), we have

$$|x - y| \geq \frac{|f(x) - f(y)|}{N} \geq \frac{\epsilon}{N}.$$

Thus, for any $\epsilon > 0$ and $x, y \in I$, taking $\delta = \epsilon/N$ yields (3.9), which shows that f is uniformly continuous on I.

We now show that (3.8) is necessary. If f is uniformly continuous on I, in view of (3.9), there is a positive integer $k \geq 1$ such that

$$k\delta \leq |x - y| \leq (k + 1)\delta.$$

Without loss of generality, we assume that $x < y$. Partitioning $[x, y]$ into $(k + 1)$ parts with partition points x_i yields

$$|x_i - x_{i-1}| = \frac{|x - y|}{k + 1} < \delta,$$

and so

$$|f(x_i) - f(x_{i-1})| < \epsilon, \quad (i = 1, 2, \ldots, k + 1).$$

Hence

$$\left| \frac{f(x) - f(y)}{x - y} \right| \leq \frac{\sum_{i=1}^{k+1} |f(x_i) - f(x_{i-1})|}{k\delta} < \frac{(k + 1)\epsilon}{k\delta} \leq \frac{2\epsilon}{\delta}.$$

It follows (3.10) with $N \geq 2\epsilon/\delta$. □

9. Let $f(x)$ be continuous on $[0, 1]$ and $f(0) = f(1)$. For every positive integer n, prove that there exists $x \in [0, 1]$ such that

$$f(x) = f\left(x + \frac{1}{n}\right).$$

Proof. Define

$$F(x) = f(x) - f\left(x + \frac{1}{n}\right). \tag{3.11}$$

Clearly, $F(x)$ is continuous on $[0, 1 - 1/n]$. Notice that

$$\begin{aligned} 0 &= f(0) - f(1) \\ &= (f(0) - f(1/n)) + (f(1/n) - f(2/n)) + \cdots + (f(1 - 1/n) - f(1)) \\ &= F(0) + F(1/n) + \cdots + F(1 - 1/n). \end{aligned}$$

Consider the set

$$A = \{F(0), F(1/n), \cdots, F(1 - 1/n)\}.$$

If $0 \in A$, then we are done. If $0 \notin A$, then there must be two elements with distinct signs. In this case, the intermediate value theorem implies that $F(x)$ has a zero on $[0, 1 - 1/n]$.

An alternative proof by contradiction is as follows:

Assume that $F(x)$ defined by (3.11) has no zero on $[0, 1 - 1/n]$. By the intermediate value theorem, $F(x)$ is either entirely positive or negative on $[0, 1 - 1/n]$. If $F(x) > 0$, then $f(x) > f(x + 1/n)$ for all $x \in [0, 1 - 1/n]$. This leads to

$$f(0) > f(1/n) > f(2/n) > \cdots > f(n/n) = f(1) = f(0),$$

which is impossible. Similarly, if $F(x) < 0$, then it follows that

$$f(0) < f(1/n) < f(2/n) < \cdots < f(n/n) = f(1) = f(0),$$

which is also impossible. □

It is interesting to see that the gap in the form of $1/n$ is necessary. For example, let

$$f(x) = 2x + \cos(5\pi x).$$

Clearly, $f(0) = f(1) = 1$. For each x, we have

$$f(x + 2/5) - f(x) = \frac{4}{5}.$$

This shows that there is no x with the gap $2/5$ such that $f(x) = f(x + 2/5)$.

10. Let $f : \mathbb{R} \to \mathbb{R}$ be a surjective continuous function that takes any value at most twice. Prove that f is strictly monotone.

Proof. Choose $a, b \in \mathbb{R}$ with $a < b$ and assume that $f(a) \leq f(b)$. We now show that

$$f(a) \leq f(x) \leq f(b), \quad \text{for all } x \in [a, b].$$

To this end, since f is continuous, $f([a, b])$ is a bounded interval, it follows that there exists $c, d \in [a, b]$ such that

$$f(c) \leq f(x) \leq f(d), \quad \text{for all } x \in [a, b].$$

First, we show that $f(c) < f(d)$. Indeed, if $f(c) = f(d)$, then f is constant on $[a, b]$, which contradicts the assumption that f takes any value at most twice. Therefore, $f(c) < f(d)$.

Next, we show that $f(c) \geq f(a)$ and $f(d) \leq f(b)$. We show the first inequality only, as the second inequality follows by similar arguments. By contradiction, assume that $f(c) < f(a)$. Then there exists $\xi \in \mathbb{R}$ such that $f(\xi) < f(a)$. Then either $\xi < a$ or $\xi > b$. Let us assume that $\xi < a$. By the intermediate value theorem, it follows that f takes a certain value at least three times on $[\xi, b]$, a contradiction. Along the same lines, we see that $\xi > b$ is also impossible. This shows that $f(c) \geq f(a)$. Thus, for all $x \in [a, b]$, we have $f(a) \leq f(x) \leq f(b)$, and so f is monotone increasing. Moreover, in view of that $f^{-1}(y)$ has at most two elements for all $y \in \mathbb{R}$, it implies that f cannot take the same value at different points. Therefore, f is strictly increasing. □

11. Let $f(x)$ be uniformly continuous on $[0, +\infty)$. For any $x \geq 0$ and positive integer n, if $\lim_{n \to \infty} f(x + n) = 0$, prove that $\lim_{x \to \infty} f(x) = 0$.

Proof. Since f is uniformly continuous on $[0, \infty)$, for any given $\epsilon > 0$, there is a $\delta > 0$ such that

$$|f(x) - f(y)| < \epsilon \quad \text{whenever} \quad x, y > 0, \ |x - y| < \delta.$$

On the other hand, since $\lim_{n \to \infty} f(x + n) = 0$ for any $x \in [0, 1]$, there exists a positive integer N_x such that $|f(n + x)| < \epsilon$ whenever $n > N_x$. Therefore, for $t \in (x - \delta, x + \delta)$ and $n > N_x$, we have

$$|f(n + t)| \leq |f(n + t) - f(n + x)| + |f(n + x)| < 2\epsilon.$$

By the Heine-Borel theorem, there exist $x_1, x_2, \ldots, x_K \in [0,1]$ such that

$$[0,1] \subset \cup_{i=1}^{K} (x_i - \delta, x_i + \delta).$$

Let $N = \max\{N_{x_1}, N_{x_2}, \ldots, N_{x_K}\}$. Then for any $x > N + 1$, there exists one x_i satisfying $x - [x] \in (x_i - \delta, x_i + \delta)$ and so

$$|f(x)| = |f([x] + t)| < 2\epsilon.$$

This proves the desired limit. $\square$

12. Determine all continuous functions $f : \mathbb{R} \to \mathbb{R}$ satisfying $f(0) = 1$ and

$$f(2x) - f(x) = x, \quad \text{for all } x \in \mathbb{R}.$$

Solution. Replacing x by $x/2$ in the given functional equation yields

$$f(x) - f(x/2) = \frac{x}{2}.$$

Repeating this process leads to

$$f(x/2^{n-1}) - f(x/2^n) = \frac{x}{2^n}, \quad \text{for all } n \in \mathbb{N}.$$

Appealing to the telescoping feature, we sum up these equations to get

$$f(x) - f(x/2^n) = \left(\frac{1}{2} + \frac{1}{2^2} + \cdots + \frac{1}{2^n}\right) x.$$

By the continuity and $f(0) = 1$, taking the limit $n \to \infty$, we find that $f(x) = x+1$ is the unique solution. $\square$

13. Let $f : \mathbb{R} \to \mathbb{R}$ be a continuous periodic function with period T. For any $\alpha > 0$, prove that there exists $x_0 \in \mathbb{R}$ such that $f(x_0) = f(x_0 + \alpha)$. In geometry, this fact can be viewed as the existence of horizontal chord of length α.

Proof. For any given $\alpha > 0$, define

$$g(x) = f(x + \alpha) - f(x).$$

Clearly, $g(x)$ is continuous on $[0, T]$. By the extreme value theorem, there are $x_1, x_2 \in [0, T]$ such that

$$f(x_1) = \min_{x \in [0,T]} f(x), \quad f(x_2) = \max_{x \in [0,T]} f(x),$$

and so

$$g(x_1) \geq 0, \quad g(x_2) \leq 0.$$

Thus, by the intermediate value theorem, there exists an $x_0 \in [0, T]$ such that $g(x_0) = 0$, the desired result follows. $\square$

Here we have seen that a continuous periodic function f has horizontal chords of arbitrary lengths. Replacing the above $g(x)$ by $f(x+\alpha) - 2f(x) + f(x-\alpha)$, along the same lines, we can prove that f has a chord of any length with midpoint on the graph of the function. We leave the details to the interested reader.

14. Let $f : \mathbb{R} \to \mathbb{R}$ be continuous and decreasing. Show that the system

$$x = f(y), \qquad y = f(z), \qquad z = f(x)$$

has a unique solution.

Proof. By the assumption that f is decreasing, we have

$$\lim_{x \to \infty} (f(x) - x) = -\infty, \quad \lim_{x \to -\infty} (f(x) - x) = \infty.$$

The intermediate value theorem implies that there is $\xi \in \mathbb{R}$ such that $f(\xi) - \xi = 0$. That is, ξ is a fixed point of f. Next, we show that ξ is the only fixed point of f. Indeed, suppose ξ_1 and ξ_2 are fixed points of f with $\xi_1 < \xi_2$. The decreasing of f implies that

$$\xi_1 = f(\xi_1) \geq f(\xi_2) = \xi_2,$$

which is impossible. Now, we see that (ξ, ξ, ξ) is a solution of the system. Let (x_0, y_0, z_0) be a solution of the system. Then

$$f(f(f(x_0))) = f(f(y_0)) = f(z_0) = x_0.$$

Since $f \circ f \circ f$ is continuous and decreasing, the argument above leads to that it has a unique fixed point. Thus, $x_0 = \xi$. This proves that the system has a unique solution $x_0 = y_0 = z_0 = \xi$. □

15. Let $f : \mathbb{R} \to \mathbb{R}$ be continuous. If there exist two numbers $a \in \mathbb{R}$ and $M > 0$ such that $|f^n(a)| \leq M$ for all $n \in \mathbb{N}$, prove that f has a fixed point. Here f^n represents the nth iteration of f.

Proof. We proceed by contradiction. Assume that f has no fixed point. Then for any $x \in \mathbb{R}$, $f(x) - x$ either positive or negative constantly. Without loss of generality, we assume that $f(x) > x$. Let $x_1 = a, x_{n+1} = f(x_n)$ for $n \geq 1$. Since $x_{n+1} = f(x_n) > x_n$, it follows that the sequence x_n is strictly increasing. Moreover,

$$x_n = f(x_{n-1}) = f^2(x_{n-2}) = \cdots = f^{n-1}(x_1) = f^{n-1}(a).$$

By assumption, we have $|x_n| = |f^{n-1}(a)| \leq M$. The monotone convergence theorem implies that x_n converges. Let $\lim_{n \to \infty} x_n = x_0$. Then

$$x_0 = \lim_{n \to \infty} x_{n+1} = \lim_{n \to \infty} f(x_n) = f(\lim_{n \to \infty} x_n) = f(x_0).$$

So x_0 is a fixed point of f. This contradicts the assumption at the beginning.

A direct existence proof can be done as follows. We separate into two cases:
(i) There is $k \in \mathbb{N}$ such that $f^k(a) = f^{k-1}(a)$. In this case, $x_0 = f^{k-1}(a)$ is a fixed point of f.
(ii) For any $n \in \mathbb{N}, f^n(a) \neq f^{n-1}(a)$. Without loss of generality, assume that $f(a) > a$. Define

$$S = \{n \in \mathbb{N} : f^n(a) < f^{n-1}(a)\}.$$

If $S = \emptyset$, the sequence $f^n(a)$ is monotonic increasing and bounded by M. The monotone convergence theorem implies that $f^n(a)$ converges. Let $\lim_{n \to \infty} f^n(a) = x_0$. Then

$$x_0 = \lim_{n \to \infty} f^n(a) = \lim_{n \to \infty} f(f^{n-1}(a)) = f(x_0),$$

which implies that x_0 is a fixed point.

If $S \neq \emptyset$, let k be the smallest integer in S. Then $k > 1$ and

$$f^k(a) < f^{k-1}(a), f^{k-1}(a) > f^{k-2}(a).$$

Let $x_1 = f^{k-2}(a), x_2 = f^{k-1}(a)$ and $F(x) = f(x) - x$. Then

$$F(x_1) = f(x_1) - x_1 = f^{k-1}(x) - f^{k-2}(x) > 0,$$

$$F(x_2) = f(x_2) - x_2 = f^k(x) - f^{k-1}(x) < 0.$$

By the intermediate value theorem, there is $x_0 \in \mathbb{R}$ such that $F(x_0) = 0$, i.e., $f(x_0) = x_0$, which shows that x_0 is a fixed point again. $\qquad\square$

16. Let $f : [a,b] \to [a,b]$ be continuous. If f satisfies that $f^n(x) = x$ for all $x \in [a,b]$ and for some $n \in \mathbb{N}$, prove that $f^2(x) = x$ for all $x \in [a,b]$.

Proof. If $f(x) = f(y)$, then $f^n(x) = f^n(y)$, and so $x = y$. Therefore, f is one-to-one. We now show that f is strictly monotone on $[a,b]$. It suffices to show that if $r, s, t \in [a,b]$ and $r < s < t$, then $f(s)$ is between $f(r)$ and $f(t)$. Indeed, suppose $f(s)$ is outside this interval, and say $f(s)$ is closer to $f(r)$. By the intermediate value property, there is $\xi \in (r,t)$ such that $f(r) = f(\xi)$, which contradicts that f is one-to-one. Therefore, for $a < x < y < b$ the only possibility are either

$$f(a) < f(x) < f(y) < f(b)$$

or

$$f(a) > f(x) > f(y) > f(b).$$

Suppose that f is increasing on $[a,b]$, we show that $f(x) = x$ for all $x \in [a,b]$. In fact, if $x > f(x)$, then

$$f(x) > f^2(x) > \cdots > f^{n-1}(x) > f^n(x).$$

Thus, $x > f^n(x) = x$, which is impossible. On the other hand, if $x < f(x)$, then

$$f(x) < f^2(x) < \cdots < f^{n-1}(x) < f^n(x),$$

so that $x < f^n(x) = x$, which is false again. Thus we must have $f(x) = x$. If f is decreasing on $[a,b]$, then for $x, y \in [a,b]$, $x < y$ implies that $f(x) > f(y)$ and $f^2(x) < f^2(x)$. Thus, f^2 is increasing. Repeating the above argument yields that $f^2(x) = x$ for all $x \in [a,b]$. $\qquad\square$

If $f^n(x) = x$ holds only for some $x \in [a,b]$ and for some $n > 2$, the following problem asserts that f still has periodic point of period 2. Since that 2 is the smallest period other than fixed point in the Sharkovsky ordering, so this problem can be view as a simple special case of Sharkovsky theorem.

17. Let $f : [a,b] \to [a,b]$ be continuous. If f satisfies that $f^n(x) = x$ for some $x \in [0,1]$ and for some positive integer $n > 2$. Show that f has a periodic point of period 2.

Proof. Here we show that if f has a periodic point of period $n > 2$, then f has a periodic point that is not fixed point and has period less than n. The desired result then follows from a descending induction.

Let $x_1 < x_2 < \cdots < x_n$ be the points on an orbit of period n. We consider the directed graph with vertices $1, 2, \ldots, n-1$ in which vertex i is joined to vertex j if and only if $f([x_i, x_{i+1}]) \supset [x_j, x_{j+1}]$. Each vertex i must be joined to at least one vertex $j \neq i$, otherwise f would have to permute the endpoints of $[x_j, x_{j+1}]$, which is impossible since these points lie on a periodic orbit for f with least period $n \geq 3$.

Starting at the vertex 1, choose an edge that joins 1 to a different vertex. Then we join this vertex to a different vertex, and so on. The path can be extended indefinitely with each edge joining two different vertices. After at most $n-1$ vertices it must return to a previously visited vertex. This gives us a loop $i_1, i_2, \ldots, i_q$ that passes through at least 2 and at most $n-1$ vertices. Set $I_k = [x_{i_k}, x_{i_k+1}]$ for $k = 1, \ldots, q$. Then $f(I_k) \supset I_{k+1}$ for $k =, \ldots, q-1$ and $f(I_q) \supset I_1$. There is a closed subinterval $I_1' \subset I_1$, such that $f(I_1') = I_2$. In order to see this, look at the intersection of the graph of f with the rectangle $I_1 \times I_2$. At least one component of this intersection must join the top and bottom edges of $I_1 \times I_2$. The interval I_1' is the projection to I_1 of such a component.

Since $f^q(I_1') = f^{q-1}(f(I_1')) = f^{q-l}(I_2) \supset I_1 \supset I_1'$, it follows from the intermediate value theorem that f^q has a fixed point $\xi \in I_1'$. Clearly ξ is a periodic point for f whose period is a factor of q and therefore less than n. We now show that ξ is not a fixed point for f. Since $\xi \in I_1' \subset I_1$, and $f(\xi) \in f(I_1') = I_2$, we can have $\xi = f(\xi)$ only if $\xi \in I_1 \cap I_2$. But I_1 and I_2 have disjoint interiors and their endpoints belong to an orbit with least period $n > 1$. □

18. (**Monthly Problem 11555, 2011**). Let f be a continuous real-valued function on $[0, 1]$ such that $\int_0^1 f(x)dx = 0$. Prove that there exists c in the interval $(0, 1)$ such that $c^2 f(c) = \int_0^c (x + x^2)f(x)dx$.

Proof. If f is identically zero, there is nothing to prove. Now, we assume that $f(x)$ is not identically zero. Since $\int_0^1 f(x)dx = 0$, there are $a, b \in [0, 1], a \neq b$ such that

$$f(a) = \max_{x \in [0,1]} f(x) > 0, \quad f(b) = \min_{x \in [0,1]} f(x) < 0.$$

Let

$$F(x) = x^2 f(x) - \int_0^x (t + t^2)f(t)dt.$$

It is clear that $F(x)$ is continues on $[0, 1]$. Moreover,

$$F(a) \geq a^2 f(a) - \int_0^a (t + t^2)f(a)dt = a^2 \left(\frac{1}{2} - \frac{a}{3} \right) f(a) > 0,$$

$$F(b) \leq b^2 f(b) - \int_0^b (t + t^2)f(b)dt = b^2 \left(\frac{1}{2} - \frac{b}{3} \right) f(b) < 0.$$

The intermediate value theorem implies that there is a number $c \in (a, b) \subset (0, 1)$ such that $F(c) = 0$, which is equivalent to $c^2 f(c) = \int_0^c (x + x^2)f(x)dx$. □

19. Let $f : [0, 1] \to \mathbb{R}$ be a strictly monotonic and continuous function such that

$$\int_0^1 f(t)\, dt = 1.$$

Prove there exist $\alpha, \beta, \gamma \in (0,1)$ with $\alpha < \beta < \gamma$ such that

$$f(\alpha)f(\beta)f(\gamma) = 1.$$

Proof. Notice that

$$\int_0^1 f(t)\, dt = \int_0^1 f(1-t)\, dt = 1.$$

Without loss of generality, we assume that f is strictly increasing. Since $\int_0^1 f(t)\, dt = 1$ and f is continues on $[0,1]$, there exists $0 < t_1 < 1$ such that

$$m = \min_{x \in [0,1]} f(x) = f(0) < f(t_1) = 1 < M = \max_{x \in [0,1]} f(x) = f(1).$$

Let $d = \max\{m, 1/M\}$. For any $0 < \delta < 1 - d$, we have

$$m < 1 - \delta < 1, \ 1 < \frac{1}{1-\delta} < M.$$

By the intermediate value theorem, there are $0 < \alpha < t_1$ and $t_1 < \gamma < 1$ such that

$$f(\alpha) = 1 - d, \ f(\gamma) = \frac{1}{1-d}.$$

Thus, letting $\beta = t_1$ yields $0 < \alpha < \beta < \gamma < 1$ and

$$f(\alpha)f(\beta)f(\gamma) = (1-d) \cdot 1 \cdot \frac{1}{1-d} = 1.$$

$\square$

This result can be generalized to any positive integer n without imposing the strictly monotonic condition. See Exercise 74.

20. Let $f : \mathbb{R} \to \mathbb{R}$ be continuous. Assume every $x \in \mathbb{R}$ is an extreme point of f, show that f is a constant function.

Proof. By contradiction, assume that f is not constant, then there are $a_1, b_1 \in \mathbb{R}, a_1 < b_1$ such that $f(a_1) \neq f(b_1)$. Without loss of generality, assume that $f(a_1) < f(b_1)$. Applying the intermediate value theorem yields that there is $c \in (a_1, b_1)$ such that

$$f(a_1) < f(c) = \frac{f(a_1) + f(b_1)}{2} < f(b_1).$$

If $b_1 - c \leq \frac{b_1 - a_1}{2}$, let $a_2 = c$ and b_2 which satisfies $a_2 = c < b_2 < b_1$ and

$$f(a_1) < f(a_2) = f(c) < f(b_2) < f(b_1).$$

If $c - a_1 \leq \frac{b_1 - a_1}{2}$, let $b_2 = c$ and a_2 which satisfies $a_1 < a_2 < c = b_2$ and

$$f(a_1) < f(a_2) < f(c) = f(b_2) < f(b_1) < f(b_1).$$

No matter which case, we always have $[a_2, b_2] \subset [a_1, b_1]$ and $f(a_2) < f(b_2)$. Repeating this process, we get

$$[a_n, b_n] \subset [a_{n-1}, b_{n-1}] \subset \cdots \subset [a_2, b_2] \subset [a_1, b_1],$$

and $0 < b_n - a_n < (b_1 - a_1)/2^{n-1}$. By the nested interval theorem, there is $x_0 \in \cap_{n=1}^\infty [a_n, b_n]$ such that

$$\lim_{n \to \infty} a_n = \lim_{n \to \infty} b_n = x_0.$$

Notice that $f(a_n)$ is increasing and $f(b_n)$ is decreasing, it follows that

$$f(a_n) < f(x_0) < f(b_n).$$

Thus, x_0 is not an extreme point, which contradicts the assumption.

Another proof is based on the fact that $R(f)$, the range of f, is at most countable. This fact can be established as follows: Let

$$U_n = \{y \in \mathbb{R} : \text{ there exists } x_0 \text{ such that } y = f(x_0) = \max_{x \in (x_0 - 1/n, x_0 + 1/n)} f(x).$$

Thus,

$$U = \cup_{n=1}^\infty U_n$$

contains all relative maximum value of f. Define $\phi_n : U_n \to \mathbb{Q}$ by

$$\phi_n : y = f(x_0) \to r \in \mathbb{Q} \cap \left(x_0 - \frac{1}{2n}, x_0 + \frac{1}{2n} \right).$$

We show that ϕ is bijective. In fact, if $\phi(y_1) = r_1 = r_2 = \phi(y_2)$, then

$$r_1 \in \left(x_1 - \frac{1}{2n}, x_1 + \frac{1}{2n} \right), \ r_2 \in \left(x_2 - \frac{1}{2n}, x_2 + \frac{1}{2n} \right).$$

Since

$$|x_1 - x_2| \le |x_1 - r_1| + |r_2 - x_2| < \frac{1}{2n} + \frac{1}{2n} = \frac{1}{n},$$

by the extreme property, we have $f(x_1) \le f(x_2)$ and $f(x_2) \le f(x_1)$. Thus, $y_1 = f(x_1) = f(x_2) = y_2$. So U_n is at most countable and therefore U is at most countable. Similarly, we can show that all relative minimum value of f is also at most countable. In summary, $R(f)$ is at most countable. Now, we show f must be a constant. If $R(f)$ contains at least two elements, say $r_1 < r_2$. The intermediate value theorem implies that $[r_1, r_2] \subset R(f)$, so $R(f)$ is uncountable since $[r_1, r_2]$ is uncountable. The contradiction shows that f is a constant function. $\square$

3.7 Exercises

Do not wait to strike till the iron is hot; but make it hot by striking.

— William Butler Yeats

1. For $x \in [0, \infty)$, define

$$f(x) = \begin{cases} \frac{1}{p+q}, & \text{if } x = p/q, \ p, q \in \mathbb{N}, \ \gcd(p,q) = 1, \\ 0, & \text{if } x = 0 \text{ or } x \text{ is irrational.} \end{cases}$$

Classify the type of discontinuities of $f(x)$.

2. (**Arnold**). Find
$$\lim_{x \to 0} \frac{\sin(\tan x) - \tan(\sin x)}{\arcsin(\arctan x) - \arctan(\arcsin x)}.$$

3. Let $a_1, a_2, \ldots, a_n$ be real numbers. Find
$$\lim_{x \to 0} \frac{1 - \cos(a_1 x)\cos(a_2 x)\cdots\cos(a_n x)}{x^2}.$$

4. Let $|x| < 1$. Find
$$\lim_{n \to \infty} \left(1 + \frac{1 + x + \cdots + x^n}{n}\right)^n.$$

5. Let $a_1, a_2, \ldots, a_n$ be positive real numbers. Prove that

   (a) $\lim_{x \to +\infty} \left[x - \sqrt[n]{(x - a_1)(x - a_2)\cdots(x - a_n)}\right] = \frac{a_1 + a_2 + \cdots + a_n}{n}$,

   (b) $\lim_{x \to 0} \left(\frac{a_1^x + a_2^x + \cdots + a_n^x}{n}\right)^{1/x} = \sqrt[n]{a_1 a_2 \cdots a_n}$.

6. Show that
$$\lim_{n \to \infty} n \sin(2\pi e n!) = 2\pi.$$

7. Let a, b, c be positive. Find the condition for which
$$\lim_{x \to \infty} \left(\frac{a^{1/x} + b}{c}\right)^x$$
   exists.

8. Let $f : (a, +\infty) \to \mathbb{R}$ be a bounded function on any finite interval. Prove that
$$\lim_{x \to +\infty} f(x)^{1/x} = \lim_{x \to +\infty} \frac{f(x + 1)}{f(x)}, \quad \text{where } f(x) \geq c > 0,$$
   provides that the right hand side limit exists.

9. (**Putnam Problem 2021-A2**). For every positive real number x, let
$$g(x) = \lim_{r \to 0} \left((x + 1)^{r+1} - x^{r+1}\right)^{1/r}.$$
   Find $\lim_{x \to \infty} \frac{g(x)}{x}$.

10. (**Putnam Problem 1998-B1**). Find the minimum value of
$$\frac{(x + 1/x)^6 - (x^6 + 1/x^6) - 2}{(x + 1/x)^3 + (x^3 + 1/x^3)}$$
    for $x > 0$.

11. (**Putnam Problem 1986-B4**). For a positive real number r, let $G(r)$ be the minimum value of $|r - \sqrt{m^2 + 2n^2}|$ for all integers m and n. Prove or disprove the assertion that $\lim_{r \to \infty} G(r)$ exists and equals 0.

12. Let $f(x) : \mathbb{R} \to [0, +\infty)$ be a function which is bounded on $[0, 1]$ and
$$f(x + y) \leq f(x)f(y), \quad \text{for all } x, y \in \mathbb{R}.$$
    Show that $\lim_{x \to +\infty} f(x)^{1/x}$ exists and is finite.

13. Let $f : (a, +\infty) \to \mathbb{R}$ be bounded on any finite interval. If

$$\lim_{x \to +\infty} \frac{f(x+1) - f(x)}{x^n} = A,$$

prove that

$$\lim_{x \to +\infty} \frac{f(x)}{x^{n+1}} = \frac{A}{n+1}.$$

14. Let $g : (0, \infty) \to (0, \infty)$ be a continuous function such that

$$\lim_{x \to \infty} \frac{g(x)}{x^{1+\lambda}} = \infty \qquad \text{for some } \lambda > 0.$$

If $f : \mathbb{R} \to (0, \infty)$ is twice-differentiable and for some $x_0 \in \mathbb{R}$,

$$f''(x) + f'(x) > \alpha g(f(x)), \qquad \text{for all } x \geq x_0 \text{ and some } \alpha > 0,$$

show that $\lim_{x \to \infty} f(x)$ exists, and find the limit.

15. Let $f : \mathbb{R} \to \mathbb{R}$ be a function satisfying

 (a) $\lim_{x \to +\infty} f(x) = 0$,

 (b) $\lim_{x \to +\infty} \frac{f(x) - f(ax)}{x} = 0$ for $a \in (0, 1)$.

 Show that $\lim_{x \to +\infty} f(x)/x = 0$.

16. (**Putnam Problem 1964-B3**). Let $f(x)$ be continuous on $(0, +\infty)$. For any $\alpha > 0$, if $\lim_{n \to \infty} f(n\alpha) = 0$, prove that $\lim_{x \to \infty} f(x) = 0$.

17. Let $n \in \mathbb{N}$. Define

$$f_1(x) = x > 0, \quad f_n(x) = x^{f_{n-1}(x)}, \qquad \text{for } n \geq 2.$$

 Find

$$\lim_{x \to 1} \frac{f_n(x) - f_{n-1}(x)}{(1-x)^n}.$$

18. Let $f : [0, \infty) \to \mathbb{R}$ be a function such that

$$f(x)e^{f(x)} = x \qquad \text{for all } x \in [0, \infty).$$

 Prove that

 (a) f is monotone.

 (b) $\lim_{x \to \infty} f(x) = \infty$.

 (c) $\lim_{x \to \infty} \frac{f(x)}{\ln x} = 1$.

19. Let f be a function satisfying $f(x) \geq \sqrt{2}/2$ and

$$f^2(x) - \ln f(x) = x \qquad \text{for all } x \geq \tfrac{1}{2}(1 + \ln 2).$$

 (a) Find $\lim_{x \to \infty} \frac{f(x)}{\sqrt{x}}$, if the limit exists.

 (b) Find $\alpha \in \mathbb{R}$ such that $\sum_{n=1}^{\infty} n^\alpha(f(n) - \sqrt{n})$ converges.

 (c) Find $\lim_{x \to \infty} \frac{\sqrt{x}f(x) - x}{\ln x}$, if the limit exists.

20. (**Putnam Problem 2012-A3**). Let $f : [-1,1] \to \mathbb{R}$ be a continuous function such that

(a) $f(x) = \frac{2-x^2}{2} f\left(\frac{x^2}{2-x^2}\right)$ for every $x \in [-1,1]$,

(b) $f(0) = f(1)$, and

(c) $\lim_{x \to 1^-} \frac{f(x)}{\sqrt{1-x}}$ exists and is finite.

Prove that f is unique, and express $f(x)$ in closed form.

21. (**Putnam Problem 2013-B3**). Let $\mathcal{C} = \cup_{N=1}^{\infty} \mathcal{C}_N$, where $\mathcal{C}_N$ denotes the set of those "cosine polynomials" of the form

$$f(x) = 1 + \sum_{n=1}^{N} a_n \cos(2\pi n x)$$

for which

(a) $f(x) \geq 0$ for every real x, and

(b) $a_n = 0$ whenever n is a multiple of 3.

Determine the maximum value of $f(0)$ as f ranges through $\mathcal{C}$, and prove that this maximum is attained.

22. (**Putnam Problem 2021-A6**). For a positive integer N, let f_N be the function defined by

$$f_N(x) = \sum_{n=0}^{N} \frac{N + 1/2 - n}{(N+1)(2n+1)} \sin((2n+1)x).$$

Determine the smallest constant M such that $f_N(x) \leq M$ for all N and for all real x.

23. Let $f : (0, +\infty) \to (0, +\infty)$ be increasing satisfying $\lim_{x \to +\infty} f(2x)/f(x) = 1$. For any $\alpha > 0$, show that

$$\lim_{x \to +\infty} \frac{f(\alpha x)}{f(x)} = 1.$$

24. Let $A_n \subset [0,1]$ be a finite element set for every $n \in \mathbb{N}$ and $A_i \cap A_j = \emptyset$ for all $i, j \in \mathbb{N}, i \neq j$. Define

$$f(x) = \begin{cases} \frac{1}{n}, & \text{if } x \in A_n; \\ 0, & \text{if } x \in [0,1] \text{ but not in any } A_n \end{cases}$$

For each $a \in [0,1]$, find $\lim_{x \to a} f(x)$.

25. Let $f, g : \mathbb{R} \to \mathbb{R}$ be periodic functions such that

$$\lim_{x \to \infty} (f(x) - g(x)) = 0.$$

Show that $f \equiv g$.

26. Let $f : \mathbb{R} \to \mathbb{R}$ be differentiable at $x = 0$ with $f(0) = 0$. For $n \in \mathbb{N}$, prove that

$$\lim_{x \to 0} \frac{1}{x} \sum_{k=1}^{n} f(x/k) = f'(0) H_n,$$

where H_n is the nth harmonic number.

27. Let $f : [0,1] \to \mathbb{R}$ be continuous. Prove that

$$\lim_{n \to \infty} \frac{1}{n} \sum_{k=1}^{n} (-1)^k f(k/n) = 0.$$

28. Let $f : [0, \infty) \to \mathbb{R}$ be continuous such that $\lim_{x \to \infty} f(x) = L$. Show that

$$\lim_{n \to \infty} \int_0^1 f(nx)\,dx = L.$$

29. Let

$$f(x) = \begin{cases} x \sin\left(\frac{1}{x}\right), & 0 < x \le 1, \\ 0, & x = 0. \end{cases}$$

Show that $f(x)$ is uniformly continuous but not Lipschitz continuous on $[0,1]$. You may prove a stronger result:

$$\frac{|f(x) - f(y)|}{|x - y|^\alpha}$$

is bounded for $x, y \in [0,1]$ if and only if $\alpha \le 1/2$.

30. Let f be well-defined on $[a,b]$. If $\lim_{x \to x_0} f(x)$ exists for every $x_0 \in [a,b]$, show that f is bounded on $[a,b]$ if only if $\lim_{x \to x_0} f(x)$ is finite for all $x_0 \in [a,b]$.

31. Let $f : \mathbb{R} \to \mathbb{R}$ attain rational at irrational points and irrational at rational points. Prove that f is not continuous.

32. Suppose that $f : \mathbb{R} \to \mathbb{R}$ is *superlinear*. i.e., if

$$\lim_{x \to \pm\infty} \frac{f(x)}{x} = +\infty.$$

Let

$$g(y) = \max_{x \in \mathbb{R}} (xy - f(x)).$$

Show that g is well-defined and also superlinear.

33. Prove that there is no function defined on $\mathbb{R}$ such that it is continuous at rationals and discontinuous at irrationals.

34. Let f be continuous on $\mathbb{R}$. Assume that $\lim_{x \to \infty} f(x)$ exists and finite, show that f attains either maximum or minimum on $\mathbb{R}$. Consider $f(x) = \frac{1+x^2}{1-x^2+x^4}$. Show that $\max_{x \in \mathbb{R}} f(x) = 1 + 2\sqrt{3}/3$ and $\inf_{x \in \mathbb{R}} f(x) = 0$, which is not achievable.

35. Let f be continuous on $[a,b]$ with a unique critical point $x_0 \in (a,b)$. Prove that if x_0 is a relative minimum (maximum) point, then f indeed attains absolute minimum (maximum) at x_0.

36. Let f be an increasing function on $[a,b]$. If $f(0) > 0, f(1) < 1$, prove that there exists a $x_0 \in (0,1)$ such that $f(x_0) = x_0^2$.

37. Is there a function defined on $\mathbb{R}$ that satisfies $f(f(x)) = e^{-x}$?

38. Let f be continuous on $[a,b]$. Let M and m be the maximum and minimum of f on $[a,b]$, respectively. Show that, for any constant c,

$$\max_{x \in [a,b]} \left| f(x) - \frac{M+m}{2} \right| \le \max_{x \in [a,b]} |f(x) - c|.$$

39. (**Chebyshev**). Let $p(x)$ any polynomial of degree n with leading coefficient 1. Show that
$$\max_{x\in[-1,1]} |p(x)| \geq \frac{1}{2^{n-1}}.$$

40. Let $f, g : [0, 1] \to [0, 1]$ be continuous functions such that
$$\max_{x\in[0,1]} f(x) = \max_{x\in[0,1]} g(x).$$

Show that there exists $c \in [0, 1]$ such that $f^2(c) + 10e^{f(c)} = g^2(c) + 10e^{g(c)}$.

41. Let $f : \mathbb{R} \to \mathbb{R}$ be a continuous function such that $f(f(x)) = x$ for all $x \in \mathbb{R}$. Prove that there exists a $x_0 \in \mathbb{R}$ such that $f(x_0) = x_0$.

42. For $\lambda \geq 0$, let
$$f(x) = \frac{x}{\sqrt{x^2 + \lambda}}.$$

Find a formula for nth iteration of $f(x)$. *Hint:* Consider two cases: $\lambda = 1$ and $\lambda \neq 1$.

43. Prove that if $f : [a, b] \to [a, b]$ is increasing, then f has a fixed point. Find a decreasing function $f : [a, b] \to [a, b]$ which has no fixed point.

44. Let $f, g : [a, b] \to [a, b]$ be continuous functions. If there exists a sequence $x_n \in [a, b]$ such that
$$g(x_n) = f(x_{n+1}), \quad \text{for all } n \in \mathbb{N},$$

show that $f(x_0) = g(x_0)$ for some $x_0 \in [a, b]$.

45. Let $f, g : [0, 1] \to [0, 1]$ be continuous and $f(g(x)) = g(f(x))$ for all $x \in [0, 1]$. Prove that

    (a) If f is decreasing, there exists a unique $a \in [0, 1]$ such that $f(a) = g(a) = a$,

    (b) If f is monotone, there exists a $a \in [0, 1]$ such that $f(a) = g(a) = a$,

    (c) If f is increasing, is the value a for which satisfies $f(a) = g(a) = a$ unique?

46. Assume that f satisfies $f(x^2) = f(x)$ for all $x \in \mathbb{R}$. If f is continuous at 0 and 1, prove that f is constant.

47. Let $f : \mathbb{R} \to \mathbb{R}$ be a continuous function such that
$$f(x) = f(x + 1) = f(x + 2\pi), \quad \text{for all } x \in \mathbb{R}.$$

Prove that f is constant.

48. Let $f : \mathbb{R} \to \mathbb{R}$ be a one-to-one continuous function. If f has a fixed point and satisfies
$$f(2x - f(x)) = x, \quad \text{for all } x \in \mathbb{R},$$

show that $f(x) \equiv x$.

49. Define $f : [1, 5] \to [1, 5]$ by
$$f(1) = 3, \quad f(3) = 4, \quad f(4) = 2, \quad f(2) = 5, \quad f(5) = 1$$

with f linear between these integers. Show that $f(x)$ has no periodic point of period 3.

50. If $f : [0, 1] \to [0, 1]$ is continuous, $f(0) = 0, f(1) = 1$ and $f(f(x)) = x$ for all $x \in [0, 1]$, show that $f(x) \equiv x$.

51. Let f be continuous on $[a, b]$. If for every $x \in [a.b]$, there exists a $t \in [a, b]$ satisfying

$$|f(t)| < \frac{1}{2}|f(x)|,$$

    show that there is a $c \in [a, b]$ such that $f(c) = 0$.

52. Let $f : [0, 1] \to \mathbb{R}$ satisfying

    (a) $f(0) > 0, f(1) < 0$,
    (b) there exists a continuous function g such that $f + g$ is monotone increasing on $[0, 1]$.

    Show that $f(c) = 0$ for some $c \in (0, 1)$.

53. Let $f : [0, 1] \to [0, 1]$ be continuous. Show that the graph of $y = f(x)$ intersects with both straight lines $y = x$ and $y = 1 - x$.

54. Let f be continuous and bounded on $[a, \infty)$. Prove that, for every $\alpha \in \mathbb{R}$, there exists a sequence $x_n \to \infty$ such that

$$\lim_{n \to \infty} (f(\alpha + x_n) - f(x_n)) = 0.$$

55. Let f be uniformly continuous on $\mathbb{R}$. Prove that there exist nonnegative constants a and b such that
$$|f(x)| \leq a|x| + b, \quad \text{for all } x \in \mathbb{R}.$$

56. Construct a function $f : \mathbb{R} \to \mathbb{R}$ which is continuous and bounded, but it is not uniformly continuous.

57. Let $f : D \to \mathbb{R}$ be uniformly continuous. Prove that if D is bounded, then f is bounded. This assertion indicates that a uniformly continuous function becomes unbounded only if its domain is unbounded.

58. Let f be continuous on $[a, b]$ with $f(a) = f(b)$. Prove that there exists a $x_0 \in [a, (a + b)/2]$ such that $f(x_0) = f(x_0 + (a + b)/2)$.

59. Let $n \in \mathbb{N}$. If f is continuous on $[0, n]$ with $f(0) = f(n)$, prove that there exist n distinct pairs (x, y) such that $|x - y| \in \mathbb{N}$ and $f(x) = f(y)$.

60. Let $f : [0, 1] \to [0, 1]$ be a function such that $|f(x) - f(y)| \leq |x - y|$ for all $x, y \in [0, 1]$. Prove that the set of all fixed points of f is either a single point or an interval.

61. Let $a, M \in \mathbb{R}$ be positive. If $f : [a, \infty) \to \mathbb{R}$ satisfies $|f(x) - f(y)| \leq M|x - y|$ for all $x, y \in [a, \infty)$, prove that $f(x)/x$ is uniformly continuous on $[a, \infty)$.

62. Let f be continuous on $[a, b]$. For every $x_1, x_2, \ldots, x_n \in [a, b]$ and positive $\lambda_1, \lambda_2, \ldots, \lambda_n$ with $\sum_{k=1}^{n} \lambda_k = 1$, prove that there exists a $c \in [a, b]$ such that

$$f(c) = \sum_{k=1}^{n} \lambda_k f(x_k).$$

63. Let $f : \mathbb{R} \to \mathbb{R}$ be continuous that maps open intervals to open intervals. Prove that f is monotone.

64. Let $f : \mathbb{R} \to \mathbb{R}$ be a function that attains neither its maximum nor its minimum on any open intervals.

(a) Prove that if f is continuous then f is monotone.

(b) Give an example of non-monotone function which satisfies the above property.

65. (**Riesz Rising Sun Lemma**). Let $f : [a, b] \to \mathbb{R}$ be a continuous function. Let

$$S = \{x \in [a, b] \,; \text{there exists } y > x \text{ such that } f(y) > f(x)\}.$$

Define $E = S \cap (a, b)$. Show that $E = \cup(a_k, b_k)$ such that $f(a_k) = f(b_k)$, unless $a_k = a \in S$ for some k, in which case $f(a) < f(b_k)$ for that one k. Furthermore, if $x \in (a_k, b_k)$, then $f(x) < f(b_k)$.

Remark. The name of this lemma derives from the following: the sun is rising from the right in a mountainous region seen in a one-dimensional profile from the side. The elevation at x is $f(x)$, and elements of E are those x values that remain in shadow at the instant the sun rises over the horizon as seen from.

66. Let $a_1, a_2, a_3 > 0$ and $b_1 < b_2 < b_3$. Show that the equation

$$\frac{a_1}{x - b_1} + \frac{a_2}{x - b_2} + \frac{a_3}{x - b_3} = 0$$

has exactly one solution in (b_1, b_2) and (b_2, b_3), respectively.

67. Let $C_n(x) = \sum_{k=0}^{n} a_k \cos kx$ with $a_i \in \mathbb{R}$ $(1 \le i \le n)$ and $\sum_{k=1}^{n-1} |a_k| < a_n$. Prove that $C_n(x)$ has at least $2n$ zeros in $[0, 2\pi)$.

68. For each $n \in \mathbb{N}$, define $T_n(x) = \cos(n \arccos x)$ for $x \in [-1, 1]$.

   (a) Show that $T_n(x)$ is an algebraic polynomial of degree n.
   (b) Find the roots of $T_n(x)$ on $[-1, 1]$ and the points where $|T_n(x)|$ attends its maximum value.
   (c) Show that among all polynomials $P_n(x)$ of degree n whose leading coefficient is 1 the polynomial $T_n(x)$ is the unique polynomial closest to zero.

69. Let $f : \mathbb{S}^1 \to \mathbb{R}$ be a continuous map, where $\mathbb{S}^1$ is the unit circle: $\{(x, y) : x^2 + y^2 = 1\}$. Prove that there exists a pair of $(x, y) \in \mathbb{S}^1$ such that $f(x, y) = f(-x, -y)$.

   Remark. This is the special case of the Borsuk-Ulam Theorem: For every continuous map $f : \mathbb{S}^n \to \mathbb{R}^n$, there exists $x \in \mathbb{S}^n$ such that $f(x) = f(-x)$. In general, the points x and $-x$ on the sphere are called *antipodal points*. A nice interpretation of this theorem is as follows: Suppose each point on the earth maps continuously to a temperature-barometric pressure pair. Then there are two antipodal points on the earth with the same temperature and barometric pressure.

70. Let $x_1, x_2, \ldots, x_n \in [0, 1]$. Prove that there exists a $x_0 \in [0, 1]$ such that

$$\frac{1}{n} \sum_{i=1}^{n} |x_0 - x_i| = \frac{1}{2}.$$

71. Let f be continuous on $[a, a + 2b]$ with $b > 0$. Show that there exists $x_0 \in [a, a + b]$ such that

$$f(x_0 + b) - f(x_0) = \frac{1}{2}(f(a + 2b) - f(a)).$$

72. Suppose that $a_0, a_1, \ldots, a_n \in \mathbb{R}$ and $x \in (0, 1)$ satisfying

$$\frac{a_0}{1 - x} + \frac{a_1}{1 - x^2} + \cdots + \frac{a_n}{1 - x^{n+1}} = 0.$$

Prove that the equation

$$a_0 + a_1 y + a_2 y^2 + \cdots + a_n y^n = 0$$

has one solution in $(0, 1)$.

73. Let f be a positive function on $[a, b]$. Define

$$F(x) = \int_a^x f(t)\, dt + \int_b^x \frac{1}{f(t)}\, dt.$$

Show that $F(x)$ has a unique zero on $[a, b]$.

74. Let $f : [0, 1] \to \mathbb{R}$ continuous and $\int_0^1 f(x) dx = 1$. For any positive integer n, prove that there exist distinct $t_1, t_2, \ldots, t_n \in (0, 1)$ such that

$$\prod_{k=1}^n f(t_k) = 1.$$

75. Let $f : [0, 1] \to \mathbb{R}$ continuous and $\int_0^1 f(x) dx = 1$. For any positive integer n, prove that there exist distinct $t_1, t_2, \ldots, t_n \in (0, 1)$ such that

   (a) $\sum_{k=1}^n f(t_k) = 1$.
   (b) $\sum_{k=1}^n \frac{1}{f(t_k)} = 1$.

76. Let f and g be continuous on $[a, b]$. If f is monotone and there exists a sequence $x_n \in [a, b]$ such that $g(x_n) = f(x_{n+1})$, prove that there exists a $\xi \in [a, b]$ satisfying $g(\xi) = f(\xi)$. Can we drop the assumption of monotone on f?

77. Let $f(x) = 4(x - 1/2)^2$, $x \in [0, 1]$. Show that for any positive integer n, $f^n(x)$ has at least one fixed point.

78. Prove that there is no continuous function $f : \mathbb{R} \to \mathbb{R}$ such that, for each $c \in \mathbb{R}$, the equation $f(x) = c$ has exactly two solutions. How about exactly 3 solutions?

79. Define the *Cantor set*

$$C = \left\{ x \in [0, 1] : x = \sum_{n=1}^\infty \frac{a_n}{3^n}, \quad \text{where } a_n = 0 \text{ or } 2. \right\}$$

   Show that C is the complement of the union of disjoint open intervals I_n, $n \in \mathbb{N}$ whose lengths add to 1. *Hint:* Use the geometric construction of C by repeated removing the middle open third from each previous intervals. Show that

$$C = [0, 1] \setminus \cup_{n=1}^\infty \cup_{k=0}^{3^{n-1}-1} \left(\frac{3k+1}{3^n}, \frac{3k+2}{3^n} \right).$$

80. Define $f(x)$ on the Cantor set $C \subset [0, 1]$ by

$$f(x) = \sum_{n=1}^\infty \frac{b_n}{2^n}, \quad \text{for } x = \sum_{n=1}^\infty \frac{a_n}{3^n},$$

   where $b_n = a_n/2$. Extend f on $[0, 1]$ continuously.

81. Construct a function $f [0, 1] \to [0, 1]$ such that $f(x) = c$ has infinitely many solutions for every $c \in [0, 1]$.

82. Let $f : [1, \infty) \to \mathbb{R}$ be a positive continuous function such that for every $x \in [1, \infty)$ holds

$$\int_0^x f(t)\, dt \le f^2(x).$$

Prove that

$$f(x) \ge \frac{x-1}{2}, \quad \text{for } x \in [1, \infty).$$

83. Prove that a continuous function $f : \mathbb{R} \to \mathbb{R}$ which maps open sets to open sets is a monotone function.

84. Find all functions $f : \mathbb{R} \to \mathbb{R}$ has the intermediate value property and $f^n(x) = -x$ for some $n \in \mathbb{N}$, where f^n is the nth iteration of f.

85. (**Lebesgue**). There exists a function $f : [0, 1] \to [0, 1]$ that has the intermediate value property and is discontinuous at every point.

86. Let $f : \mathbb{R} \to \mathbb{R}$ be monotone increasing with $f(-\infty) = 0$ and $f(\infty) = 1$. Define a function $g(x)$ on $(0, 1)$ by

$$g(x) = \inf\{t : f(t) > x\}.$$

Prove that $g(x) = g(x^+)$, i.e., g is continuous from the right.

87. Let $f : \mathbb{R} \to \mathbb{R}$ be a continuous periodic function with period 1. Prove that

(a) f is bounded and achieves its maximum and minimum,
(b) f is uniformly continuous on $\mathbb{R}$, and
(c) there exists $x_0 \in \mathbb{R}$ such that $f(x_0 + \pi) = f(x_0)$.

88. If $f : \mathbb{R} \to \mathbb{R}$ takes every value exactly twice, show that f is not continuous.

89. A function f is said to be *upper semicontinuous* at x_0 if given $\epsilon > 0$, there exists a $\delta > 0$ such that if $|x - x_0| < \delta$,, then $f(x) < f(x_0) + \epsilon$. Prove that an upper semicontinuous function f on $[a, b]$ is bounded above and achieves its maximum on $[a, b]$. Give an example for which an upper semicontinuous function does not satisfy the intermediate value property.

90. Let $f : [0, 1] \to [0, 1]$ be a continuous function. Define the iteration sequence $x_0 \in [0, 1], x_{n+1} = f(x_n)$ and $A = \{x_0, x_1, x_2, \ldots\}$. Prove that if A is a closed set then A has only finitely many elements.

91. Prove that there exists no function $f : (0, +\infty) \to (0, +\infty)$ such that

$$f^2(x) \ge f(x+y)(f(x) + y) \text{ for all } x, y \in (0, +\infty).$$

92. Let $1 < a < b$ and $m, n \in \mathbb{R}$ with $m \ne 0$. Find all continuous function $f : [0, \infty) \to \mathbb{R}$ such that for $x \ge 0$, $f(a^x) + f(b^x) = mx + n$.

93. Prove that there exists no continuous function $f : [0, 1] \to \mathbb{R}$ such that

$$f(x) + f(x^2) = x \text{ for all } x \in [0, 1].$$

94. Let $f : [0, 1] \to [0, 1]$ be an increasing function. Prove that there exists an $x \in [0, 1]$ such that $f(x) = x$. Find a decreasing function $f : [0, 1] \to [0, 1]$ which has no fixed point.

95. Find all continuous functions $f : \mathbb{R} \to \mathbb{R}$ such that $f(x) - f(y)$ is rational for all reals x and y such that $x - y$ is rational.

96. Let $f : \mathbb{R} \to \mathbb{R}$ be a surjective continuous function that takes any value at most twice. Prove that f is strictly monotone.

97. Let f be a function from $[a, b]$ into $[a, b]$. Let $S \subset [a, b]$ be the set of all periodic point of f with period n. Prove that if S has exactly n elements and $n > 1$, then there is no function $g : [a, b] \to [a, b]$ such that $g^n = f$.

98. (**Monthly Problem 10818, 2000**). Let $g : \mathbb{R} \to \mathbb{R}$ be a continues function such that $\lim_{x \to \infty} (g(x) - x) = \infty$ and such that the set $\{x : g(x) = x\}$ is finite and nonempty. Prove that if $f : \mathbb{R} \to \mathbb{R}$ be continues and $f \circ g = f$, then f is constant.

99. (**Monthly Problem 11872, 2015**). Let f be a continuous function from $[0, 1]$ into $\mathbb{R}$ such that $\int_0^1 f(x)dx = 0$. Prove that for all positive integers n there exists $c \in (0, 1)$ such that

$$n \int_0^c x^n f(x)dx = c^{n+1} f(c).$$

100. Let k be a real number. Find all continuous functions $f : \mathbb{R} \to \mathbb{R}$ such that

$$f(x + y) = f(x) + f(y) + kxy, \quad x, y \in \mathbb{R}.$$

101. Find all continuous functions $f : \mathbb{R} \to \mathbb{R}$ such that

$$f(x + y) + f(x - y) = 2f(x)f(y), \quad \text{for all } x, y \in \mathbb{R}.$$

102. Let n be a positive integer. Find all continuous functions f such that

$$f(x + y^n) = f(x) + (f(y))^n, \quad \text{for all } x, y \in \mathbb{R}.$$

103. (**Putnam Problem 1971-B2**). Let $f(x)$ be a real function defined on $\mathbb{R}$ except $x = 0$ and $x = 1$ and satisfying the functional equation

$$f(x) + f\left(\frac{x - 1}{x}\right) = 1 + x.$$

Find all functions $f(x)$ satisfying these conditions.

104. (**Putnam Problem 1988-A5**). Prove there exists a unique function $f : \mathbb{R}^+ \to \mathbb{R}^+$ such that

$$f(f(x)) = 6x - f(x)$$

and $f(x) > 0$ for all $x > 0$.

105. Let $f(x)$ be a real function defined on $\mathbb{R} \setminus \{-3, -1, 0, 1, 3\}$ and satisfying the functional equation

$$f(x) + f\left(\frac{13 + 3x}{1 - x}\right) = ax + b$$

for some given constants a and b. Find all functions $f(x)$ satisfying these conditions.

106. (**Putnam Problem 1990-B5**). Is there an infinitely sequence $a_0, a_1, a_2, \ldots$ of nonzero real numbers such that for $n = 1, 2, 3, \ldots$ the polynomial

$$P_n(x) = a_0 + a_1 x + \cdots + a_n x^n$$

has exactly n distinct real roots?

107. (**Putnam Problem 1996-A6**). Let $c \geq 0$ be a constant. Given a complete description, with proof, of the set of all continuous function $f : \mathbb{R} \to \mathbb{R}$ such that $f(x) = f(x^2 + c)$ for all $x \in \mathbb{R}$.

108. (**Putnam Problem 2014-B6**). Let $f : [0,1] \to \mathbb{R}$ be a function for which there is a constant $K > 0$ such that $|f(x) - f(y)| \leq K|x - y|$ for all $x, y \in [0,1]$. Suppose also that for each rational number $r \in [0,1]$, there exist integers a and b such that $f(r) = a + br$. Prove that there are finitely many intervals $I_1, I_2, \ldots, I_n$ such that f is a linear function on each I_i and $[0,1] = \cup_{i=1}^{n} I_i$.

109. Define $f : [0,1] \to [0,1]$ by

$$f(x) = \begin{cases} 0.0a_1 0a_2 0a_3 \cdots, & \text{if } x = 0.a_1 a_2 a_3 \cdots, \\ 1, & \text{if } x = 1. \end{cases}$$

Here $x = 0.a_1 a_2 a_3 \cdots$ is the decimal representation of x. When the non-uniqueness happens, choose the finite representation (for example, if $x = 0.1 = 0.0999 \cdots$, use $x = 0.1$). Study the continuity of $f(x)$.

110. Let $f : [a,b] \to \mathbb{R}$ be continuous on $[a,b]$ and differentiable on (a,b). Suppose that f has infinitely many zeros, but there is no $x \in (a,b)$ with $f(x) = f'(x) = 0$.

(a) Prove that either $f(a)$ or $f(b)$ equals zero.

(b) Given an example of such function on $[0,1]$.

4

Differentiation

The calculus was the first achievement of modern mathematics, and it is difficult to overestimate its importance.

— John von Neumann

A hundred years ago such a function would have been considered an outrage on common sense.

— Henri Poincaré on Weierstrass's nowhere differentiable continuous function

The notion of the derivative comes from the intuitive concepts of velocity and the slope of the tangent line. It is the main theme of differential calculus—one of the major discoveries in mathematics. In this chapter we give a rigorous treatment of differentiation and related concepts based on limits. First, we focus on two fundamental theorems of derivatives. The first proves that a derivative has an intermediate value property. The second is the mean value theorem which plays a central role in analysis. As an application of the mean value theorem, we present L'Hôpital's rule, which provides us with a powerful tool to compute limits of some indeterminate forms. Next, we study some basic properties of convex functions including Jensen's inequality which offers an effective approach to establish inequalities. Finally, we extend the mean value theorem to the Taylor theorem, which is considered to be the pinnacle of the differential calculus. As usual, we conclude this chapter with some interesting problems.

4.1 Derivatives

And I dare say that this is not only the most useful and most general problem in geometry that I know, but even that I ever desired to know.

— René Descartes on the calculation of the angles between two curves

The concept of derivative has been originated from various practical problems. We begin with the following formal definition:

Definition 4.1. *Let $f : I \to \mathbb{R}$, where I is an interval in $\mathbb{R}$. For any $x_0 \in I$, we say that f is differentiable at x_0 if the limit*

$$\lim_{x \to x_0} \frac{f(x) - f(x_0)}{x - x_0} := f'(x_0) \tag{4.1}$$

exists. The limit $f'(x_0)$ is called the derivative of f at x_0.

The geometric motivation for the derivative reveals the impetus behind the definition (4.1). Here the difference $(f(x) - f(x_0))/(x - x_0)$ is the slope of the chord through two points $(x, f(x))$ and $(x_0, f(x_0))$. Taking the limit as x approaches x_0 yields the slope of the tangent line at $x = x_0$.

Let $x = x_0 + h$. Then $x \to x_0 \iff h \to 0$. The definition (4.1) is rephrased as the familiar form

$$f'(x_0) = \lim_{h \to 0} \frac{f(x_0 + h) - f(x_0)}{h}.$$

Now we state the basic operation laws of derivatives with which you are all familiar.

Theorem 4.1. *Let f and g be differentiable at x_0. Then the differentiation operation is closed under the four basic operations. i.e.,*

  1. $(f \pm g)'(x_0) = f'(x_0) \pm g'(x_0)$,
  2. $(fg)'(x_0) = f'(x_0)g(x_0) + f(x_0)g'(x_0)$,
  3. $(f/g)'(x_0) = (f'(x_0)g(x_0) - f(x_0)g'(x_0))/g^2(x_0)$ *if $g(x_0) \neq 0$.*

The last two rules are usually refereed as the *product rule* and *quotient rule*, respectively.

Theorem 4.2 (Chain Rule). *Let g be differentiable at x_0 and f be differentiable at $g(x_0)$. Then $f \circ g$ is differentiable at x_0 with*

$$(f \circ g)'(x_0) = f'(g(x_0))\, g'(x_0).$$

Let $f^{(k)}$ denote the kth derivative of f. The product rule can be extended to the higher derivatives.

Theorem 4.3 (Leibniz's Rule). *Suppose that both f and g have nth derivatives. Then*

$$(fg)^{(n)}(x) = \sum_{k=0}^{n} \binom{n}{k} f^{(k)}(x) g^{(n-k)}(x).$$

It is well-known that $f(x) = |x|$ is continuous but fails to be differentiable at $x = 0$. Thus, a continuous function is not necessarily differentiable. However, the converse is true, i.e., the differentiability implies the continuity.

Theorem 4.4. *If f is differentiable at $x_0 \in I$, then f is continuous at x_0.*

Proof. We show that $\lim_{x \to x_0} f(x) = f(x_0)$. To this end, by the algebraic limit theorem and Definition 4.1, we have

$$\lim_{x \to x_0} (f(x) - f(x_0)) = \lim_{x \to x_0} \frac{f(x) - f(x_0)}{x - x_0} \cdot (x - x_0) = f'(x_0) \cdot 0 = 0$$

as desired. $\square$

Next, observe that there exist functions which are continuous and differentiable everywhere on $\mathbb{R}$, but the derivatives need not be continuous.

Example 4.1. *Let*

$$f(x) = \begin{cases} x^2 \sin(1/x), & \text{if } x \neq 0, \\ 0, & \text{if } x = 0. \end{cases}$$

If $x \neq 0$, $f'(x) = 2x\sin(1/x) - \cos(1/x)$. So $f'(x)$ exists for all $x \neq 0$. Moreover, by Definition (4.1), we have

$$f'(0) = \lim_{x \to 0} \frac{f(x) - f(0)}{x} = \lim_{x \to 0} x \sin \frac{1}{x} = 0.$$

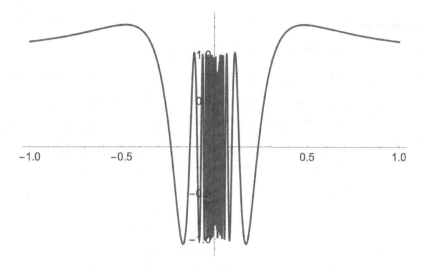

FIGURE 4.1
Graph of $f'(x)$.

Since $\lim_{x \to 0} \cos(1/x)$ does not exist, consequently, $f'(x)$ has no limit at $x = 0$, so $f'(x)$ is not continuous at $x = 0$ (see Figure 4.1).

It is always necessary to be careful to confirm what appears to be clear from a geometric viewpoint. For example, if $f'(0) > 0$, is it true that there is neighborhood of 0 on which f is increasing? Answer is negative!

Example 4.2. *Let*

$$f(x) = \begin{cases} x^2 \sin(1/x^2) + x, & \text{if } x \neq 0, \\ 0, & \text{if } x = 0. \end{cases}$$

Here $f(x)$ is differentiable everywhere and satisfies $f'(0) = 1$, but, for $x \neq 0$, $f'(x)$ densely oscillates near the origin and $f(x)$ oscillates strongly around the straight line $y = x$ (see Figure 4.2). Thus, f is not increasing on any interval near the origin. This phenomenon is caused by the fact that $f'(x)$ is discontinuous at $x = 0$.

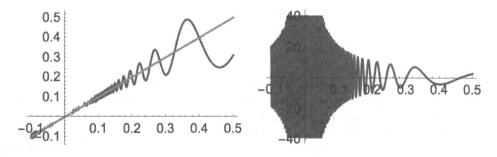

FIGURE 4.2
f is not monotone in the neighborhood of $f'(x_0) \neq 0$.

4.2   Fundamental Theorems of Differentiation

The truth always turns out to be simpler than you thought. — Richard Feynman

We have seen that continuous functions enjoy many nice properties (see Section 3.3). In this section, we will look at the consequences of differentiability. Equipped with this stronger condition, we should expect to gain even richer rewards. We begin with the classical maxima and minimum problem. As one learnt in basic calculus course, the derivative of a function is directly relevant to determining its maxima and minima. Recall that

Definition 4.2. $f(x)$ *has a local maximum (minimum) at* $x = x_0$ *if there exists* $\delta > 0$ *such that*

$$f(x) \le \ (\ge) \ f(x_0) \quad \text{for all} \ \ x \in (x_0 - \delta, x_0 + \delta).$$

The connection between local extrema and derivatives is the following *Fermat's theorem*, which displays a geometric intuitive fact: a differentiable function attains extrema at a point where the graph has a horizontal tangent line.

Theorem 4.5 (Fermat's Theorem). *Let* $f : (a,b) \to \mathbb{R}$. *If* f *attains a local maximum (or local minimum) at* $x_0 \in (a,b)$, *and if* $f'(x_0)$ *exists, then* $f'(x_0) = 0$.

Proof. Without loss of generality, we assume that f attains local maximum at x_0. Choose δ based on Definition 4.2, so that

$$a < x_0 - \delta < x_0 < x_0 + \delta < b.$$

If $x_0 - \delta < x < x_0$, then

$$\frac{f(x) - f(x_0)}{x - x_0} \ge 0.$$

Letting $x \to x_0$ yields $f'(x_0) \ge 0$. On the other side, if $x_0 < x < x_0 + \delta$, then

$$\frac{f(x) - f(x_0)}{x - x_0} \le 0,$$

which shows that $f'(x_0) \le 0$. Therefore, we find that $f'(x_0) = 0$ as desired.   □

Recall Definition 3.5, we see that $f'(x)$ in Example 4.1 has an essential discontinuity at $x = 0$. It is interesting to know what about a derivative with a simple jump discontinuity. For example, is there a function $f(x)$ such that

$$f'(x) = \begin{cases} -1, & \text{if } x \le 0, \\ 1, & \text{if } x > 0. \end{cases}$$

A quick answer may bring to mind the absolute value function $|x|$. However, $|x|$ is not differentiable at $x = 0$ and so is not differentiable everywhere. As an application of Fermat's theorem, the following surprising theorem due to Darboux tells us that, without sacrificing differentiability at any point, there is no function for which its slopes jump from -1 to 1. In other words, *not every function is a derivative!* A function that is a derivative of some function must be special.

Theorem 4.6 (Darboux's Theorem). *Let* f *be differentiable on* $[a,b]$. *Then for any* α *between* $f'(a)$ *and* $f'(b)$, *there exists a point* $c \in (a,b)$ *such that* $f'(c) = \alpha$. *In other words,* $f'(x)$ *exhibits the intermediate value property.*

Proof. First, we treat the special case by assuming that $f'(a)$ and $f'(b)$ have opposite sign. For example, $f'(a) > 0$ and $f'(b) < 0$. Since $f(x)$ is continuous on $[a, b]$, the extreme value theorem yields that there exists a $c \in [a, b]$ such that $f(c) = \max_{x \in [a,b]} f(x)$. By the assumption and Definition 4.1, we have that $f(x) > f(a)$ for $x \in (a, a + \delta_1)$ and $f(x) > f(b)$ for $x \in (b - \delta_2, b)$, and so $a < c < b$. It follows that $f'(c) = 0$ by Fermat's Theorem.

Next, by introducing an auxiliary function, we can always transfer the general case to the above special case. To this end, for any α between $f'(a)$ and $f'(b)$, Let

$$F(x) = f(x) - \alpha x, \quad x \in [a, b].$$

Clearly, $F(x)$ is differentiable on $[a, b]$ and $F'(x) = f'(x) - \alpha$. By the assumption of α, $F'(a)$ and $F'(b)$ have opposite sign. The preceding argument guarantees that there exists a $c \in (a, b)$ with $F'(c) = 0$, so $f'(c) = \alpha$. □

The following proof of Darboux's theorem is due to Nadler [70]. This elegant proof shows systematically and conceptually why the theorem is true. Let $\mathcal{D}$ be the set of all values of the derivative of f and $\mathcal{C}$ be the set of all slopes of chords joining distinct points of the graph of f. We now prove that $\mathcal{D}$ is an interval.

Proof. The mean value theorem says that $\mathcal{C} \subset \mathcal{D}$, and the definition of the derivative implies that $\mathcal{D} \subset \overline{\mathcal{C}}$. Thus, it suffices to prove that $\mathcal{C}$ is an interval. To this end, we prove that any two points of $\mathcal{C}$ can be connected by an interval in $\mathcal{C}$. To prove this, fix $p, q \in \mathcal{C}$, where

$$p = \frac{f(x_1) - f(x_2)}{x_1 - x_2}, \quad q = \frac{f(x_3) - f(x_4)}{x_3 - x_4}$$

with $x_1 < x_2, x_3 < x_4$, and $x_i \in [a, b]$ $(i = 1, 2, 3, 4)$. We define a continuous function $g : [0, 1] \to \mathcal{C}$ by

$$g(t) = \frac{f((1 - t)x_1 + tx_3) - f((1 - t)x_2 + tx_4)}{((1 - t)x_1 + tx_3) - ((1 - t)x_2 + tx_4)}.$$

Geometrically, g is the slope of the chords one obtains by sliding the points $(x_1, f(x_1))$ and $(x_2, f(x_2))$ linearly with respective to the x-axis along the graph of f to the points $(x_3, f(x_3))$ and $(x_4, f(x_4))$. By the intermediate value theorem, $g([0, 1])$ is an interval in $\mathcal{C}$, also $p = g(0)$ and $q = g(1)$, and this concludes the proof. □

Similar to the intermediate value theorem for continuous functions, Darboux's theorem indeed offers us an intermediate value theorem for derivatives. Consequently, we have

(1) A derivative cannot pose jump discontinuity. Therefore, if $f'(x)$ is monotone on (a, b), then $f'(x)$ is continuous on (a, b).

(2) A derivative cannot change sign in an interval without taking the value 0. Therefore, for example, if $f'(x) > 0$ on (a, b), then f is strictly increasing on (a, b).

The second fact provides us a very useful approach in proving inequalities. We illustrate this by two examples.

Example 4.3. *Consider $f(x) = x^\alpha - \alpha x$ on $[0, \infty)$, where $0 < \alpha < 1$.*

Since

$$f'(x) = \alpha(x^{\alpha-1} - 1) = \begin{cases} > 0, & \text{if } 0 < x < 1, \\ 0, & \text{if } x = 1, \\ < 0, & \text{if } x > 1, \end{cases}$$

it follows that f attains the maximum at $x = 1$, so

$$x^\alpha - \alpha x \le f(1) = 1 - \alpha \quad \text{for all } x \in (0, \infty). \tag{4.2}$$

We now derive some classical inequalities from (4.2). For example, for any positive real numbers a and b, let $x = a/b, \beta = 1 - \alpha$. Then (4.2) becomes

$$a^{\alpha} b^{\beta} \leq \alpha a + \beta b, \tag{4.3}$$

which is often called as the *weighted AM-GM inequality*. In particular, when $\alpha = \beta = 1/2$, we obtain the well-known *AM-GM inequality*

$$\sqrt{ab} \leq \frac{a+b}{2}.$$

Furthermore, let $p = 1/\alpha, q = 1/\beta$, where $p, q > 1, 1/p + 1/q = 1$. Replacing a and b in (4.3) by a^p and b^q, respectively, yields *Young's inequality*

$$ab \leq \frac{1}{p} a^p + \frac{1}{q} b^q. \tag{4.4}$$

Substituting a by a_k and b by b_k in (4.4) and summing over $1 \leq k \leq n$ gives

$$\sum_{k=1}^{n} a_k b_k \leq \frac{1}{p} \sum_{k=1}^{n} a_k^p + \frac{1}{q} \sum_{k=1}^{n} b_k^q. \tag{4.5}$$

By a standard normalized variable approach, applying

$$a_k \to \frac{a_k}{\left(\sum_{k=1}^{n} a_k^p\right)^{1/p}}, \quad b_k \to \frac{b_k}{\left(\sum_{k=1}^{n} b_k^q\right)^{1/q}}$$

in (4.5), we finally obtain the classical *Hölder's inequality*

$$\sum_{k=1}^{n} a_k b_k \leq \left(\sum_{k=1}^{n} a_k^p\right)^{1/p} \left(\sum_{k=1}^{n} b_k^q\right)^{1/q}. \tag{4.6}$$

Example 4.4. *Let $x_i \, (i = 1, 2, \ldots, n)$ be positive real numbers with $\prod_{i=1}^{n} x_i = 1$. For $0 \leq s \leq t$, prove that*

$$\sum_{i=1}^{n} x_i^s \leq \sum_{i=1}^{n} x_i^t.$$

Proof. We prove that function $f(x) = \sum_{i=1}^{n} x_i^x$ is increasing for $x \geq 0$. Indeed, we have

$$f'(x) = \sum_{i=1}^{n} x_i^x \ln x_i.$$

In particular, $f'(0) = \sum_{i=1}^{n} \ln x_i = \ln \left(\prod_{i=1}^{n} x_i\right) = 0$. Clearly,

$$f''(x) = \sum_{i=1}^{n} x_i^x (\ln x_i)^2 > 0.$$

This implies that $f'(x)$ is increasing and $f'(x) \geq f'(0) = 0$. Hence f itself is increasing for $x \geq 0$. $\qquad \square$

We next turn to the *mean value theorem*, which is fundamental to the theory of derivatives. First we treat a special case of the mean value theorem. Geometrically, it provides a sufficient condition for the existence of horizontal tangent line.

Theorem 4.7 (Rolle's Theorem). *Let $f : [a, b] \to \mathbb{R}$. Suppose that*

    1. f is continuous on $[a, b]$,

    2. f is differentiable on (a, b), and

    3. $f(a) = f(b) = 0$.

Then there is a point $x_0 \in (a, b)$ at which

$$f'(x_0) = 0.$$

The popular procedure of proving Rolle's theorem runs as follows. First the extreme value theorem implies that f attains a maximum and a minimum on $[a, b]$. Then the assumption derives that the extreme occurs at an interior. Otherwise, f is a constant function and $f'(x) = 0$ for all $x \in (a, b)$. Finally the conclusion follows from Fermat's theorem. Notice that Rolle's theorem is just another existence theorem. It guarantees the existence of a certain number x_0 without giving any information about how to find x_0 . Here we subsequently give two direct proofs around Fermat's theorem. Samelson's proof [83] is based on the intermediate value theorem and the nested interval theorem. The point x_0 found is not necessarily the absolute (or even a relative) extrema. Abian's proof [1] uses the nested interval theorem only. Both proofs are highly constructive in the sense that it can be readily computerized and x_0 can be computed within any desired degree of accuracy. Moreover, Abian's proof also shows the extreme value theorem constructively without invoking the boundedness of f on $[a, b]$.

Samelson's proof. Let

$$F(x) = f(x) - f\left(x + \frac{b - a}{2}\right).$$

Since $f(a) = f(b)$, we have

$$F(a) = f(a) - f\left(\frac{a + b}{2}\right), \quad \text{and} \quad F\left(\frac{a + b}{2}\right) = f\left(\frac{a + b}{2}\right) - f(b) = -F(a).$$

The intermediate value theorem ensures that there exists $a_1 \in [a, (a + b)/2]$ such that $F(a_1) = 0$. Let $b_1 = a_1 + \frac{1}{2}(b - a)$. Then

$$f(a_1) = f(b_1) \text{ and } b_1 - a_1 = \frac{1}{2}(b - a).$$

Continuing in this way, we obtain a sequence of nested intervals $[a_n, b_n]$ such that

$$f(a_n) = f(b_n) \text{ and } b_n - a_n = \frac{1}{2^n}(b - a), \quad n \in \mathbb{N}.$$

The nested interval theorem implies that there is a unique number $x_0 \in [a_n, b_n]$ for every n and

$$\lim_{n \to \infty} a_n = \lim_{n \to \infty} b_n = x_0.$$

Since $f(x)$ is differentiable at x_0, we have

$$\lim_{n \to \infty} \frac{f(b_n) - f(a_n)}{b_n - a_n} = f'(x_0).$$

It follows that $f'(x_0) = 0$ from the difference quotients all vanish.   □

Abian's proof. Let

$$a_1 = a, b_1 = b, m_1 = \frac{a_1 + b_1}{2}, p_1 = \frac{m_1 + a_1}{2}, q_1 = \frac{b_1 + m_1}{2}.$$

Without loss of generality, we assume that $f(m_1) \geq 0$. Define

$$f(m_2) = \max\{f(m_1), f(p_1), f(q_1)\}.$$

Of above three values, if two or more of them attain the maximum, we choose m_2 the first according to the order of m_1, p_1, q_1 from left to right. Next, let

$$a_2 = m_2 - \frac{b_1 - a_1}{4}, b_2 = m_2 + \frac{b_1 - a_1}{4}, p_2 = \frac{m_2 + a_2}{2}, q_2 = \frac{b_2 + m_2}{2}$$

and

$$f(m_3) = \max\{f(m_2), f(p_2), f(q_2)\}.$$

This process can be repeated as often as we like. Consequently, we obtain a sequence of nested intervals $[a_n, b_n]$ with

$$0 \leq f(m_1) \leq f(m_2) \leq \cdots \leq f(m_n) \leq \cdots$$

and

$$f(m_n) \geq f(a_n), \quad f(m_n) \geq f(b_n).$$

Let x_0 be the unique common point of $[a_n, b_n]$'s, which is asserted by the nested interval theorem. Clearly, $x_0 \in [a, b]$ and

$$\lim_{n \to \infty} a_n = \lim_{n \to \infty} b_n = \lim_{n \to \infty} m_n = x_0.$$

Since f is continuous, we have

$$\lim_{n \to \infty} f(a_n) = \lim_{n \to \infty} f(b_n) = \lim_{n \to \infty} f(m_n) = f(x_0)$$

and

$$f(x_0) \geq f(a_n), \quad f(x_0) \geq f(b_n) \quad \text{for all } n \in \mathbb{N}.$$

We claim that $x_0 \in (a, b)$. Indeed, if $f(m_i) > 0$ for some $i \in \mathbb{N}$, clearly, $f(x_0) > 0$ and so $x_0 \in (a, b)$. On the other hand, if $f(m_i) = 0$ for all $i \in \mathbb{N}$, the selection of m_i yields that $x_0 = m_1 \in (a, b)$. Finally, since

$$\frac{f(x_0) - f(a_n)}{x_0 - a_n} \geq 0, \quad \frac{f(b_n) - f(x_0)}{b_n - x_0} \leq 0,$$

taking $n \to \infty$ concludes that $f'(x_0) = 0$, as desired. $\qquad\square$

In general, if the graph of a differentiable function connects the points $(a, f(a))$ and $(b, f(b))$, we can find an auxiliary function that enables us to reduce this case to Rolle's Theorem. The precise statement and proof are as follows.

Theorem 4.8 (Lagrange's Mean Value Theorem). *Let $f : [a, b] \to \mathbb{R}$. Suppose*

    1. f is continuous on $[a, b]$,

    2. f is differentiable in (a, b).

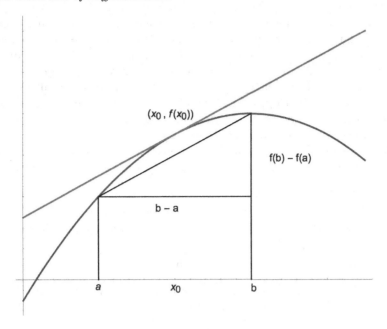

FIGURE 4.3
The tangent line at x_0 is parallel to the secant line.

Then there is a point $x_0 \in (a, b)$ where (see Figure 4.3)

$$f'(x_0) = \frac{f(b) - f(a)}{b - a}.$$

Proof. By the point-slope form, the equation of the line through $(a, f(a))$ and $(b, f(b))$ is given by

$$y = T(x) := f(a) + \frac{f(b) - f(a)}{b - a}(x - a).$$

Let $g(x) = f(x) - T(x)$. Observe that $g(x)$ is continuous on $[a, b]$, differentiable on (a, b) and satisfies $g(a) = g(b) = 0$. Applying Rolle's theorem to g yields that there exists a point $x_0 \in (a, b)$ such that

$$g'(x_0) = f'(x_0) - \frac{f(b) - f(a)}{b - a} = 0.$$

Thus

$$f'(x_0) = \frac{f(b) - f(a)}{b - a}.$$

$\square$

The mean value theorem has a more general form due to Cauchy. It follows by directly applying Lagrange's mean value theorem to the auxiliary function

$$h(x) = (f(b) - f(a))g(x) - (g(b) - g(a))f(x).$$

Theorem 4.9 (Cauchy's Mean Value Theorem). *Let $f, g : [a, b] \to \mathbb{R}$ be continuous on $[a, b]$ and differentiable on (a, b). If $g'(x) \neq 0$ for all $x \in (a, b)$, then there exists a point $x_0 \in (a, b)$ such that*

$$\frac{f'(x_0)}{g'(x_0)} = \frac{f(b) - f(a)}{g(b) - g(a)}.$$

Beside its geometrically plausible assertion, the mean value theorem can be stated as many special designations. For example, in Physics, the mean value theorem assures that there is a time at which the instantaneous velocity equals the average velocity on $[a, b]$. In general, the mean value theorem is the cornerstone for almost every major theorems related to differentiation, we will have many opportunities to apply it in the subsequent sections. For example, we will use it to prove L'Hôpital's rules, to analyze the behavior of infinite series, and to establish the *fundamental theorem of calculus*.

The mean value theorem is also rich in applications of differentiation, especially when it relates the behavior of f' to f in an intricate way. Here we comprise some immediately consequences from the mean value theorem. The proofs are left to the interested reader.

1. If $f : (a, b) \to \mathbb{R}$ is differentiable and $f'(x) = 0$ for all $x \in (a, b)$, then $f(x)$ is a constant.

2. *A criterion for the identity.* If $f, g : (a, b) \to \mathbb{R}$ are differentiable and $f'(x) = g'(x)$ for all $x \in (a, b)$, then
$$f(x) = g(x) + C.$$

3. *A criterion for selecting maximizers and minimizers.* Suppose that $f : (a, b) \to \mathbb{R}$ has a second derivative and $f'(x_0) = 0$.

    - If $f''(x_0) > 0$, then x_0 is a local minimum point of f.
    - If $f''(x_0) < 0$, then x_0 is a local maximum point of f.

4. *A criterion for proving inequality.* If $|f'(x)| \leq M$ for all $x \in (a, b)$, then
$$|f(x) - f(y)| \leq M\, |x - y| \quad \text{for all } x, y \in (a, b).$$

    A function satisfies the above inequality is called *Lipschitz* on (a, b).

5. If f' exists and is bounded on some interval I, then f is Lipschitz and uniform continues on I.

Finally, we show that Darboux's theorem and the mean value theorem turn out also to be equivalent to the cut principle (and hence completeness per se). This fact is based on the Corollary 4 in Chapter 3. We prove the contrapositive. Recall that If the cut principle does not hold, then there exists a cut of open sets A and B. Clearly, if $f, g : \mathbb{R} \to \mathbb{R}$ are differentiable, then the function
$$F(x) = \begin{cases} f(x), & x \in A, \\ g(x), & x \in B \end{cases}$$
is also differentiable on $\mathbb{R}$.

For the case of Darboux's theorem, let
$$F(x) = \begin{cases} x, & x \in A \cap [a, b], \\ a - 1, & x \in B \cap [a, b]. \end{cases}$$

$F(x)$ is differentiable on $[a, b]$. However, since $F'([a, b]) = \{0, 1\}$, the derivative clearly does not have the intermediate value property. Therefore, Darboux's theorem implies the cut principle.

For the case of the mean value theorem, let
$$F(x) = \begin{cases} 0, & x \in A \cap [a, b], \\ 1, & x \in B \cap [a, b]. \end{cases}$$

Here F is differentiable on (a, b) and satisfies $F'(x) = 0$. However,

$$\frac{F(b) - F(a)}{b - a} = \frac{1}{b - a},$$

which violates the conclusion of the mean value theorem. Thus, the mean value theorem implies the cut principle.

The connection between these two fundamental theorems of differentiations and the completeness of real numbers may offer a philosophical "explanation" as to why they almost seem to be too obvious. After all, the completeness of real numbers itself appears completely self-evident. This sheds light on the landscape of mathematical theorems and structures.

4.3   L'Hôpital's Rules

When the only tool you own is a hammer, every problem begins to resemble a nail.
— Abraham Maslow

In this section, we use Cauchy's mean value theorem to establish a collection of familiar theorems usually called *L'Hôpital's rules*. They are favorite tools for evaluating the limits of indeterminate forms.

Theorem 4.10 (L'Hôpital's Rule I). *Suppose $f, g : I \to \mathbb{R}$ are differentiable and $g'(x) \neq 0$ for all x near a. If*

$$f(a) = g(a) = 0 \quad or \quad \lim_{x \to a} g(x) = +\infty$$

and

$$\lim_{x \to a} \frac{f'(x)}{g'(x)} = L,$$

then

$$\lim_{x \to a} \frac{f(x)}{g(x)} = L.$$

Proof. We begin with the case where $f(a) = g(a) = 0$. Observe that Cauchy's mean value theorem implies that

$$\frac{f(x) - f(a)}{g(x) - g(a)} = \frac{f'(x_0)}{g'(x_0)}$$

for some x_0 between a and x. Therefore,

$$\left| \frac{f(x)}{g(x)} - L \right| = \left| \frac{f(x) - f(a)}{g(x) - g(a)} - L \right| = \left| \frac{f'(x_0)}{g'(x_0)} - L \right|$$

For any given $\epsilon > 0$, there exists a δ that works for $f'(x)/g'(x)$. Since x_0 lies between a and x,

$$0 < |a - x_0| < |a - x| < \delta,$$

the same δ will work for $f(x)/g(x)$.

Next, Let $\lim_{x \to a} g(x) = +\infty$. By assumption, for any $\epsilon > 0$, there exists a $\delta_1 > 0$ such that

$$\left| \frac{f'(x)}{g'(x)} - L \right| < \frac{\epsilon}{2}$$

for all $a < x < a + \delta_1$. Next, applying Cauchy's mean value theorem on $[x, a + \delta_1]$ yields

$$\frac{f(x) - f(a + \delta_1)}{g(x) - g(a + \delta_1)} = \frac{f'(x_0)}{g'(x_0)}$$

for some $x_0 \in (x, a + \delta_1)$. Our choice of δ_1 then implies

$$L - \frac{\epsilon}{2} \leq \frac{f(x) - f(a + \delta_1)}{g(x) - g(a + \delta_1)} \leq L + \frac{\epsilon}{2} \tag{4.7}$$

for all $x \in (a, a + \delta_1)$.

To isolate the ratio $f(x)/g(x)$, we multiply (4.7) by $(g(x) - g(a + \delta_1))/g(x)$. Since $\lim_{x \to a} g(x) = +\infty$, there exists a $\delta_2 > 0$ so that $g(x) > g(a + \delta_1)$ for all $x \in (a, a + \delta_2)$. Thus $g(a + \delta_1)/g(x) < 1$ and

$$\left(L - \frac{\epsilon}{2}\right)\left(1 - \frac{g(a + \delta_1)}{g(x)}\right) \leq \frac{f(x) - f(a + \delta_1)}{g(x)} \leq \left(L + \frac{\epsilon}{2}\right)\left(1 - \frac{g(a + \delta_1)}{g(x)}\right).$$

It follows that, after some algebraic manipulations,

$$\left(L - \frac{\epsilon}{2}\right) - \left(\frac{g(a + \delta_1)(L - \epsilon/2) - f(a + \delta_1)}{g(x)}\right) \leq \frac{f(x)}{g(x)} \leq \left(L + \frac{\epsilon}{2}\right)$$
$$+ \left(\frac{g(a + \delta_1)(L + \epsilon/2) + f(a + \delta_1)}{g(x)}\right).$$

Since L, ϵ, and δ_1 are all fixed, there exists a $\delta_3 > 0$ such that

$$\frac{g(a + \delta_1)(L - \epsilon/2) - f(a + \delta_1)}{g(x)} \leq \frac{\epsilon}{2},$$

$$\frac{g(a + \delta_1)(L + \epsilon/2) + f(a + \delta_1)}{g(x)} \leq \frac{\epsilon}{2}.$$

Putting all these inequalities together and choosing $\delta = \min\{\delta_1, \delta_2, \delta_3\}$ guarantee that

$$\left|\frac{f(x)}{g(x)} - L\right| < \epsilon$$

for all $a < x < a + \delta$. $\square$

Notice that there is no assumption on f in the second case. Similarly, when $x \to \infty$, we have

Theorem 4.11 (L'Hôpital's Rule II). *Suppose $f, g : [a, \infty) \to \mathbb{R}$ are differentiable and $g'(x) \neq 0$ for all $x \in [a, \infty)$. If*

$$\lim_{x \to \infty} g(x) = +\infty$$

and

$$\lim_{x \to \infty} \frac{f'(x)}{g'(x)} = L,$$

then

$$\lim_{x \to \infty} \frac{f(x)}{g(x)} = L.$$

Proof. Here we give a short proof based on Problem 2 in Section 3.6, which serves as a bridge between Stolz-Cesàro theorem and L'Hôpital's rules. To see this, we only need to verify the conditions in the Problem 2. Here $T = 1$. Since $g'(x) \neq 0$, Darboux's theorem implies that $g'(x)$ does not change sign in $[a, \infty)$. In view of the fact that $\lim_{x \to \infty} g(x) = +\infty$, we have $g'(x) > 0$ for $x \in [a, \infty)$ and so $g(x + 1) > g(x)$ for all $x \in [a, \infty)$. Since f and g are differentiable on $[a, \infty)$, clearly they are bounded on any finite interval in $[a, \infty)$. Finally, for any $x \in [a, \infty)$, Cauchy's mean value theorem gives

$$\frac{f(x + 1) - f(x)}{g(x + 1) - g(x)} = \frac{f'(x + \theta_x)}{g'(x + \theta_x)},$$

where $\theta_x \in (0, 1)$. Thus,

$$\lim_{x \to \infty} \frac{f'(x + \theta_x)}{g'(x + \theta_x)} = \lim_{x \to \infty} \frac{f'(x)}{g'(x)} = L,$$

and so

$$\lim_{x \to \infty} \frac{f(x + 1) - f(x)}{g(x + 1) - g(x)} = L.$$

By the result of Problem 2 in Section 3.6, we find

$$\lim_{x \to \infty} \frac{f(x)}{g(x)} = \lim_{x \to \infty} \frac{f(x + 1) - f(x)}{g(x + 1) - g(x)} = L.$$

$\square$

By taking the logarithm, the indeterminate forms of

$$0^0, 1^\infty, \infty^0$$

can be transformed into cases of either $0/0$ or ∞/∞, so that L'Hôpital's rule still can be applied.

There are two comments related to the applicability of L'Hôpital's rules.

(1) We cannot draw any conclusion about $\lim_{x \to a} f(x)/g(x)$ if $\lim_{x \to a} f'(x)/g'(x)$ does not exist. Indeed, consider $f(x) = x + \sin x, g(x) = x + \cos x$. Clearly,

$$\lim_{x \to \infty} \frac{f'(x)}{g'(x)} = \lim_{x \to \infty} \frac{1 + \cos x}{1 - \sin x}$$

does not exist, but

$$\lim_{x \to \infty} \frac{f(x)}{g(x)} = \lim_{x \to \infty} \frac{1 + \sin x/x}{1 + \cos x/x} = 1.$$

(2) The requirement that g' is nonzero for x near a is crucial. If g' has zeros in every neighborhood of a, the zeros of g' may cancel the zeros of f'. In this case, we will see that $\lim_{x \to a} \frac{f'(x)}{g'(x)}$ exists but $\lim_{x \to a} \frac{f(x)}{g(x)}$ does not exists. For example, let

$$f(x) = \frac{x}{2} + \frac{\sin 2x}{4}, \quad g(x) = f(x)e^{\sin x}.$$

Hence

$$\lim_{x \to \infty} \frac{f'(x)}{g'(x)} = \lim_{x \to \infty} \frac{2 \cos x}{(x + 2 \cos x + \sin x \cos x)e^{\sin x}} = 0.$$

However, $\lim_{x \to \infty} \frac{f(x)}{g(x)} = \lim_{x \to \infty} e^{-\sin x}$ does not exist. This is due to the fact that $g'(x) = e^{\sin x} \cos x(f(x) + \cos x)$ has zero in every neighborhood of ∞. Consequently, we are not entitled to apply L'Hôpital's rule here.

We now end this section by introduce a general method for proving the monotonicity of a large class of quotients. Because of the similarity to the hypotheses to those of L'Hôpital's rule, we refer to the following theorem as the *L'Hôpital's monotone rule*.

Theorem 4.12 (L'Hôpital's Monotone Rule (LMR)). *Let f, $g : [a, b] \to \mathbb{R}$ be continuous functions that are differentiable on (a, b) with $g' \neq 0$ on (a, b). If f'/g' is increasing (decreasing) on (a, b), then the functions*

$$\frac{f(x) - f(b)}{g(x) - g(b)} \quad and \quad \frac{f(x) - f(a)}{g(x) - g(a)}$$

are likewise increasing (decreasing) on (a, b).

Proof. We may assume that $g'(x) > 0$ and $f'(x)/g'(x)$ is increasing for all $x \in (a, b)$. By the Cauchy mean value theorem, for each given $x \in (a, b)$ there exists $c \in (a, x)$ such that

$$\frac{f(x) - f(a)}{g(x) - g(a)} = \frac{f'(c)}{g'(c)} \leq \frac{f'(x)}{g'(x)},$$

and so

$$f'(x)(g(x) - g(a)) - g'(x)(f(x) - f(a)) \geq 0.$$

Therefore,

$$\left(\frac{f(x) - f(a)}{g(x) - g(a)} \right)' = \frac{f'(x)(g(x) - g(a)) - g'(x)(f(x) - f(a))}{(g(x) - g(a))^2} \geq 0.$$

This shows that $(f(x) - f(a))/(g(x) - g(a))$ is increasing on (a, b) as desired. Along the same lines, we can show that $(f(x) - f(b))/(g(x) - g(b))$ is increasing. $\square$

Here we assume that a and b are finite. Along the same path, this rule can be extended easily to the case where a or b is infinity. The LMR first appeared in Gromov's work for volume estimation in differential geometry. Since then, the LMR and its variants have been used in approximation theory, quasi-conformal theory, and probability. But, for most of readers, the LMR is not as well known as it should be. We illustrate this method by proving *Wilker's inequality*

$$\left(\frac{\sin x}{x} \right)^2 + \frac{\tan x}{x} > 2, \quad \text{for } 0 < x < \tfrac{\pi}{2}.$$

Proof. The proof is almost algorithmic in nature. Set

$$F(x) = \begin{cases} \left(\frac{\sin x}{x} \right)^2 + \frac{\tan x}{x}, & \text{if } x \neq 0, \\ 2, & \text{if } x = 0. \end{cases}$$

Let $f(x) = \sin^2 x + x \tan x$, $g(x) = x^2$. Now

$$f'(x) = 2 \sin x \cos x + x \sec^2 x + \tan x$$

and

$$\frac{f''(x)}{g''(x)} = \cos^2 x - \sin^2 x + \sec^2 x + x \tan x \sec^2 x.$$

Since

$$\left(\frac{f''(x)}{g''(x)} \right)' = 3 \tan x \sec^2 x (1 - \cos^4 x) + \sin x \left(\frac{x}{\sin x} \frac{1}{\cos^4 x} - \cos x \right) + 2x \tan^2 x \sec^2 x > 0,$$

$f''(x)/g''(x)$ is increasing on $(0, \pi/2)$. Using the LMR twice and noticing that $f(0) = f'(0) = g(0) = g'(0) = 0$, we deduce that $f'(x)/g'(x)$ and then $F(x) = f(x)/g(x)$ is strictly increasing on $(0, \pi/2)$. Wilker's inequality now follows from $F(x) > F(0)$. $\square$

The interested reader can find more applications of the LMR in [25, Chapter 4].

4.4 Convex Functions

Nothing is built on stone; all is built in sand. But we must build as if the sand were stone. — Jorge Borges

The convex functions consist of a special class of functions and often appear in many contexts in mathematics. Their importance in various fields of analysis is steadily growing. In this section, we introduce the basic properties of convex functions and examine the connections between the convexity and continuity/differentiability.

Definition 4.3. *Let* $f : I \to \mathbb{R}$, *where* I *is an interval.* f *is called convex if for every* $a, b \in I$ *and every* α *with* $0 < \alpha < 1$,

$$f(\alpha a + (1 - \alpha)b) \le \alpha f(a) + (1 - \alpha)f(b).$$

If $a < b$, for each $x \in (a, b)$, we have $x = \alpha a + (1 - \alpha)b$ with $\alpha = (b - x)/(b - a)$ and $1 - \alpha = (x - a)/(b - a)$. Thus, the definition can be restated as

$$f(x) \le \frac{b - x}{b - a} f(a) + \frac{x - a}{b - a} f(b). \tag{4.8}$$

Geometrically, this indicates that the graph of f, between $(a, f(a))$ and $(b, f(b))$, lies below the chord joining these two points. While outside the interval (a, b), the graph of f lies above this chord. A change of notation in (4.8) yields

Theorem 4.13. *If* f *is convex on* I *and* $a < b < c \in I$, *then*

$$f(x) \ge \frac{b - x}{b - a} f(a) + \frac{x - a}{b - a} f(b) \ \text{ for } b < x < c \tag{4.9}$$

and

$$f(x) \ge \frac{c - x}{c - b} f(b) + \frac{x - b}{c - b} f(c) \ \text{ for } a < x < b. \tag{4.10}$$

Consequently, we have

Theorem 4.14. *If* f *is convex on an open interval* I, *then* f *is continuous on* I.

Proof. For any $b \in I$, we show that $f(b^-) = f(b^+) = f(b)$. To this end, choose $a, c \in I$ with $a < b < c$. When $b < x < c$, the inequalities (4.9) and (4.10) yield

$$\frac{b - x}{b - a} f(a) + \frac{x - a}{b - a} f(b) \le f(x) \le \frac{c - x}{c - b} f(b) + \frac{x - b}{c - b} f(c),$$

from which it follows that $f(b^+) = f(b)$ from $x \to b^+$. Similarly, when $a < x < b$, we have

$$\frac{c - x}{c - b} f(b) + \frac{x - b}{c - b} f(c) \le f(x) \le \frac{x - a}{b - a} f(b) + \frac{b - x}{b - a} f(a),$$

which implies that $f(b^-) = f(b)$ from $x \to b^-$. This proves that f is continuous at $x = b$. $\quad\square$

Remark. The hypothesis that I is open is necessary. For example, the function

$$f(x) = \begin{cases} 1, & \text{if } x = 0, \\ 0, & \text{if } 0 < x \le 1 \end{cases}$$

is convex on $[0, 1]$, but not continuous.

We next turn to the connection between the convexity and differentiability. Subtracting $f(a)$ and $f(b)$ from both sides of (4.8), respectively, we get inequalities among the difference quotients

$$\frac{f(x) - f(a)}{x - a} \le \frac{f(b) - f(a)}{b - a} \le \frac{f(b) - f(x)}{b - x}, \quad a < x < b, \tag{4.11}$$

which is also known as the *three chord lemma*. Geometrically, inequalities (4.11) shows that the slope of the chord joining $(a, f(a))$ and $(b, f(b))$ for a convex function is increased if b is increased, or if a is increased.

Theorem 4.15. *Let f be convex on an open interval I. Then for any $c \in I$, both $f'(c^+)$ and $f'(c^-)$ exist. In particular, if $a < c < b$, we have*

$$f'(a^+) \le f'(c^-) \le f'(c^+) \le f'(b^-).$$

Proof. Let $c \in I$. For any $t, u \in I$ with $t < c < u$, (4.11) gives

$$\frac{f(c) - f(t)}{c - t} \le \frac{f(u) - f(c)}{u - c},$$

so that

$$\sup_{t<c} \frac{f(c) - f(t)}{c - t} \le \inf_{u>c} \frac{f(u) - f(c)}{u - c}. \tag{4.12}$$

Now assume that $s, v \in I$ with $s < t < c < u < v$, by the left-hand inequality of (4.11), we have

$$\frac{f(u) - f(c)}{u - c} \le \frac{f(v) - f(c)}{v - c},$$

so the function

$$u \to \frac{f(u) - f(c)}{u - c}$$

is increasing on $I \cap (c, \infty)$. Hence,

$$f'(c^+) = \lim_{u \to c^+} \frac{f(u) - f(c)}{u - c} = \inf_{u>c} \frac{f(u) - f(c)}{u - c}$$

exists. Similarly, the right-hand inequality of (4.11) gives

$$\frac{f(c) - f(s)}{c - s} \le \frac{f(c) - f(t)}{c - t},$$

so the function

$$t \to \frac{f(c) - f(t)}{c - t}$$

is increasing on $I \cap (-\infty, c)$. Therefore,

$$f'(c^-) = \lim_{t \to c^-} \frac{f(t) - f(c)}{c - t} = \sup_{t<c} \frac{f(c) - f(t)}{c - t}$$

exists. We have seen that $f'(c^-) \leq f'(c^+)$ in (4.12). Finally, let $a < c < b$. For $t \in (a, c)$, we have

$$f'(a^+) \leq \frac{f(t) - f(a)}{t - a} \leq \frac{f(c) - f(t)}{c - t} \leq f'(c^-).$$

Similarly, for $u \in (c, b)$ we get

$$f'(c^+) \leq \frac{f(u) - f(c)}{u - c} \leq \frac{f(b) - f(u)}{b - u} \leq f'(b^-).$$

This proves that

$$f'(a^+) \leq f'(c^-) \leq f'(c^+) \leq f'(b^-)$$

as desired. $\qquad\qquad\qquad\qquad\qquad\qquad\qquad\qquad\qquad\qquad\qquad\qquad\qquad\qquad\qquad\quad\square$

Appealing to Theorem 3.13, this theorem implies that

Theorem 4.16. *If f is convex on an open interval I, then the set of points where f is not differentiable is at most countable.*

Let $c \in I$. For any $f'(c^-) \leq m \leq f'(c^+), x \in I$, we have

$$f(x) \geq f(c) + m(x - c).$$

Thus, at each point $(c, f(c))$ on the graph, there exists a straight line passing through that point, and lying below the graph of f. Such a line is called a *support line* of f (see Figure 4.4).

If $f'(c)$ exist, then $m = f'(c)$ and so there is only one support line. Otherwise, m can be any number between $f'(c^-)$ and $f'(c^+)$.

Based on this property, we have the following simple tests of convexity on differentiable functions.

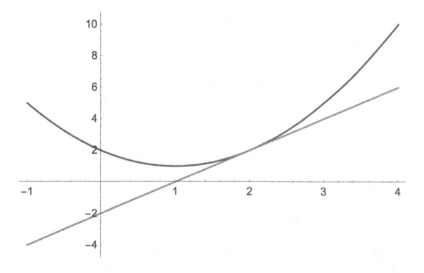

FIGURE 4.4
A convex function has a support line.

Theorem 4.17. *Let f be convex and differentiable on an open interval I.*

1. *f is convex if and only if $f(x) \geq f(a) + f'(a)(x-a)$ for any $x, a \in I$.*

2. *If f is twice differentiable, then f is convex if and only if $f''(x) \geq 0$ for all $x \in I$.*

Monotonicity and convexity are two basic tools for establishing inequalities. We have illustrated the applications of monotonicity in previous section. As an application of convexity, we now end this section by proving the following master inequality.

Theorem 4.18 (Jensen's Inequality). *Let $f : I \to \mathbb{R}$ be a convex function. If all $x_i \in I, p_i \geq 0, i = 1, 2, \ldots, n$ with $p_1 + p_2 + \cdots + p_n = 1$, then*

$$f(p_1 x_1 + p_2 x_2 + \cdots + p_n x_n) \leq p_1 f(x_1) + p_2 f(x_2) + \cdots + x_n f(x_n). \tag{4.13}$$

Proof. We prove (4.13) by induction. When $n = 2$ we see that Jensen's inequality (4.13) is same with the definition of convexity. So the basis for induction is clearly true. For the induction step, without loss of generality, we assume that $0 < p_{n+1} < 1$ and rewrite

$$\sum_{i=1}^{n+1} p_i x_i = p_{n+1} x_{n+1} + (1 - p_{n+1}) \sum_{i=1}^{n} \frac{p_i}{1 - p_{n+1}} x_i.$$

Now, by this expression, the definition of convexity, and the induction hypothesis—all applied in that order—we get

$$\begin{aligned}
f\left(\sum_{i=1}^{n+1} p_i x_i\right) &\leq p_{n+1} f(x_{n+1}) + (1 - p_{n+1}) f\left(\sum_{i=1}^{n} \frac{p_i}{1 - p_{n+1}} x_i\right) \\
&\leq p_{n+1} f(x_{n+1}) + (1 - p_{n+1}) \sum_{i=1}^{n} \frac{p_i}{1 - p_{n+1}} f(x_i) \\
&= \sum_{i=1}^{n+1} p_i f(x_i).
\end{aligned}$$

This proves (4.13). $\qquad \square$

Jensen's inequality provides straightforward answers to many problems. For example, let $f(x) = -\ln x$ for $x > 0$. Since $f''(x) = 1/x^2 > 0$, f is convex for $x > 0$. Let $p_i = 1/n$. Jeasen's inequality gives

$$-\ln\left(\frac{x_1 + x_2 + \cdots + x_n}{n}\right) \leq -\frac{1}{n}(\ln x_1 + \ln x_2 + \cdots + \ln x_n).$$

This yields the well-known *AM-GM inequality*

$$\sqrt[n]{x_1 x_2 \cdots x_n} \leq \frac{x_1 + x_2 + \cdots + x_n}{n}.$$

However, to treat more challenge problems, we may need to make some deliberate preparations. The following Putnam problem provides a nice example for this situation.

Example 4.5 (Putnam 2003-A2). *Let $a_1, a_2, \ldots, a_n$ and $b_1, b_2, \ldots, b_n$ be nonnegative real numbers. Show that*

$$(a_1 a_2 \cdots a_n)^{1/n} + (b_1 b_2 \cdots b_n)^{1/n} \leq [(a_1 + b_1)(a_2 + b_2) \cdots (a_n + b_n)]^{1/n}.$$

Proof. If $a_i = 0$ for some $1 \leq i \leq n$, the inequality is self-evident. So we assume that $a_i > 0$ for all $1 \leq i \leq n$. Set $t_i = b_i/a_i$. Dividing the given inequality by $(a_1 a_2 \cdots a_n)^{1/n}$ yields an equivalent inequality

$$1 + (t_1 t_2 \cdots t_n)^{1/n} \leq (1 + t_1)^{1/n}(1 + t_2)^{1/n} \cdots (1 + t_n)^{1/n}.$$

Let $t_i = e^{x_i}$. Taking the logarithm of both sides gives

$$\ln\left(1 + e^{(x_1 + x_2 + \cdots + x_n)/n}\right) \leq \frac{1}{n}\sum_{i=1}^{n} \ln(1 + e^{x_i}). \tag{4.14}$$

In view of this form and Jensen's inequality, if we choose $f(x) = \ln(1 + e^x)$, then (4.14) is equivalent to

$$f\left(\frac{x_1 + x_2 + \cdots + x_n}{n}\right) \leq \frac{1}{n}(f(x_1) + f(x_2) + \cdots + f(x_n)).$$

This will follow from Jensen's inequality if we can verify that f is convex. Indeed, we have that $f''(x) = e^{2x}/(1 + e^x)^2 > 0$. Thus, f is convex and Jensen's inequality concludes the desired result. $\square$

4.5 Taylor's Theorem

You can see a lot just by looking. — Yogi Berra

In this section, we present the most important theorem in differential calculus—Taylor's theorem, which is a generalization of the mean value theorem and allows us to approximate a differentiable function by the Taylor polynomials. Qualitatively, Taylor's theorem answers the following question: to what extent do the derivatives of a function at a single point dictate the behavior of the function at nearby points? It also answers a practical concern: if we wish to use Taylor polynomials to actually approximate a function, how much accuracy can we assure ourselves in this approximation?

Definition 4.4. *Let $f : (a,b) \to \mathbb{R}$ be nth differentiable at $x_0 \in (a,b)$. The Taylor polynomial of f at x_0 is given by*

$$T_n(x) = \sum_{k=0}^{n} \frac{f^{(k)}(x_0)}{k!} (x - x_0)^k. \tag{4.15}$$

In particular, if $x_0 = 0$, the Taylor polynomial is called the Maclaurin polynomial.

For $n = 1$, we have $T_1(x) = f(x_0) + f'(x_0)(x - x_0)$, whose graph is the tangent line to f at x_0. Applying L'Hôpital's rule yields

$$\lim_{x \to x_0} \frac{f(x) - T_1(x)}{x - x_0} = 0.$$

Hence

$$f(x) = T_1(x) + o(x - x_0). \tag{4.16}$$

In general, we have

Theorem 4.19 (Taylor's Theorem with Peano's Remainder). *Let f be nth differentiable at x_0. Then*

$$f(x) = T_n(x) + o((x - x_0)^n),$$

where $T_n(x)$ is given by (4.15).

Here, as usual, $f(x) = o(g(x))$ means that $\lim_{x \to x_0} f(x)/g(x) = 0$.

Proof. Let $R_n(x) = R_n(f, x) = f(x) - T_n(x)$, which is called the *Taylor remainder*. It suffices to show that

$$R_n(x) = o((x - x_0)^n). \tag{4.17}$$

We prove (4.17) by induction. The base step $n = 1$ holds by (4.16). Now, assume that (4.17) is true for $n = k$. When $n = k + 1$, applying the induction hypothesis to $f'(x)$ yields

$$R'_{k+1}(f, x) = R_k(f', x) = o((x - x_0)^k).$$

Thus, by L'Hôpita's rule, we find that

$$\lim_{x \to x_0} \frac{R_{k+1}(f, x)}{(x - x_0)^{k+1}} = \lim_{x \to x_0} \frac{R'_{k+1}(f, x)}{(k + 1)(x - x_0)^k} = 0,$$

which implies that (4.17) holds for $n = k + 1$. $\square$

The condition (4.17) actually characterizes the Taylor polynomial $T_n(x)$ completely. Indeed, by induction again, we have the following uniqueness theorem.

Theorem 4.20. *Let f be nth differentiable at x_0. If $Q_n(x)$ is a polynomial such that $f(x) - Q_n(x) = o((x - x_0)^n)$, then $Q_n(x) = T_n(x)$, where $T_n(x)$ is given by (4.15).*

This theorem is very useful for calculating the Taylor polynomial $T_n(x)$. It shows that using the formula $f^{(k)}(x_0)/k!$ is not the only way to calculate the coefficients of $T_n(x)$. In fact, if by *any* means we can find a polynomial Q of degree $\leq n$ such that $f(x) = Q(x) + o((x - x_0)^n)$, then $Q(x)$ must be $T_n(x)$. Observe that if $T_n(f, x)$ and $T_n(g, x)$ are nth order Taylor polynomials of f and g, respectively, then

$$
\begin{aligned}
f(x) \cdot g(x) &= (T_n(f, x) + o((x - x_0)^n))(T_n(g, x) + o((x - x_0)^n)) \\
&= [\text{the terms of degree} \leq n \text{ in } T_n(f, x) \cdot T_n(g, x)] + o((x - x_0)^n)).
\end{aligned}
$$

Thus, to find the nth degree Taylor polynomial of $f(x)g(x)$, simply multiply the nth Taylor polynomials of f and g together, discarding all terms of degree large than n. Here we illustrate an application of this fact by determining the Maclaurin polynomials of $\sec x$ without computing $(\sec x)^{(n)}$. Since $\sec x$ is even, let

$$\sec x = \sum_{k=0}^{n} (-1)^k \frac{E_{2k}}{(2k)!} x^{2k} + o(x^{2n+1}),$$

where E_{2k} is called the *Euler number*. It is well-known that

$$\cos x = \sum_{k=0}^{n} (-1)^k \frac{1}{(2k)!} x^{2k} + o(x^{2n+1}).$$

By the identity $\sec x \cdot \cos x = 1$, we have

$$\left(E_0 - \frac{E_2}{2!} x^2 + \cdots + (-1)^n \frac{E_{2n}}{(2n)!} x^{2n} + o(x^{2n+1}) \right)$$
$$\cdot \left(1 - \frac{1}{2!} x^2 + \cdots + (-1)^n \frac{1}{(2n)!} x^{2n} + o(x^{2n+1}) \right) = 1,$$

and so

$$E_0 - \frac{1}{2!}(E_2 + E_0)x^2 + \frac{1}{4!}\left(E_4 + \binom{4}{2}E_2 + E_0\right)x^4 + \cdots$$

$$+(-1)^n \frac{1}{(2n)!}\left(E_{2n} + \binom{2n}{2}E_{2n-2} + \binom{2n}{4}E_{2n-4} + \cdots + E_0\right)x^{2n} + o(x^{2n+1}) = 1.$$

Matching the coefficients of x^{2k} for $k = 0, 1, \ldots, n$, we obtain the following recursive formulas:

$$E_0 = 1, \ E_2 + E_0 = 0, \ E_4 + \binom{4}{2}E_2 + E_0 = 0;$$

$$E_{2n} + \binom{2n}{2}E_{2n-2} + \binom{2n}{4}E_{2n-4} + \cdots + E_0 = 0.$$

This gives that, for example, the first six Euler numbers are

$$E_0 = 1, E_2 = -1, E_4 = 5, E_6 = -61, E_8 = 1385, E_{10} = 50521.$$

Hence,

$$\sec x = 1 + \frac{1}{2}x^2 + \frac{5}{24}x^4 + \frac{61}{720}x^6 + \frac{277}{8064}x^8 + \frac{50521}{3628800}x^{10} + o(x^{11}).$$

To establish a more precise quantitative estimate of the remainder, one usually imposes slightly stronger conditions on f. The estimates will involve bounds on the derivative $f^{(n+1)}$. Here we present two remainders that are most often encountered.

Theorem 4.21 (Cauchy's Remainder and Lagrange's Remainder). *Let f be $(n+1)$th differentiable on (a, b). For $x_0 \in (a, b)$, let $T_n(x)$ be given by (4.15) and*

$$f(x) = T_n(x) + R_n(x).$$

Then for each $x \in (a, b)$, there is a point c between x_0 and x such that

$$R_n(x) = \frac{f^{(n+1)}(c)}{n!}(x - c)^n(x - x_0), \quad (\text{Cauchy's Remainder})$$

and

$$R_n(x) = \frac{f^{(n+1)}(c)}{(n+1)!}(x - x_0)^{n+1}. \quad (\text{Lagrange's Remainder})$$

Proof. Motivated by (4.15), we introduce

$$F(t) = f(t) + \sum_{k=1}^{n} \frac{f^{(k)}(t)}{k!}(x - t)^k, \quad t \in (a, b).$$

By the assumption, $F(t)$ is differentiable on (a, b) and

$$F'(t) = f'(t) + \sum_{k=1}^{n}\left(\frac{f^{(k+1)}(t)}{k!}(x - t)^k - \frac{f^{(k)}(t)}{(k-1)!}(x - t)^{k-1}\right).$$

Telescoping yields

$$F'(t) = \frac{1}{n!}(x - t)^n f^{(n+1)}(t).$$

Since

$$F(x) - F(x_0) = f(x) - T_n(x) = R_n(x),$$

the mean value theorem implies that there exists c between x and x_0 such that

$$R_n(x) = F'(c)(x - x_0) = \frac{f^{(n+1)}(c)}{n!}(x - c)^n(x - x_0),$$

which is precisely Cauchy's remainder.

On the other hand, let $G(t) = -(x - t)^{n+1}$. Applying Cauchy's mean value theorem yields

$$\frac{R_n(x)}{(x - x_0)^{n+1}} = \frac{F(x) - F(x_0)}{G(x) - G(x_0)} = \frac{F'(c)}{G'(c)} = \frac{f^{(n+1)}(c)}{(n+1)!},$$

which implies the claimed Lagrange's remainder. □

Remark. In addition, if $f^{(n+1)}(x)$ is continuous, by the fundamental theorem of calculus, we find that

$$R_n(x) = F(x) - F(x_0) = \int_{x_0}^{x} F'(t)dt = \frac{1}{n!}\int_{x_0}^{x}(x - t)^n f^{(n+1)}(t)\, dt, \qquad (4.18)$$

which gives us another remainder form of Taylor's theorem—the *integral remainder*.

If f has derivatives of all orders, this leads naturally into the notion of the Taylor series.

Definition 4.5. *Let $f : (a, b) \to \mathbb{R}$ be infinitely differentiable and $x_0 \in (a, b)$. The Taylor series generated by f at $x = x_0$ is*

$$\sum_{k=0}^{\infty} \frac{f^{(k)}(x_0)}{k!}(x - x_0)^k = f(x_0) + f'(x_0)(x - x_0) + \cdots + \frac{f^{(n)}(x_0)}{n!}(x - x_0)^n + \cdots.$$

In particular, if $x_0 = 0$, the Taylor series is called the Maclaurin series.

Based on his remainder, Lagrange once asserted that if $f^{(n)}(x)$ exists for all $n \in \mathbb{N}$ and $x, x_0 \in (a, b)$, then the Taylor series always converges to f itself, namely,

$$f(x) = \sum_{k=0}^{\infty} \frac{f^{(k)}(x_0)}{k!}(x - x_0)^k.$$

But, in 1823, Cauchy discovered the following counterexample:

$$f(x) = \begin{cases} e^{-1/x^2}, & \text{if } x \neq 0 \\ 0, & \text{if } x = 0. \end{cases}$$

Here f has derivatives of all orders at $x_0 = 0$ with $f^{(n)}(0) = 0$ for all $n \in \mathbb{N}$. Thus, the Maclaurin series converges to zero for every x, but converges to $f(x)$ itself only at $x = 0$. A more surprising result was given by Borel: Let a_n be any real sequence. Then there are infinitely many differentiable function f such that $f^{(n)}(0) = a_n$ for all $n \in \mathbb{N}$.

Recall that

$$f(x) = T_n(x) + R_n(x).$$

Thus, the Taylor series generated by f does converge to f itself on (a, b) is equivalent to show that

$$\lim_{n \to \infty} R_n(x) = 0 \quad \text{for all } x \in (a, b). \qquad (4.19)$$

In practice, it may be quite difficult to deal with this limit without knowing the value of c in either Cauchy's or Lagrange's remainder. One popular sufficient condition for (4.19) will definitely hold is: there exists a positive constant M such that

$$|f^{(n)}(x)| \leq M^n, \quad \text{for all } n \in \mathbb{N}, x \in (a, b) .$$

Another sufficient condition for (4.19) valids is due to Bernstein.

Theorem 4.22 (Bernstein Theorem). *Let f and all its derivatives be nonnegative on* $[x_0, a]$. *Then, for any* $x \in (x_0, a)$,

$$f(x) = \sum_{k=0}^{\infty} \frac{f^{(k)}(x_0)}{k!} (x - x_0)^k.$$

Proof. By a translation we can assume that $x_0 = 0$. For any $x \in (0, a)$, we show that

$$0 \leq R_n(x) \leq \left(\frac{x}{a}\right)^{n+1} f(a). \tag{4.20}$$

This implies that $R_n(x) \to 0$ as $n \to \infty$ since $(x/a)^{n+1} \to 0$ for $x \in (0, a)$.

To prove (4.20), we apply the integral remainder (4.18). The change of variable $t = (1 - u)x$ with $x_0 = 0$ gives

$$R_n(x) = \frac{x^{n+1}}{n!} \int_0^1 u^n f^{(n+1)}((1 - u)x) du.$$

By the assumption that all derivations are nonnegative, it follows that $f^{(n+1)}(x)$ is increasing on $(0, a]$. Thus,

$$\begin{aligned}
\frac{R_n(x)}{x^{n+1}} &= \frac{1}{n!} \int_0^1 u^n f^{(n+1)}((1 - u)x) du \\
&\leq \frac{1}{n!} \int_0^1 u^n f^{(n+1)}((1 - u)a) du \quad \text{(use } t = (1 - u)a) \\
&= \frac{1}{a^{n+1}n!} \int_0^a (a - t)^n f^{(n+1)}(t)\, dt = \frac{R_n(a)}{a^{n+1}},
\end{aligned}$$

and so

$$R_n(x) \leq \left(\frac{x}{a}\right)^{n+1} R_n(a). \tag{4.21}$$

Recall that

$$f(x) = T_n(x) + R_n(x) = \sum_{k=0}^{n} \frac{f^{(k)}(0)}{k!} x^k + R_n(x).$$

It follows that $R_n(a) \leq f(a)$ since the terms in $T_n(a)$ are nonnegative. This fact, together with (4.21), now proves (4.20) as desired, in turn, it completes the proof. $\square$

4.6 Worked Examples

The intelligence is proved not by ease of learning, but by understanding what we learn.
— Joseph Whitney

1. Prove that Thomae's function $T(x)$ (see Example 3.3) is nowhere differentiable.

   Proof. Recall that $T(x)$ is discontinuous at every rational point. Thus $T(x)$ is not differentiable at every rational point. We now prove that $T(x)$ is not differentiable at every irrational point by contradiction. Assume that $T(x)$ is differentiable at $x_0 \in \mathbb{R} \setminus \mathbb{Q}$. Then

$$T'(x_0) = \lim_{x \to x_0, x \in \mathbb{R} \setminus \mathbb{Q}} \frac{T(x) - T(x_0)}{x - x_0} = \lim_{x \to x_0, x \in \mathbb{R} \setminus \mathbb{Q}} \frac{0 - 0}{x - x_0} = 0.$$

By the definition of the derivative, for any $\epsilon \in (0, 1)$, there exists a $\delta > 0$ such that

$$\left| \frac{T(x) - T(x_0)}{x - x_0} \right| = \left| \frac{T(x)}{x - x_0} \right| < \epsilon, \quad \text{whenever } 0 < |x - x_0| < \delta. \tag{4.22}$$

On the other hand, choose $q \in \mathbb{N}$ such that $q > 1/\delta$. Let $p = \lfloor x_0 q \rfloor$. We have

$$p < x_0 q < p + 1 \quad \text{and} \quad \frac{p}{q} < x_0 < \frac{p}{q} + \frac{1}{q}.$$

Thus, $0 < x_0 - p/q < 1/q < \delta$, and

$$\left| \frac{T(p/q) - T(x_0)}{p/q - x_0} \right| = \left| \frac{1/q}{x_0 - p/q} \right| = \frac{1}{x_0 q - p} > 1 > \epsilon.$$

This contradicts (4.22), from which it concludes our proof. $\qquad\qquad \square$

2. It is well-known that $f'(x) > 0$ on (a, b) implies that f is strictly increasing on (a, b). Find an example for which f is strictly increasing on (a, b) with $f'(x) = 0$ infinitely many times.

Solution. Let

$$f(x) = \begin{cases} x(2 - \cos(\ln x) - \sin(\ln x)), & \text{if } x \in (0, 1], \\ 0, & \text{if } x = 0. \end{cases}$$

Then $f'(x) = 2(1 - \cos(\ln x)) \geq 0$. For any $x_1, x_2 \in [0, 1]$ with $x_1 < x_2$, we have

$$f(x_2) - f(x_1) = \int_{x_1}^{x_2} f'(t) \, dt > 0.$$

This implies that f is strictly increasing on $(0, 1)$. Moreover, we find that $f'(x_n) = 0$ where $x_n = e^{-2n\pi} \in [0, 1]$ for every $n \in \mathbb{N}$. See Figure 4.5 of the function nearby $x_1 = e^{-2\pi} = 0.0018734\ldots$.

$\qquad\qquad \square$

3. Another proof of Darboux's theorem.

Proof. This proof is due to Olsen [74]. Assume that α lies strictly between $f'(a)$ and $f'(b)$. Define

$$F_a(t) = \begin{cases} f'(a), & \text{if } t = a, \\ \frac{f(t) - f(a)}{t - a}, & \text{if } t \neq a. \end{cases}$$

and

$$F_b(t) = \begin{cases} f'(b), & \text{if } t = b, \\ \frac{f(t) - f(b)}{t - b}, & \text{if } t \neq b. \end{cases}$$

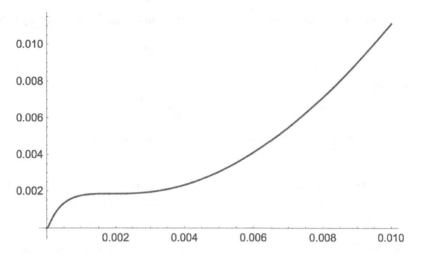

FIGURE 4.5
A strictly increasing function with infinitely many zero derivatives.

Then both $F_a(t)$ and $F_b(t)$ are continuous on $[a, b]$. Since $F_a(b) = F_b(a)$, either α lies between $F_a(a)$ and $F_a(b)$, or α lies between $F_b(a)$ and $F_b(b)$. In the first case, the intermediate value theorem implies that there exists $s \in (a, b]$ such that

$$\alpha = F_a(s) = \frac{f(s) - f(a)}{s - a}.$$

By Lagrange's mean value theorem, there exists $x_1 \in (a, s)$ such that

$$\alpha = \frac{f(s) - f(a)}{s - a} = f'(x_1).$$

In the second case, along the same lines, there exist $s \in [a, b)$ and $x_2 \in (s, b)$ such that

$$\alpha = \frac{f(s) - f(b)}{s - b} = f'(x_2).$$

This completes the proof. $\qquad\qquad\qquad\qquad\qquad\qquad\qquad\qquad\qquad\quad$ $\square$

4. Let f be continuously differentiable on $[a, b]$. If $f'(c) = 0$ for some $c \in (a, b)$, prove that there exists $x_0 \in (a, b)$ such that

$$f'(x_0) = \frac{f(x_0) - f(a)}{b - a}.$$

Proof. Define $g : [a, b] \to \mathbb{R}$ by

$$g(x) = f'(x) - \frac{f(x) - f(a)}{b - a}.$$

If $f(c) = f(a)$, just choosing $x_0 = c$ yields $f'(c) = 0 = \frac{f(c) - f(a)}{b - a}$. Without loss of generality, we assume that $f(c) > f(a)$. Let $x_1 \in (a, c]$ be such that $f(x_1) = \max\{f(x) \,;\, x \in [a, c]\}$. Then

$$f(x_1) - f(a) > 0 \geq (b - a)f'(x_1),$$

which implies that $g(x_1) < 0$.

On the other hand, by Lagrange's mean value theorem, there exists $x_2 \in (a, x_1)$ such that

$$f'(x_2) = \frac{f(x_1) - f(a)}{x_1 - a}.$$

Therefore,

$$g(x_2) = \frac{f(x_1) - f(a)}{x_1 - a} - \frac{f(x_2) - f(a)}{b - a} \geq 0 \geq g(x_1).$$

Using the intermediate value theorem, we find that there exists $x_0 \in [x_2, x_1)$ such that $g(x_0) = 0$. This ends the proof. □

5. Let f be a three times differentiable function such that f has at least four distinct real zeros. Prove that $f'''(x) + 3af''(x) + 3a^2 f'(x) + a^3 f(x)$ has at least one zero for every $a \in \mathbb{R}$.

Proof. Let $F(x) = e^{ax} f(x)$. By the assumption, $F(x)$ has at least four distinct real zeros as well. Applying Rolle's theorem repeatedly yields that $F'(x), F''(x)$ and $F'''(x)$ have at least $3, 2, 1$ distinct real zeros, respectively. Since, by Leibniz's rule (Theorem 4.3),

$$F'''(x) = e^{ax}(f'''(x) + 3af''(x) + 3a^2 f'(x) + a^3 f(x)),$$

and e^{ax} is never zero, it follows the desired result. □

6. Suppose that $f(0) = 0$ and $f'(0)$ exists and is finite. Define

$$x_n = \sum_{k=1}^{n} f\left(\frac{k}{n^2}\right).$$

Prove that $\lim_{n \to \infty} x_n = f'(0)/2$.

Proof. Note that

$$f'(0) = \lim_{x \to 0} \frac{f(x) - f(0)}{x - 0} = \lim_{x \to 0} \frac{f(x)}{x}.$$

Thus, by definition, for any $\epsilon > 0$, there exists a $\delta > 0$ such that

$$f'(0) - \epsilon < \frac{f(x)}{x} < f'(0) + \epsilon \quad \text{whenever} \ \ 0 < x < \delta.$$

Choose $n \in \mathbb{N}$ such that $N > 1/\delta$. When $n > N$, we have $k/n^2 \leq 1/n < 1/N < \delta$, and so

$$(f'(0) - \epsilon) \cdot \frac{k}{n^2} < f\left(\frac{k}{n^2}\right) < (f'(0) + \epsilon) \cdot \frac{k}{n^2}.$$

Summing k from 1 to n yields

$$(f'(0) - \epsilon) \cdot \sum_{k=1}^{n} \frac{k}{n^2} < \sum_{k=1}^{n} f\left(\frac{k}{n^2}\right) < (f'(0) + \epsilon) \cdot \sum_{k=1}^{n} \frac{k}{n^2}.$$

Since $\sum_{k=1}^{n} k = n(n+1)/2$, we have

$$(f'(0) - \epsilon) \frac{n+1}{2n} < \sum_{k=1}^{n} f\left(\frac{k}{n^2}\right) < (f'(0) + \epsilon) \cdot \frac{n+1}{2n}.$$

Letting $n \to \infty$ gets

$$\frac{1}{2}(f'(0) - \epsilon) < \lim_{n \to \infty} x_n < \frac{1}{2}(f'(0) + \epsilon).$$

Since ϵ is arbitrary, this yields the proposed limit by the squeeze theorem. □

Remark. Letting $f(x) = \sin x$ and $f(x) = \ln(1 + x)$, respectively, we find that

$$\lim_{n \to \infty} \sum_{k=1}^{n} \sin\left(\frac{k}{n^2}\right) = 1/2, \quad \lim_{n \to \infty} \prod_{k=1}^{n}\left(1 + \frac{k}{n^2}\right) = e^{1/2}.$$

7. Let $f_n(x) = x^n \ln x$ with $n \in \mathbb{N}$. Find $\lim_{n \to \infty} f_n^{(n)}(1/n)/n!$.

Solution. Observe that

$$f_n'(x) = nx^{n-1} \ln x + x^{n-1} = nf_{n-1} + x^{n-1}.$$

Differentiating both sides $n - 1$ times yields

$$f_n^{(n)}(x) = nf_{n-1}^{(n-1)}(x) + (n - 1)!.$$

Dividing both sides by $n!$ gives

$$\frac{f_n^{(n)}(x)}{n!} - \frac{f_{n-1}^{(n-1)}(x)}{(n-1)!} = \frac{1}{n}, \quad n = 1, 2, \ldots.$$

Telescoping these sums yields

$$\frac{f_n^{(n)}(x)}{n!} = \ln x + 1 + \frac{1}{2} + \cdots + \frac{1}{n}.$$

Let $x = 1/n$. In terms of the nth harmonic number H_n, we find that

$$\lim_{n \to \infty} \frac{f_n^{(n)}(1/n)}{n!} = \lim_{n \to \infty} (H_n - \ln n) = \gamma,$$

which is the Euler-Mascheroni constant. □

8. If $p(x)$ is a non-constant polynomial with only real roots, prove that $(p'(x))^2 \geq p(x)p''(x)$ for all $x \in \mathbb{R}$.

Proof. Let $p(x) = a(x - x_1)(x - x_2) \cdots (x - x_n)$. The logarithmic differentiation gives

$$p'(x) = p(x)\left(\frac{1}{x - x_1} + \frac{1}{x - x_2} + \cdots + \frac{1}{x - x_n}\right).$$

When $x \neq x_i$ $(i = 1, 2, \ldots, n)$, we have

$$\frac{p'(x)}{p(x)} = \frac{1}{x - x_1} + \frac{1}{x - x_2} + \cdots + \frac{1}{x - x_n}.$$

Differentiating this yields

$$\frac{p''(x)p(x) - (p'(x))^2}{p^2(x)} = -\left(\frac{1}{(x - x_1)^2} + \frac{1}{(x - x_2)^2} + \cdots + \frac{1}{(x - x_n)^2}\right) < 0.$$

It follows that $(p'(x))^2 \geq p(x)p''(x)$ for all $x \neq x_i$ $(i = 1, 2, \ldots, n)$. When $x = x_i$ $(i = 1, 2, \ldots, n)$. The proposed inequality has nothing to prove since $p(x_i) = 0$. □

9. Let $p(z)$ be a polynomial with distinct roots $\alpha_1, \alpha_2, \ldots, \alpha_n$ and with the degree $n \geq 2$. Prove that

$$\sum_{k=1}^{n} \frac{1}{p'(\alpha_k)} = 0.$$

Proof. Let

$$p(z) = C \prod_{k=1}^{n} (z - \alpha_k),$$

where C is a nonzero constant. By partial fraction decomposition, we have

$$\frac{1}{p(z)} = \sum_{k=1}^{n} \frac{\beta_k}{z - \alpha_k},$$

where $\beta_k \, (k = 1, 2, \ldots, n)$ is a constant to be determined. To find β_k, appealing to

$$\frac{z - \alpha_k}{p(z)} = \beta_k + \sum_{1 \leq i \leq n, i \neq k} \frac{\beta_i (z - \alpha_k)}{z - \alpha_i},$$

then letting $z \to \alpha_k$ and applying L'Hôpital's rule, we get

$$\beta_k = \lim_{z \to \alpha_k} \frac{z - \alpha_k}{p(z)} = \frac{1}{p'(\alpha_k)},$$

where $p'(\alpha_k) \neq 0$ since α_k is a single root. Therefore,

$$\frac{1}{p(z)} = \sum_{k=1}^{n} \frac{1}{p'(\alpha_k)} \frac{1}{z - \alpha_k},$$

or equivalently

$$1 = \sum_{k=1}^{n} \frac{1}{p'(\alpha_k)} \frac{p(z)}{z - \alpha_k}. \tag{4.23}$$

Note that, for each $k = 1, 2, \ldots, n$,

$$\frac{p(z)}{z - \alpha_k} = C \left(z^{n-1} - \left(\sum_{1 \leq i \leq n, i \neq k} \alpha_i \right) z^{n-2} + \cdots + (-1)^{n-1} \prod_{1 \leq i \leq n, i \neq k} \alpha_i \right).$$

Matching the coefficient of z^{n-1} in (4.23), we find that

$$\sum_{k=1}^{n} \frac{C}{p'(\alpha_k)} = 0.$$

Since $C \neq 0$, it follows that

$$\sum_{k=1}^{n} \frac{1}{p'(\alpha_k)} = 0.$$

$\square$

10. Let $f(x) = (1 + \sqrt{x})^{2(n+1)}$. Find $f^{(n)}(1)$.

Solution. Let $g(x) = (1 - \sqrt{x})^{2(n+1)}$. By induction, it is easy to show that $g^{(n)}(1) = 0$. Applying the binomial theorem yields

$$f(x) + g(x) = \sum_{k=0}^{2(n+1)} \binom{2(n+1)}{k} [x^{k/2} + (-1)^k x^{k/2}] = \sum_{k=0}^{n+1} 2 \binom{2(n+1)}{2k} x^k.$$

Hence

$$(f(x) + g(x))^{(n)} = 2 \left(n! \binom{2(n+1)}{2n} + (n+1)! x \right) = 2(n+1)!(2n+1+x),$$

and so $f^{(n)}(1) = 4(n+1) \cdot (n+1)!$. $\qquad\square$

11. Let n be positive integer and $f(x) = \sin x / x$ for $x > 0$. Prove that

$$|f^{(n)}(x)| < \frac{1}{n+1},$$

where $f^{(n)}(x)$ means the nth derivative of f.

Proof. We offer two different proofs. The first one is based on the auxiliary function defined by

$$g(x) = x^{n+1} \left(f^{(n)}(x) - \frac{1}{n+1} \right).$$

Clearly, $g(0) = 0$ and

$$\begin{aligned}
g'(x) &= (n+1)x^n \left(f^{(n)}(x) - \frac{1}{n+1} \right) + x^{n+1} f^{(n+1)} \\
&= x^n [(n+1)f^{(n)}(x) + x f^{(n+1)} - 1] \\
&= x^n [(xf)^{(n+1)}(x) - 1]
\end{aligned}$$

where Leibniz's rule is applied in the last equation. Using that

$$(xf)^{(n+1)}(x) = (\sin x)^{(n+1)} = \sin \left(x + \frac{\pi}{2}(n+1) \right) \le 1,$$

we find that $g'(x) \le 0$ for all $x > 0$ and consequently $g(x) < 0$ for $x \in (0, \pi/2)$. Hence the desired strict inequality holds.

The second proof is based on the observation:

$$\begin{aligned}
f^{(n)}(x) &= \frac{d^n}{dx^n} \left(\int_0^1 \cos(xt)\, dt \right) \\
&= \int_0^1 \frac{\partial^n}{\partial x^n} (\cos(xt))\, dt \\
&= \int_0^1 t^n F_n(xt)\, dt,
\end{aligned}$$

where $F_n(x)$ equals one of $\{\pm \sin(xt), \pm \cos(xt)\}$. Thus, $|F_n(xt)| \le 1$ and so

$$|f^{(n)}(x)| \le \int_0^1 t^n |F_n(xt)|\, dt < \int_0^1 t^n\, dt = \frac{1}{n+1}.$$

$\qquad\square$

12. Let $f : [0,1] \to \mathbb{R}$ be twice differentiable and satisfy $f(0) = 2, f'(0) = -2$ and $f(1) = 1$. Prove that there exists a real number $c \in (0,1)$ such that $f(c)f'(c) + f''(c) = 0$.

Proof. We introduce the function

$$g(x) = \frac{1}{2}f^2(x) + f'(x).$$

Since $g(0) = 0$ and

$$g'(x) = f(x) \cdot f'(x) + f''(x),$$

by Rolle's theorem, it suffices to prove that there exists a real number $\alpha \in (0,1)$ such that $g(\alpha) = 0$. To this end, we separate into three cases:

(1) If f is never zero in $(0,1)$, let

$$h(x) = \frac{1}{2}x - \frac{1}{f(x)}.$$

Since $h(0) = h(1) = -1/2$, it follows that there exists $\alpha \in (0,1)$ such that $h'(\alpha) = 0$. Thus, $g(\alpha) = f^2(\alpha)h'(\alpha) = 0$, as desired.

(2) If f has exact one zero $\alpha \in (0,1)$, then $f(\alpha) = f'(\alpha) = 0$. In this case, clearly, $g(\alpha) = 0$.

(3) If f has at least two zeros in $(0,1)$, let x_1 be the first one and x_2 be the last. Then

$$f(x) > 0 \text{ when } x \in [0, x_1) \cup (x_2, 1].$$

This implies that $f'(x_1) \le 0$ and $f'(x_2) \ge 0$. Notice that

$$g(x_1) = f'(x_1) \le 0, \ g(x_2) = f'(x_2) \ge 0.$$

By the intermediate value theorem, there is $\alpha \in [x_1, x_2]$ such that $g(\alpha) = 0$. □

13. Let f be differentiable on $[a, \infty)$ with $f(a) = 0$. If $|f'(x)| \le |f(x)|$ for all $x > a$, prove that $f(x) \equiv 0$.

Proof. Since the differentiability implies the continuity, we see that f is continuous on $[a, a + 1/2]$. By the extreme value theorem, there exists an $x_0 \in [a, a + 1/2]$ such that

$$M = \max_{x \in [a, a+1/2]} |f(x)| = |f(x_0)|.$$

Applying the mean value theorem, we deduce that there exists $x_1 \in (a, x_0)$ such that

$$M = |f(x_0)| = |f(x_0) - f(a)| = |f'(x_1)(x_0 - a)| \le \frac{1}{2}|f(x_1)| \le \frac{1}{2}M.$$

It follows that $M = 0$, i.e., $f(x) \equiv 0$ for all $x \in [a, a + 1/2]$. Along the same lines, we can prove that $f(x) \equiv 0$ for all $x \in [a + (n-1)/2, a + n/2], n \in \mathbb{N}$. Therefore, $f(x) \equiv 0$ for all $x \ge a$. □

14. Let f be continuous on $[0,1]$ and differentiable in $(0,1)$. Assume that $f(0) = 0, 0 < f'(x) \le 1$. Prove that

$$\left(\int_0^1 f(x)\,dx \right)^2 \ge \int_0^1 f^3(x)\,dx.$$

Proof. Let

$$F(x) = \left(\int_0^x f(t)\, dt \right)^2, \quad G(x) = \int_0^x f^3(t)\, dt.$$

We need to show that $F(1) \geq G(1)$. By the assumptions that $f(0) = 0$ and $0 < f'(x) \leq 1$, we have $f(x) = \int_0^x f'(t)\, dt > 0$ for $x \in (0, 1)$. Then $G'(x) = f^3(x) > 0$ for $x \in (0, 1)$. By Cauchy's mean value theorem, there is a $c_1 \in (0, 1)$ such that

$$\frac{F(1)}{G(1)} = \frac{F(1) - F(0)}{G(1) - G(0)} = \frac{F'(c_1)}{G'(c_1)} = \frac{2 \int_0^{c_1} f(t)\, dt}{f^2(c_1)}.$$

Since

$$\frac{2 \int_0^{c_1} f(t)\, dt}{f^2(c_1)} = \frac{2 \int_0^{c_1} f(t)\, dt - 0}{f^2(c_1) - f^2(0)},$$

applying Cauchy's mean value theorem again yields

$$\frac{F(1)}{G(1)} = \frac{2 \int_0^{c_1} f(t)\, dt - 0}{f^2(c_1) - f^2(0)} = \frac{1}{f'(c_2)}$$

for some $c_2 \in (0, c_1)$. Now the desired inequality $F(1) \geq G(1)$ follows from the assumption that $f'(x) \leq 1$. $\qquad \Box$

15. Let f be differentiable on $[a, b]$. Prove that there exists an $x_0 \in [a, b]$ such that

$$|f'(x_0)| \geq \left| \frac{f(b) - f(a)}{b - a} \right|.$$

Proof. Without loss of generality, we assume that $[a, b] = [0, 1]$. Define $D = |f(1) - f(0)|$ and

$$A = \{0, 1, 2, \cdots, 9\}.$$

First, we claim that there exists at least one number $p_1 \in A$ such that

$$\left| \frac{f(0.(p_1 + 1)) - f(0.p_1)}{0.(p_1 + 1) - 0.p_1} \right| \geq D.$$

Indeed, otherwise,

$$D = |f(1) - f(0)| \leq |f(1) - f(0.9)| + \cdots + |f(0.1) - f(0)| < 10 \cdot \frac{D}{10} = D,$$

which is impossible.

Next, let $[a_1, b_1] = [0.p_1, 0.(p_1 + 1)]$. We repeat the same argument, there exists $p_2 \in A$ such that

$$\left| \frac{f(0.p_1(p_2 + 1)) - f(0.p_1 p_2)}{0.p_1(p_2 + 1) - 0.p_1 p_2} \right| \geq D.$$

Let $[a_2, b_2] = [0.p_1 p_2, 0.p_1(p_2 + 1)]$. In general, we have

$$\left| \frac{f(b_n) - f(a_n)}{b_n - a_n} \right| \geq D,$$

where $a_n = 0.p_1p_2 \cdots p_n$, $b_n = 0.p_1p_2 \cdots (p_n + 1)$, $(n = 1, 2, \ldots,)$. In view of the fact that $b_n - a_n = 1/10^n \to 0$ as $n \to \infty$, by the nested interval theorem, there is $x_0 \in [a_n, b_n]$ for all $n \in \mathbb{N}$. Now we only need to show that

$$\lim_{n \to \infty} \frac{f(b_n) - f(a_n)}{b_n - a_n} = f'(x_0).$$

To this end, let $\alpha_n = (b_n - x_0)/(b_n - a_n)$. Then $0 < \alpha_n < 1$. The desired limit follows from

$$\frac{f(b_n) - f(a_n)}{b_n - a_n} - f'(x_0)$$

$$= \alpha_n \left(\frac{f(b_n) - f(x_0)}{b_n - x_0} - f'(x_0) \right) + (1 - \alpha_n) \left(\frac{f(a_n) - f(x_0)}{a_n - x_0} - f'(x_0) \right)$$

and the definition of the derivative of f at x_0. $\qquad\square$

Remark. This result can be viewed as a weaker version of Lagrange's mean value theorem. The nice ingredient of the argument is bypassing the application of Rolle's theorem. Moreover, the above argument implies that there exist $x_1, x_2 \in (a, b)$ such that

$$f'(x_1) \le \frac{f(b) - f(a)}{b - a} \le f'(x_2).$$

16. Let $f : I \to \mathbb{R}$. If f is bounded above on any closed subinterval of I and satisfies that

$$f\left(\frac{x_1 + x_2}{2}\right) \le \frac{f(x_1) + f(x_2)}{2}, \quad \text{for any } x_1, x_2 \in I,$$

prove that f is convex on I.

Proof. By contradiction. Assume that f is not convex on I. By Definition 4.3, there exist $x_1, x_2 \in I, 0 < \alpha < 1/2$ such that

$$f(\alpha x_1 + (1 - \alpha)x_2) - \alpha f(x_1) - (1 - \alpha)f(x_2) = L > 0.$$

Define

$$F(x) = f(x) - f(x_1) - \frac{f(x_2) - f(x_1)}{x_2 - x_1}(x - x_1).$$

Clearly, $F(x)$ is also bounded above on any closed interval of I. Moreover, direct calculation yields

$$F(x_1) = F(x_2) = 0, \quad F(\alpha x_1 + (1 - \alpha x_2)) = L > 0,$$

and

$$F\left(\frac{a + b}{2}\right) \le \frac{F(a) + F(b)}{2}, \quad \text{for } a, b \in I.$$

Hence

$$\begin{aligned}
2L &= 2F[\alpha x_1 + (1 - \alpha)x_2] \\
&= 2F\left[\frac{2\alpha x_1 + (1 - 2\alpha)x_2 + x_2}{2}\right] \\
&\le F[2\alpha x_1 + (1 - 2\alpha)x_2] + F(x_2) \\
&= F[2\alpha x_1 + (1 - 2\alpha)x_2].
\end{aligned}$$

Let $\alpha_1 = 2\alpha$. Then $F[\alpha_1 x_1 + (1 - \alpha_1)x_2] = 2L > 0$. If $\alpha_1 < 1/2$, we repeat the above process. If $\alpha_1 > 1/2$, we replace α_1 by $1 - \alpha_1$, x_1 by x_2, and x_2 by x_1, then repeat the above precess. Therefore, there exist $\alpha_1, \alpha_2, \ldots, \alpha_n$ such that

$$F(\alpha_n x_1 + (1 - \alpha_n)x_2) = 2^n L.$$

This will contradict the assumption for which F is bounded above. $\qquad\square$

17. (**Monthly Problem 11369, 2008**). Prove that for all real t, and all $\alpha \geq 2$,

$$e^{\alpha t} + e^{-\alpha t} - 2 \leq (e^t + e^{-t})^\alpha - 2^\alpha.$$

Proof. First, observe that the proposed inequality is invariant when t is replaced by $-t$. Hence, it suffices to show the inequality holds for $t \geq 0$. Next, let $x = e^t$. Then the proposed inequality is equivalent to

$$\frac{(x^2 + 1)^\alpha - x^{2\alpha} - 1}{x^\alpha} \geq 2^\alpha - 2. \qquad (4.24)$$

To apply the LMR (Theorem 4.12), we choose that $f(x) = (x^2 + 1)^\alpha - x^{2\alpha} - 1$ and $g(x) = x^\alpha$. Then

$$\frac{f'(x)}{g'(x)} = \frac{2x(x^2 + 1)^{\alpha-1} - 2x^{2\alpha-1}}{x^{\alpha-1}}.$$

Define

$$F(x) = \frac{f'(x)}{g'(x)} = 2x\left(x + \frac{1}{x}\right)^{\alpha-1} - 2x^\alpha.$$

To show that $f'(x)/g'(x)$ is increasing for $x > 1$, it suffices to prove that $F'(x) \geq 0$ for $x > 1$. Indeed,

$$F'(x) = 2\left(x + \frac{1}{x}\right)^{\alpha-1}\left[\alpha - \frac{2(\alpha - 1)}{x^2 + 1}\right] - 2\alpha x^{\alpha-1}.$$

Using Bernoulli's inequality

$$(1 + t)^p \geq 1 + pt \qquad \text{for } t \geq 0,\, p \geq 1,$$

which can be proved via the LMR by considering $(1 + t)^p/(1 + pt)$ as well, we have

$$\left(x + \frac{1}{x}\right)^{\alpha-1} = x^{\alpha-1}\left(1 + \frac{1}{x^2}\right)^{\alpha-1} \geq x^{\alpha-1} + (\alpha - 1)x^{\alpha-3}.$$

Therefore, for $x > 1$ and $\alpha \geq 2$,

$$F'(x) \geq \frac{2(\alpha - 1)(\alpha - 2)x^{\alpha-3}(x^2 - 1)}{x^2 + 1} \geq 0.$$

Now, the LMR deduces that

$$\frac{f(x) - f(1)}{g(x) - g(1)} = \frac{(x^2 + 1)^\alpha - x^{2\alpha} - 2^\alpha + 1}{x^\alpha - 1}$$

is increasing for $x > 1$. In particular, applying L'Hôpital's rule yields

$$\lim_{x \to 1} \frac{f(x) - f(1)}{g(x) - g(1)} = \lim_{x \to 1} \frac{(x^2 + 1)^\alpha - x^{2\alpha} - 2^\alpha + 1}{x^\alpha - 1} = 2^\alpha - 2.$$

Thus,
$$\frac{f(x) - f(1)}{g(x) - g(1)} = \frac{(x^2 + 1)^\alpha - x^{2\alpha} - 2^\alpha + 1}{x^\alpha - 1} \geq 2^\alpha - 2.$$

This proves the desired (4.24) and therefore concludes the proposed inequality.
□

The following two examples demonstrate that Taylor's formula is also an instrument for revealing quantitative information for derivatives.

18. Let f be twice differentiable on $[0, 1]$ and $f(0) = f(1) = 0$. Show that if $\min_{x \in [0,1]} f(x) = -1$, then there exist $c_1, c_2 \in (0, 1)$ such that

$$f''(c_1) \geq 8, \quad f''(c_2) \leq 8.$$

Proof. Let $f(x_0) = \min_{x \in [0,1]} f(x)$. By the assumption that $f(x_0) = -1 < 0 = f(0) = f(1)$, it follows that $x_0 \in (0, 1)$. Moreover, Fermat's theorem implies that $f'(x_0) = 0$. Applying Taylor's theorem with Lagrange's remainder to $f(1)$ and $f(0)$ at x_0, respectively, we obtain

$$\begin{aligned}
f(1) &= f(x_0) + f'(x_0)(1 - x_0) + \frac{1}{2}f''(c_1)(1 - x_0)^2 \\
&= f(x_0) + \frac{1}{2}(1 - x_0)^2 f''(c_1); \\
f(0) &= f(x_0) + f'(x_0)(0 - x_0) + \frac{1}{2}f''(c_2)(0 - x_0)^2 \\
&= f(x_0) + \frac{1}{2}x_0^2 f''(c_2),
\end{aligned}$$

where $c_1 \in (x_0, 1), c_2 \in (0, x_0)$. In view of $f(0) = f(1) = 0, f(x_0) = -1$, we find that

$$f''(c_1) = \frac{2}{(1 - x_0)^2}, \quad f''(c_2) = \frac{2}{x_0^2}.$$

If $x_0 \in (0, 1/2]$, then $1/2 \leq 1 - x_0 < 1$. In this case,

$$f''(c_1) = \frac{2}{(1 - x_0)^2} \leq \frac{2}{(1/2)^2} = 8, \quad f''(c_2) = \frac{2}{x_0^2} \geq \frac{2}{(1/2)^2} = 8.$$

If $x_0 \in (1/2, 1)$, then $0 \leq 1 - x_0 < 1/2$. In this case,

$$f''(c_1) = \frac{2}{(1 - x_0)^2} > \frac{2}{(1/2)^2} = 8, \quad f''(c_2) = \frac{2}{x_0^2} < \frac{2}{(1/2)^2} = 8.$$

□

19. Let $f(x)$ be twice differentiable on $\mathbb{R}$. Let $M_k = \sup_{x \in \mathbb{R}} |f^{(k)}(x)|, k = 0, 1, 2$. If M_0 and M_2 are bounded, prove that $M_1^2 \leq 2M_0 M_2$, so M_1 is also bounded.

Proof. First observe that if $M_2 = 0$, then the only bounded functions satisfying the hypotheses are the constant functions. So, without loss of generality, we assume that $M_2 > 0$. For any fixed $h \in \mathbb{R}$, by Taylor's formula, we have

$$\begin{aligned}
f(x + h) &= f(x) + f'(x)h + \frac{1}{2}f''(c_1)h^2, \quad \text{where } c_1 \text{ is between } x \text{ and } x + h, \\
f(x - h) &= f(x) - f'(x)h + \frac{1}{2}f''(c_2)h^2, \quad \text{where } c_2 \text{ is between } x \text{ and } x - h.
\end{aligned}$$

Subtracting these two equations yields

$$f(x + h) - f(x - h) = 2f'(x)h + \frac{1}{2}h^2(f''(c_1) - f''(c_2)).$$

That is,

$$2f'(x)h = f(x + h) - f(x - h) - \frac{1}{2}h^2(f''(c_1) - f''(c_2)).$$

Applying the hypothesis gives

$$2|f'(x)|h \le |2f'(x)h| \le 2M_0 + M_2h^2.$$

It follows that, for any $h \in \mathbb{R}$, the trinomial

$$M_2h^2 - 2|f'(x)|h + 2M_0 \ge 0.$$

Thus the discriminant $b^2 - 4ac = 4(|f'(x)|^2 - 2M_0M_2) \le 0$, which concludes that $M_1^2 \le 2M_0M_2$. $\qquad\square$

This result asserts that if f itself is bounded together with its second derivative, then the derivative of f is also bounded. Landau and Kolmogorov generalized the above inequality to the following *interpolation inequalities* between different derivatives: Let $M_k = \sup_{x \in \mathbb{R}} |f^{(k)}(x)|, k = 0, 1, \ldots, n$. Then

$$M_k \le 2^{k(n-k)/2} M_0^{1-k/n} M_n^{k/n}, \quad \text{for } 1 \le k < n.$$

A variant of this kind of inequalities in the Hölder's spaces are called the *Sobolev embedding inequalities*.

We now conclude this section with the proof of Stirling's formula.

20. (**Stirling's Formula**). Prove that, for sufficiently large n,

$$n! = \sqrt{2\pi n} \left(\frac{n}{e}\right)^n e^{\theta_n/12n}, \quad \text{for some } 0 < \theta_n < 1.$$

Proof. First, we find an estimate for $\sqrt[n]{n!}$. By the AM-GM inequality, we have

$$\sqrt[n+1]{\left(1 + \frac{1}{n}\right)^n \cdot 1} < \frac{1}{n+1}\left[n\left(1 + \frac{1}{n}\right) + 1\right] = 1 + \frac{1}{n+1}.$$

This equivalent to

$$\left(1 + \frac{1}{n}\right)^n < \left(1 + \frac{1}{n+1}\right)^{n+1}.$$

On the other hand, since

$$\left(\frac{1 + \frac{1}{n-1}}{1 + \frac{1}{n}}\right)^n = \left(1 + \frac{1}{n^2 - 1}\right)^n > 1 + \frac{n}{n^2 - 1} > 1 + \frac{1}{n},$$

it follows that

$$\left(1 + \frac{1}{n-1}\right)^n > \left(1 + \frac{1}{n}\right)^{n+1}.$$

In summary, we find that

$$\left(1 + \frac{1}{n}\right)^n \text{ is strictly increasing and } \left(1 + \frac{1}{n}\right)^{n+1} \text{ is strictly decreasing.}$$

Recall that $e = \lim_{n\to\infty}(1 + 1/n)^n$. We find that, for all $k \in \mathbb{N}$,

$$\left(1 + \frac{1}{k}\right)^k < e < \left(1 + \frac{1}{k}\right)^{k+1}.$$

Multiplying this inequality for k from 1 to n yields

$$\frac{(n+1)^n}{n!} < e^n < \frac{(n+1)^{n+1}}{n!},$$

which is equivalent to

$$\left(\frac{n+1}{e}\right)^n < n! < e\left(\frac{n+1}{e}\right)^{n+1}.$$

Therefore, by the squeeze theorem, we have

$$\lim_{n\to\infty} \frac{\sqrt[n]{n!}}{n} = \frac{1}{e},$$

and so $\sqrt[n]{n!} \sim n/e$. To establish an estimate of $n!$ itself, we introduce $a_n = n!(e/n)^n$. Then

$$\frac{a_{n+1}}{a_n} = \frac{e}{\left(1 + \frac{1}{n}\right)^n} = e^{1-n\ln(1+1/n)} = \frac{\sqrt{n+1}}{\sqrt{n}} e^{1-(n+1/2)\ln(1+1/n)}.$$

Since

$$\ln\left(1 + \frac{1}{n}\right) = \ln\left(1 + \frac{1}{2n+1}\right) - \ln\left(1 - \frac{1}{2n+1}\right),$$

using the Maclaurin series of $\ln(1 + x)$, we find that

$$\ln\left(1 + \frac{1}{n}\right) = 2\left(\frac{1}{2n+1} + \frac{1}{3}\frac{1}{(2n+1)^3} + \frac{1}{5}\frac{1}{(2n+1)^5} + \cdots\right).$$

This yields

$$0 < (n+1/2)\ln\left(1 + \frac{1}{n}\right) - 1 < \frac{1}{3}\left(\frac{1}{(2n+1)^2} + \frac{1}{(2n+1)^4} + \cdots\right) = \frac{1}{12}\left(\frac{1}{n} - \frac{1}{n+1}\right).$$

Thus $n^{-1/2}a_n$ is strictly decreasing and $n^{-1/2}e^{-1/12n}a_n$ is strictly increasing. Moreover, they converge to the same limit, denote this limit as C. Then

$$n^{-1/2}e^{-1/12n}a_n < C < n^{-1/2}a_n,$$

and so

$$n! = C\sqrt{n}\left(\frac{n}{e}\right)^n e^{\theta_n/12n}, \quad (0 < \theta_n < 1). \tag{4.25}$$

Finally, we determine the value of C. To this end, by (4.25), we have

$$\lim_{n\to\infty} \sqrt{n}\frac{(2n)!!}{(2n+1)!!} = \lim_{n\to\infty} \frac{\sqrt{n}}{2n+1}\frac{(2^n n!)^2}{(2n)!} = \frac{C}{2\sqrt{2}}. \tag{4.26}$$

On the other hand, by iteration, we have

$$\int_0^1 (1-x^2)^{n+1/2}dx = \frac{(2n+1)!!}{(2n+2)!!}\frac{\pi}{2}.$$

Since

$$\int_0^1 (1-x^2)^{n+1}dx < \int_0^1 (1-x^2)^{n+1/2}dx < \int_0^1 (1-x^2)^n dx,$$

it follows that

$$\frac{(2n+2)!!}{(2n+3)!!} < \frac{(2n+1)!!}{(2n+2)!!}\frac{\pi}{2} < \frac{(2n)!!}{(2n+1)!!}.$$

Multiplying this inequality by $\sqrt{n}$ and letting $n \to \infty$ yields

$$\frac{C}{2\sqrt{2}} \le \frac{\sqrt{2}}{C}\frac{\pi}{2} \le \frac{C}{2\sqrt{2}}.$$

Thus, $C = \sqrt{2\pi}$, in turn, we obtain Stirling's formula as desired. $\qquad \square$

Remark. Applying Wallis's formula (2.22) in (4.26) will lead to a shorter proof of $C = \sqrt{2\pi}$.

4.7 Exercises

Be not afraid of going slowly; be afraid only of standing still. — Chinese Proverb

1. Let f be differentiable on (a, b). If f is unbounded, show that $f'(x)$ is also unbounded. Determine whether the converse is also true.

2. Let f be differentiable on (a, b). If $f'(x)$ is monotone on (a, b), show that $f'(x)$ is continuous on (a, b).

3. (**Putnam 1986-A1**). Find, with explanation, the maximum value of $f(x) = x^3 - 3x$ on the set of all real numbers x satisfying $x^4 + 36 \le 13x^2$.

4. (**Putnam 2010-A2**). Find all differentiable functions $f : \mathbb{R} \to \mathbb{R}$ such that

$$f'(x) = \frac{f(x+n) - f(x)}{n}$$

for all real numbers x and all positive integers n.

5. (**Di Bruno's Formula**). Let g be nth differentiable at a, and let f be nth differentiable at $b = g(a)$. Prove that

$$(f(g(x))^{(n)})(a) = n! \sum \frac{f^{(k)}(b) \prod_{j=1}^n [g^{(j)}(a)]^{k_j}}{\prod_{j=1}^n [k_j!(j!)^{k_j}]},$$

where the summation is over a sequence of nonnegative integers $\{k_j\}$ $(j = 1, 2, \ldots, n)$ satisfying $\sum_{j=1}^n jk_j = n$, and k is defined by $k = \sum_{j=1}^n k_j$. For example, for $n = 3$, we have

$$(f(g(x)))^{(3)}(a) = f^{(3)}(b)g'^3(a) + 3f''(b)g'(a)g''(a) + f'(b)g^{(3)}(a).$$

6. Let f be differentiable on $[0, \infty)$ such that $\lim_{x \to \infty} \frac{f(x)}{x} = 0$. Prove that

$$\liminf |f'(x)| = 0.$$

7. Let $f(x)$ be well-defined on $(-1, 1)$ such that $f'(0)$ exists. If the sequences a_n and b_n satisfy

$$-1 < a_n < 0 < b_n < 1, \qquad \lim_{n \to \infty} a_n = \lim_{n \to \infty} b_n = 0,$$

show that

$$\lim_{n \to \infty} \frac{f(b_n) - f(a_n)}{b_n - a_n} = f'(0).$$

8. Suppose that f is continuously differentiable on $(-\delta, \delta)$ for some $\delta > 0$. If both $f'(0)$ and $f''(0)$ exist, prove that

$$\lim_{x \to 0} \frac{f(x) - f(\ln(1 + x))}{x^3} = \frac{1}{2} f''(0).$$

9. Let $a_i > 0$ for $1 \le i \le n$. Compute

$$\lim_{x \to \infty} \left(\frac{a_1^{1/x} + a_2^{1/x} + \cdots + a_n^{1/x}}{n} \right)^{nx}.$$

10. Let $x, y > 0$. Show that $x^y + y^x > 1$.

11. For $x \ge 0$, show that

$$\frac{2x}{2 + x} \le \ln(1 + x) \le \frac{x}{\sqrt{x + 1}}.$$

12. Let $a_1 \ge a_2 \ge \cdots \ge a_n \ge 0$ and let $f'(0) \ge 0$ and $f''(x) \ge 0$ for $x \ge 0$. Show that

$$\sum_{k=1}^{n} (-1)^{k+1} f(a_k) \ge f \left(\sum_{k=1}^{n} (-1)^{k+1} a_k \right).$$

13. Let f be continuous on $[0, 3]$ and differentiable on $(0, 3)$ with $f(0) = 1, f(1) = 2, f(3) = 2$.

   (a) Show that f has at least one fixed point in $[0, 3]$.
   (b) Let $\alpha \in (0, 1)$. Show that there exists a $c \in [0, 3]$ such that $f'(c) = \alpha$.

14. Let $f : \mathbb{R} \to [0, \infty)$ be twice differentiable with $f''(x) \le 0$. Prove that f is a constant.

15. Prove that there is no differentiable function $f(x)$ such that $f(f(x)) = x^2 - 3x + 3$.

16. Prove that there is no twice differentiable function $f : \mathbb{R} \to \mathbb{R}$ such that

$$f(x) > 0, \quad f'(x) > 0, \quad f''(x) < 0.$$

17. Let $p(x)$ be a polynomial satisfying

$$p'''(x) - p''(x) - p'(x) + p(x) \ge 0 \quad \text{for all } x \in \mathbb{R}.$$

Show that $p(x) \ge 0$ for all $x \in \mathbb{R}$.

18. Let $p(x)$ be a polynomial with all real zeros $x_1, x_2, \ldots, x_n$ Show that

$$\exp\left(\int_a^b \frac{p'''(x)p(x)}{p'(x)}\, dx\right) < \left|\frac{p(a)^2 p'(b)^3}{p'(a)^3 p(b)^2}\right|,$$

where $a < b < \min\{x_1, x_2, \ldots, x_n\}$.

19. Let
$$f(x) = \frac{ax^2 + bx + c}{\alpha x^2 + \beta x + \gamma} \quad (\alpha \neq 0).$$
Show that if f has three inflection points then these three points stay on the same line.

20. Let f be third differentiable on $[-1, 1]$ with $f(0) = f'(0) = f(-1) = 0$ and $f(1) = 1$. Show that there exists a $c \in (-1, 1)$ such that $f''(c) \geq 3$.

21. Let $f : \mathbb{R} \to \mathbb{R}$ be twice differentiable with $f(0) = 2, f'(0) = -2$ and $f(1) = 1$. Show that there exists $c \in (-1, 1)$ such that $f(c)f'(c) + f''(c) = 0$.

22. Let f and g be continuously differentiable on $[a, b]$ with $f(a) = f(b) = 0$. If
$$W(f, g) = \left|\begin{array}{cc} f(x) & g(x) \\ f'(x) & g'(x) \end{array}\right| \neq 0 \text{ for each } x \in [a, b],$$
show that $g(x)$ has zero in (a, b).

23. Let $f(x) = \arctan x$. Find $f^{(n)}(0)$. *Hint:* f satisfies the differential equation $(1 + x^2)y'' + 2xy' = 0$.

24. Let $f(x) = \frac{2x}{1+e^{2x}}$. Show that $f^{(n)}(0) \in \mathbb{Z}$ for all $n \in \mathbb{N}$.

25. Let $n \in \mathbb{N}$. Show that, for $x > 0$,
$$\left|\frac{d^n}{dx^n}\cos(\sqrt{x})\right| \leq \frac{n!}{(2n)!}.$$

Hint. Find an integral representation for the above nth derivative. See Problem 11 in Section 4.6.

26. Show that the sequence a_n defined by
$$a_n = \left(1 + \frac{1}{n}\right)^{n+x}$$
is decreasing if and only if $x \geq 1/2$.

27. For $n \in \mathbb{N}$, show that
$$\left(1 + \frac{1}{2n+1}\right)\left(1 + \frac{1}{n}\right)^n < e < \left(1 + \frac{1}{2n}\right)\left(1 + \frac{1}{n}\right)^n.$$

28. (**Putnam 2002-B3**). Show that, for all integers $n > 1$,
$$\frac{1}{2ne} < \frac{1}{e} - \left(1 - \frac{1}{n}\right)^n < \frac{1}{ne}.$$

29. (**Putnam 2014-A1**). Prove that every nonzero coefficient of the Taylor series of
$$(1 - x + x^2)e^x$$
about $x = 0$ is a rational number whose numerator (in lowest terms) is either 1 or a prime number.

30. Let $f : [0,1] \to \mathbb{R}$ be a differentiable function such that there is no x with $f(x) = f'(x) = 0$. Prove that the set $\{x : f(x) = 0\}$ is finite.

31. Let f be a differentiable function such that $f(x) + f'(x) \leq 1$ for all x, and $f(0) = 0$. Show that $f(x) \leq 1 - e^{-x}$.

32. Let $x_1 = a > 0, x_{n+1} = (n+1)\ln(1 + x_n/n)$. Find $\lim_{n\to+\infty} x_n$.

33. Show that $\sum_{n=1}^{\infty}(-1)^{n+1}\sin(\ln n)/n^p$ converges if and only if $p > 0$. *Hint:* Let $f(x) = \sin(\ln x)/x^p$. Observe that $\sum_{n=1}^{\infty}(-1)^{n+1}f(n) = \sum_{n=1}^{\infty}(f(2n-1)-f(2n))$. Then use the mean value theorem.

34. Let f be continuous on $[0,1]$ and differentiable on $(0,1)$. If $|f'(x)| < 1$ for all $x \in (0,1)$ and $f(0) = f(1)$, prove that there exist $x_1, x_2 \in (0,1)$ such that

$$|f(x_1) - f(x_2)| < \frac{1}{2}.$$

35. Let $f : [0,1] \to [0,1]$ be non-constant continuous function. Prove that there exist $x_1, x_2 \in (0,1), x_1 \neq x_2$, such that

$$|f(x_1) - f(x_2)| = |x_1 - x_2|^2.$$

36. Let f be twice differentiable on $[0,1]$. If $|f''(x)| < 1$ for all $x \in [0,1]$ and $f(0) = f(1)$, prove that

$$|f'(x)| \leq \frac{1}{2} \text{ for all } x \in [0,1].$$

37. Let f be twice differentiable on $(-1,1)$ and $f(0) = f'(0) = 0$. If

$$|f''(x)| \leq |f(x)| + |f'(x)| \text{ for all } x \in (-1,1),$$

prove that there is $\delta > 0$ such that $f(x) = 0$ on $(-\delta, \delta)$.

38. Assume that f has bounded derivatives on $(a, +\infty)$. Prove that

$$\lim_{x\to+\infty} \frac{f(x)}{x \ln x} = 0.$$

39. Let f be continuous on $[a,b]$ and twice differentiable on (a,b). If $L(x)$ is the linear function which passes through $(a, f(a))$ and $(b, f(b))$, show that

$$|f(x) - L(x)| \leq \frac{1}{2}(b-x)(x-a)M,$$

where $M = \max_{x \in [a,b]} |f''(x)|$.

40. Let f be twice differentiable on $[x, x+h]$ and $\alpha \in [0,1]$. Show that there exists a $\theta \in (0,1)$ such that

$$f(x + \alpha h) = \alpha f(x + h) + (1 - \alpha)f(x) + \frac{1}{2}h^2 \alpha(1 - \alpha)f'(x + \theta h).$$

41. **(Optimal Property of the Taylor polynomial)**. Let $T_n(x)$ be the Taylor polynomial of $f(x)$ at $x = a$. Show that there is $\delta > 0$ such that

$$|f(x) - T_n(x)| \leq |f(x) - p(x)|$$

for every polynomial $p(x)$ with $\deg(p(x)) \leq n$ on $(a - \delta, a + \delta)$.

42. Show that

$$\sum_{n=0}^{\infty} \frac{1}{n!} \left\lfloor \frac{n!}{e} \right\rfloor x^n = -\frac{1}{2} \left(e^x + \frac{x+1}{x-1} e^{-x} \right).$$

43. For $0 < x < \frac{\pi}{2}$, show that $2 \sin x + \tan x \geq 3x$.

44. Prove the following inequalities:

(a) If $0 < x < y < \pi/2$, then $x \tan y > y \tan x$.

(b) If $x, y > 0$, then $m(x, y, t) = (x^t + y^t)^{1/t}$ is decreasing with respect to t.

(c) If $p \geq 2, x \in [0, 1]$, then

$$\left(\frac{1+x}{2} \right)^p + \left(\frac{1-x}{2} \right)^p \leq \frac{1}{2}(1 + x^p).$$

45. Let $0 < x_i < \pi$, $(i = 1, 2, \ldots n)$ and $x = (x_1 + x_2 + \cdots + x_n)/n$. Prove that

$$\prod_{i=1}^{n} \frac{\sin x_i}{x_i} \leq \left(\frac{\sin x}{x} \right)^n.$$

46. Let $a_i, p_i > 0$ for $i = 1, 2, \ldots, n$. Show that

$$\frac{p_1 + p_2 + \cdots + p_n}{\frac{p_1}{a_1} + \frac{p_2}{a_2} + \cdots + \frac{p_n}{a_n}} \leq \exp \left(\frac{p_1 \ln a_1 + p_2 \ln a_2 + \cdots + p_n \ln a_n}{p_1 + p_2 + \cdots + p_n} \right)$$
$$\leq \frac{p_1 a_1 + p_2 a_2 + \cdots + p_n a_n}{p_1 + p_2 + \cdots + p_n}.$$

47. Let $a > 0$. Show that

$$\lim_{n \to \infty} \frac{1}{(a+1)^n} \sum_{k=0}^{n} \binom{n}{k} a^k \ln \left(\frac{k+1}{n} \right) = \ln \frac{a}{a+1}.$$

48. (**Putnam 1999-B4**). Let f be a real function with a continuous third derivatives such that

$$f(x), f'(x), f''(x), f'''(x) > 0, \qquad \text{for all } x.$$

Suppose that $f'''(x) \leq f(x)$ for all x. Show that $f'(x) < 2f(x)$ for all x.

Remark. Based on the total scores of the top 205 Putnam competition writers, this problem is viewed as one of the hardest problems in many years. Under the conditions of the problem, let C be the constant such that $f'(x) \leq Cf(x)$. The third solution in [59] improved C from 2 to $2^{1/6} = 1.12246\ldots$. The best possible constant C in the inequality seems still open.

49. (**Monthly Problem 10261, 1992**). Let $0 < x < \pi/4$. Prove that

$$(\sin x)^{\sin x} < (\cos x)^{\cos x}.$$

50. (**Monthly Problem 10604, 1997**). Determine positive numbers c and C such that if $0 < a < b$ then

$$c \left(1 - \frac{a}{b} \right) \leq \sup_{x>0} \left| \frac{\sin(ax)}{ax} - \frac{\sin(bx)}{bx} \right| \leq C \left(1 - \frac{a}{b} \right).$$

Can you find the best possible constants such the above inequalities hold whenever $0 < a < b$?

51. (**Monthly Problem 11957, 2017**). Let k and n be two integers with $n \geq k \geq 2$. Let $S(n, k)$ be the Stirling number of the second kind, i.e., the number of ways to partition a set of n objects into k nonempty subsets. Show that

$$n^k S(n, k) \geq k^n \binom{n}{k}.$$

52. (**Dobinski's Formula**). For $n \in \mathbb{N}$, define the nth Bell number by

$$B_n = \sum_{k=0}^{n} S(n, k),$$

where $S(n, k)$ is the Stirling number of the second kind. Show that

$$\sum_{k=0}^{\infty} \frac{k^n}{k!} = e\, B_n.$$

53. Let f be differentiable on $(a, +\infty)$. If $\lim_{x\to\infty} (f(x) + f'(x)) = L$, show that $\lim_{x\to\infty} f(x) = L$.

54. Let f be continuously differentiable on $\mathbb{R}$. If $\lim_{x\to\infty} [(f'(x))^2 + f^3(x)] = 0$, show that $\lim_{x\to\infty} f(x) = 0$.

55. Let $p(x)$ be continuously differentiable on $\mathbb{R}$. Let $f(x)$ be a solution of

$$y'' + p(x)y' - y = 0.$$

If f has more than one zero, show that $f(x) \equiv 0$ on $\mathbb{R}$.

56. Let f satisfy that $f''(x) + f^3(x) = 0$ on $(0, b)$ with $f(0) = f(b) = 0$. Prove that there exists a unique function f such that $f(x) > 0$ on $(0, b)$ and

$$\int_0^b f(x)\,dx = \frac{\pi}{\sqrt{2}}.$$

57. Let f be continuously differentiable with $f(0) = 1$. If $f'(x) + e^x f(x) + 1 \leq 0$, show that $f(x)$ has a zero on $[0, 3/4]$.

58. Let f be twice differentiable on $[a, b]$. If $f^2(x) + (f'')^2(x) = k^2 > 0$, show that, for any $x \in [a, b]$,

$$|f'(x)| \leq k\left(\frac{2}{b-a} + \frac{b-a}{2}\right).$$

59. For every $n \in \mathbb{N}$, show that

$$x^{2n} - 2x^{2n-1} + 3x^{2n-2} + \cdots - 2nx + (2n+1) = 0$$

has no real solution.

60. Let f be continuous on $(a, +\infty)$. Let x_n be distinct roots of $f(x)$ with $f'(x_n) \neq 0$. Prove that $\lim_{n\to\infty} x_n = +\infty$.

61. Let f be twice differentiable on $[a, b]$. If $f(a) = f(b) = 0$ and $f(c) > 0$ for some $c \in (a, b)$, prove that there exists $x_0 \in (a, b)$ such that $f''(x_0) < 0$.

62. Let f be continuous on $[a, b]$ and satisfy $f(a) = f(b) = 0, f'(a)f'(b) > 0$. Prove that there exists $c \in (a, b)$ such that $f(c) = 0$.

63. Let f be differentiable on $[a, b]$ and satisfy $f(a) = f(b) = 0, f'(a)f'(b) > 0$. Prove that $f'(x) = 0$ has at least two roots.

64. Let f be a real-valued function with continuous nonnegative derivative and let L be the arc length of f on $[0, 1]$. If $f(0) = 0, f(1) = 1$, show that

$$\sqrt{2} \leq L < 2.$$

65. Let f be continuous on $[a, b]$ and differentiable on (a, b). If $f(a) = 0$ and $f(x) > 0$ on (a, b), for every $\alpha > 0$, prove that there exists $x_1, x_2 \in (a, b)$ such that

$$\frac{f'(x_1)}{f(x_1)} = \alpha \frac{f'(x_2)}{f(x_2)}.$$

66. Let f be continuous on $[a, b]$ and differentiable on (a, b). If $f(a) = f(b) = 0$, for any real number α, prove that there exists $c \in (a, b)$ such that $f'(c) = \alpha f(c)$.

67. Let f be continuous on $[a, b]$ and differentiable on (a, b). If $f'(a) = f'(b)$, prove that there exists $c \in (a, b)$ such that

$$f'(c) = \frac{f(c) - f(a)}{c - a}.$$

68. (**Monthly Problem 10739, 1999**). Suppose that $f : [0, 1] \to \mathbb{R}$ has a continuous second derivative with $f''(x) > 0$ on $(0, 1)$, and suppose that $f(0) = 0$. Choose $a \in (0, 1)$ such that $f'(a) < f(1)$. Show that there is a unique $b \in (0, 1)$ such that $f'(a) = f(b)/b$.

69. Let f be twice differentiable on (a, b). If $f''(a) = f''(b)$, prove that there exists $c \in (a, b)$ such that

$$f(c) - f(a) = f'(c)(c - a) - \frac{1}{2}(c - a)^2 f''(c).$$

70. Let $P(x)$ be a polynomial of odd degree and $a \in \mathbb{R}$ such that $P''(a) \neq 0$. Prove that for any $t \in (0, 1/2)$ there exists $b \neq a$ such that

$$\frac{P(b) - P(a)}{b - a} = P'(tb + (1 - t)a).$$

71. Let f be continuous on $[0, 1]$ and differentiable on $(0, 1)$ with $f(0) = 0, f(1) = 1$. For any given positive numbers $c_1, c_2, \ldots, c_n$ with $\sum_{k=1}^{n} c_k = 1$, prove that there exist distinct numbers $x_1, x_2, \ldots, x_n \in [0, 1]$ such that

$$\sum_{k=1}^{n} \frac{c_k}{f'(x_k)} = 1.$$

72. Let f be continuous on $[0, 1]$ and differentiable on $(0, 1)$ with $f(0) = 0, f(1) = 1$. Show that for each $n \in \mathbb{N}$ there exist distinct numbers $x_1, x_2, \ldots, x_n \in [0, 1]$ such that

$$\prod_{k=1}^{n} f'(x_k) = 1.$$

73. Let $f_n(x) = \cos x + \cos^2 x + \cdots + \cos^n x$.

    (a) For any given $n \in \mathbb{N}$, prove that $f_n(x) = 1$ has a unique solution on $[0, \pi/3]$.

    (b) Let $x_n \in [0, \pi/3]$ be the unique solution of $f_n(x) = 1$. Prove that

$$\lim_{n \to \infty} x_n = \frac{\pi}{3}.$$

74. Let $f(x) = 1 - x + \frac{x^2}{2} + \cdots + (-1)^n \frac{x^n}{n}$. Prove that $f(x) = 0$ has exactly one real root when n is odd while $f(x) = 0$ has no real root at all when n is even.

75. Let $p_n(x) = 1 + x + \frac{x^2}{2} + \cdots + \frac{x^n}{n}$. Prove that

    (a) $p_n(x) > 0$ for all $x \in \mathbb{R}$ when n is even.

    (b) $p_n(x) = 0$ has a unique solution in $\mathbb{R}$ when n is odd.

    (c) $p_{2n}(x) > e^x > p_{2n+1}(x)$ when $x < 0$

    (d) $e^x > p_n(x) \geq (1 + x/n)^n$ when $x > 0$.

76. Let $f(x)$ be twice continuously differentiable on $[0, a]$ and satisfy the functional equation

$$f\left(\frac{x}{2}\right) + f\left(a - \frac{x}{2}\right) = \frac{1}{2} f(x).$$

Find $f(x)$.

77. Find a function which satisfies the functional equation

$$2f^2(x) - f(2x) = 1.$$

78. (**Putnam Problem 2005-B3**). Find all differential function $f : (0, \infty) \to (0, \infty)$ for which there is a positive number a such that

$$f'\left(\frac{a}{x}\right) = \frac{x}{f(x)}$$

for all $x > 0$.

79. (**Putnam Problem 2010-B5**). Is there a strictly increasing function $f : \mathbb{R} \to \mathbb{R}$ such that $f'(x) = f(f(x))$ for all x?

80. (**Putnam Problem 2009-B5**). Let $f : (0, \infty) \to \mathbb{R}$ be differentiable such that

$$f'(x) = \frac{x^2 - f^2(x)}{x^2(1 + f^2(x))}$$

for all $x > 1$. Prove that $\lim_{x \to \infty} f(x) = \infty$.

81. Let $f(x) = \sum_{k=1}^n c_k e^{\alpha_k x}$, where α_k $(k = 1, 2, \ldots, n)$ are distinct real numbers and $c_1, c_2, \ldots, c_n$ are not likewise zero. How many real roots can $f(x)$ have?

82. Let f be differentiable on $[a, b]$ with $f(a) = 0$, and let g be bounded on $[a, b]$. If for any nonzero constant λ,

$$|f(x)g(x) + \lambda f'(x)| \leq |f(x)|, \text{ for } x \in [a, b],$$

show that $f(x) \equiv 0$ on $[a, b]$.

83. Let $|f''(x)| \leq M$ on $[0, a]$. If f attains the maximum in $(0, a)$, prove that

$$|f'(0)| + |f'(a)| \leq Ma.$$

84. Let $f : [0,1] \to \mathbb{R}$ be continuous and differentiable on $(0,1)$. If $f(0) = 0, f(x) \neq 0$ for all $x \in (0,1)$, prove that, for any $n \in \mathbb{N}$, there exists an $x_0 \in (0,1)$ such that
$$\frac{nf'(x_0)}{f(x_0)} = \frac{f'(1-x_0)}{f(1-x_0)}.$$

85. Let $f : \mathbb{R} \to \mathbb{R}$ be twice differentiable. If $|f(x)| \leq 1$ and
$$f^2(0) + (f'(0))^2 = 4,$$
prove that there exists $c \in (-2, 2)$ such that $f(c) + f''(c) = 0$.

86. Let f be $(n+1)$th differentiable on $[a,b]$. Assume that $f^{(k)}(a) = f^{(k)}(b) = 0$ for $k = 0, 1, \ldots, n$. Prove that there exists $c \in (a,b)$ such that $f^{(n+1)}(c) = f(c)$.

87. Let f be differentiable on $[a,b]$ and $ab > 0$. Prove that there exists an $x_0 \in (a,b)$ such that
$$\frac{1}{b-a} \begin{vmatrix} a & b \\ f(a) & f(b) \end{vmatrix} = f(x_0) - x_0 f'(x_0).$$

88. (**Monthly Problem, 11892, 2016**). Let f be a real-valued continuously differentiable function on $[a,b]$ with positive derivative on (a,b). Prove that, for all pairs (x_1, x_2) with $a \leq x_1 < x_2 \leq b$ and $f(x_1)f(x_2) > 0$, there exists $\xi \in (x_1, x_2)$ such that
$$\frac{x_1 f(x_2) - x_2 f(x_1)}{f(x_2) - f(x_1)} = \xi - \frac{f(\xi)}{f'(\xi)}.$$

89. Let f be convex on $[0, \infty)$ with $f(0) = 0, f(x) \geq 0$. Prove that $f(x)/x$ is monotone increasing on $(0, +\infty)$.

90. Let f be convex on (a,b), and $L(x)$ be a linear function with $L(x) \leq f(x)$. Find $L(x)$ such that
$$\int_a^b (f(x) - L(x))\, dx$$
is minimal.

91. Let f_1, f_2 be nonnegative convex on $[0, \infty)$ and $f_1(0) = f_2(0)$. Prove that $f_1(x)f_2(x)/x$ is convex on $(0, +\infty)$.

92. Let f be continuous on $[a,b]$ and twice differentiable in (a,b). For any $x_i \in [a,b], p_i \geq 0 \, (i = 1, 2, \ldots, n)$ with $\sum_{i=1}^n p_i = 1$, show that there exists $c \in (a,b)$ such that
$$f\left(\sum_{i=1}^n p_i x_i\right) - \sum_{i=1}^n p_i f(x_i) = -\frac{f''(c)}{2} \sum_{i<j} p_i p_j (x_i - x_j)^2.$$

Clearly, Jensen's inequality is the consequence of this result. Moreover, it gives the error estimates of Jensen's inequality as well.

93. Let f be differentiable on (a,b) with $f'(x) \neq 0$. Show that there exist $c, d \in (a,b)$ such that
$$\frac{f'(c)}{f'(d)} = \frac{e^b - e^a}{b-a} e^{-d}.$$

94. Let $f : [a,b] \to \mathbb{R}$ be continuous and differentiable on (a,b). Prove that there exists $c \in (a,b)$ such that
$$\frac{2}{a-c} < f'(c) < \frac{2}{b-c}.$$

95. Let f be twice differentiable on $[a, b]$. If $f'(a) = f'(b) = 0$, show that there exists $c \in (a, b)$ such that
$$|f''(c)| \geq \frac{4}{(b-a)^2} |f(b) - f(a)|.$$

96. Let f be twice differentiable on $[a, b]$ and $f''((a+b)/2) = 0$.

    (a) Show that there exists $c \in (a, b)$ such that
    $$|f''(c)| \geq \frac{4}{(b-a)^2} |f(b) - f(a)|.$$

    (b) Show the constant 4 in the above inequality is the best possible one.
    (c) Show that, if f is not constant, then there exists $c \in (a, b)$ such that
    $$|f''(c)| > \frac{4}{(b-a)^2} |f(b) - f(a)|.$$

97. Let f be continuous on $[0, 1]$ and differentiable on $(0, 1)$. If $f(1) - f(0) = 1$, prove that, for every $k = 0, 1, \ldots, n-1$, there exists $x_k \in (0, 1)$ such that
$$f'(x_k) = \frac{n!}{k!(n-1-k)!} x_k^k (1 - x_k)^{n-1-k}.$$

98. Let f be twice differentiable on $\mathbb{R}$ and $f''(x) > 0$. If there exist x_0 such that $f(x_0) < 0$, $\lim_{x \to -\infty} f'(x) = \alpha < 0$ and $\lim_{x \to \infty} f'(x) = \beta > 0$. Prove that $f(x) = 0$ has exactly two solutions.

99. Let f be continuous on $[a, b]$ and $f(a) = f(b) = 0$. If f is twice differentiable and $f''(x) \leq 0$, show that $f(x) \geq 0$ on $[a, b]$.

100. Let $f_0(x) = 1$, $f_{n+1}(x) = x f_n(x) - f_n'(x)$ Prove that

    (a) $f_n(x)$ is an nth degree polynomial with the leading coefficient 1.
    (b) $f_n(x)$ has n distinct real roots, which are distributed symmetrically about $x = 0$.

101. Let $f : \mathbb{R} \to (0, \infty)$ be an increasing differentiable function for which $\lim_{x \to \infty} f(x) = \infty$ and $f'(x)$ is bounded. Let $F(x) = \int_0^x f(t)dt$. Define the sequence a_n by
$$a_0 = 1, \quad a_{n+1} = a_n + \frac{1}{f(a_n)}, \quad (\text{ for } n \geq 0$$
and the sequence $b_n = F^{-1}(n)$. Prove that $\lim_{n \to \infty} (a_n - b_n) = 0$.

102. Let $f : \mathbb{R} \to \mathbb{R}$ be continuously differentiable function that satisfies that $f'(x) > f(f(x))$ for all $x \in \mathbb{R}$. Prove that $f(f(f(x))) \leq 0$ for all $x \geq 0$.

103. Let $f(x) = \sum_{n=0}^{\infty} \frac{\sin(2^n x)}{n!}$. Show that the Taylor series of $f(x)$ diverges everywhere except $x = 0$.

104. The nth *Legendre polynomial* is defined by
$$P_n(x) = \frac{1}{2^n n!} \frac{d^n}{dx^n} (x^2 - 1)^n, \quad (n = 1, 2, \ldots,)$$

    For example, $P_1(x) = x, P_2(x) = \frac{1}{2}(3x^2 - 1), P_3(x) = \frac{1}{2}(5x^3 - 3x)$. Prove that $P_n(x)$ has n distinct roots in $(-1, 1)$.

105. (**Monthly Problem 6585, 1988**). Prove that if $-1 < x < 1$, then

$$\sum_{k=1}^{n} \frac{\sin(k \arccos x)}{k} = \frac{\sqrt{1-x}}{2} \int_{-1}^{x} \frac{1 - P_n(t)}{1-t} \frac{dt}{\sqrt{x-t}},$$

where $P_n(x)$ is the nth Legendre polynomial.

Remark. Since $|P_n(t)| < 1$ for all $t \in (-1,1)$, this provides a proof of the Fejer-Jackson inequality

$$\sum_{k=1}^{n} \frac{\sin(k\theta)}{k} > 0, \quad 0 < \theta < \pi.$$

106. (**Monthly Problem 11641, 2012**). Let f be a convex function from $\mathbb{R}$ into $\mathbb{R}$ and suppose that $f(x+y) + f(x-y) - 2f(x) \le y^2$ for all real x and y.

   (a) Show that f is differentiable.
   (b) Show that for all real x and y,

   $$|f'(x) - f'(y)| \le |x - y|.$$

107. Let f have derivatives of all orders on $\mathbb{R}$ and $|f^{(k)}(x)| \le M$ for all $x \in \mathbb{R}, k = 0, 1, \dots$. Let $E \subset \mathbb{R}$ be a infinite but bounded set. If $f(x) = 0$ on E, show that $f(x) \equiv 0$ for all $x \in \mathbb{R}$.

108. Let $\mathbf{A} = (a_{ij})$ be an $n \times n$ matrix and

$$\sum_{i=1}^{n} a_{ij} = \sum_{j=1}^{n} a_{ij} = 1.$$

Given $(x_1, x_2, \dots, x_n)$ with $x_i > 0$, define $y_i = \sum_{j=1}^{n} a_{ij} x_j$. Prove that

$$\prod_{i=1}^{n} y_i \ge \prod_{i=1}^{n} x_i.$$

109. Show that, for any $x \in \mathbb{R}$.

$$\left| \frac{\pi}{4} - \arctan x \right| \le \frac{\pi}{4} \frac{|x-1|}{\sqrt{x^2+1}}.$$

110. For $0 < x \le \pi$, show that

$$1 - \frac{x}{3} < \frac{\sin x}{x} < 1.1 - \frac{x}{4}.$$

111. Jordan's inequality claims that $\sin x \ge \frac{2}{\pi} x$ for $x \in [0, \pi/2]$. Prove the following improved Jordan's inequality: If $x \in [0, \pi/2]$, then

$$\frac{2}{\pi}x + \frac{\pi-2}{\pi^2}x(\pi - 2x) \quad \le \sin x \le \quad \frac{2}{\pi}x + \frac{2}{\pi^2}x(\pi - 2x),$$

$$\frac{2}{\pi}x + \frac{1}{\pi^3}x(\pi^2 - 4x^2) \quad \le \sin x \le \quad \frac{2}{\pi}x + \frac{\pi-2}{\pi^2}x(\pi^2 - 4x^2),$$

$$\frac{2}{\pi}x + \frac{1}{2\pi^5}x(\pi^4 - 16x^4) \quad \le \sin x \le \quad \frac{2}{\pi}x + \frac{\pi-2}{\pi^5}x(\pi^4 - 16x^4),$$

where the coefficients are all the best possible.

112. Let f be a function that has a continuous third derivative on $[0,1]$. Assume that $f(0) = f'(0) = f''(0) = f'(1) = f''(1) = 0$ and $f(1) = 1$. Prove that there exists $c \in [0,1]$ such that $f'''(c) \geq 24$.

113. Let a_i be positive for all $1 \leq i \leq n$. Prove that

$$\sqrt[n]{\prod_{i=1}^{n} a_i} \leq \ln\left(1 + \sqrt[n]{\prod_{i=1}^{n}(e^{a_i} - 1)}\right) \leq \frac{\sum_{i=1}^{n} a_i}{n}.$$

114. Let a_i, b_i be positive for all $1 \leq i \leq n$ with $\sum_{i=1}^{n} a_i = \sum_{i=1}^{n} b_i$. Show that

$$\sum_{i=1}^{n} a_i \ln a_i \geq \sum_{i=1}^{n} a_i \ln b_i.$$

115. Let $f : I \to (0, \infty)$ such that $e^{cx} f(x)$ is convex in I for all real number c. Prove that $\ln f(x)$ is convex in I.

116. Let $f^{(n+1)}(x_0) \neq 0$ and

$$f(x_0 + h) = f(x_0) + f'(x_0)h + \cdots + \frac{f^{(n-1)}(x_0)}{(n-1)!} + \frac{h^n}{n!} f^{(n)}(x_0 + \theta(h)h),$$

where $\theta(h) \in (0, 1)$. Prove that

$$\lim_{h \to 0} \theta(h) = \frac{1}{n+1}.$$

117. Let f be n times continuously differentiable on $(x_0 - \delta, x_0 + \delta)$. Let

$$f''(x_0) = f'''(x_0) = \cdots = f^{(n-1)}(x_0) = 0,$$

but $f^{(n)}(x_0) \neq 0$. For $0 < h < \delta$, if

$$f(x_0 + h) - f(x_0) = hf'(x_0 + \theta(h)h),$$

where $\theta(h) \in (0, 1)$. Prove that

$$\lim_{h \to 0} \theta(h) = \frac{1}{\sqrt[n-1]{n}}.$$

118. Let

$$\left(1 + \frac{1}{x}\right)^x = e \sum_{n=0}^{\infty} \frac{a_n}{x^n}.$$

Show that

$$a_n = (-1)^n \sum \frac{1}{k_1! \cdots k_n!} \left(\frac{1}{2}\right)^{k_1} \cdots \left(\frac{1}{n+1}\right)^{k_n}.$$

119. Consider the differential equation $x''(t) + a(t)x^3(t) = 0$ on $[0, \infty)$, where $a(t)$ is continuously differentiable and $a(t) \geq k > 0$. If $a'(t)$ has only finitely many changes of sign, prove that any solution $x(t)$ is bounded.

120. (**Extension of Pólya's Problem** (See Example 1.23)). Let x_n be a positive sequence satisfying $x_{n+1} = f(x_n)$ and $x_n \to 0$ as $n \to \infty$. Assume that

$$f(x) = x + Ax^k + o(x^k) \quad (x \to 0),$$

where $k \in \mathbb{N}, k > 1$ and $A \neq 0$. Show that there is a constant $\alpha > 0$ such that $\lim_{n \to \infty} nx_n^{\alpha}$ exists and determine the limit.

5

Integration

I will not define time, space, place, and motion, as being well known to all.

— Isaac Newton

The art of doing mathematics consists in finding that special case which contains all the germs of generality.

— David Hilbert

In this chapter we study the Riemann integral on a finite closed interval. We assume familiarity with this concept from a first-year calculus sequence, but try to develop the theory in a more precise way than typical calculus textbooks. We establish the criteria of integrability and the sense in which differentiation is the inverse of integration.

5.1   The Riemann Integral

Our first question is therefore: what meaning should we give to $\int_a^b f(x)dx$?

— Bernhard Riemann

In most calculus textbooks, the common approach to introduce the concept of integration was defined as the inverse process of differentiation, often called "antiderivatives." Here it is implicitly assumed that such an antiderivative exists and is unique up to an additive constant. Recall Darboux's theorem (Theorem 4.6), we see that any function with a jump discontinuity cannot be a derivative. In other words, if we define integration in terms of antiderivative, it results in a very limited number of functions that can be integrated. For example, a function as simple as the floor function is not integrable. Thus, we need a careful development of integration theory. In particular, we expect that is independent of differentiation. Finally, around 1850, Cauchy, following the work of Riemann, motivated by computing areas between the function and the horizontal axis, rigorously defined the integration as the limit of a *Riemann sum*.

Throughout this chapter, we assume that $[a, b]$ is a finite interval. First, to conform with the most modern approach, we introduce some notation and terminology.

Definition 5.1. *Let $a < b$ and $n \in \mathbb{N}$. If there are a finite set of points $x_0, x_1, \ldots, x_n$ with*

$$a = x_0 < x_1 < \cdots < x_{n-1} < x_n = b,$$

then $P = \{x_0, \ldots, x_n\}$ is called a partition of $[a, b]$. A partition P^* of $[a, b]$ is called a refinement of P if it contains all the points of P, i.e., $P \subseteq P^*$.

Definition 5.2. *Let $f : [a, b] \to \mathbb{R}$ be an arbitrary function. Given a partition $P = \{x_0, \ldots, x_n\}$ of $[a, b]$ and a set of tags $c_i \in [x_{i-1}, x_i]$ for $i = 1, 2, \ldots, n$, let $\Delta x_i = x_i - x_{i-1}$. The Riemann sum $R(f, P)$ is then given by*

$$R(f, P) = \sum_{i=1}^n f(c_i) \Delta x_i.$$

DOI: 10.1201/9781003304135-5

Definition 5.3. *Let $f : [a, b] \to \mathbb{R}$ be an arbitrary function. f is Riemann integrable on $[a, b]$ if there exists a real number I satisfying the following property: for every $\epsilon > 0$, there is a $\delta > 0$ such that for every partition $P = \{x_0, \ldots, x_n\}$ with $\|P\| = \max_{1 \leq i \leq n} \Delta x_i < \delta$ and for any set of tags $c_i \in [x_{i-1}, x_i]$, we have*

$$|R(f, P) - I| = \left| \sum_{i=1}^{n} f(c_i) \Delta x_i - I \right| < \epsilon.$$

The number I is called the definite integral of f on $[a, b]$ and is denoted by $\int_a^b f(x) dx$.

The significance of this definition is that the integral is a limit and its existence has nothing to do with antiderivatives. Moreover, this definition also ensures that the definite integral of f is unique. Suppose that both I_1 and I_2 are the definite integrals of f. By this definition, for every $\epsilon > 0$, there is a $\delta > 0$ such that for every partition $P = \{x_0, \ldots, x_n\}$ with $\|P\| = \max_{1 \leq i \leq n} \Delta x_i < \delta$ and for any set of tags $c_i \in [x_{i-1}, x_i]$, we have

$$|R(f, P) - I_k| = \left| \sum_{i=1}^{n} f(c_i) \Delta x_i - I_k \right| < \frac{\epsilon}{2}$$

for $k = 1, 2$. Then

$$
\begin{aligned}
|I_2 - I_1| &= \left| I_2 - \sum_{i=1}^{n} f(c_i) \Delta x_i + \sum_{i=1}^{n} f(c_i) \Delta x_i - I_1 \right| \\
&\leq \left| I_2 - \sum_{i=1}^{n} f(c_i) \Delta x_i \right| + \left| \sum_{i=1}^{n} f(c_i) \Delta x_i - I_1 \right| \\
&< \frac{\epsilon}{2} + \frac{\epsilon}{2} = \epsilon.
\end{aligned}
$$

Since ϵ is arbitrary, we must have $I_1 = I_2$.

Furthermore, this definition implies that if f is integrable, then f is bounded.

Theorem 5.1 (The Necessary Condition for Integrability). *If f is integrable on $[a, b]$, then f is bounded on $[a, b]$.*

Proof. Assume that f is integrable. Let $\epsilon = 1$. Then there exists $\delta > 0$, for any partition $P = \{x_0, \ldots, x_n\}$ with $\|P\| < \delta$ and for any set of tags $c_i \in [x_{i-1}, x_i]$ $(1 \leq i \leq n)$, such that

$$\left| \sum_{i=1}^{n} f(c_i) \Delta x_i - I \right| < 1. \tag{5.1}$$

We show that $f(x)$ is bounded on $[a, b]$ by contradiction. Suppose that $f(x)$ is unbounded on $[a, b]$. Then $f(x)$ is unbounded on some $[x_{j-1}, x_j]$ with $1 \leq j \leq n$. We now choose $\xi_j \in [x_{j-1}, x_j]$ such that

$$|f(\xi_j) \Delta x_j| > 1 + \left| \sum_{i \neq j}^{n} f(c_i) \Delta x_i \right| + |I|. \tag{5.2}$$

On the other hand, by (5.1), we have

$$
\begin{aligned}
|f(\xi_j)\Delta x_j| &= \left| \left(\sum_{i \neq j}^{n} f(c_i)\Delta x_i + f(\xi_j)\Delta x_j \right) - I - \sum_{i \neq j}^{n} f(c_i)\Delta x_i + I \right| \\
&\leq \left| \left(\sum_{i \neq j}^{n} f(c_i)\Delta x_i + f(\xi_j)\Delta x_j \right) - I \right| + \left| \sum_{i \neq j}^{n} f(c_i)\Delta x_i \right| + |I| \\
&\leq 1 + \left| \sum_{i \neq j}^{n} f(c_i)\Delta x_i \right| + |I|.
\end{aligned}
$$

This contradicts (5.2). Thus, f is bounded on $[a, b]$. $\qquad\square$

However, a bounded function is not necessarily integrable.

Example 5.1. *Consider the Dirichlet's function which is defined on $[0, 1]$ by*

$$
D(x) = \begin{cases} 1, & \text{if } x \text{ is rational,} \\ 0, & \text{if } x \text{ is irrational.} \end{cases}
$$

Then D is not integrable on $[0, 1]$.

Indeed, for any partition $P = \{x_0, x_1, \ldots, x_n\}$ of $[0, 1]$, by choosing $c_i \in [x_{i-1}, x_i]$ rational and irrational, respectively, we obtain

$$
R(f, P) = 1 \text{ and } R(f, P) = 0
$$

correspondingly. The integrability of $D(x)$ will violate the uniqueness of the definite integral.

One difficulty with Definition 5.3 is handling the variability of the set of tags c_i since $c_i \in [x_{i-1}, x_i]$ is arbitrary. Similar to study the convergence of bounded sequence via the supremum and infimum, Darboux used the supremum and infimum of the set $\{f(x) : x \in [x_{i-1}, x_i]\}$ around the set of tags.

Definition 5.4 (Darboux Sums). *Let f be bounded on $[a, b]$ and $P = \{x_0, \ldots, x_n\}$ be a partition of $[a, b]$. For $1 \leq i \leq n$, let*

$$
m_i = \inf\{f(x) : x \in [x_{i-1}, x_i]\}, \quad M_i = \sup\{f(x) : x \in [x_{i-1}, x_i]\}.
$$

The upper and lower Darboux sums associated to the P are defined by

$$
U(f, P) = \sum_{i=1}^{n} M_i \Delta x_i \quad \text{and} \quad L(f, P) = \sum_{i=1}^{n} m_i \Delta x_i,
$$

respectively.

Clearly, for a given partition, every Riemann sum is always squeezed by its upper and lower Darboux sums. Moreover, for any $p \in (x_{i-1}, x_i)$ for $1 \leq i \leq n$, we have

$$
m_i \Delta x_i \leq \inf\{f(x) : x \in [x_{i-1}, p]\}(p - x_{i-1}) + \inf\{f(x) : x \in [p, x_i]\}(x_i - p);
$$

$$
M_i \Delta x_i \geq \sup\{f(x) : x \in [x_{i-1}, p]\}(p - x_{i-1}) + \sup\{f(x) : x \in [p, x_i]\}(x_i - p).
$$

Consequently, we obtain

Theorem 5.2. *Let $f : [a, b] \to \mathbb{R}$ be bounded.*

1. *For every partition P of $[a, b]$, if $m \leq f(x) \leq M$, then*

$$m(b - a) \leq L(f, P) \leq R(f, P) \leq U(f, P) \leq M(b - a).$$

2. *If P^* is a refinement of P (i.e., $P \subseteq P^*$), then*

$$L(f, P) \leq L(f, P^*) \leq U(f, P^*) \leq U(f, P).$$

Next, we show that the inequality $L(f, P) \leq U(f, P)$ also holds for different partitions.

Theorem 5.3. *Let P_1 and P_2 be any two partitions of $[a, b]$. Then*

$$L(f, P_1) \leq U(f, P_2).$$

Proof. Using the refinement partition $P^* = P_1 \cup P_2$ as a bridge, by Theorem 5.2, we have

$$L(f, P_1) \leq L(f, P*) \leq U(f, P*) \leq U(f, P_2)$$

$\square$

Thus, a reasonable definition of $\int_a^b f(x)dx$ must be some number between $L(f, P)$ and $U(f, P)$ for any partition P. As the partition gets refined, we see that the upper Darboux sums get smaller while the lower Darboux sums get larger. Therefore, it makes sense to define

Definition 5.5.

$$\begin{cases} L(f) = \underline{\int}_a^b f \, dx = \sup\{L(f, P) : \ P \text{ is a partition of } [a, b])\} & (\text{lower Darboux integral}) \\ U(f) = \overline{\int}_a^b f \, dx = \inf\{U(f, P) : \ P \text{ is a partition of } [a, b])\} & (\text{upper Darboux integral}) \end{cases}$$

Darboux used the definition that f is integrable on $[a, b]$ provided $L(f) = U(f)$. Moreover, he established the following result on *one partition sufficing criterion* for integrability.

Theorem 5.4 (Darboux Theorem). *Let $f : [a, b] \to \mathbb{R}$ be bounded. f is Douboux integrable on $[a, b]$ if and only if for every $\epsilon > 0$ there exists a partition P of $[a, b]$ such that*

$$U(f, P) - L(f, P) = \sum_{i=1}^{n} (M_i - m_i)\Delta x_i < \epsilon. \tag{5.3}$$

Proof. "$\Longrightarrow$" Suppose that f is Douboux integrable. For every $\epsilon > 0$, there are partitions P_1 and P_2 of $[a, b]$ such that

$$L(f, P_1) > L(f) - \frac{\epsilon}{2} \quad \text{and} \quad U(f, P_1) < U(f) + \frac{\epsilon}{2}.$$

Let $P = P_1 \cup P_2$. By Theorem 5.2, we have

$$U(f, P) - L(f, P) \leq U(f, P_2) - L(f, P_1)$$
$$\leq U(f) + \frac{\epsilon}{2} - \left(L(f) - \frac{\epsilon}{2}\right)$$
$$= U(f) - L(f) + \epsilon = \epsilon. \quad (\text{use } U(f) = L(f))$$

This proves (5.3) as desired.

"$\Longleftarrow$" Suppose that for every $\epsilon > 0$ there is some partition P such that (5.3) holds. Then

$$U(f) \leq U(f, P) = U(f, P) - L(f, P) + L(f, P)$$
$$\leq \epsilon + L(f, P) \leq \epsilon + L(f).$$

Since ϵ is arbitrary, we obtain that $U(f) \leq L(f)$. This, together with the fact that $L(f) \leq U(f)$ from Theorem 5.3, yields that $U(f) = L(f)$. Thus f is Darboux integrable. $\square$

Remark. The characterization of Darboux integrability (5.3) for some partition can be replaced by: For every $\epsilon > 0$ and any partition P of $[a, b]$, there is a $\delta > 0$ such that

$$U(f, P) - L(f, P) < \epsilon \quad \text{whenever } \|P\| < \delta. \tag{5.4}$$

We leave the proof to the reader.

In contrast to Definition 5.3 on the Riemann integral, where in order to show the integrability by that definition, one must first have a candidate for the number I (the value of the definite integral) in mind, the Darboux theorem characterizes the integrability in terms of the upper and lower sums getting closer to each other. There is no need to know what number the upper and lower sums approaches.

We now show that both Riemann's and Darboux's definitions of integrability indeed are equivalent.

Theorem 5.5. *Let $f : [a, b] \to \mathbb{R}$ be bounded. f is Riemann integrable on $[a, b]$ if and only if it is Darboux integrable, in which case the values of the integrals are same.*

Proof. Suppose that f is Darboux integrable on $[a, b]$. For every Riemann sum $R(f, P)$ with $\|P\| < \delta$, we have

$$L(P, f) \leq R(P, f) \leq U(P, f).$$

By (5.4), we also have

$$U(f, P) < L(f, P) + \epsilon \leq L(f) + \epsilon$$

and

$$L(f, P) > U(f, P) - \epsilon \geq U(f) - \epsilon.$$

Since $U(f) = L(f)$, combining these three inequalities yields

$$|R(f, P) - U(f)| < \epsilon$$

for every partition P of $[a, b]$ with $\|P\| < \delta$. This proves that f is Riemann integrable and its value of the integral is $U(f)$.

Next suppose that f is Riemann integrable in the sense of Definition 5.3. For any partition $P = \{a = x_0, x_1, \ldots, x_n = b\}$ with $\|P\| < \delta$, we choose $c_i \in [x_{i-1}, x_i], 1 \leq i \leq n$ such that

$$f(c_i) \leq m_i + \epsilon = \inf\{f(x) : x \in [x_{i-1}, x_i]\} + \epsilon.$$

The Riemann sum $R(f, P)$ associated with this choice of c_i's satisfies

$$R(f, P) \leq L(f, P) + \epsilon(b - a)$$

and

$$|R(f, P) - I| < \epsilon,$$

where I is the value of the Riemann integral. Then

$$L(f) \geq L(f, P) \geq R(f, P) - \epsilon(b - a) > I - \epsilon - \epsilon(b - a).$$

Since ϵ is arbitrary, we obtain that $L(f) \geq I$. Along the same lines, we can show that $U(f) \leq I$. Hence we have $L(f) = U(f) = I$, i.e., f is Darboux integrable by the definition. $\square$

5.2   Classes of Integrable Functions

By one of those insights of which only the greatest minds are capable, the famous ge-
ometer (Riemann) generalizes the concept of the definite integral. — Gaston Darboux

Combining of Darboux theorem and Theorem 5.5 gives us a simple test on Riemann
integrability: For each $\epsilon > 0$, we just need to find one partition P for which $U(f,P) -$
$L(f,P) < \epsilon$. Based on this fact, we study which classes of functions are integrable. The
following well-known result was asserted by Cauchy in 1823. But it was proved rigorously
only some 50 years later with the notion of uniform continuity.

Theorem 5.6. *If f is continuous on $[a,b]$, then f is integrable.*

Proof. The uniform continuity of f implies that for every $\epsilon > 0$ there exists a $\delta > 0$ such
that

$$|f(x) - f(y)| < \frac{\epsilon}{b-a} \qquad \text{whenever } |x - y| < \delta. \tag{5.5}$$

Let $P = \{x_0, \ldots, x_n\}$ be a partition of $[a,b]$ with $\|P\| < \delta$. By the extreme value theorem,
there are $u_i, v_i \in [x_{i-1}, x_i]$ such that

$$m_i = f(u_i), \quad M_i = f(v_i).$$

It follows from (5.5) that

$$f(v_i) - f(u_i) = M_i - m_i < \frac{\epsilon}{b-a}$$

for all $1 \le i \le n$. Hence

$$U(f,P) - L(f,P) = \sum_{i=1}^{n} M_i(x_i - x_{i-1}) - \sum_{i=1}^{n} m_i(x_i - x_{i-1})$$

$$\le \sum_{i=1}^{n} \frac{\epsilon}{b-a}(x_i - x_{i-1}) = \frac{\epsilon}{b-a} \sum_{i=1}^{n}(x_i - x_{i-1}) = \epsilon.$$

This shows that f is integrable by the Darboux theorem.   □

The following theorem shows that monotone functions are also integrable.

Theorem 5.7. *Every monotone function $f : [a,b] \to \mathbb{R}$ is integrable.*

Proof. Without loss of generality, we assume that $f(x)$ is increasing. Since $f(a) \le f(x) \le$
$f(b)$ for all $x \in [a,b]$, f is bounded on $[a,b]$. Let $P = \{x_0, \ldots, x_n\}$ be the uniform partition
of $[a,b]$ such that

$$\Delta x_i = x_i - x_{i-1} = \frac{b-a}{n}, \quad \frac{b-a}{n}(f(b) - f(a)) < \epsilon.$$

Then

$$U(f, P_n) - L(f, P_n) = \sum_{i=1}^{n}(f(x_i) - f(x_{i-1}))\frac{b-a}{n}$$

$$= \frac{b-a}{n} \sum_{i=1}^{n}(f(x_i) - f(x_{i-1}))$$

$$= \frac{b-a}{n}(f(b) - f(a)) < \epsilon.$$

The integrability of f now follows from the Darboux theorem.   □

What about a discontinuous function? If f is discontinuous, for every partition, then there are subintervals that contains the discontinuity points where $M_i - m_i$ cannot be arbitrarily small. To have integrability, if we choose a small bound for $U(f, P) - L(f, P)$, then we need an even smaller bound on the total lengths of these subintervals which contain the discontinuity points. The following theorem demonstrates how to control the lengths of the subintervals so that they contain a finite number of discontinuity points.

Theorem 5.8. *If f is bounded and continuous at all but a finite number of points in $[a, b]$, then f is integrable.*

Proof. Let $M = \sup_{x \in [a,b]} f(x), m = \inf_{x \in [a,b]} f(x)$. Assume that f is discontinuous at $\bar{x}_i$ $(i = 1, 2, \ldots, N)$. For every $\epsilon > 0$, take $0 < \rho < \frac{\epsilon}{4(M-m+1)N}$ such that $(\bar{x}_i - \rho, \bar{x}_i + \rho)$ $(i = 1, 2, \ldots, N)$ are disjoint. By assumption, f is continuous on $[a, b] \setminus \cup_{i=1}^{N} (\bar{x}_i - \rho, \bar{x}_i + \rho) = \cap_{i=1}^{n} ([a, b] \setminus (\bar{x}_i - \rho, \bar{x}_i + \rho))$. Since

$$[a, b] \setminus (\bar{x}_i - \rho, \bar{x}_i + \rho) = [a, \bar{x}_i - \rho] \cup [\bar{x}_i + \rho, b],$$

by the uniform continuity of f on these sets, there exists a partition P' with $M_i - m_i < \frac{\epsilon}{2(b-a)}$. Now, P', adding $(\bar{x}_i - \rho, \bar{x}_i + \rho)$ $(i = 1, 2, \ldots, N)$ forms a partition of $[a, b]$, we denoted it by P. Then

$$U(f, P) - L(f, P) \leq \frac{\epsilon}{2(b - a)}(b - a) + (M - m) \sum_{i=1}^{N} 2\rho \leq \frac{\epsilon}{2} + (M - m)\frac{2N\epsilon}{4(M - m + 1)N} < \epsilon.$$

We conclude that f is integrable by the Darboux theorem. □

Definition 5.6. *If there is a partition $P = \{x_0, \ldots, x_n\}$ of $[a, b]$ such that f is a constant on each $(x_{i-1}, x_i]$, $(i = 1, 2, \ldots, n)$, then f is called a step function.*

In this case, f has at most a finite number of discontinuity points. Therefore,

Theorem 5.9. *A step function is integrable.*

The next example shows that an integrable function may have infinitely many discontinuity points. Recall that Riemann's function is discontinuous at every rational point. But, we have

Example 5.2. *Riemann's function $R : [0, 1] \to \mathbb{R}$ defined by*

$$R(x) = \begin{cases} 0, & \text{if } x \text{ is irrational in } [0, 1], \\ \frac{1}{n}, & \text{if } x = \frac{m}{n}, (m, n) = 1. \end{cases}$$

is integrable and $\int_0^1 R(x)dx = 0$.

Proof. To see this, for every $\epsilon > 0$, choose $n_0 \in \mathbb{N}$ such that $n_0 > 2/\epsilon$. Let

$$A = \left\{ x \in [0, 1] : R(x) > \frac{1}{n_0} \right\}.$$

Then $x \in A$ if and only if $x = m/n$ with $(m, n) = 1$ and $1 \leq m \leq n < n_0$. Since $1 \in A$, A is a finite nonempty set. Let M be the number of members in A. Clearly, $M \geq 1$.

Let $\delta = \frac{\epsilon}{2M}$. For any partition $P = \{x_0, x_1, \ldots, x_n\}$ of $[0,1]$ with $\|P\| < \delta$, we have $L(R, P) = 0$ and

$$
U(R, P) = \sum_{i=1}^{n} M_i (x_i - x_{i-1}) = \sum_{i=1}^{n} \sup_{x \in [x_{i-1}, x_i]} R(x)(x_i - x_{i-1})
$$

$$
= \sum_{\{i=1,\ldots,n; x \in A\}} \sup_{x \in [x_{i-1}, x_i]} R(x)(x_i - x_{i-1}) + \sum_{\{i=1,\ldots,n; x \notin A\}} \sup_{x \in [x_{i-1}, x_i]} R(x)(x_i - x_{i-1})
$$

$$
\leq \sum_{\{i=1,\ldots,n; x \in A\}} 1 \cdot (x_i - x_{i-1}) + \sum_{\{i=1,\ldots,n; x \notin A\}} \frac{1}{n_0} \cdot (x_i - x_{i-1})
$$

$$
< M \cdot 1 \cdot \delta + \frac{1}{n_0} \sum_{i=1}^{n} (x_i - x_{i-1}) = M\delta + \frac{1}{n_0} < \frac{\epsilon}{2} + \frac{\epsilon}{2} = \epsilon.
$$

We conclude that $R(x)$ is integrable by the Darboux theorem.   □

This example demonstrates that the obstacle to the integrability of a function is not only its discontinuity, but also how the values of the function vary in some subintervals. For a function to be integral, it seems that the set where the function is discontinuous (for example, the set A above) cannot be "too big." Motivated by this example, let

$$
\omega_i = \omega_i(f, [x_{i-1}, x_i]) = M_i - m_i, \tag{5.6}
$$

which is called the *oscillation* of f on $[x_{i-1}, x_i]$. This leads to the following Riemann's test for integrability.

Theorem 5.10 (Riemann Theorem). *Let $f : [a, b] \to \mathbb{R}$ be bounded. f is integrable on $[a, b]$ if and only if for every $\epsilon, \rho > 0$ there exists a partition of $[a, b]$ such that*

$$
\sum_{\omega_i \geq \rho} \Delta x_i < \epsilon. \tag{5.7}
$$

Proof. "$\Longrightarrow$" Assume f is integrable, by the Darboux theorem, for every $\epsilon, \rho > 0$, there exists a partition $P = \{x_0, \ldots, x_n\}$ of $[a, b]$ such that

$$
U(f, P) - L(f, P) = \sum_{i=1}^{n} \omega_i \Delta x_i < \epsilon \cdot \rho.
$$

Thus,

$$
\rho \cdot \sum_{\omega_i \geq \rho} \Delta x_i \leq \sum_{i=1}^{n} \omega_i \Delta x_i < \epsilon \cdot \rho,
$$

and so (5.7) holds.

"$\Longleftarrow$" By assumption, for every $\epsilon > 0$, there exists a partition P of $[a, b]$ such that

$$
\sum_{\omega_i \geq \frac{\epsilon}{2(b-a)}} \Delta x_i \leq \frac{\epsilon}{2(M - m + 1)}.
$$

For this partition, we have

$$U(f, P) - L(f, P) = \sum_{i=1}^{n} \omega_i \Delta x_i$$

$$= \sum_{\omega_i < \frac{\epsilon}{2(b-a)}} \omega_i \Delta x_i + \sum_{\omega_i \geq \frac{\epsilon}{2(b-a)}} \omega_i \Delta x_i$$

$$\leq \frac{\epsilon}{2(b-a)} \sum_{\omega_i < \frac{\epsilon}{2(b-a)}} \Delta x_i + (M - m) \sum_{\omega_i \geq \frac{\epsilon}{2(b-a)}} \Delta x_i$$

$$\leq \frac{\epsilon}{2(b-a)} (b-a) + (M - m) \frac{\epsilon}{2(M - m + 1)} < \epsilon.$$

This implies that f is integrable by the Darboux theorem. □

Drawing on Theorem 5.10, Lebesgue began working on infinite sets of discontinuities and finally revealed the actual continuity character of Riemann integrable functions. In short, he proved that Riemann integrable functions must be *continuous almost everywhere*.

The journey begins with focusing the oscillation (5.6) onto a single point. The oscillation of f at x_0 is defined by

$$\omega(f, x_0) = \inf_{\delta > 0} \omega(f, [x_0 - \delta, x_0 + \delta] \cap [a, b]).$$

Thus, a function f is continuous at x_0 if and only if $\omega(f, x_0) = 0$. For every $\delta > 0$ we define

$$D_\delta = \{x \in [a, b] : \omega(f, x) \geq \delta\}.$$

Clearly, f has a discontinuity at each point in D_δ. Let

$$D(f) = \{x \in [a, b] : f \text{ is discontinuous at } x\}.$$

It follows that

$$D(f) = \{x \in [a, b] : \omega(f, x) > 0\} = \cup_{\delta > 0} D_\delta.$$

Since $0 < \delta_1 < \delta_2$ implies that $D_{\delta_2} \subset D_{\delta_1}$, it follows that

$$D(f) = \bigcup_{n=1}^{\infty} D_{1/n}. \tag{5.8}$$

The following definition made the meaning of, a set is not "too big," precise.

Definition 5.7. *A set of real numbers A has measure zero if for every $\epsilon > 0$ there exists a collection of bounded intervals $\{(a_i, b_i); 1 \leq i \leq \infty\}$ such that*

$$A \subseteq \cup_{i=1}^{\infty} (a_i, b_i) \quad and \quad \sum_{i=1}^{\infty} (b_i - a_i) < \epsilon.$$

Definition 5.8. *A property holds almost everywhere, if it holds for all x in the domain except for a set of measure zero.*

The laws of set operations allow us to establish many examples of sets of measure zero.

Example 5.3. *The following sets all have measure zero:*

    1. A set with finite number of elements.

    2. A set with countable infinitely many elements.

    3. A subset of a set of measure zero.

    4. Countable union of sets of measure zero.

Equipped with Definitions 5.7 and 5.8, Lebesgue succeeded in yielding a simple characterization of Riemann integrable functions.

Theorem 5.11 (Lebesgue Theorem). *Let $f : [a,b] \to \mathbb{R}$ be bounded. f is Riemann integrable on $[a,b]$ if and only if f is continuous almost everywhere.*

Proof. "$\Longrightarrow$" Suppose that f is integrable on $[a,b]$. Fix $\delta > 0$, for every $\epsilon > 0$, there exists a partition $P = \{x_0, x_1, \ldots, x_n\}$ of $[a,b]$ such that

$$\sum_{i=1}^{n} \omega_i \Delta x_i < \epsilon \cdot \frac{\delta}{2}.$$

If $x \in D_\delta \cap [x_{i-1}, x_i]$, then $\omega_i \geq \omega(f, x) \geq \delta$, and so

$$\sum_{D_\delta \cap [x_{i-1}, x_i] \neq \emptyset} \Delta x_i < \frac{\epsilon}{2}.$$

Clearly,

$$D_\delta \subset \bigcup_{D_\delta \cap [x_{i-1}, x_i] \neq \emptyset} (x_{i-1}, x_i) \bigcup_{i=0}^{n} \left(x_i - \frac{\epsilon}{4(n+1)}, x_i + \frac{\epsilon}{4(n+1)} \right).$$

Since

$$\sum_{D_\delta \cap [x_{i-1}, x_i] \neq \emptyset} \Delta x_i + \frac{2\epsilon}{4(n+1)} \cdot (n+1) < \frac{\epsilon}{2} + \frac{\epsilon}{2} = \epsilon,$$

by Definition 5.7, this implies that D_δ is a set of measure zero. In view of (5.8), $D(f)$ is a countable union of sets of measure zero. Hence $D(f)$ also has measure zero.

"$\Longleftarrow$" Let $|f(x)| \leq M$. The fact that $D(f)$ has measure zero means that for every $\epsilon > 0$, there are open intervals $\{(\alpha_j, \beta_j) : j \in \mathbb{N}\}$ such that $D(f) \subset \cup_j (\alpha_j, \beta_j)$ and

$$\sum_j (\beta_j - \alpha_j) \leq \frac{\epsilon}{4M + 1}.$$

When $x \in [a,b] \setminus \cup_j (\alpha_j, \beta_j)$, since f is continuous at x, it follows that there exists a open interval I_x, where $x \in I_x$ and $\omega(f, I_x) < \frac{\epsilon}{2(b-a)}$. Thus, $\{(\alpha_j, \beta_j), I_x\}$ constitutes an open cover of $[a,b]$. By the Heine-Borel theorem, there is a finite subcovering of $[a,b]$. Select a partition P of $[a,b]$ such that either $[x_{i-1}, x_i] \subset (\alpha_j, \beta_j)$ for some j or $[x_{i-1}, x_i] \subset I_x$ for some x. Hence,

$$
\begin{aligned}
\sum_{i=1}^{n} \omega_i \Delta_i &\leq \sum_{[x_{i-1},x_i] \subset (\alpha_j, \beta_j)} \omega_i \Delta_i + \sum_{[x_{i-1},x_i] \subset I_x} \omega_i \Delta_i \\
&\leq 2M \sum_{[x_{i-1},x_i] \subset (\alpha_j, \beta_j)} \Delta_i + \frac{\epsilon}{2(b-a)} \sum_{[x_{i-1},x_i] \subset I_x} \Delta_i \\
&\leq 2M \sum_j (\beta_j - \alpha_j) + \frac{\epsilon}{2(b-a)} (b-a) \\
&\leq 2M \frac{\epsilon}{4M+1} + \frac{\epsilon}{2} < \epsilon.
\end{aligned}
$$

This implies that f is integrable by the Riemann theorem. $\qquad\square$

The Lebesgue theorem demonstrates how discontinuous function can be but still be Riemann integrable. It provides an intuitive idea of integrability in terms of the discontinuity set. Based on this theorem, a function that is discontinuous only at the rational numbers is not very discontinuous. It finally enables us to leave the Riemann integration as a historical subcase.

5.3 The Mean Value Theorem

All intelligent thoughts have already been thought; what is necessary is only to try to think them again. — Johann Goethe

Let f be continuous on $[a, b]$. Geometrically, the inequalities

$$\min_{x \in [a,b]} \{f(x)\}(b - a) \leq \int_a^b f(x)dx \leq \max_{x \in [a,b]} \{f(x)\}(b - a),$$

can be viewed as the area of a region under a curve is greater than the area of an inscribed rectangle and less the area of a circumscribed rectangle. The following theorem states that somewhere "between" the inscribed and circumscribed rectangles there is a rectangle whose area is precisely equal to the area of the region under the curve.

Theorem 5.12 (The First Mean Value Theorem for Integrals). *If $f : [a, b] \to \mathbb{R}$ is continuous, then there exists $x_0 \in [a, b]$ such that*

$$\int_a^b f(x)dx = f(x_0)(b - a).$$

Proof. Since $\min_{x \in [a,b]}\{f(x)\} \leq f(x) \leq \max_{x \in [a,b]}\{f(x)\}$ for all $x \in [a, b]$, we have

$$\min_{x \in [a,b]} \{f(x)\} \leq \frac{1}{b - a} \int_a^b f(x)dx \leq \max_{x \in [a,b]} \{f(x)\}.$$

The present theorem follows from the intermediate value theorem. □

In view of the above proof, if $g : [a, b] \to \mathbb{R}$ is a nonnegative integrable function, then

$$\min_{x \in [a,b]} \{f(x)\} \int_a^b g(x)dx \leq \int_a^b f(x)g(x)dx \leq \max_{x \in [a,b]} \{f(x)\} \int_a^b g(x)dx.$$

Thus, there exists $x_0 \in [a, b]$ such that

$$\int_a^b f(x)g(x)dx = f(x_0) \int_a^b g(x)dx.$$

If, in addition, assume that $g(x)$ is monotone, by using Abel's summation formula (Theorem 2.24), Bonnet and Weierstrass established

Theorem 5.13 (The Second Mean Value Theorem for Integrals). *Let $f : [a, b] \to \mathbb{R}$ be continuous.*

1. *If $g : [a, b] \to \mathbb{R}$ is decreasing and $g(x) \geq 0$ on $[a, b]$, then there exists $\xi \in [a, b]$ such that*

$$\int_a^b f(x)g(x)dx = g(a) \int_a^\xi f(x)dx. \tag{5.9}$$

2. *If $g : [a, b] \to \mathbb{R}$ is increasing and $g(x) \geq 0$ on $[a, b]$, then there exists $\eta \in [a, b]$ such that*

$$\int_a^b f(x)g(x)dx = g(b) \int_\eta^b f(x)dx. \tag{5.10}$$

3. *If $g : [a, b] \to \mathbb{R}$ is monotone, then there exists $\xi \in [a, b]$ such that*

$$\int_a^b f(x)g(x)dx = g(a) \int_a^\xi f(x)dx + g(b) \int_\xi^b f(x)dx. \tag{5.11}$$

Proof. (1) By the assumption that g is monotone, Theorem 5.7 implies that g is integrable on $[a, b]$. Thus, for every $\epsilon > 0$, there exists a partition $P = \{x_0, \ldots, x_n\}$ of $[a, b]$ such that

$$\sum_{i=1}^n \omega_i(g)\Delta x_i < \epsilon.$$

Let $F(x) = \int_a^x f(t)dt$. Then F is continuous. Furthermore,

$$\begin{aligned}
\int_a^b f(x)g(x)dx &= \sum_{i=1}^n \int_{x_{i-1}}^{x_i} f(x)g(x)dx \\
&= \sum_{i=1}^n \int_{x_{i-1}}^{x_i} [g(x) - g(x_{i-1})]f(x)dx + \sum_{i=1}^n g(x_{i-1}) \int_{x_{i-1}}^{x_i} f(x)dx \\
&\leq \sum_{i=1}^n \int_{x_{i-1}}^{x_i} |g(x) - g(x_{i-1})||f(x)|dx + \sum_{i=1}^n g(x_{i-1})[F(x_i) - F(x_{i-1})] \\
&\leq M \sum_{i=1}^n \omega_i(g)\Delta x_i + \sum_{i=1}^{n-1} F(x_i)[g(x_{i-1}) - g(x_i)] + F(b)g(x_{n-1}) \\
&\leq M\epsilon + \max_{x \in [a,b]} F(x) \cdot \sum_{i=1}^{n-1} [g(x_{i-1}) - g(x_i)] + \max_{x \in [a,b]} F(x) \cdot g(x_{n-1}) \\
&= M\epsilon + g(a) \cdot \max_{x \in [a,b]} F(x).
\end{aligned}$$

Here Abel's summation formula is used in the second inequality above. Replacing f by $-f$, notice that $\max_{x \in [a,b]}(-F(x)) = -\min_{x \in [a,b]} F(x)$, the argument above shows that

$$\int_a^b -f(x)g(x)dx \leq M\epsilon - g(a) \cdot \min_{x \in [a,b]} F(x).$$

In summary, we find that

$$g(a) \cdot \min_{x \in [a,b]} F(x) - M\epsilon \leq \int_a^b f(x)g(x)dx \leq g(a) \cdot \max_{x \in [a,b]} F(x) + M\epsilon.$$

Since ϵ is arbitrary, we must have

$$g(a) \cdot \min_{x \in [a,b]} F(x) \leq \int_a^b f(x)g(x)dx \leq g(a) \cdot \max_{x \in [a,b]} F(x).$$

If $g(a) = 0$, together with g is decreasing and nonnegative, g must be zero. Thus, $\int_a^b f(x)g(x)dx = 0$. If $g(a) \neq 0$, then

$$\min_{x \in [a,b]} F(x) \leq \frac{\int_a^b f(x)g(x)dx}{g(a)} \leq \max_{x \in [a,b]} F(x).$$

(5.9) now follows from applying the intermediate value theorem to $F(x)$.

(2) The proof of (5.10) is similar except replacing $F(x)$ by $\int_x^b f(x)dx$.

(3) First, we assume that g is decreasing. Let $G(x) = g(x) - g(b)$. Then G is monotone and $G(x) \geq 0$. By (1), there exists $\xi \in [a, b]$ such that

$$\int_a^b f(x)G(x)dx = G(a)\int_a^\xi f(x)dx.$$

Plugging $G(x) = g(x) - g(b)$ yields (5.11) as desired. When g is increasing, (5.11) follows by applying (2) to $G(x) = g(x) - g(a)$. □

We close this section with an application of the second mean value theorem for integrals.

Example 5.4. *Let $\beta \geq 0, b > a > 0$. Then*

$$\left| \int_a^b e^{-\beta x} \frac{\sin x}{x} dx \right| \leq \frac{2}{a}.$$

Indeed, let $f(x) = \sin x, g(x) = \frac{e^{-\beta x}}{x}$. Here $g(x)$ is decreasing and positive on $[a, b]$. By (5.9), there is $\xi \in [a, b]$ such that

$$\int_a^b e^{-\beta x} \frac{\sin x}{x} dx = \frac{e^{-\beta a}}{a} \int_a^\xi \sin x dx = \frac{e^{-\beta a}}{a}(\cos a - \cos \xi).$$

Therefore,

$$\left| \int_a^b e^{-\beta x} \frac{\sin x}{x} dx \right| \leq 2 \frac{e^{-\beta a}}{a} \leq \frac{2}{a}.$$

5.4 The Fundamental Theorems of Calculus

"This link was confirmed by Newton and Leibniz. For this achievement, we honor them as the discovers of calculus." — David Perkins

There are two versions of the *fundamental theorems of calculus*. Roughly speaking, each says that differentiation and integration are inverse operations. The first version

$$\int_a^b f(x)dx = F(b) - F(a)$$

is the central result of all the integral computations. It enables us to evaluate an integral that does not require the explicit evaluation of upper and lower Darboux sums. The combination of the two versions yields

$$F(x) = F(a) + \int_a^x F'(t)dt.$$

Thus, with the help of a limit process (definite integral is a limit), one can recover the function itself based on its given derivative. The above approach provides the crucial step on the road to studying general differential equations.

Definition 5.9. *Let $f : I \to \mathbb{R}$. $F(x) : I \to \mathbb{R}$ is an antiderivative of f if $F'(x) = f(x)$ for all $x \in I$.*

Theorem 5.14 (The First Fundamental Theorem of Calculus). *Suppose that f is integrable on $[a, b]$, and $F(x)$ is any antiderivative of f on (a, b). Then*

$$\int_a^b f(x)dx = F(b) - F(a).$$

Proof. Let $P = \{x_0, \ldots, x_n\}$ be a partition of $[a, b]$. By the mean value theorem, there exists $\xi_i \in (x_{i-1}, x_i)$ such that

$$F(x_i) - F(x_{i-1}) = F'(\xi_i)(x_i - x_{i-1}) = f(\xi_i)\Delta x_i, \ \ i = 1, 2, \ldots, n.$$

Thus,

$$F(b) - F(a) = \sum_{i=1}^{n} (F(x_i) - F(x_{i-1})) = \sum_{i=1}^{n} f(\xi_i)\Delta x_i.$$

The assumption of integrability of f yields

$$F(b) - F(a) = \lim_{\|P\| \to 0} \sum_{i=1}^{n} f(\xi_i)\Delta x_i = \int_a^b f(x)dx. \qquad \square$$

The following example shows the necessity of the integrability of f in this theorem.

Example 5.5. *Define*
$$F(x) = \begin{cases} x^2 \sin\left(\frac{1}{x^2}\right), & x \neq 0, \\ 0, & x = 0. \end{cases}$$

Here $F(x)$ is differentiable on $[0, 1]$ and

$$F'(x) = \begin{cases} 2x \sin\left(\frac{1}{x^2}\right) - \frac{2}{x} \cos\left(\frac{1}{x^2}\right), & x \neq 0, \\ 0, & x = 0. \end{cases}$$

$F'(x)$ is unbounded on $[0, 1]$ and therefore $f(x) = F'(x)$ is not integrable.

Smooth action. Let $f : [a, b] \to \mathbb{R}$ be integrable. Then the function

$$F(x) = \int_a^x f(t)dt$$

is continuous on $[a, b]$. If we strengthen the condition of integrability with the condition of continuity, we have another cornerstone of analysis.

Theorem 5.15 (The Second Fundamental Theorem of Calculus). *Let $f : [a, b] \to \mathbb{R}$ be continuous. Then the function*

$$F(x) = \int_a^x f(t)dt$$

is differentiable on (a, b) and satisfies $F'(x) = f(x)$. Hence, F is an antiderivative of $f(x)$.

Proof. Suppose f is continuous at $x_0 \in (a, b)$. Note that

$$\frac{F(x) - F(x_0)}{x - x_0} = \frac{1}{x - x_0} \int_{x_0}^x f(t)dt \ \ (x \neq x_0)$$

and

$$f(x_0) = \frac{1}{x - x_0} \int_{x_0}^{x} f(x_0)dt.$$

Thus,

$$\frac{F(x) - F(x_0)}{x - x_0} - f(x_0) = \frac{1}{x - x_0} \int_{x_0}^{x} (f(t) - f(x_0))dt. \tag{5.12}$$

For every $\epsilon > 0$, the continuity of f implies that there exists $\delta > 0$ such that

$$|f(t) - f(x_0)| < \epsilon$$

whenever $|t - x_0| < \delta, t \in (a, b)$. It follows from (5.12) that

$$\left| \frac{F(x) - F(x_0)}{x - x_0} - f(x_0) \right| < \epsilon$$

for $x \in (a, b)$ satisfying $|x - x_0| < \delta$. This shows that

$$F'(x_0) = \lim_{x \to x_0} \frac{F(x) - F(x_0)}{x - x_0} = f(x_0).$$

$\square$

This theorem asserts that every continuous function has an antiderivative. By the Lebesgue theorem (Theorem 5.11), if f is Riemann integrable, then f is continuous almost everywhere. Therefore we have

$$\frac{d}{dx} \int_{a}^{x} f(t)dt = f(x) \tag{5.13}$$

almost everywhere. On the other hand, the following example shows that a Riemann integrable function has no antiderivative.

Example 5.6. *Consider*

$$f(x) = \begin{cases} 1, & \text{if } 1 \leq x < 2, \\ 2, & \text{if } 2 \leq x \leq 3. \end{cases}$$

As a step function, $f(x)$ is integrable on $[1, 3]$. Recall that a derivative has no jump discontinuity, $f(x)$ is not the derivative of any function defined on $[1, 3]$. So f has no antiderivative! Thus, there is a class of integrable functions for which

$$\int_{a}^{b} f(x)dx = F(b) - F(a) \tag{5.14}$$

is not directly applicable. The question naturally arises: when F is an antiderivative? The search for a concise characterization of all possible derivatives remains largely unsuccessful. Even if we extend the Riemann integral to the Lebesgue integral, (5.14) is true only when F is *absolute continuous*. For details, see Royden's *Real Analysis* [81, Chapter 5].

Coupled with the chain rule, if f is continuous, $u(x)$ and $v(x)$ are differentiable, (5.14) can be extended to

$$\frac{d}{dx} \int_{v(x)}^{u(x)} f(t)dt = f(u(x))u'(x) - f(v(x))v'(x). \tag{5.15}$$

The next result, known as *integration by parts*, is very useful in dealing with integrals.

Theorem 5.16. *Let* $u, v : [a, b] \to \mathbb{R}$ *be differentiable. If* $u'(x)$ *and* $v'(x)$ *are integrable, then*

$$\int_a^b u(x)v'(x)dx = u(x)v(x) \mid_a^b - \int_a^b u'(x)v(x)dx. \tag{5.16}$$

Another important technique of integration is known as "substitution—change of variable." This process is the reverse of the chain rule.

Theorem 5.17. *Let* $f : [a, b] \to \mathbb{R}$ *be continuous. If* $x = \phi(t)$ *is continuously differentiable on* $[\alpha, \beta]$ *such that* $\phi([\alpha, \beta]) \subset [a, b]$, $\phi(\alpha) = a, \phi(\beta) = b$, *then*

$$\int_a^b f(x)dx = \int_\alpha^\beta f(\phi(t))\phi'(t)dt. \tag{5.17}$$

5.5   Worked Examples

The greatest challenge to any thinker is stating the problem in a way that will allow a solution. — Bertrand Russell

In this section, we present 24 examples to illustrate the applications of the integral theorems. The first four examples demonstrate some nonstandard tricks for computing integrals.

1. Evaluate $\int \frac{dx}{1+x^4}$.

Solution. The standard approach is to use partial fractions. But it is much simpler to introduce

$$I = \int \frac{dx}{1 + x^4} \quad \text{and} \quad J = \int \frac{x^2}{1 + x^4} dx.$$

Since

$$I + J = \int \frac{1 + x^2}{1 + x^4} dx = \int \frac{1 + 1/x^2}{x^2 + 1/x^2} dx$$

$$= \int \frac{d(x - 1/x)}{(x - 1/x)^2 + 2} dx = \frac{1}{\sqrt{2}} \arctan^{-1} \left(\frac{x - 1/x}{\sqrt{2}} \right) + C;$$

$$I - J = \int \frac{1 - x^2}{1 + x^4} dx = - \int \frac{1 - 1/x^2}{x^2 + 1/x^2} dx$$

$$= - \int \frac{d(x + 1/x)}{(x + 1/x)^2 - 2} dx = - \frac{1}{2\sqrt{2}} \ln \frac{x + 1/x + \sqrt{2}}{x + 1/x - \sqrt{2}} + C,$$

Solving the system yields that

$$\int \frac{dx}{1 + x^4} = \frac{1}{2\sqrt{2}} \arctan \frac{x^2 - 1}{\sqrt{2}\,x} - \frac{1}{4\sqrt{2}} \ln \frac{x^2 + \sqrt{2}x + 1}{x^2 - \sqrt{2}x + 1} + C.$$

□

Remark. Along these same lines, we can evaluate a class of integrals such as

$$\int \frac{e^x}{e^x + e^{-x}} dx, \quad \int \frac{\sin x\, dx}{a \sin x + b \cos x} \ (a \neq b), \quad \text{and} \quad \int_0^{\pi/4} \ln(\sin x)\, dx.$$

2. Let n be nonnegative integer. Evaluate $\int_0^\pi \frac{1-\cos nx}{1-\cos x}\,dx$.

   Solution. Denote the proposed integral by I_n. For $n \geq 1$, we have

   $$\frac{1}{2}(I_{n+1} + I_{n-1}) = \int_0^\pi \frac{2 - \cos(n+1)x - \cos(n-1)x}{2(1 - \cos x)}\,dx$$

   $$= \int_0^\pi \frac{1 - \cos nx \cos x}{1 - \cos x}\,dx$$

   $$= \int_0^\pi \frac{(1 - \cos nx) + \cos nx(1 - \cos x)}{1 - \cos x}\,dx$$

   $$= I_n.$$

   This implies that I_n is an arithmetic sequence. Since $I_0 = 0$ and $I_1 = \pi$, we find that $I_n = n\pi$. $\qquad\square$

3. For $n \in \mathbb{N}$, evaluate

   $$I_n = \int_{-\pi}^\pi \frac{\sin nx}{(1 + 2^x)\sin x}\,dx.$$

   Solution. Splitting the integral into two parts, we have

   $$I_n = \int_{-\pi}^0 \frac{\sin nx}{(1 + 2^x)\sin x}\,dx + \int_0^\pi \frac{\sin nx}{(1 + 2^x)\sin x}\,dx.$$

   Making the substitution $u = -x$ in the first integral yields

   $$I_n = \int_0^\pi \frac{\sin nu}{(1 + 2^{-u})\sin u}\,du + \int_0^\pi \frac{\sin nx}{(1 + 2^x)\sin x}\,dx$$

   $$= \int_0^\pi \frac{(1 + 2^x)\sin nx}{(1 + 2^x)\sin x}\,dx$$

   $$= \int_0^\pi \frac{\sin nx}{\sin x}\,dx.$$

   For $n \geq 2$, it follows that

   $$I_n - I_{n-2} = \int_0^\pi \frac{\sin nx - \sin(n-2)x}{\sin x}\,dx = 2\int_0^\pi \cos(n-1)x\,dx = 0.$$

   Using the facts that $I_0 = 0, I_1 = \pi$, we conclude that

   $$I_n = \begin{cases} \pi, & n \text{ is odd,} \\ 0, & n \text{ is even.} \end{cases}$$

   $\qquad\square$

4. For $|x| \leq \pi/2$, let $\phi(x) = -\int_0^x \ln\cos t\,dt$. Prove that

   $$\phi(x) = -x\ln 2 + 2\phi\left(\frac{\pi}{4} + \frac{x}{2}\right) - 2\phi\left(\frac{\pi}{4} - \frac{x}{2}\right).$$

Proof. Note that

$$\left(-x\ln 2 + 2\phi\left(\frac{\pi}{4}+\frac{x}{2}\right) - 2\phi\left(\frac{\pi}{4}-\frac{x}{2}\right)\right)' = -\ln 2 + \phi'\left(\frac{\pi}{4}+\frac{x}{2}\right) + \phi'\left(\frac{\pi}{4}-\frac{x}{2}\right)$$

$$= -\ln 2 - \ln\cos\left(\frac{\pi}{4}+\frac{x}{2}\right) - \ln\cos\left(\frac{\pi}{4}-\frac{x}{2}\right)$$

$$= -\ln\left[2\cos\left(\frac{\pi}{4}+\frac{x}{2}\right)\cos\left(\frac{\pi}{4}-\frac{x}{2}\right)\right]$$

$$= -\ln(\cos\pi/2 + \cos x) = -\ln\cos x = \phi'(x).$$

Thus, the claimed identity follows from $\phi(0) = 0$.   □

5. Let $f, g : [a, b] \to \mathbb{R}$ be Riemann integrable, and let $P = \{x_0, x_1, \ldots, x_n\}$ be a partition of $[a, b]$. Prove that, for every set of tags $\xi_i, \eta_i \in [x_{i-1}, x_i]$,

$$\lim_{\|P\|\to 0}\sum_{i=1}^{n} f(\xi_i)g(\eta_i)\Delta x_i = \int_a^b f(x)g(x)dx.$$

Proof. The integrability of f and g implies that $f(x)g(x)$ is integrable on $[a, b]$. Thus, for every $\epsilon > 0$, there exists a $\delta_1 > 0$ and a partition $P = \{x_0, x_1, \ldots, x_n\}$ of $[a, b]$ satisfying $\|P\| < \delta_1$ such that

$$\left|\sum_{i=1}^{n} f(\xi_i)g(\xi_i)\Delta x_i - \int_a^b f(x)g(x)dx\right| < \frac{\epsilon}{2},$$

where $\xi_i \in [x_{i-1}, x_i]$. Let $M = \sup_{x\in[a,b]} |f(x)|$. Since g is integrable, there exists $\delta_2 > 0$ such that

$$\sum_{i=1}^{n} \omega_i(g)\Delta x_i < \frac{\epsilon}{2(M+1)}$$

whenever $\|P\| < \delta_2$. Now, when $\|P\| < \delta = \min\{\delta_1, \delta_2\}, \xi_i, \eta_i \in [x_{i-1}, x_i]$, we have

$$\left|\sum_{i=1}^{n} f(\xi_i)g(\eta_i)\Delta x_i - \sum_{i=1}^{n} f(\xi_i)g(\xi_i)\Delta x_i\right| \le \sum_{i=1}^{n} |f(\xi_i)||g(\eta_i) - g(\xi_i)|\Delta x_i$$

$$\le M\sum_{i=1}^{n} |g(\eta_i) - g(\xi_i)|\Delta x_i \le M\sum_{i=1}^{n} \omega_i(g)\Delta x_i$$

$$< M\frac{\epsilon}{2(M+1)} < \frac{\epsilon}{2}.$$

Therefore, by the triangle inequality, we have

$$\left|\sum_{i=1}^{n} f(\xi_i)g(\eta_i)\Delta x_i - \int_a^b f(x)g(x)dx\right| \le \left|\sum_{i=1}^{n} f(\xi_i)g(\eta_i)\Delta x_i - \sum_{i=1}^{n} f(\xi_i)g(\xi_i)\Delta x_i\right|$$

$$+ \left|\sum_{i=1}^{n} f(\xi_i)g(\xi_i)\Delta x_i - \int_a^b f(x)g(x)dx\right| < \frac{\epsilon}{2} + \frac{\epsilon}{2} = \epsilon.$$

This proves the claimed result.   □

6. Let $f : [a, b] \to \mathbb{R}$ be Riemann integrable.

   (a) Prove that there exists $x_0 \in (a, b)$ such that f is continuous at x_0.

   (b) If f is nonnegative and Riemann integrable on $[a, b]$, show the following statements are equivalent

       (1) $\int_a^b f(x)dx = 0$.

       (2) $f(x) = 0$ where x is a continuous point of f.

       (3) $f = 0$ almost everywhere.

   (c) Let f be nonnegative and bounded. If $f = 0$ almost everywhere, does $\int_a^b f(x)dx = 0$?

Proof. (a) By the Lebesgue theorem, f is continuous almost everywhere. Since (a, b) is not measure zero, it follows that f is continuous at some $x_0 \in (a, b)$. Here we give a different proof—using the nested interval theorem to find the point x_0. To this end, first, for every $\epsilon > 0$, we show that if $P = \{x_0, x_1, \ldots, x_n\}$ is a partition of $[a, b]$ such that $U(f, p) - L(f, P) < \epsilon(b - a)$, then there exists some subinterval $[x_{i-1}, x_i]$ on which $\omega_i(f) < \epsilon$. By contradiction, if $\omega_i(f) \geq \epsilon$ for all $i \, (1 \leq i \leq n)$, then

$$\epsilon(b - a) > U(f, P) - L(f, P) = \sum_{i=1}^{n} \omega_i(f)\Delta x_i \geq \sum_{i=1}^{n} \epsilon \Delta x_i = \epsilon(b - a).$$

This contradiction shows that $\omega_i(f) < \epsilon$ for some $1 \leq i \leq n$. In particular, let $\epsilon = 1$ and $[a_1, b_1] \subset [x_{i-1}, x_i] \subset [a, b]$ such that $b_1 - a_1 < \frac{b-a}{2}$. Then

$$\omega(f, [a_1, b_1]) \leq \omega(f, [x_{i-1}, x_i]) < 1.$$

Next, let $\epsilon = 1/2$. Since f is integrable on $[a_1, b_1]$, it follows that there exists $[a_2, b_2] \subset [a_1, b_1]$ such that $b_2 - a_2 < \frac{b-a}{2^2}$ and

$$\omega(f, [a_2, b_2]) < \frac{1}{2}.$$

Repeat the process above to obtain a sequence of closed intervals that satisfy

- $[a_n, b_n] \subset [a_{n-1}, b_{n-1}]$ and $b_n - a_n < \frac{b-a}{2^n}$
- $\omega(f, [a_n, b_n]) < \frac{1}{n}$.

By the nested interval theorem, there exists $x_0 \in \cap_{n=1}^{\infty}[a_n, b_n]$ such that

$$0 \leq \omega(f, x_0) = \lim_{n \to \infty} \omega(f, [a_n, b_n]) \leq \lim_{n \to \infty} \frac{1}{n} = 0.$$

Thus, $\omega(f, x_0) = 0$, which implies that f is continuous at x_0. In fact, such x_0 is dense in $[a, b]$. For any $t \in [a, b]$, let (α, β) be an arbitrary neighborhood of t. The integrability of f on $(\alpha, \beta) \cap [a, b]$ asserts that there is $x_0 \in (\alpha, \beta) \cap [a, b]$ at which f is continuous. So the set of continuity points of f is dense in $[a, b]$.

(b) (1) $\Rightarrow$ (2) By contradiction, assume that $f(x_0) \neq 0$ where x_0 is a continuous point of f. Without loss of generality, assume that $f(x_0) > 0$ and $x_0 \in (a, b)$. The continuity of f at x_0 implies that there exists $(x_0 - \delta, x_0 + \delta) \subset (a, b)$ such that

$$f(x) > \frac{1}{2}f(x_0), \quad \text{for } x \in (x_0 - \delta, x_0 + \delta).$$

Thus, we have

$$0 = \int_a^b f(x)dx \geq \int_{x_0-\delta}^{x_0+\delta} f(x)dx > \frac{1}{2}\int_{x_0-\delta}^{x_0+\delta} f(x_0)dx = \delta f(x_0) > 0.$$

The contradiction shows that $f(x_0) = 0$.

$(2) \Rightarrow (3)$ By the Lebesgue theorem, f is continuous almost everywhere. The assumption of $f(x) = 0$ at continuous points concludes that $f(x) = 0$ almost everywhere.

$(3) \Rightarrow (1)$ Observe that $f = 0$ almost everywhere implies that $f(x) = 0$ has solution in every $(\alpha, \beta) \subset [a, b]$. Let $P = \{x_0, x_1, \ldots, x_n\}$ be a partition of $[a, b]$. Taking $\xi_i \in [x_{i-1}, x_i]$ such that $f(\xi_i) = 0$ yields

$$\int_a^b f(x)dx = \lim_{\|P\|\to 0} \sum_{i=1}^n f(\xi_i)\Delta x_i = 0.$$

(c) Dirichlet's function provides us an example that a nonnegative, bounded and zero almost everywhere does not necessarily have zero integral value.   □

7. Let $f : [a, b] \to \mathbb{R}$ be continuous. If

$$\int_a^b f(x)dx = 0, \quad \int_a^b xf(x)dx = 0,$$

show that $f(x)$ has at least two zeros in (a, b).

Proof. We give two proofs.

Proof I If f has no zero at all in (a, b), by its continuity and the intermediate value property, we find that either $f(x) > 0$ or $f(x) < 0$ for any $x \in (a, b)$. This yields that either $\int_a^b f(x)dx > 0$ or $\int_a^b f(x)dx < 0$, which contradicts the assumption.

If f has only one zero $x_0 \in (a, b)$, then f has opposite sign on (a, x_0) and (x_0, b). Thus, $(x - x_0)f(x)$ does not change sign on (a, b). Therefore,

$$\int_a^b (x - x_0)f(x)dx \neq 0. \tag{5.18}$$

On the other hand,

$$\int_a^b (x - x_0)f(x)dx = \int_a^b xf(x)dx - x_0 \int_a^b f(x)dx = 0.$$

This contradicts (5.18). Hence f must have at least two zeros in (a, b).

Proof II The assumptions of the continuity and that $\int_a^b f(x)dx = 0$ imply that f has at least one zero in (a, b). By contradiction, we assume that f has only one zero at x_0. Appealing to

$$0 = \int_a^b f(x)dx = \int_a^{x_0} f(x)dx + \int_{x_0}^b f(x)dx,$$

we have

$$\int_a^{x_0} f(x)dx = -\int_{x_0}^b f(x)dx \neq 0. \tag{5.19}$$

Similarly, we rewrite

$$\int_a^b xf(x)dx = \int_a^{x_0} xf(x)dx + \int_{x_0}^b xf(x)dx. \tag{5.20}$$

Applying the first mean value theorem for integrals gives

$$\int_a^{x_0} xf(x)dx = \xi_1 \int_a^{x_0} f(x)dx \ \text{ and } \ \int_{x_0}^b xf(x)dx = \xi_2 \int_{x_0}^b f(x)dx,$$

where $a < \xi_1 < x_0 < \xi_2 < b$. In view of (5.19) and (5.20) , we have

$$0 = \int_a^b xf(x)dx = (\xi_2 - \xi_1) \int_{x_0}^b f(x)dx \neq 0.$$

The contradiction shows that f has at least two zeros in (a, b). $\qquad\square$

Remark. Based on this result, by induction, we can show that if

$$\int_a^b x^k f(x)dx = 0 \ \text{ for } k = 0, 1, \ldots, n$$

then f has at least $n + 1$ zeros in (a, b).

8. **(Riemann Lemma)**. Let $f : [a, b] \to \mathbb{R}$ be Riemann integrable. Show that

$$\lim_{p \to \infty} \int_a^b f(x) \sin px \, dx = \lim_{p \to \infty} \int_a^b f(x) \cos px \, dx = 0.$$

Proof. Due to the similarity, we prove that $\lim_{p\to\infty} \int_a^b f(x) \sin px dx = 0$ only. By the assumption, for every $\epsilon > 0$, there exists a partition $P = \{x_0, x_1, \ldots, x_n\}$ of $[a, b]$ such that

$$\left| \int_a^b f(x)dx - \sum_{i=1}^n m_i \Delta x_i \right| < \frac{\epsilon}{2}, \tag{5.21}$$

where $m_i = \inf_{x \in [x_{i-1}, x_i]} f$ for $1 \leq i \leq n$. Let $M = \sup_{x \in [a,b]} |f(x)|$. Choose $p > \frac{4nM}{\epsilon}$. Using that $f(x) - m_i \geq 0$ on $[x_{i-1}, x_i]$ for $1 \leq i \leq n$, we have

$$\left| \int_a^b f(x) \sin px \, dx \right| \leq \left| \sum_{i=1}^n \int_{x_{i-1}}^{x_i} (f(x) - m_i) \sin px \, dx \right| + \left| \sum_{i=1}^n \int_{x_{i-1}}^{x_i} m_i \sin px \, dx \right|$$

$$\leq \sum_{i=1}^n \int_{x_{i-1}}^{x_i} (f(x) - m_i) |\sin px| \, dx + \left| \sum_{i=1}^n \int_{x_{i-1}}^{x_i} m_i \sin px \, dx \right|$$

$$\leq \left(\int_a^b f(x) \, dx - \sum_{i=1}^n m_i \Delta x_i \right) + \frac{1}{p} \left| \sum_{i=1}^n m_i [\cos(px_i) - \cos(px_{i-1})] \right|$$

$$< \frac{\epsilon}{2} + \frac{1}{p} 2nM < \frac{\epsilon}{2} + \frac{\epsilon}{2} = \epsilon.$$

This proves that $\lim_{p\to\infty} \int_a^b f(x) \sin px \, dx = 0$, as desired. $\qquad\square$

Remark. If, in addition, assume that f is continuously differentiable, integrating by parts will yield a brief proof. On the other hand, For each partition $P = \{x_0, x_1, \ldots, x_n\}$ of $[a, b]$, define a step function

$$S_n(x) = \begin{cases} m_i, & x \in [x_{i-1}, x_i), i = i = 1, 2, \ldots, n-1 \\ m_n, & x \in [x_{n-1}, x_n]. \end{cases} \tag{5.22}$$

(5.21) becomes

$$\int_a^b (f(x) - S_n(x))dx < \frac{\epsilon}{2}.$$

Thus, f is approximated by a step function. The following problem provides a stronger version—that f can be approximated by a continuous function.

9. Let $f : [a, b] \to \mathbb{R}$ be Riemann integrable. Prove that for every $\epsilon > 0$, there exists a continuous function $\phi : [a, b] \to \mathbb{R}$ such that

$$\int_a^b |f(x) - \phi(x)|dx < \epsilon.$$

Proof. Since f is integrable, for every $\epsilon > 0$, there exists a partition $P = \{x_0, x_1, \ldots, x_n\}$ of $[a, b]$ such that

$$\int_a^b (f(x) - S_n(x))dx < \frac{\epsilon}{2}.$$

where $S_n(x)$ is given by (5.22). For each $S_n(x)$, define

$$\phi(x) = \begin{cases} m_i, & x \in [x_{i-1}, x_i - \delta] \\ m_i + \frac{m_{i+1} - m_i}{\delta}(x - x_i), & x \in (x_i - \delta, x_i], \ i = 1, 2, \ldots, n-1 \\ m_n, & x \in (x_{n-1}, x_n], \end{cases}$$

where $0 < \delta < \min\{\frac{\epsilon}{4nM}, \min_{1 \le i \le n} \Delta x_i\}$, $M = \sup_{x \in [a,b]} |f(x)|$. Clearly, $\phi(x)$ is continuous on $[a, b]$. Thus,

$$\begin{aligned} \int_a^b |f(x) - \phi(x)|dx &\le \int_a^b |f(x) - S_n(x)|dx + \int_a^b |S_n(x) - \phi(x)|dx \\ &\le \frac{\epsilon}{2} + \sum_{i=1}^{n-1} \int_{x_i-\delta}^{x_i} \left| m_i + \frac{m_{i+1} - m_i}{\delta}(x - x_i) \right| dx \\ &\le \frac{\epsilon}{2} + \sum_{i=1}^{n-1} 2M\delta < \frac{\epsilon}{2} + 2nM\delta < \frac{\epsilon}{2} + \frac{\epsilon}{2} = \epsilon. \end{aligned}$$

$\square$

Remark. By the Weierstrass approximation theorem in Chapter 6, which asserts that every continuous function on $[a, b]$ is a uniform limit of polynomials, the above result can be further strengthened to: For every $\epsilon > 0$, there exists a polynomial $P(x)$ such that $\int_a^b |f(x) - P(x)|dx < \epsilon$.

10. Let $f : [0, \pi] \to \mathbb{R}$ be Riemann integrable and $n \in \mathbb{N}$. Show that

$$\lim_{n \to \infty} \int_0^\pi f(x)|\sin nx| \, dx = \frac{2}{\pi} \int_0^\pi f(x) \, dx.$$

Proof. Let $P = \{x_0, x_1, \ldots, x_n\}$ be the equal partition of $[a, b]$. Then, for $1 \le i \le n$,

$$m_i \int_{\frac{i-1}{n}\pi}^{\frac{i}{n}\pi} |\sin nx| dx \le \int_{\frac{i-1}{n}\pi}^{\frac{i}{n}\pi} f(x) |\sin nx| dx \le M_i \int_{\frac{i-1}{n}\pi}^{\frac{i}{n}\pi} |\sin nx| dx,$$

where

$$m_i = \inf\{f(x) : x \in [(i-1)\pi/n, i\pi/n]\}, \quad M_i = \sup\{f(x) : x \in [(i-1)\pi/n, i\pi/n]\}.$$

Summing up these inequalities on i from 1 to n yields

$$\sum_{i=1}^n m_i \int_{\frac{i-1}{n}\pi}^{\frac{i}{n}\pi} |\sin nx| dx \le \int_0^\pi f(x) |\sin nx| dx \le \sum_{i=1}^n M_i \int_{\frac{i-1}{n}\pi}^{\frac{i}{n}\pi} |\sin nx| dx. \quad (5.23)$$

Since

$$\int_{\frac{i-1}{n}\pi}^{\frac{i}{n}\pi} |\sin nx| dx = \frac{2}{n},$$

it follows from (5.23) that

$$\frac{2}{\pi} \sum_{i=1}^n m_i \frac{\pi}{n} \le \int_0^\pi f(x) |\sin nx| dx \le \frac{2}{\pi} \sum_{i=1}^n M_i \frac{\pi}{n}. \quad (5.24)$$

The integrability of f implies that

$$\lim_{n \to \infty} \sum_{i=1}^n m_i \frac{\pi}{n} = \lim_{n \to \infty} \sum_{i=1}^n M_i \frac{\pi}{n} = \int_0^\pi f(x) dx.$$

The desired statement now follows from (5.24) and the squeeze theorem. $\square$

Remark. We assume that f is continuous on $[0, \pi]$, the statement can be proved by using the first mean value theorem for integrals.

11. Let $f : [-1, 1] \to \mathbb{R}$ be continuous. Show that

$$\lim_{h \to 0} \frac{1}{\pi} \int_{-1}^1 \frac{h}{h^2 + x^2} f(x) dx = f(0).$$

Proof. Since f is continuous, it follows that, for every $\epsilon > 0$, there exists a $\eta > 0$ such that

$$|f(x) - f(0)| < \frac{\epsilon}{4\pi} \quad \text{whenever } |x| < \eta.$$

Let $M = \max_{x \in [-1,1]} |f(x)|$. For such fixed η, there exists a $\delta > 0$, if $|h| < \delta$, we have

$$\left| \int_{-1}^{-\eta} \frac{h}{h^2 + x^2} f(x) dx \right| \le M \left| \int_{-1}^{-\eta} \frac{h}{h^2 + x^2} dx \right| < \frac{\epsilon}{4},$$

$$\left| \int_{\eta}^{1} \frac{h}{h^2 + x^2} f(x) dx \right| \le M \left| \int_{\eta}^{1} \frac{h}{h^2 + x^2} dx \right| < \frac{\epsilon}{4},$$

$$\left| 2 \arctan \frac{\eta}{h} - \pi \right| < \frac{\epsilon}{4M}.$$

Hence,

$$\left|\int_{-1}^{1}\frac{h}{h^2+x^2}f(x)dx - \pi f(0)\right| \leq \left|\int_{-1}^{-\eta}\frac{h}{h^2+x^2}f(x)dx\right| + \left|\int_{\eta}^{1}\frac{h}{h^2+x^2}f(x)dx\right|$$

$$+ \left|\int_{-\eta}^{\eta}\frac{h}{h^2+x^2}f(x)dx - \pi f(0)\right|$$

$$< \frac{\epsilon}{4} + \frac{\epsilon}{4} + \left|f(\xi_\eta)\int_{-\eta}^{\eta}\frac{h}{h^2+x^2}dx - \pi f(0)\right|$$

$$= \frac{\epsilon}{2} + \left|2f(\xi_\eta)\arctan\frac{\eta}{h} - \pi f(0)\right|$$

$$\leq \frac{\epsilon}{2} + 2|f(\xi_\eta) - f(0)|\arctan\frac{\eta}{h} + \left|2\arctan\frac{\eta}{h} - \pi\right||f(0)|$$

$$< \frac{\epsilon}{2} + \frac{\epsilon}{4} + \frac{\epsilon}{4} = \epsilon.$$

This proves the statement.   □

Remark. Here $g_h(x) := \frac{1}{\pi}\frac{h}{h^2+x^2}$ is called the *Lorentzian kernel*. It is singly peaked at $x = 0$ but with a total area of unity. Let

$$\delta(x) := \lim_{h\to 0} g_h(x).$$

It is often called *Dirac's delta function* and it has the fundamental property:

$$\int_{-\infty}^{\infty} f(x)\delta(x)\,dx = f(0)$$

for every function $f(x)$. The Delta function can also be defined as the following limits:

$$\delta(x) = \lim_{n\to\infty}\frac{1}{2\pi}\frac{\sin[(n+1/2)x]}{\sin(x/2)} = \lim_{\sigma\to 0}\frac{1}{\sqrt{2\pi}\sigma}e^{-x^2/2\sigma^2}.$$

The function in the first limit is called the *Fejér kernel*, which is used to express the effect of Cesàro summation on Fourier series. The function in the second limit is named the *Gaussian kernel* that is widely used in probability and statistics, differential equations, and other areas of mathematics.

12. Let $f : [0,1] \to \mathbb{R}$ be continuous. If $\lim_{x\to 0+}\frac{f(x)}{x}$ exists and is finite, show that

$$\lim_{n\to\infty} n\int_0^1 f(x^n)dx = \int_0^1\frac{f(x)}{x}.$$

Proof. Define a function $g : [0,1] \to \mathbb{R}$ by

$$g(x) = \begin{cases} \frac{f(x)}{x}, & x \in (0,1], \\ \lim_{t\to 0+}\frac{f(t)}{t} & x = 0. \end{cases}$$

Clearly, $g(x)$ is continuous on $[0,1]$. Let $G(x) = \int_0^x g(t)dt$. Using integration by parts, we have

$$n\int_0^1 f(x^n)dx = n\int_0^1 x^n g(x^n)dx = xG(x^n)|_0^1 - \int_0^1 G(x^n)dx$$

$$= G(1) - \int_0^1 G(x^n)dx = \int_0^1\frac{f(t)}{t}dt - \int_0^1 G(x^n)dx.$$

Now, it suffices to show that $\lim_{n\to\infty}\int_0^1 G(x^n)dx = 0$. To this end, let $M = \max_{x\in[0,1]}|G(x)|$. Observe that, for any $0 < a < 1$,

$$\left|\int_0^1 G(x^n)dx\right| \le \int_0^1 |G(x^n)|dx = \int_0^a |G(x^n)|dx + \int_a^1 |G(x^n)|dx. \qquad (5.25)$$

Applying the first mean value theorem of integrals yields

$$\left|\int_0^a G(x^n)dx\right| = \left|G(\xi_n^n)\int_0^a dx\right| \le a|G(\xi_n^n)|,$$

where $0 < \xi_n < a$. Thus, for every $\epsilon > 0$, choosing $a > 1 - \frac{\epsilon}{2M}$, we have

$$\lim_{n\to\infty} G(\xi_n^n) = G\left(\lim_{n\to\infty} \xi_n^n\right) = G(0) = 0.$$

This implies that there is a positive integer N such that $|G(\xi_n^n)| < \frac{\epsilon}{2}$ for all $n > N$. Thus, in view of (5.25), we have

$$\left|\int_0^1 G(x^n)dx\right| < \frac{\epsilon}{2} + (1 - a)M < \frac{\epsilon}{2} + \frac{\epsilon}{2} = \epsilon$$

and the conclusion follows. $\qquad\square$

13. Let $f : [0,1] \to \mathbb{R}$ be integrable. Show that if f is positive on $[0,1]$, then

$$\frac{3}{4}\left(\int_0^1 f(x)dx\right)^2 \le \frac{1}{16} + \int_0^1 f^3(x)dx.$$

Proof. Notice that

$$f^3 - \frac{3}{4}f^2 = \frac{1}{4}\left((2f - 1)^2(f + 1/4) - \frac{1}{4}\right).$$

Integrating the above identity on $[0,1]$ yields

$$\int_0^1 f^3(x)\,dx + \frac{1}{16} \ge \frac{3}{4}\int_0^1 f^2(x)\,dx.$$

The desired inequality now follows from the Cauchy-Schwarz inequality

$$\int_0^1 f^2(x)\,dx \cdot \int_0^1 1^2\,dx \ge \left(\int_0^1 f(x)\,dx\right)^2. \qquad\square$$

Remark. With the concave assumption, we can show that

$$\int_0^1 f^3(x)\,dx \le 2\left(\int_0^1 f(x)\,dx\right)^3.$$

The following problem presents another lower bound for $\int_0^1 f^3(x)\,dx$.

14. (**Monthly Problem 11819, 2015**). Let f be a continuous, nonnegative function on $[0,1]$. Show that

$$\int_0^1 f^3(x)\,dx \ge 4\left(\int_0^1 x^2 f(x)\,dx\right)\left(\int_0^1 xf^2(x)\,dx\right).$$

Proof. Since $\int_0^1 x^3 dx = 1/4$, the proposed inequality is equivalent to

$$\int_0^1 x^3\,dx \int_0^1 f^3(x)\,dx \geq \int_0^1 x^2 f(x)\,dx \int_0^1 xf^2(x)\,dx. \qquad (5.26)$$

To prove this inequality, for $0 \leq t \leq 1$, let

$$F(t) = \int_0^t x^3\,dx \int_0^t f^3(x)\,dx - \int_0^t x^2 f(x)\,dx \int_0^t xf^2(x)\,dx.$$

Then (5.26) becomes that $F(1) \geq 0$. To this end, applying the product rule, we find that

$$\begin{aligned}
F'(t) &= t^3 \int_0^t f^3(x)\,dx + f^3(t) \int_0^t x^3\,dx - t^2 f(t) \int_0^t xf^2(x)\,dx - tf^2(t) \int_0^t x^2 f(x)\,dx \\
&= \int_0^t t^2 f^2(x)(tf(x) - xf(t))\,dx + \int_0^t f^2(t)x^2(xf(t) - tf(x))\,dx \\
&= \int_0^t (t^2 f^2(x) - x^2 f^2(t))(tf(x) - xf(t))\,dx \\
&= \int_0^t (tf(x) + xf(t))(tf(x) - xf(t))^2\,dx.
\end{aligned}$$

This implies that $F'(t) \geq 0$ for all $0 \leq t \leq 1$. Hence $F(t)$ is increasing on $[0,1]$. In particular, we obtain that $F(1) \geq F(0) = 0$. $\qquad\square$

15. (**Monthly Problem 11780, 2014**). Let $f : [-1,1] \to \mathbb{R}$ be continuously differentiable up to the order of $2n + 2$. If $f(0) = f''(0) = \cdots = f^{(2n+2)}(0) = 0$, prove that
$$\frac{(4n+5)((2n+2)!)^2}{2} \left(\int_{-1}^1 f(x)dx\right)^2 \leq \int_{-1}^1 (f^{(2n+2)}(x))^2\,dx.$$

Proof. Applying Taylor's theorem with the integral remainder, we have

$$f(x) = T_{2n+1}(x) + R_{2n+1}(x),$$

where

$$T_{2n+1}(x) = \sum_{k=0}^{2n+1} \frac{f^{(k)}(0)}{i!} x^k \quad \text{and} \quad R_{2n+1}(x) = \int_0^x \frac{f^{(2n+2)}(t)}{(2n+1)!}(x-t)^{2n+1}dt.$$

Since $f(0) = f''(0) = \cdots = f^{(2n+2)}(0) = 0$, it implies that $T_{2n+1}(x)$ is an odd function and so
$$\int_{-1}^1 T_{2n+1}(x)dx = 0.$$

Therefore,

$$\int_{-1}^{1} f(x)dx = \int_{-1}^{1} R_{2n+1}(x)dx$$

$$= \int_{-1}^{0} \int_{x}^{0} \frac{f^{(2n+2)}(t)}{(2n+1)!}(t-x)^{2n+1}dt\,dx + \int_{0}^{1} \int_{0}^{x} \frac{f^{(2n+2)}(t)}{(2n+1)!}(x-t)^{2n+1}dt\,dx$$

$$= \int_{-1}^{0} \int_{-1}^{t} \frac{f^{(2n+2)}(t)}{(2n+1)!}(t-x)^{2n+1}dx\,dt + \int_{0}^{1} \int_{t}^{1} \frac{f^{(2n+2)}(t)}{(2n+1)!}(x-t)^{2n+1}dx\,dt$$

$$= \int_{-1}^{0} \frac{f^{(2n+2)}(t)}{(2n+1)!}\left[-\frac{(t-x)^{2n+2}}{2n+2}\right]\bigg|_{x=-1}^{t} dt + \int_{0}^{1} \frac{f^{(2n+2)}(t)}{(2n+1)!}\left[\frac{(x-t)^{2n+2}}{2n+2}\right]\bigg|_{x=t}^{1} dt$$

$$= \int_{-1}^{0} \frac{f^{(2n+2)}(t)}{(2n+2)!}(t+1)^{2n+2}\,dt + \int_{0}^{1} \frac{f^{(2n+2)}(t)}{(2n+2)!}(1-t)^{2n+2}dt$$

$$= \int_{-1}^{1} \frac{f^{(2n+2)}(t)}{(2n+2)!}\phi(t)\,dt,$$

where

$$\phi(t) = \begin{cases} (1+t)^{2n+2}, & t \in [-1,0), \\ (1-t)^{2n+2}, & t \in [0,1]. \end{cases}$$

Since

$$\int_{-1}^{1} \phi^2(t)\,dt = \int_{-1}^{0} (1+t)^{4n+4}\,dx + \int_{0}^{1} (1-t)^{4n+4}\,dx = \frac{2}{4n+5},$$

by the Cauchy-Schwarz inequality, we find that

$$\left(\int_{-1}^{1} f(x)\,dx\right)^2 \le \int_{-1}^{1} \phi^2(x)\,dx \cdot \int_{-1}^{1} \left[\frac{f^{(2n+2)}(x)}{(2n+2)!}\right]^2 dx$$

$$= \frac{2}{4n+5} \cdot \frac{1}{((2n+2)!)^2} \int_{-1}^{1} \left(f^{(2n+2)}(x)\right)^2 dx,$$

which is equivalent to the desired inequality. $\qquad\square$

16. Let f be continuous on $[0, \sqrt{a^2+b^2}]$. Show that

$$\int_{0}^{2\pi} f(a\cos\theta + b\sin\theta)\,d\theta = 2\int_{0}^{\pi} f(\sqrt{a^2+b^2}\cos t)\,dt.$$

Proof. Let

$$\cos\phi = \frac{a}{\sqrt{a^2+b^2}}, \quad \sin\phi = \frac{b}{\sqrt{a^2+b^2}}.$$

Then

$$a\cos\theta + b\sin\theta = \sqrt{a^2+b^2}(\cos\theta\cos\phi + \sin\theta\sin\phi) = \sqrt{a^2+b^2}\cos(\theta-\phi).$$

Since $\cos(\theta-\phi)$ is a periodic function with the period 2π, it follows that

$$\int_{0}^{2\pi} f(a\cos\theta+b\sin\theta)d\theta = \int_{\phi-\pi}^{\phi+\pi} f(\sqrt{a^2+b^2}\cos(\theta-\phi))d\theta$$

$$= \int_{-\pi}^{\pi} f(\sqrt{a^2+b^2}\cos t)dt \quad (\text{use } t = \theta - \phi)$$

$$= 2\int_{0}^{\pi} f(\sqrt{a^2+b^2}\cos t)dt.$$

$\qquad\square$

17. (**Gauss Formula**). Let $a > b > 0, a_1 = \frac{a+b}{2}$, and $b_1 = \sqrt{ab}$. Prove that

$$G(a,b) = \int_0^{\pi/2} \frac{d\phi}{\sqrt{a^2 \cos^2 \phi + b^2 \sin^2 \phi}} = \int_0^{\pi/2} \frac{d\theta}{\sqrt{a_1^2 \cos^2 \theta + b_1^2 \sin^2 \theta}}.$$

Proof. Let

$$\sin \phi = \frac{2a \sin \theta}{(a+b) + (a-b)\sin^2 \theta}. \tag{5.27}$$

Here ϕ varies from 0 to $\pi/2$ as θ changes from 0 to $\pi/2$. Differentiating (5.27) gives

$$\cos \phi \, d\phi = 2a \frac{(a+b) - (a-b)\sin^2 \theta}{[(a+b) + (a-b)\sin^2 \theta]^2} \cos \theta \, d\theta.$$

On the other hand,

$$\cos \phi = \sqrt{1 - \sin^2 \phi} = \frac{\sqrt{(a+b)^2 - (a-b)^2 \sin^2 \theta}}{(a+b) + (a-b)\sin^2 \theta} \cos \theta.$$

Thus,

$$d\phi = 2a \frac{(a+b) - (a-b)\sin^2 \theta}{(a+b) + (a-b)\sin^2 \theta} \frac{d\theta}{\sqrt{(a+b)^2 - (a-b)^2 \sin^2 \theta}}.$$

Since

$$\sqrt{a^2 \cos^2 \phi + b^2 \sin^2 \phi} = a \frac{(a+b) - (a-b)\sin^2 \theta}{(a+b) + (a-b)\sin^2 \theta},$$

it follows that

$$\frac{d\phi}{\sqrt{a^2 \cos^2 \phi + b^2 \sin^2 \phi}} = \frac{d\theta}{\sqrt{a_1^2 \cos^2 \theta + b_1^2 \sin^2 \theta}}. \tag{5.28}$$

Integrating both sides of (5.28) from 0 to $\pi/2$ arrives at the Gauss formula. □

Remark. The substitution (5.27) that Gauss used is remarkable. Gauss must have known something which lead him rationally along the path. But he didn't reveal any guidance of a proof which involves a great deal of algebra. Here we give an alternative proof with a simpler substitution. The idea is motivated by **Putnam Problem 1968-B4**: Let $f : \mathbb{R} \to \mathbb{R}$ be continuous. Show that

$$\int_{-\infty}^{\infty} f\left(x - \frac{1}{x}\right) dx = \int_{-\infty}^{\infty} f(x) \, dx.$$

First, using the substitution $u = b \tan \phi$, we represent G as

$$G(a,b) = \frac{1}{2} \int_{-\infty}^{\infty} \frac{du}{\sqrt{(u^2 + a^2)(u^2 + b^2)}}.$$

Let $u = \frac{1}{2}(x - ab/x)$. We have

$$du = \frac{1}{2x^2}(x^2 + ab)dx,$$

$$u^2 + a_1^2 = \frac{1}{4}\left(x - \frac{ab}{x}\right)^2 + \frac{1}{4}(a+b)^2$$

$$= \frac{1}{4x^2}(x^4 - 2abx^2 + a^2b^2 + a^2x^2 + 2abx^2 + b^2x^2)$$

$$= \frac{1}{4x^2}(x^2 + a^2)(x^2 + b^2),$$

$$u^2 + b_1^2 = \frac{1}{4}\left(x - \frac{ab}{x}\right)^2 + ab$$

$$= \frac{1}{4x^2}(x^4 - 2abx^2 + a^2b^2 + 4abx^2)$$

$$= \frac{1}{4x^2}(x^2 + ab)^2.$$

Consequently, similar to (5.28), we find another invariance:

$$\frac{du}{\sqrt{(u^2 + a_1^2)(u^2 + b_1^2)}} = \frac{2x}{\sqrt{(u^2 + a^2)(u^2 + b^2)}} \frac{2x}{x^2 + ab} \frac{x^2 + ab}{2x^2} dx$$

$$= \frac{2dx}{\sqrt{(x^2 + a^2)(x^2 + b^2)}}.$$

Integrating both sides yields that $G(a_1, b_1) = G(a, b)$ again.

The Gauss formula is significant. Its end result is very beautiful and striking. Let

$$a_n = \frac{a_{n-1} + b_{n-1}}{2}, b_n = \sqrt{a_{n-1}b_{n-1}}.$$

Repeatedly using the Gauss formula yields

$$G(a, b) = \int_0^{\pi/2} \frac{d\phi}{\sqrt{a_n^2 \cos^2 \phi + b_n^2 \sin^2 \phi}}.$$

Recall that

$$\lim_{n \to \infty} a_n = \lim_{n \to \infty} b_n := \mathrm{agm}(a, b),$$

which is called the *Gauss arithmetic-geometric mean*. The Gauss formula finally yields

$$G(a, b) = \int_0^{\pi/2} \frac{d\phi}{\sqrt{(\mathrm{agm}(a, b))^2 \cos^2 \phi + (\mathrm{agm}(a, b))^2 \sin^2 \phi}} = \frac{\pi}{2\,\mathrm{agm}(a, b)}.$$

18. Let $0 < \alpha, \beta < 1$. Evaluate

$$\int_{-1}^{1} \frac{dx}{\sqrt{1 - 2\alpha x + \alpha^2}\sqrt{1 - 2\beta x + \beta^2}}.$$

Solution. Let

$$ax^2 + bx + c = (1 - 2\alpha x + \alpha^2)(1 - 2\beta x + \beta^2). \tag{5.29}$$

Clearly, $a = 4\alpha\beta > 0$. Differentiating (5.29) yields

$$ax + \frac{b}{2} = -\alpha(1 - 2\beta x + \beta^2) - \beta(1 - 2\alpha x + \alpha^2).$$

Recall that

$$\int \frac{dx}{\sqrt{ax^2 + bx + c}} = \frac{1}{\sqrt{a}} \ln \left| ax + \frac{b}{2} + \sqrt{a}\sqrt{ax^2 + bx + c} \right| + C. \tag{5.30}$$

When $x = 1$, we have

$$\begin{aligned}
ax + \frac{b}{2} + \sqrt{a}\sqrt{ax^2 + bx + c} &= -\alpha(1 - \beta)^2 - \beta(1 - \alpha)^2 + 2\sqrt{\alpha\beta}(1 - \alpha)(1 - \beta) \\
&= -[\sqrt{\alpha}(1 - \beta) - \sqrt{\beta}(1 - \alpha)]^2 \\
&= -(\sqrt{\alpha} - \sqrt{\beta})^2 (1 + \sqrt{\alpha\beta})^2.
\end{aligned}$$

Similarly, when $x = -1$, we have

$$ax + \frac{b}{2} + \sqrt{a}\sqrt{ax^2 + bx + c} = -(\sqrt{\alpha} - \sqrt{\beta})^2 (1 - \sqrt{\alpha\beta})^2.$$

Using (5.30), we conclude that

$$\int_{-1}^{1} \frac{dx}{\sqrt{1 - 2\alpha x + \alpha^2}\sqrt{1 - 2\beta x + \beta^2}} = \frac{1}{\sqrt{\alpha\beta}} \ln \frac{1 + \sqrt{\alpha\beta}}{1 - \sqrt{\alpha\beta}}. \qquad \square$$

19. Find a nth degree polynomial $P_n(x)$ such that

$$\int_a^b P_n(x)Q(x)\,dx = 0$$

for any polynomial $Q(x)$ with the degree up to $n - 1$.

Solution. Observe that every nth degree polynomial $P_n(x)$ can be obtained as the nth derivative of a $(2n)$th degree polynomial $R(x)$, and $R(x)$ can be determined by integrating $P_n(x)$ n times. For each integration, in particular, we select the arbitrary integral constants such that

$$R(a) = R'(a) = \cdots = R^{(n-1)}(a) = 0. \tag{5.31}$$

We now transform the problem as follows: Find a $(2n)$th degree polynomial $R(x)$, subject to (5.31), such that

$$\int_a^b R^{(n)}(x)Q(x)\,dx = 0.$$

Thus, using integration by parts, we have

$$\begin{aligned}
\int_a^b R^{(n)}(x)Q(x)\,dx = [Q(x)R^{(n-1)}(x) - Q'(x)R^{(n-2)}(x) + \cdots \\
\pm Q^{(n-1)}(x)R(x)]\,\big|_a^b \mp \int_a^b Q^{(n)}(x)R(x)\,dx.
\end{aligned}$$

By (5.31) and $Q^{(n)}(x) = 0$, this leads to

$$Q(b)R^{(n-1)}(b) - Q'(b)R^{(n-2)}(b) + \cdots \pm Q^{(n-1)}(b)R(b) = 0. \qquad (5.32)$$

Since Q is arbitrary, this implies that $\{Q(b), Q'(b), \cdots, Q^{(n-1)}(b)\}$ are arbitrary too. Thus, (5.32) implies

$$R(b) = R'(b) = \cdots = R^{(n-1)}(b) = 0. \qquad (5.33)$$

By (5.31) and (5.33), we conclude that $R(x)$ has a and b as nth repeat roots. Therefore, we can choose $R(x) = c(x-a)^n(x-b)^n$ with $c \neq 0$ and so

$$P_n(x) = c\,\frac{d^n}{dx^n}[(x-a)^n(x-b)^n]$$

is one of the required nth degree polynomials. $\qquad \square$

Remark. Let $a = -1, b = 1$. We obtain the well-known *Legendre polynomials* (see Exercise 101 in Chapter 4):

$$P_n(x) = \frac{1}{2^n\,n!}\frac{d^n}{dx^n}[(x^2-1)^n],$$

which are frequently encountered in physics and other technical fields. Here the coefficient is *normalized* such that $P_n(1) = 1$. Moreover, $\{P_n(x)\}_{n=0}^{\infty}$ are orthogonal with respect to the L^2–inner product:

$$\int_{-1}^{1} P_n(x)P_m(x)\,dx = \frac{2}{2n+1}\delta_{nm},$$

where δ_{nm} is the *Kronecker delta*, which equals to 1 if $m = n$ and to 0 otherwise. An alternative derivation of the Legendre polynomials is by carrying out the *Gram–Schmidt process* on the polynomials $\{1, x, x^2, ...\}$ with respect to the L^2–inner product.

20. Let $f : [-1,1] \to \mathbb{R}$ be continuous. If $\int_0^1 f(\sin(x+t))dx = 0$ for all $t \in \mathbb{R}$, show that $f(x) = 0$ for all $x \in [-1,1]$.

Proof. Let $u = x + t$. Then the given integral becomes

$$\int_t^{t+1} f(\sin u)\,du = 0 \qquad (5.34)$$

for all $t \in \mathbb{R}$. Differentiating (5.34) yields $f(\sin(t+1)) = f(\sin t)$ for all $t \in \mathbb{R}$. In particular, for every $n \in \mathbb{N}$, we have

$$f(\sin n) = f(\sin(n-1)) = \cdots = f(0),$$

which indicates that $f(\sin n)$ is a constant for all $n \in \mathbb{N}$. Let $f(\sin n) = C$. Recall that the set $\{\sin n, n \in \mathbb{N}\}$ is dense in $[-1.1]$. The continuity of f implies that $f(x) = C$ for all $x \in [-1,1]$. Substituting f into (5.34) yields that $C = 0$. $\qquad \square$

21. Let f be differentiable on $[-1,1]$ and $M = \max_{-1 \le x \le 1}|f'(x)|$. If there exists $a \in (0,1)$ such that $\int_{-a}^{a} f(x)\,dx = 0$, prove that

$$\left|\int_{-1}^{1} f(x)\,dx\right| \le M(1-a^2).$$

Proof. By applying the first mean value theorem of integral, there are $\xi_1 \in [-a, 0]$ and $\xi_2 \in [0, a]$ such that

$$f(\xi_1) = \frac{1}{a} \int_{-a}^{0} f(x)\, dx,$$

$$f(\xi_2) = \frac{1}{a} \int_{0}^{a} f(x)\, dx,$$

respectively. On the other hand, the Lagrange mean value theorem implies that

$$f(x) - f(\xi_1) = f'(\eta_1)(x - \xi_1), \quad \text{for } x, \eta_1 \in (-1, -a),$$
$$f(x) - f(\xi_2) = f'(\eta_2)(x - \xi_2), \quad \text{for } x, \eta_2 \in (a, 1).$$

Using the assumption that $\int_{-a}^{a} f(x)\, dx = 0$, we have $f(\xi_1) + f(\xi_2) = 0$. Hence

$$\left| \int_{-1}^{1} f(x)\, dx \right| = \left| \int_{-1}^{-a} f(x)\, dx + \int_{a}^{1} f(x)\, dx \right|$$

$$= \left| \int_{-1}^{-a} f(x)\, dx - f(\xi_1)(-a + 1) - f(\xi_2)(1 - a) + \int_{a}^{1} f(x)\, dx \right|$$

$$= \left| \int_{-1}^{-a} (f(x) - f(\xi_1))\, dx + \int_{a}^{1} (f(x) - f(\xi_2))\, dx \right|$$

$$= \left| \int_{-1}^{-a} f'(\eta_1)(x - \xi_1)\, dx + \int_{a}^{1} f'(\eta_2)(x - \xi_2)\, dx \right|$$

$$\leq M \left(\int_{-1}^{a} (\xi_1 - x)\, dx + \int_{a}^{1} (x - \xi_2)\, dx \right)$$

$$= \frac{1}{2} M(2 - 2a^2 + 2(a - 1)(\xi_2 - \xi_1))$$

$$= M(1 - a^2 + (a - 1)(\xi_2 - \xi_1))$$

$$\leq M(1 - a^2).$$

$\square$

22. **(Putnam Problem 2007-B2).** Suppose that $f : [0, 1] \to \mathbb{R}$ has a continuous derivative and that $\int_{0}^{1} f(x)\, dx = 0$. Prove that for every $a \in (0, 1)$,

$$\left| \int_{0}^{a} f(x)\, dx \right| \leq \frac{1}{8} \max_{0 \leq x \leq 1} |f'(x)|.$$

Proof. Let $M = \max_{0 \leq x \leq 1} |f'(x)|$ and $F(x) = \int_{0}^{x} f(t)\, dt$. Since $F(x)$ is continuous on $[0, 1]$, and

$$F(0) = 0, F(1) = \int_{0}^{1} f(t)\, dt = 0,$$

by the extreme value theorem, there is $x_0 \in (0, 1)$ such that

$$\max_{0 \leq x \leq 1} |F(x)| = |F(x_0)| = \left| \int_{0}^{x_0} f(x)\, dx \right|.$$

Hence it suffices to prove that

$$|F(x_0)| \leq \frac{1}{8} M.$$

By Fermat's theorem, we also have $F'(x_0) = f(x_0) = 0$. To proceed, we represent $F(x_0)$ in two different forms. Using integration by parts, we first have

$$F(x_0) = tf(t)\big|_0^{x_0} - \int_0^{x_0} tf'(t)\, dt = -\int_0^{x_0} tf'(t)\, dt.$$

On the other hand, using the assumption that $\int_0^1 f(x)\, dx = 0$, then integrating by parts, we have

$$F(x_0) = \int_0^1 f(t)\, dt - \int_{x_0}^1 f(t)\, dt = -\int_{x_0}^1 f(t)\, dt$$

$$= f(t)(1-t)\big|_{x_0}^1 - \int_{x_0}^1 (1-t)f'(t)\, dt$$

$$= -\int_{x_0}^1 (1-t)f'(t)\, dt.$$

Using these two forms, respectively, we find that

$$|F(x_0)| = \left|\int_0^{x_0} tf'(t)\, dt\right| \le \int_0^{x_0} t|f'(t)|\, dt \le \int_0^{x_0} Mt\, dt = \frac{1}{2}Mx_0^2,$$

$$|F(x_0)| = \left|\int_{x_0}^1 (1-t)f'(t)\, dt\right| \le \int_{x_0}^1 (1-t)|f'(t)|\, dt \le \int_{x_0}^1 M(1-t)\, dt = \frac{1}{2}M(1-x_0)^2.$$

Therefore,

$$|F(x_0)| \le \frac{1}{2}M \min\{x_0^2, (1-x_0)^2\} \le \frac{1}{2}M \cdot \frac{1}{4} = \frac{1}{8}M. \qquad \square$$

23. **(Monthly Problem 11812, 2015)**. Let f be a twice continuously differentiable function from $[0,1]$ to $\mathbb{R}$. Let p be an integer greater than 1. Given that $\sum_{k=1}^{p-1} f(k/p) = -\frac{1}{2}(f(0) + f(1))$, prove that

$$\left(\int_0^1 f(x)\, dx\right)^2 \le \frac{1}{5!\, p^4} \int_0^1 (f''(x))^2\, dx.$$

Proof. For $1 \le k \le p$, integrating by parts twice gives

$$\int_{(k-1)/p}^{k/p} \left(t - \frac{k-1}{p}\right)\left(\frac{k}{p} - t\right) f''(x)\, dx = \frac{1}{p}\left(f\left(\frac{k-1}{p}\right) + f\left(\frac{k}{p}\right)\right) - 2\int_{(k-1)/p}^{k/p} f(x)\, dx.$$

Hence,

$$\int_{(k-1)/p}^{k/p} f(x)\, dx - \frac{1}{2p}\left(f\left(\frac{k-1}{p}\right) + f\left(\frac{k}{p}\right)\right)$$

$$= -\frac{1}{2}\int_{(k-1)/p}^{k/p} \left(t - \frac{k-1}{p}\right)\left(\frac{k}{p} - t\right) f''(x)\, dx.$$

Summing k from 1 to p yields

$$\int_0^1 f(x)\, dx - \frac{1}{2p}\sum_{k=1}^p \left(f\left(\frac{k-1}{p}\right) + f\left(\frac{k}{p}\right)\right)$$

$$= -\frac{1}{2}\sum_{k=1}^p \int_{(k-1)/p}^{k/p} \left(t - \frac{k-1}{p}\right)\left(\frac{k}{p} - t\right) f''(x)\, dx. \qquad (5.35)$$

By the assumption, we have

$$\sum_{k=1}^{p} \left(f\left(\frac{k-1}{p}\right) + f\left(\frac{k}{p}\right) \right) = f(0) + 2\sum_{k=1}^{p-1} f(k/p) + f(1) = 0.$$

This simplifies (5.35) as

$$\int_0^1 f(x)\,dx = -\frac{1}{2}\sum_{k=1}^{p}\int_{(k-1)/p}^{k/p} \left(t - \frac{k-1}{p}\right)\left(\frac{k}{p} - t\right) f''(x)\,dx. \qquad (5.36)$$

Applying the Cauchy-Schwarz inequality (the sequence form) yields

$$S := \left(\sum_{k=1}^{p}\int_{(k-1)/p}^{k/p} \left(t - \frac{k-1}{p}\right)\left(\frac{k}{p} - t\right) f''(x)\,dx \right)^2$$

$$\leq p \sum_{k=1}^{p} \left(\int_{(k-1)/p}^{k/p} \left(t - \frac{k-1}{p}\right)\left(\frac{k}{p} - t\right) f''(x)\,dx \right)^2.$$

Applying the Cauchy-Schwarz inequality (the integral from), we find that

$$S \leq p \sum_{k=1}^{p} \left(\int_{(k-1)/p}^{k/p} \left[\left(t - \frac{k-1}{p}\right)\left(\frac{k}{p} - t\right) \right]^2 dx \right) \left(\int_{(k-1)/p}^{k/p} (f''(x))^2\,dx. \right)$$

For $1 \leq k \leq p$, direct calculation gives

$$\int_{(k-1)/p}^{k/p} \left[\left(t - \frac{k-1}{p}\right)\left(\frac{k}{p} - t\right) \right]^2 dx = \frac{1}{30\,p^5}.$$

Thus,

$$S \leq \frac{1}{30\,p^4} \sum_{k=1}^{p} \int_{(k-1)/p}^{k/p} (f''(x))^2\,dx = \frac{1}{30\,p^4}\int_0^1 (f''(x))^2\,dx. \qquad (5.37)$$

Combining (5.36) and (5.37), we now arrive at

$$\left(\int_0^1 f(x)\,dx \right)^2 = \frac{1}{4} S \leq \frac{1}{5!\,p^4}\int_0^1 (f''(x))^2\,dx,$$

as desired. $\square$

Remark. The form of $\sum_{k=1}^{p-1} f(k/p)$ in the assumption also hints to us to use the Euler-Maclaurin formula:

$$\sum_{k=1}^{p-1} g(k) = \int_0^p g(x)\,dx + \frac{1}{2}(g(p) - g(0)) + \frac{1}{12}(g'(p) - g'(0)) - \frac{1}{2}\int_0^p B_2(\{x\})g''(x)\,dx$$

$$(5.38)$$

where $B_2(x) = x^2 - x + 1/6$ is the second Bernoulli polynomial and $\{x\} = x - \lfloor x \rfloor$ is the fraction part of x. This leads to a shorter proof. To this end, let $g(x) = f(x/p)$. By the assumption, (5.38) becomes

$$\int_0^1 f(x)\,dx = \frac{1}{2p^2} \int_0^1 B_2(\{nx\})f''(x)\,dx - \frac{1}{12p^2}(f'(1) - f'(0))$$

$$= \frac{1}{2p^2} \int_0^1 \left(B_2(\{nx\}) - \frac{1}{6} \right) f''(x)\,dx$$

$$= \frac{1}{2p^2} \int_0^1 (\{nx\}^2 - \{nx\})f''(x)\,dx.$$

Applying the Cauchy-Schwarz inequality and using

$$\int_0^1 (\{nx\}^2 - \{nx\})^2\,dx = \frac{1}{30},$$

we conclude the proposed inequality again.

24. (**Monthly Problem 12308, 2022**). What is the minimum value of $\int_0^1 (f'(x))^2\,dx$ over all continuously differentiable functions $f : [0,1] \to \mathbb{R}$ such that $\int_0^1 f(x)\,dx = \int_0^1 x^2 f(x)\,dx = 1$?

Proof. The minimum value is $105/2$. To this end, integrating by parts, then using the assumptions of f yields

$$\int_0^1 x(1 - x^2)f'(x)\,dx = x(1 - x^2)f(x)|_0^1 - \int_0^1 f(x)(1 - 3x^2)\,dx$$

$$= -\int_0^1 f(x)\,dx + 3 \int_0^1 x^2 f(x)\,dx = 2$$

Applying the Cauchy-Schwarz inequality, we have

$$4 = \left(\int_0^1 x(1 - x^2)f'(x)\,dx \right)^2 \le \left(\int_0^1 x^2(1 - x^2)^2\,dx \right) \left(\int_0^1 (f'(x))^2\,dx \right).$$

Since

$$\int_0^1 x^2(1 - x^2)^2\,dx = \frac{8}{105},$$

we conclude

$$\int_0^1 (f'(x))^2\,dx \ge \frac{105}{2}.$$

We now find a function $f(x)$ that satisfies the assumptions and attains the value of $105/2$. Let

$$f(x) = \alpha \left(\frac{x^2}{2} - \frac{x^4}{4} \right) + \beta,$$

where α and β are constants to be determined. Using

$$\int_0^1 f(x)\,dx = \frac{7}{60}\alpha + \beta = 1,$$

$$\int_0^1 x^2 f(x)\,dx = \frac{9}{140}\alpha + \frac{\beta}{3} = 1,$$

we find $\alpha = 105/4, \beta = -33/16$. Hence

$$f(x) = \frac{105}{4}\left(\frac{x^2}{2} - \frac{x^4}{4}\right) - \frac{33}{16} = \frac{210x^2 - 105x^4 - 33}{16}.$$

This gives

$$\int_0^1 (f'(x))^2 \, dx = \int_0^1 \left(\frac{105}{4}(x - x^3)\right)^2 dx = \frac{105}{2}. \qquad \Box$$

5.6   Exercises

When I was young I observed that nine out of every ten things I did were failures, so I did ten times more work. — George Shaw

1. Evaluate
$$\int_{-1}^1 \frac{dx}{(e^x + 1)(x^2 + 1)}.$$

2. Evaluate
$$\int_0^{2\pi} \sqrt{1 + \sin x} \, dx.$$

3. Let $f : [-a, a] \to \mathbb{R}$ be continuous. Prove that
$$\int_{-a}^a f(x) \, dx = \int_0^a (f(x) + f(-x)) \, dx.$$

   Use this formula to find
$$\int_{-\pi/2}^{\pi/2} \frac{dx}{1 + \sin x}.$$

4. For $n \in \mathbb{N}$, find
$$\int \frac{x^n}{1 + x + \frac{x^2}{2!} + \cdots + \frac{x^n}{n!}} \, dx.$$

5. For $n \in \mathbb{N}$, show that
$$\int_0^{\pi/2} \cos(2nx) \ln(\cos x) \, dx = \frac{(-1)^{n-1}\pi}{4n}.$$

6. Show that
$$F(x) = -\frac{1}{3} \arctan\left(\frac{3x(x^2 - 1)}{x^4 - 4x^2 + 1}\right)$$

   is an antiderivative of $\frac{x^4+1}{x^6+1}$. Then evaluate
$$\int_0^1 \frac{x^4 + 1}{x^6 + 1} \, dx.$$

7. Let f be a continuous function on $[0, 1]$. Show that
$$\int_0^{\pi/2} f(\sin^2 x) \cos x \, dx = \int_0^{\pi/2} f(\cos^2 x) \sin x \, dx.$$

8. Let $k \in \mathbb{N}$ and $P_k(x) = 1 + x + \cdots + x^{k-1}$. Show that

$$\int_0^1 \frac{P_k'(x)}{P_k(x)} x^{kn} \, dx = \ln k - (H_{kn} - H_n),$$

where H_n is the nth harmonic number.

9. Show that

$$\int_0^{1/\sqrt{2}} \frac{4\sqrt{2} - 8x^3 - 4\sqrt{2}x^4 - 8x^5}{1 - x^8} = \pi.$$

Remark. Computing the integral in terms of series yields

$$\pi = \sum_{n=0}^{\infty} \frac{1}{16^n} \left(\frac{4}{8n+1} - \frac{2}{8n+4} - \frac{1}{8n+5} - \frac{1}{8n+6} \right).$$

In 1996, based on this remarkable series, Bailey, Borwein, and Plouffe found an algorithm for computing individual digits of π. The series is now referred to as a BBP-type formula. Since then, a large number of BBP-type formulas for other mathematical constants have been discovered [10].

10. Consider the *lemniscate* defined by

$$x(t) = \frac{\cos t}{1 + \sin^2 t}, \quad y(t) = \frac{\sin t \cos t}{1 + \sin^2 t}, \quad 0 \le t \le 2\pi.$$

Let L be its arclength. Show that

$$2\pi = L \cdot \mathrm{agm}(1, \sqrt{2}),$$

where $\mathrm{agm}(a, b)$ is the Gauss arithmetic-geometric mean.

11. Let $I_n = \int_0^{\pi/2} \sin^n x \, dx$. Show that

$$I_{2n} = \frac{(2n-1)!!}{(2n)!!} \frac{\pi}{2}, \quad I_{2n+1} = \frac{(2n)!!}{(2n+1)!!}.$$

Then use $I_{2n+1} \le I_{2n} < I_{2n-1}$ to prove the Wallis's formula

$$\frac{\pi}{2} = \lim_{n \to \infty} \frac{1}{2n+1} \left(\frac{(2n)!!}{(2n-1)!!} \right)^2.$$

12. Let f be a decreasing continuous function on $[a, b]$. Show that

$$\int_a^b f(x) \, dx + \int_{f(a)}^{f(b)} f^{-1}(x) \, dx = bf(b) - af(a).$$

In particular, if $f(a) = b$, $f(b) = a$, then

$$\int_a^b f(x) \, dx = \int_a^b f^{-1}(x) \, dx.$$

13. Let f, g be integrable on $[a, b]$. Suppose that f is decreasing and $0 \le g(x) \le 1$, prove that

$$\int_{b-\alpha}^b f(x) \, dx \le \int_a^b f(x)g(x) \, dx \le \int_a^{a+\alpha} f(t) \, dt,$$

where $\alpha = \int_a^b g(x) \, dx$.

14. Let f be continuous and strictly increasing on $[a, b]$. By the first mean value theorem for integral, for each $n \in \mathbb{N}$, there is $x_n \in [a, b]$ such that

$$(f(x_n))^n = \frac{1}{b-a} \int_a^b (f(x))^n \, dx.$$

Show that $\lim_{n \to \infty} x_n = b$.

15. (**Monthly Problem 11961, 2017**). Evaluate

$$\int_0^{\pi/2} \frac{\sin x}{1 + \sqrt{\sin(2x)}} \, dx.$$

16. Let f be continuous on $[0, 1]$ and differentiable on $(0, 1)$. Suppose that $\int_{7/8}^1 f(x) \, dx = \frac{1}{8} f(0)$. Show that there exists a $c \in (0, 1)$ such that $f'(c) = 0$.

17. Let $J_n = \int_0^{\pi/2} \frac{\sin^2 nt}{\sin t} \, dt$ for all $n \in \mathbb{N}$. Determine

$$\lim_{n \to \infty} \frac{J_n}{\ln n}.$$

18. Determine the existence of the limit

$$\lim_{t \to 0} \left(\ln t + \int_0^1 \frac{dx}{(x^4 + t^4)^{1/4}} \right).$$

19. Let f be a differentiable function satisfying $f(1) = 1$ and

$$f'(x) = \frac{1}{x^2 + f^2(x)}.$$

Prove that $\lim_{x \to \infty} f(x)$ exists and is less that $1 + \pi/4$.

20. Let

$$f(x) = \begin{cases} \frac{1}{x} - \lfloor \frac{1}{x} \rfloor, & x \in (0, 1], \\ 0, & x = 0. \end{cases}$$

Show that f is integrable on $[0, 1]$.

21. (a) Find an example that f and g both are integrable on $[0, 1]$, but $(f \circ g)(x)$ is not integrable.

   (b) If f is integrable and g is continuous on $[0, 1]$, prove that $(f \circ g)(x)$ is integrable.

22. Let f be integrable on $[a, b]$ and $\int_a^b f(x) \, dx > 0$. Prove that there are subinterval $[c, d] \subset [a, b]$ and $\mu > 0$ such that $f(x) \geq \mu$ on $[c, d]$.

23. Let f be convex on $[a, b]$. Show that

$$f \left(\frac{a+b}{2} \right) \leq \frac{1}{b-a} \int_a^b f(x) \, dx \leq \frac{f(a) + f(b)}{2}.$$

24. (**Marcel Chirita**). Let $n \geq 2$ be an integer. Determine all continuous functions $f : [1, \infty) \to \mathbb{R}$ such that

$$\int_x^{x^n} f(t) \, dt = \int_1^x (t^{n-1} + t^{n-2} + \cdots + t) f(t) \, dt$$

for every $x \in [1, \infty)$.

25. Let $f : [0, 1] \to \mathbb{R}$ be convex and decreasing. Define

$$S_r(n) = \frac{1}{n} \sum_{k=1}^{n} f\left(\frac{k}{n}\right) \quad \text{and} \quad S_l(n) = \frac{1}{n} \sum_{k=0}^{n-1} f\left(\frac{k}{n}\right).$$

Show that $S_r(n)$ is increasing and $S_l(n)$ is decreasing as n increasing.

26. Let $f : [0, 1] \to \mathbb{R}$ be convex on $[0, c]$ $(0 < c < 1)$, concave on $[c, 1]$, and decreasing on $[0, 1]$. Define

$$S_r(n) = \frac{1}{n} \sum_{k=1}^{n} f\left(\frac{k}{n}\right) \quad \text{and} \quad S_l(n) = \frac{1}{n} \sum_{k=0}^{n-1} f\left(\frac{k}{n}\right).$$

Show that $S_r(n)$ is increasing and $S_l(n)$ is decreasing as n increasing.

27. Let $f : [0, 1] \to \mathbb{R}$ be decreasing. Define

$$S_r(n) = \frac{1}{n} \sum_{k=1}^{n} f\left(\frac{k}{n}\right).$$

If $(f(x) + f(1 - x))/2$ is concave, show that $S_r(n)$ is increasing as n increasing.

28. Let both $f''(x)$ and $f'''(x)$ be nonnegative on $[0, 1]$. Define

$$a_n = \frac{1}{n} \sum_{k=1}^{n} f\left(\frac{2k - 1}{2n}\right).$$

Show that a_n is increasing as n increasing.

29. Let $f : [a, b] \to \mathbb{R}$ be a function such that

$$\lim_{x \to x_0} f(x)$$

exists for every $x_0 \in [a, b]$. Show that f is Riemann integrable on $[a, b]$.

30. Let f be a nth degree polynomial such that $\int_0^1 x^k f(x)\, dx = 0$ for $k = 1, 2, \ldots, n$. Show that

$$\int_0^1 f^2(x)\, dx = (n + 1)^2 \left(\int_0^1 f(x)\, dx \right)^2.$$

31. Let $\alpha > 0, A > 0$, and $0 \le a < b$. Prove that

$$\left| \int_a^b \sin\left(nt - \frac{A}{t^\alpha}\right) dt \right| < \frac{2}{n}.$$

32. Let $\alpha > 0$. Prove that

$$\left| \int_\alpha^\infty \cos(t^2)\, dt \right| < \frac{1}{\alpha}.$$

33. (**van der Corput's inequality**). Let f be twice differentiable with $f''(x) \ge \lambda > 0$ on $[a, b]$. Show that

$$\left| \int_a^b e^{if(x)}\, dx \right| \le \frac{8}{\sqrt{\lambda}}.$$

34. Let f be a continuous function on $[0,1]$ such that $\int_0^1 f(x)\,dx = 0$. Show that there exists $c \in (0,1)$ such that

(a) $f(c) = \int_0^c f(x)\,dx$.

(b) $(1-c)f(c) = c\int_0^c f(x)\,dx$.

(c) $\int_0^c xf(x)\,dx = 0$.

(d) $cf(c) = f'(c)\int_0^c xf(x)\,dx$ if in addition assume that f is differentiable on $(0,1)$.

35. (**Monthly Problem 12046, 2018**). Suppose that $f : [0,1] \to \mathbb{R}$ has a continuous and nonnegative third derivative, and suppose $\int_0^1 f(x)\,dx = 0$. Prove

$$10\int_0^1 x^3 f(x)\,dx + 6\int_0^1 xf(x)\,dx \geq 15\int_0^1 x^2 f(x)\,dx.$$

36. Let $f : [a,b] \to \mathbb{R}$ be twice continuously differentiable function with $\int_a^b f(x)dx = 0$. Prove

$$\int_a^b (f''(x))^2\,dx \geq \frac{30(f(a)+f(b))^2}{(b-a)^3},$$

where 30 is the best possible constant.

37. (**Monthly Problem 11872, 2015**). Let f be a continuous function on $[0,1]$ such that $\int_0^1 f(x)\,dx = 0$. Prove that for all positive integers n there exists $c \in (0,1)$ such that

$$n\int_0^c x^n f(x)\,dx = c^{n+1} f(c).$$

38. (**Monthly Problem 11933, 2016**). For positive integer n, let $H_n = \sum_{k=1}^n 1/k$. Prove

$$\int_0^1 \frac{1}{x+1}\,dx \cdot \int_0^1 \frac{x+1}{x^2+x+1}\,dx \cdots \int_0^1 \frac{x^{n-2}+\cdots+x+1}{x^{n-1}+\cdots+x+1}\,dx \geq \frac{1}{H_n}.$$

39. (**Monthly Problem 12205, 2020**). Find the minimum value of

$$\frac{\int_0^1 x^2(f'(x))^2\,dx}{\int_0^1 x^2(f(x))^2\,dx}$$

over all nonzero continuously differentiable function $f : [0,1] \to \mathbb{R}$ with $f(1) = 0$.

40. (**Putnam 2006-B5**). For each continuous function $f : [0,1] \to \mathbb{R}$, let $I(f) = \int_0^1 x^2 f(x)dx$ and $J(x) = \int_0^1 x(f(x))^2 dx$. Find the maximum value of $I(f) - J(f)$ over all such functions f.

41. (**Monthly Problem 12318, 2022**). Let a be a positive real number, and let S_a be the set of functions $f : [-a,a] \to \mathbb{R}$ such that $\int_{-a}^a (f(x))^2 dx = 1$. Let $A(f) = \int_{-a}^a f(x)dx, B(f) = \int_{-a}^a xf(x)dx$, and $C(f) = \int_{-a}^a x^2 f(x)dx$.

(a) What is $\sup\{A(f)^2 + B(f)^2 : f \in S_a\}$?

(b) What is $\sup\{A(f)^2 + B(f)^2 + C(f)^2 : f \in S_a\}$?

42. Evaluate

$$(1) \quad \int_0^1 \frac{\ln(1+x)}{1+x^2}\,dx, \qquad (2) \quad \int_0^1 \frac{\arctan x}{1+x}\,dx.$$

43. Evaluate the *Poisson integral*

$$\int_0^\pi \ln(1 - 2x \cos\theta + x^2)\, d\theta.$$

44. Evaluate

$$\int_{\pi/4}^{\pi/2} \frac{1 + \sin x}{1 - \cos x} e^x\, dx.$$

45. Evaluate $\int_0^{\pi/2} \ln\sin x\, dx$. Then use this result to calculate

$$\int_0^\pi x \ln\sin x\, dx, \quad \int_0^{\pi/2} x \cot x\, dx, \quad \int_0^1 \frac{\ln x}{\sqrt{1 - x^2}}\, dx, \quad \text{and} \quad \int_0^1 \frac{\arcsin x}{x}\, dx.$$

46. Let $n \in \mathbb{N}$. Evaluate

(1) $\displaystyle\int_0^{\pi/2} \frac{\sin(2n - 1)x}{\sin x}\, dx$ (Dirichlet); (2) $\displaystyle\int_0^{\pi/2} \left(\frac{\sin nx}{\sin x}\right)^2 dx$ (Fejer).

47. Show that

$$\int_0^1 [(\ln(\sqrt{1 + x} - \sqrt{1 - x}))^2 - x]\, dx = \frac{1}{4}\ln^2 2 - \frac{1}{2}\ln 2 + \frac{\pi}{4} + G,$$

where G is the *Catalan constant* defined by

$$G = \sum_{n=0}^\infty \frac{(-1)^n}{(2n + 1)^2}.$$

48. (**Monthly Problem 12221, 2020**). Prove that

$$\int_0^1 \frac{\ln(1 + x^6)}{1 + x^2}\, dx = \frac{\pi}{2}\ln 6 - 3G,$$

where G is the *Catalan constant*.

Remark. Similarly, you should be able to show that

$$\int_0^\infty \frac{\ln(1 + x^6)}{1 + x^2}\, dx = \pi\ln 6.$$

49. (**Monthly Problem 11152, 2005**). Evaluate

$$\int_0^1 \frac{\ln(\cos(\pi x/2))}{x(1 + x)}\, dx.$$

50. (**Putnam Problem 1980-A3**). Evaluate

$$\int_0^{\pi/2} \frac{dx}{1 + \tan^{\sqrt{2}} x}.$$

51. (**Putnam Problem 2016-A3**). Suppose that f is a function from $\mathbb{R}$ to $\mathbb{R}$ such that

$$f(x) + f\left(1 - \frac{1}{x}\right) = \arctan x$$

for all real $x \neq 0$. Find

$$\int_0^1 f(x)\,dx.$$

52. (**Monthly Problem 11924, 2016**). Calculate

$$\int_0^{\pi/2} \frac{\{\tan x\}}{\tan x}\,dx,$$

where $\{u\}$ denotes $u - \lfloor u \rfloor$.

53. (**Monthly Problem 11457, 2009**). For real numbers a and b with $0 \leq a \leq b$, find

$$\int_a^b \arccos\left(\frac{x}{\sqrt{(a+b)x - ab}}\right)\,dx.$$

54. Let $f : [a, b] \to \mathbb{R}$ be continuous and let g be a periodic function with period T. Show that

$$\lim_{n\to\infty} \int_a^b f(t)g(nt)\,dt = \frac{1}{T} \int_a^b f(t)\,dt \cdot \int_0^T g(t)\,dt.$$

Remark. This is an extension of Problem 9 in Section 5.5.

55. Let f be a periodic function with period T. If f is integrable on $[0, T]$, show that

$$\lim_{x\to\infty} \frac{1}{x} \int_0^x f(t)dt = \frac{1}{T} \int_0^T f(t)dt.$$

56. Let f be a integrable periodic function. Show that

$$F(x) := \int_0^x f(t)dt$$

can be represented as the sum of one periodic function and a linear function.

57. Let $f_n(x)$ be continuous on $[0, 1]$ with $\int_0^1 f_n(x)\,dx = 1$ for each $n \in \mathbb{N}$. For every positive constant M, show that there exists a positive integer N and constants $c_i, (i = 1, 2, \ldots, N)$ such that

$$\sum_{i=1}^N c_i^2 = 1, \quad \max_{x\in[0,1]} \left|\sum_{i=1}^N c_i f_i(x)\right| > M.$$

Hint: Use the mean value theorem of integral.

58. Let f be continuously differentiable on $[a, b]$. If there are positive constants M and α such that

$$\left|\int_c^d f(x)\,dx\right| \leq M(d-c)^{1+\alpha} \text{ for each } [c, d] \subseteq [a, b],$$

show that $f(x) = 0$ on $[a, b]$.

59. Let $f : [0,1] \to \mathbb{R}$ be continuously differentiable. If $f(0) = 0$, $f(1) = 1$, show that

$$\int_0^1 |f(x) - f'(x)| \, dx \geq \frac{1}{e}.$$

60. Let $f : [a,b] \to \mathbb{R}$ be continuous. If

$$\int_0^\pi f(x) \cos x \, dx = 0, \quad \int_0^\pi f(x) \sin x \, dx = 0,$$

show that $f(x)$ has at least two distinct zeros in $(0, \pi)$.

61. Let $f : [a,b] \to \mathbb{R}$ be Riemann integrable. If

$$\int_a^b x^n f(x) \, dx = 0, \quad \text{for } n = 0, 1, 2, \ldots,$$

show that $f(x) = 0$ where x is a continuous point of f. In particular, if f is continuous, then $f(x) \equiv 0$.

62. Let $f : [0,1] \to \mathbb{R}$ be continuous. If

$$\int_0^1 x^k f(x) \, dx = 0, \quad \text{for } k = 0, 1, \cdots, n-1, \quad \int_0^1 x^n f(x) \, dx = 1,$$

show that there exists $c \in (0,1)$ such that $|f(c)| \geq 2^n(n+1)$.

63. Let $f : [a,b] \to \mathbb{R}$ be continuous. If for every continuous function $g(x)$ satisfying $\int_a^b g(x) dx = 0$,

$$\int_a^b f(x) g(x) dx = 0,$$

show that $f(x)$ is a constant on $[a,b]$.

64. Let f be continuous on $[-1,1]$. If for any even function $g : [-1,1] \to \mathbb{R}$ satisfying $\int_{-1}^1 f(x) g(x) \, dx = 0$, show that f is odd on $[-1,1]$.

65. Let $f : [0,1] \to \mathbb{R}$ be continuous. Show that

$$\lim_{n \to \infty} \int_0^1 x^n f(x) dx = 0 \quad \text{and} \quad \lim_{n \to \infty} n \int_0^1 x^n f(x) dx = f(1).$$

66. Let $f : [0,1] \to \mathbb{R}$ be continuous and let $g : [0,1] \to \mathbb{R}$ be continuously differentiable. Prove that

$$\lim_{n \to \infty} n \int_0^1 x^n f(x^n) g(x) \, dx = g(1) \int_0^1 f(x) \, dx.$$

67. Let $f : [0,1] \to \mathbb{R}$ be continuous. If $\lim_{x \to 0^+} \frac{f(x)}{x}$ exists and is finite, show that for any continuous differentiable function $g : [0,1] \to \mathbb{R}$,

$$\lim_{n \to \infty} n \int_0^1 g(x) f(x^n) dx = g(1) \int_0^1 \frac{f(x)}{x} \, dx.$$

Remark. A class of limits are direct consequences of this principle. For example,

$$\lim_{n \to \infty} n \int_0^1 f(x) \ln(1 + x^n) \, dx = \frac{\pi^2}{2} f(1), \quad \lim_{n \to \infty} n \int_0^1 \frac{x^{n-2}}{x^{2n} + x^n + 1} \, dx = \frac{\pi}{3\sqrt{3}}.$$

68. Let $f : [0,1] \to \mathbb{R}$ be continuous. Show that

$$\lim_{n\to\infty} \int_0^n \frac{f(x/n)}{1 + n^2 \cos^2 x} = \int_0^1 f(x)dx.$$

69. Let $f : \mathbb{R} \to \mathbb{R}$ be a continuous periodic function with period 1. Prove that, for irrational number α, we have

$$\lim_{n\to\infty} \frac{1}{n} (f(\alpha) + f(2\alpha) + \cdots + f(n\alpha)) = \int_0^1 f(x)\,dx.$$

70. Let $f : [0,1] \to \mathbb{R}$ be integrable. Prove that

$$\lim_{n\to\infty} \int_0^1 (1 - x^n)^n f(x)\,dx = \int_0^1 f(x)\,dx.$$

71. Let $f : [0,1] \to (0,\infty)$ be continuous. Find

$$\lim_{n\to\infty} \sqrt[n]{\int_0^1 (1 + x^n)^n f(x)\,dx}.$$

72. Let $f : [1,\infty] \to \mathbb{R}$ be a continuous function such that $\lim_{x\to\infty} xf(x)$ exists and is finite. Show that $\lim_{t\to\infty} \int_1^t \frac{f(x)}{x} dx$ exists and for any $a > 1$,

$$\lim_{n\to\infty} n \int_0^a f(x^n)dx = \lim_{t\to\infty} \int_1^t \frac{f(x)}{x}dx.$$

73. If f and g are nonnegative and integrable on $[a,b]$, prove that

$$\lim_{n\to\infty} \int_a^b \sqrt[n]{f^n(x) + g^n(x)}\,dx = \int_a^b \max\{f(x), g(x)\}\,dx.$$

74. Let f be a positive continuous function on $[0,1]$ with $f(0) = 1$. Let $g(x)$ be continuous on $[0,1]$. Prove that

$$\lim_{n\to\infty} n^2 \left(\int_0^1 \sqrt[n]{f(x^n)}g(x)\,dx - \int_0^1 g(x)\,dx \right) = g(1) \int_0^1 \frac{\ln f(x)}{x}dx.$$

75. **(Monthly Problem 11941, 2016).** Let $L = \lim_{n\to\infty} \int_0^1 \sqrt[n]{x^n + (1-x)^n}\,dx$.

    (a) Find L.

    (b) Find

    $$\lim_{n\to\infty} n^2 \left(\int_0^1 \sqrt[n]{x^n + (1-x)^n}\,dx - L \right).$$

76. Let $L = \lim_{n\to\infty} \int_0^{\pi/2} \sqrt[n]{\sin^n x + \cos^n x}\,dx$.

    (a) Find L.

    (b) Find

    $$\lim_{n\to\infty} n^2 \left(\int_0^{\pi/2} \sqrt[n]{\sin^n x + \cos^n x}\,dx - L \right).$$

77. (**Monthly Problem 11535, 2010**). Let f be continuously differential function on $[0,1]$. Let $A = f(1)$ and let $B = \int_0^1 x^{-1/2} f(x) dx$. Evaluate

$$\lim_{n\to\infty} n \left(\int_0^1 f(x) dx - \sum_{k=1}^{n} \left(\frac{k^2}{n^2} - \frac{(k-1)^2}{n^2} \right) f\left(\frac{(k-1)^2}{n^2} \right) \right)$$

in terms of A and B.

78. (**Frullani Integral**). Let $f : [0,1] \to \mathbb{R}$ be continuous and let $0 < a < b \le 1$. Show that

$$\int_0^1 \frac{f(ax) - f(bx)}{x} dx = f(0) \ln \frac{b}{a} - \int_a^b \frac{f(x)}{x} dx.$$

79. (**Monthly Problem 11548, 2011**). Let f be a twice differentiable real-value function with continuous second derivative, and suppose that $f(0) = 0$. Show that

$$\int_{-1}^{1} (f''(x))^2 dx \ge 10 \left(\int_{-1}^{1} f(x) dx \right)^2.$$

80. (**Monthly Problem 11517, 2010**). Let $f : [a,b] \to \mathbb{R}$ be three-times differentiable with $f(a) = f(b)$. Prove that

$$\left| \int_a^{(a+b)/2} f(x) dx - \int_{(a+b)/2}^b f(x) dx \right| \le \frac{(b-a)^4}{192} \sup_{x\in[a,b]} |f'''(x)|.$$

81. (**Monthly Problem 11417, 2009**). Let $f : [0,1] \to \mathbb{R}$ be continuously differentiable such that $\int_{1/3}^{2/3} f(x) dx = 0$. Show that

$$\int_0^1 (f'(x))^2 dx \ge 27 \left(\int_0^1 f(x) dx \right)^2.$$

82. (**Monthly Problem 11946, 2016**). Let $f : [0,1] \to \mathbb{R}$ be twice differentiable with f'' continuous on $[0,1]$ and $\int_{1/3}^{2/3} f(x) dx = 0$. Prove

$$4860 \left(\int_0^1 f(x) dx \right)^2 \le 11 \int_0^1 (f''(x))^2 dx.$$

83. (**Monthly Problem 12229, 2021**). Let $f : [0,1] \to \mathbb{R}$ be a function that has a continuous second derivate and that satisfies $f(0) = f(1)$ and $\int_0^1 f(x) dx = 0$. Prove

$$30240 \left(\int_0^1 x f(x) dx \right)^2 \le \int_0^1 (f''(x))^2 dx.$$

84. (**Monthly Problem 11133, 2005**). Let f be a nonnegative, continuous, concave function on $[0,1]$ with $f(0) = 1$. Prove that

$$2 \int_0^1 x^2 f(x) dx + \frac{1}{12} \le \left(\int_0^1 f(x) dx \right)^2.$$

85. (**Putnam Problem 1940-A3**). Find $f(x)$ such that

$$\int f^n(x)\, dx = \left(\int f(x)\, dx\right)^n,$$

when constant of integration are suitably chosen.

86. (**Putnam Problem 1968-A1**). Prove that

$$\frac{22}{7} - \pi = \int_0^1 \frac{x^4(1-x)^4}{1+x^2}\, dx.$$

87. (**Putnam Problem 1987-B1**). Evaluate

$$\int_2^4 \frac{\sqrt{\ln(9-x)}}{\sqrt{\ln(9-x)} + \sqrt{\ln(x+3)}}\, dx.$$

88. (**Putnam Problem 1991-A5**). For $0 \le y \le 1$, find the maximum value of

$$\int_0^y \sqrt{x^4 + (y-y^2)^2}\, dx.$$

89. (**Putnam Problem 2014-B2**). Suppose that $f : [1,3] \to \mathbb{R}$ is a function such that $-1 \le f(x) \le 1$ for all x and $\int_1^3 f(x)\, dx = 0$. How large can $\int_1^3 \frac{f(x)}{x}\, dx$ be?

90. Find all continuous functions $f : \mathbb{R} \to \mathbb{R}$ such that

$$n^2 \int_x^{x+1/n} f(t)\, dt = nf(x) + \frac{1}{2},$$

for all $x \in \mathbb{R}, n \in \mathbb{N}$.

91. Find all continuous function f on $(0, \infty)$ such that

$$\int_x^{x^2} f(t)\, dt = \int_1^x f(t)\, dt \quad \text{for all } x > 0.$$

92. Let $a \in [0,1]$. Find all continuous functions $f : [0,1] \to [0,\infty)$ such that

$$\int_0^1 f(x)\, dx = 1, \quad \int_0^1 xf(x)\, dx = a, \quad \int_0^1 x^2 f(x)\, dx = a^2.$$

93. (**Monthly Problem 11981, 2017**). Suppose that $f : [0,1] \to \mathbb{R}$ is differentiable function with continuous derivative and with

$$\int_0^1 f(x)\, dx = \int_0^1 xf(x)\, dx = 1.$$

Prove

$$\int_0^1 |f'(x)|^3\, dx \ge \left(\frac{128}{3\pi}\right)^2.$$

Hint. First show that $\int_0^1 x(1-x)f'(x)\, dx = 1$.

94. Let f be twice continuously differentiable and positive on $(0, 1)$. If $f(0) = f(1) = 0$, show that

$$\int_0^1 \left| \frac{f''(x)}{f(x)} \right| dx \geq 4.$$

Remark. An elegant solution due to Larson can be found in [63, pp. 238–239]. Similar to the solution of Problem 22 in this Chapter, you may give another proof. Hint: Let $f(x_0) = \max_{x \in [0,1]} f(x)$ for some $x_0 \in (0, 1)$. Then $f'(x_0) = 0$. Furthermore, show that

$$f(x_0) \leq x_0 \int_0^{x_0} |f''(x)| \, dx \quad \text{and} \quad f(x_0) \leq (1 - x_0) \int_{x_0}^1 |f''(x)| \, dx.$$

95. **(Monthly Problem 12193, 2020)**. Suppose that $f : [0, 1] \to \mathbb{R}$ has a continuous third derivative and $f(0) = f(1)$. Prove

$$\left| \int_0^1 f'(x) x^{k-1}(1 - x)^{k-1} \, dx \right| \leq \frac{(k-1)k!(k-1)!}{6(2k+1)!} \max_{0 \leq x \leq 1} |f'''(x)|,$$

where k is a positive integer.

96. Suppose that $f : [0, 1] \to \mathbb{R}$ is differentiable. If $a, b \in (0, 1), a < b$ such that

$$\int_0^a f(x) \, dx = \int_b^1 x f(x) \, dx = 0,$$

show that

$$\left| \int_0^1 f(x) \, dx \right| \leq \frac{1 - a + b}{4} M,$$

where $M = \max_{x \in (0,1)} |f'(x)|$.

97. Let f be twice continuously differentiable on $[a, b]$. Prove that there exists $\xi \in (a, b)$ such that

$$\int_a^b f(x) \, dx = (b - a) f \left(\frac{a + b}{2} \right) + \frac{(b - a)^3}{24} f''(\xi).$$

98. Let f be differentiable on $[a, b]$. If $f(a) = f(b) = 0$, prove that there is a $x_0 \in (a, b)$ such that

$$|f'(x_0)| \geq \frac{4}{(b - a)^2} \left| \int_a^b f(x) \, dx \right|.$$

In addition, if $f(x)$ is not identical zero on $[a, b]$, show that there is a $x_0 \in (a, b)$ such that

$$|f'(x_0)| > \frac{4}{(b - a)^2} \int_a^b |f(x)| \, dx.$$

99. Let f be differentiable on $[a, b]$. If $f((a+b)/2) = 0$, prove that there is a $x_0 \in (a, b)$ such that

$$|f'(x_0)| \geq \frac{4}{(b - a)^2} \int_a^b |f(x)| \, dx \geq \frac{4}{(b - a)^2} \left| \int_a^b f(x) \, dx \right|.$$

Find an example that there is no $x_0 \in (a, b)$ such that

$$|f'(x_0)| > \frac{4}{(b - a)^2} \int_a^b |f(x)| \, dx.$$

100. Let f be twice continuously differentiable on $[a, b]$. If $f(a) = f(b) = 0$, prove that

$$\left| \int_a^b f(x) \, dx \right| \leq \frac{(b-a)^3}{12} \max_{x \in [a,b]} |f''(x)|.$$

Hint: First show that

$$\int_a^b f(x) \, dx = \frac{1}{2} \int_a^b (x-a)(x-b) f''(x) \, dx.$$

101. Let $f : [a, b] \to \mathbb{R}$ be continuously differentiable. Show that

$$\max_{x \in [a,b]} |f(x)| \leq \frac{1}{b-a} \left| \int_a^b f(x) \, dx \right| + \int_a^b |f'(x)| \, dx.$$

102. Let $f : [0, 1] \to \mathbb{R}$ be a continuous function which satisfies $xf(y) + yf(x) \leq 1$ for all $x, y \in [0, 1]$. Show that

$$\int_0^1 f(x) \, dx \leq \frac{\pi}{4}.$$

Find a function satisfying the equality.

103. (**Niven's proof on irrationality of** π). For $a, b, n \in \mathbb{N}$, let $f(x) = \frac{1}{n!} x^n (a - bx)^n$. Prove that

   (a) $f(a/b - x) = f(x)$,
   (b) $f^{(k)}(x) \in \mathbb{Z}$ for all $0 \leq k \leq 2n$ and $x = 0, \frac{a}{b}$,
   (c) Assume that $\pi = \frac{a}{b}$, a and b are relatively prime. Show that

$$\int_0^\pi f(x) \sin x \, dx \in \mathbb{Z}.$$

   (d) Show that

$$1 \leq \int_0^\pi f(x) \sin x \, dx \leq \frac{(a\pi)^n}{n!} \pi < 1 \quad \text{(for sufficiently large } n).$$

104. Show that e is irrational.

105. Let f be continuously differentiable on $(0, 1)$. If $f(0) = f(1) = 0$, prove that there exists a positive constant M such that

$$\int_0^1 f^2(x) \, dx \leq M \int_0^1 (f'(x))^2 \, dx.$$

Try to determine the best possible constant M in the above inequality.

106. Let f be twice continuously differentiable on $(0, 1)$. If $f(0) = f(1) = f'(0) = 0$, $f'(1) = 1$, show that

$$\int_0^1 (f''(x))^2 \, dx \geq 4.$$

Find the condition for which function(s) the equality holds.

107. Let f be continuous and increasing on $[a, b]$. Show that

$$\int_a^b x f(x)\, dx \geq \frac{a+b}{2} \int_a^b f(x)\, dx.$$

108. Let f be integrable on $[0, 1]$. If $0 < m \leq f(x) \leq M$ for all $x \in [0, 1]$, show that

$$1 \leq \int_0^1 f(x)\, dx \cdot \int_0^1 \frac{dx}{f(x)} \leq \frac{(m+M)^2}{4mM}.$$

109. Let f be continuously differentiable on $[0, 2]$. If $f(0) = f(2) = 1$ and $|f'(x)| \leq 1$ for all $x \in [0, 2]$, show that

$$1 \leq \int_0^2 f(x)\, dx \leq 3.$$

110. Let f be continuous on $[0, 1]$ and $0 \leq f(x) < 1$ for all $x \in [0, 1]$. Show that

$$\int_0^1 \frac{f(x)}{1 - f(x)}\, dx \geq \frac{\int_0^1 f(x)\, dx}{1 - \int_0^1 f(x)\, dx}.$$

111. Show that there is no integrable function f that satisfies

$$\int_0^\pi (f(x) - \sin x)^2\, dx \leq \frac{3}{4} \quad \text{and} \quad \int_0^\pi (f(x) - \cos x)^2\, dx \leq \frac{3}{4}.$$

112. (**Bellman Inequality**). Let f, g, and x be nonnegative continuous function on $[t_0, t_1]$. If

$$x(t) \leq g(t) + \int_{t_0}^t f(s)x(s)\, ds, \quad t_0 \leq t \leq t_1,$$

prove that

$$x(t) \leq g(t) + \int_{t_0}^t f(s)g(s)e^{\int_s^t f(\xi)\, d\xi}\, ds, \quad t_0 \leq t \leq t_1.$$

113. Let f be continuously differentiable on $[a, b]$ and let g be convex on $[a, b]$ such that $f(a) = g(a)$, $f(b) = g(b)$, and $f(x) \leq g(x)$ for all $x \in (a, b)$. Prove that

$$\int_a^b \sqrt{1 + (f'(x))^2}\, dx \geq \int_a^b \sqrt{1 + (g'(x))^2}\, dx.$$

114. Prove that

$$\int_0^\pi e^{\sin x}\, dx > \pi\, e^{2/\pi}.$$

Find a sharper lower bound for this integral.

115. Let f be continuously differentiable on $[a, b]$ such that $f(a) = f(b) = 0$. Prove that

$$\int_a^b \frac{f^2(x)}{4d^2(x)}\, dx \leq \int_a^b (f'(x))^2\, dx,$$

where $d(x) = \min\{x - a, b - x\}$.

6

Sequences and Series of Functions

It is not once nor twice but times without number that the same ideas make their appearance in the world.

— Aristotle

All intelligent thoughts have already been thought; what is necessary is only to try to think them again.

— Johann Goethe

Since many important functions are defined by using infinite sequences or series, in this chapter, we extend the study of numerical sequences and series to the study of sequences and series of functions. Given a sequence or series of functions, it is often important to know whether or not certain desirable properties of these functions carry over to the limit function. In analysis parlance, we want to know if each function in a sequence or series of functions is continuous or differentiable or integrable, can we assume then that the limit function will be continuous or differentiable or integrable? The intuitive answer is "Yes, why not?" For example, for the case of integration, if f_n is very close to f, then its integral (the area under the curve) should be close to that of f. However, the Examples 6.1–6.4 in the following section show that these desirable properties are not inherited by the pointwise limit functions. Hence pointwise limits do not behave nicely enough for analysis. In a manner reminiscent of uniform continuity, we introduce Weierstrass's stronger form of convergence—uniform convergence, and illustrate its nature and significance.

6.1 Pointwise and Uniform Convergence

Not every soil can bear all things. — Virgil (19 B.C.)

Recall that, in Chapter 2, the convergence of infinite series is defined in terms of the associated sequence of the partial sums. Along the same lines, we initially study the behavior and properties of converging sequences of functions, the results from our study of sequences can be immediately applied to the series of functions via the n partial sums.

Let $f_n(x)$ be a sequence of functions each defined on an interval $I \subseteq \mathbb{R}$. Observe that $f_n(x_0)$ is a numerical sequence for each given $x_0 \in I$.

Definition 6.1. *For each $n \in \mathbb{N}$, the sequence of functions $f_n(x) : I \to \mathbb{R}$ converges pointwise to $f(x)$ on I if for each $x_0 \in I$ the numerical sequence $f_n(x_0)$ converges to $f(x_0)$.*

Thus, in $\epsilon - N$ language, for every $\epsilon > 0, x_0 \in I$, there exists an $N(x_0, \epsilon) \in \mathbb{N}$ such that

$$|f_n(x_0) - f(x_0)| < \epsilon \qquad \text{whenever } n > N.$$

Example 6.1. *Consider $f_n(x) = x^n$ on $[0,1]$ (see Figure 6.1). We have*

$$\lim_{n \to \infty} f_n(x) = f(x) = \begin{cases} 0, & \text{if } 0 \le x < 1, \\ 1, & \text{if } x = 1, \end{cases} \quad \text{pointwise.}$$

DOI: 10.1201/9781003304135-6

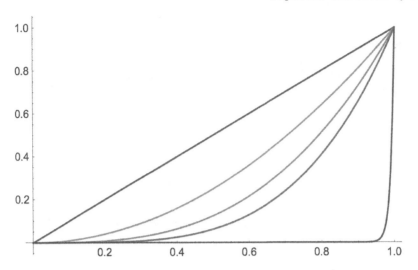

FIGURE 6.1
$f_n(x) = x^n$ for $n = 1, 2, 3, 4, 100$.

Let $x_0 \in (0, 1)$. For every $\epsilon > 0$, choose $N > (\ln \epsilon / \ln x_0)$ yields that, if $n > N$,

$$|f_n(x_0) - f(x_0)| = x_0^n < x_0^N < \epsilon.$$

The choice of N above depends on both ϵ and x_0. For example, let $x_0 = 1/2, \epsilon = 1/3$. To achieve

$$|f_n(1/2) - f(1/2)| < \frac{1}{3},$$

we have $N = 2$. But if $x_0 = 9/10, \epsilon = 1/3$,

$$|f_n(9/10) - f(9/10)| < \frac{1}{3}$$

holds only for $n > N = 10$. In general, there is no N such that $|f_n(x) - f(x)| < \epsilon$ when $n > N$ for every $x \in (0, 1)$. In fact, if $1 > x_n > 1/\sqrt[n]{3}$, we have $|f_n(x_n) - f(x_n)| = x_n^n \geq 1/3 = \epsilon$.

This example shows that the pointwise convergence does not necessarily carry over the continuity to the limit function.

The following generic Mathematica code enables us to animate the convergence:

```
Animate[Plot[x^n, {x, 0, 1},
  Prolog -> {Red, PointSize[0.01], Point[{1/n, 1/n^n}]},
  PlotRange -> {{0, 1}, {-0.1, 1}}], {n, 1, 15, 1, Appearance -> "Labeled"}]
```

Example 6.2. *Consider $f_n(x) = 2n^2 x e^{-n^2 x^2}$ on $[0, 1]$ (see Figure 6.2). We have*

$$\lim_{n \to \infty} f_n(x) = f(x) \equiv 0 \; pointwise.$$

Thus $\int_0^1 f(x) \, dx = 0$. On the other hand,

$$\lim_{n \to \infty} \int_0^1 f_n(x) \, dx = \lim_{n \to \infty} (1 - e^{-n^2}) = 1.$$

This example shows that, for pointwise convergence, in general,

$$\lim_{n \to \infty} \int_a^b f_n(x) \, dx \neq \int_a^b \lim_{n \to \infty} f_n(x) \, dx.$$

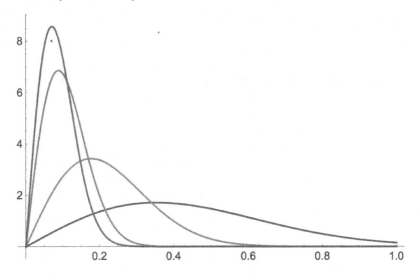

FIGURE 6.2
$f_n(x) = 2n^2 x e^{-n^2 x^2}$ on $[0,1]$ for $n = 2, 4, 8, 10$.

Example 6.3. *Consider $f_n(x) = x^{1+1/(2n-1)}$ on $[-1,1]$ (see Figure 6.3). We have*

$$\lim_{n \to \infty} f_n(x) = f(x) = |x| \ pointwise.$$

Since $|x|$ is not differentiable at $x = 0$, this example shows that the piontwise limit function does not inherit the differentiability from the approximating sequence.

Example 6.4. *Consider $f_n(x) = \frac{\sin nx}{\sqrt{n}}$ on $\mathbb{R}$. We have*

$$\lim_{n \to \infty} f_n(x) = f(x) \equiv 0 \ pointwise.$$

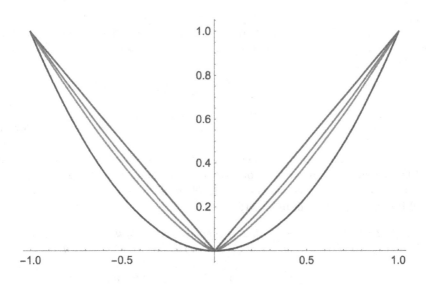

FIGURE 6.3
$f_n(x) = x^{1+1/(2n-1)}$ on $[-1,1]$ for $n = 2, 3, 4, \infty$.

So $f(x)$ is differentiable and $f'(x) = 0$. But,

$$f_n'(x) = \sqrt{n} \cos nx \not\to 0, \qquad x \in \mathbb{R}.$$

This example shows that even though the limit function exists and is differentiable,

$$\lim_{n \to \infty} f_n'(x) \neq \left(\lim_{n \to \infty} f_n(x) \right)' = f'(x).$$

Examples 1–4 convinces us that pointwise convergence does not interact well with the concepts of analysis. Let us first try to gain an insight into a difficulty we must overcome if we wish to transfer continuity from the individuals f_n to the limit function f. Assume that each $f_n(x)$ is continuous and $f_n(x) \to f(x)$ pointwise. To argue that f is continuous at x_0, let $\epsilon > 0$, we need to find a $\delta > 0$ such that

$$|f(x) - f(x_0)| < \epsilon \qquad \text{whenever } |x - x_0| < \delta.$$

By the typical "$\epsilon/3$ argument," the critical inequality becomes

$$|f(x) - f(x_0)| \leq |f(x) - f_n(x)| + |f_n(x) - f_n(x_0)| + |f_n(x_0) - f(x_0)| < \epsilon. \qquad (6.1)$$

The pointwise convergence asserts that, for $n > N$,

$$|f_n(x_0) - f(x_0)| < \frac{\epsilon}{3}.$$

Now for N fixed, the continuity of f_n implies that

$$|f_n(x) - f_n(x_0)| < \frac{\epsilon}{3} \qquad \text{whenever } |x - x_0| < \delta.$$

But here is the problem. We also need

$$|f_n(x) - f(x)| < \frac{\epsilon}{3} \qquad \text{whenever } |x - x_0| < \delta.$$

The values of x depend on δ, which depends on the choice of N. Thus, we cannot go back and choose a different N. As Example 6.1 indicates, the number N that works for x_0 may not work for $x \neq x_0$. To resolve this drawback, we would require that for each $\epsilon > 0$, there is an $N \in \mathbb{N}$ such that

$$|f_n(x) - f(x)| < \frac{\epsilon}{3} \qquad \text{for all } x \in I \text{ and } n > N.$$

Here N would depend on ϵ but not on x. Thus, the values f_n would be "uniformly" close to the values $f(x)$.

By 1841 Weierstrass recognized these phenomena exhibited in Examples 1–4 and first proposed the following stronger form of convergence—uniform convergence. As we will see in the next section, uniform convergence will allow us to transfer the key properties from individual functions to their limit function.

Definition 6.2. *For each $n \in \mathbb{N}$, the sequence of functions $f_n(x) : I \to \mathbb{R}$ converges uniformly to $f(x)$ on I if for every $\epsilon > 0$, there exists an $N \in \mathbb{N}$ such that*

$$|f_n(x) - f(x)| < \epsilon \qquad \text{for all } x \in I \text{ and for all } n > N.$$

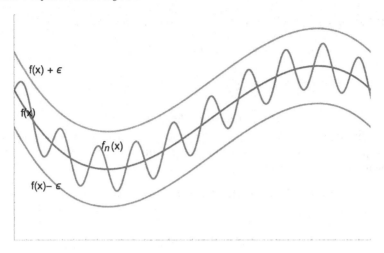

FIGURE 6.4
$f_n(x)$ is within a ϵ band of f for $n > N$.

Similar to the phrase "uniformly continuous" in Chapter 3, the term *uniformly* emphasize the fact that the response N to a prescribed ϵ can be chosen to work simultaneously for all $x \in I$, i.e., one N fits all x in the domain of the functions f_n. Geometrically, construct a band of radius ϵ around the limit function f, if $f_n \to f$ uniformly, then for every $x \in I$, there exists an $N \in \mathbb{N}$ such that $f_n(x)$ is entirely contained in this ϵ band whenever $n > N$ (see Figure 6.4).

Return to Example 6.1, in view of Figure 6.1, no matter how large n is, $f_n(x) = x^n$ never completely falls within the band $0 < y < \epsilon < 1$. So x^n does not converge to 0 uniformly.

Let $f_n(x) \to f(x)$ pointwise. Define

$$\omega_n = \sup_{x \in I} |f_n(x) - f(x)|.$$

This offers a typical procedure to test the uniform convergence for a large class of sequences.

Theorem 6.1. $f_n \to f$ *uniformly on I if and only if*

$$\lim_{n \to \infty} \omega_n = 0.$$

Proof. "$\Longrightarrow$" Suppose that $f_n(x) \to f(x)$ uniformly on I. Then for every $\epsilon > 0$ there is an $N = N(\epsilon) \in \mathbb{N}$ such that

$$|f_n(x) - f(x)| < \epsilon \quad \text{for all } x \in I \text{ whenever } n > N.$$

Thus

$$\omega_n = \sup_{x \in I} |f_n(x) - f(x)| \le \epsilon.$$

This shows that $\lim_{n \to \infty} \omega_n = 0$.

"$\Longleftarrow$" Suppose that $\lim_{n \to \infty} \omega_n = 0$. Then for every $\epsilon > 0$ there exists an $N = N(\epsilon) \in \mathbb{N}$ such that

$$\omega_n = \sup_{x \in I} |f_n(x) - f(x)| < \epsilon \quad \text{whenever } n > N.$$

Since

$$|f_n(x) - f(x)| \le \omega_n < \epsilon \quad \text{for all } x \in I \text{ whenever } n > N,$$

it follows that $f_n(x) \to f(x)$ uniformly on I by the definition. $\qquad \square$

Example 6.5. *Does $f_n(x) = 2n^2xe^{-n^2x^2}$ converges uniformly on $[0,1]$?*

In Example 6.2, we already knew that $f_n(x) \to 0$ pointwise. Since

$$\omega_n = \sup_{x \in [0,1]} 2n^2xe^{-n^2x^2} \geq f_n(1/n) = 2ne^{-1} \not\to 0,$$

it follows that the convergence is not uniform from Theorem 6.1.

When the limit function $f(x)$ cannot be found explicitly, which is especially apparent when f_n is a partial sum of a series of functions, uniform convergence can often be tested by the following Cauchy criterion.

Theorem 6.2 (Cauchy Criterion). *A sequence of functions $f_n(x) : I \to \mathbb{R}$ converges uniformly on I if and only if for every $\epsilon > 0$ there exists an $N \in \mathbb{N}$ such that*

$$|f_n(x) - f_m(x)| < \epsilon$$

for all $m, n \geq N$ and for all $x \in I$.

Proof. "$\Rightarrow$" Suppose that $f_n(x) \to f(x)$ uniformly on I. Then for every $\epsilon > 0$ there is an $N \in \mathbb{N}$ such that $|f_n(x) - f(x)| < \epsilon/2$ for all $x \in I$ whenever $n > N$. Therefore, if $m, n > N$, we have

$$|f_n(x) - f_m(x)| \leq |f_n(x) - f(x)| + |f(x) - f_m(x)| < \frac{\epsilon}{2} + \frac{\epsilon}{2} = \epsilon$$

for all $x \in I$.

"$\Leftarrow$" Suppose that given every $\epsilon > 0$ there exists an $N \in \mathbb{N}$ such that $|f_n(x) - f_m(x)| < \epsilon$ for all $m, n \geq N$ and for all $x \in I$. Then for each $x_0 \in I$, $f_n(x_0)$ is a numerical Cauchy sequence and hence converges. Let its limit be $f(x_0)$. In this way a function $f(x)$ is characterized by domain I and $f_n(x) \to f(x)$ pointwise. We have to prove the convergence is uniform. Let $\epsilon > 0$. There is an $N \in \mathbb{N}$ such that $|f_n(x) - f_m(x)| < \epsilon/2$ for every $x \in I$ and for all $n, m > N$. For each $x \in I$ there is an $N_x > N$ for which $|f_{N_x}(x) - f(x)| < \epsilon/2$. Thus, if $n > N$, we have

$$|f_n(x) - f(x)| \leq |f_n(x) - f_{N_x}(x)| + |f_{N_x}(x) - f(x)| < \frac{\epsilon}{2} + \frac{\epsilon}{2} = \epsilon$$

for all $x \in I$. Therefore $f_n(x)$ converges uniformly to $f(x)$ on I. $\square$

Next we study the convergence of a series of functions by looking at the sequence of its partial sums. The pointwise and uniform convergence of a series are defined as follows.

Definition 6.3. *The series $\sum_{n=1}^{\infty} f_n(x)$ converges pointwise to $f(x)$ on I if the numerical series $\sum_{n=1}^{\infty} f_n(x_0)$ converges to $f(x_0)$ for each $x_0 \in I$.*

Remark. It is clear that the series $\sum_{n=1}^{\infty} f_n(x) \to f(x)$ pointwise $\Longleftrightarrow$ the sequence $S_n(x) = \sum_{k=1}^{n} f_k(x) \to f(x)$ pointwise.

Definition 6.4. *The series $\sum_{n=1}^{\infty} f_n(x)$ converges uniformly to $f(x)$ on I if the sequence $S_n(x) = \sum_{k=1}^{n} f_k(x)$ converges uniformly to $f(x)$ on I.*

Using the Cauchy criterion (Theorem 6.2), we have the following important applicable test for uniform convergence.

Theorem 6.3 (Weierstrass M-test). *Suppose that $|f_n(x)| \leq M_n$ for all $x \in I$ and for every $n \in \mathbb{N}$. If $\sum_{n=1}^{\infty} M_n$ converges, then $\sum_{n=1}^{\infty} f_n(x)$ converges uniformly on I.*

Proof. If $\sum_{n=1}^{\infty} M_n$ converges, then, for every $\epsilon > 0$, there is an $N \in \mathbb{N}$ such that $\sum_{k=m+1}^{n} M_k < \epsilon$ whenever $n > m > N$. Therefore, for $n > m > N$ and for all $x \in I$, we have

$$|S_n(x) - S_m(x)| = \left| \sum_{k=m+1}^{n} f_k(x) \right| \le \sum_{k=m+1}^{n} M_k < \epsilon.$$

Uniform convergence now follows from the Cauchy criterion. $\qquad\square$

Clearly, the series on which the Weierstrass M-test is applicable must converge absolutely. However, there does exist series which converges uniformly but not absolutely.

Example 6.6. *Consider the series $\sum_{n=1}^{\infty} \frac{(-1)^{n-1}}{x^2+n}$ on $\mathbb{R}$.*

By the Alternating series test (Theorem 2.8), this series converges pointwise on $\mathbb{R}$. Appealing to the fact that an alternating series remainder $|R_n| \le a_{n+1}$, we have

$$|R_n(x)| = \left| \sum_{k=n+1}^{\infty} \frac{(-1)^{k-1}}{x^2+k} \right| \le \frac{1}{x^2+n+1} \le \frac{1}{n+1}.$$

So the series indeed converges uniformly by the Cauchy criterion. Comparing with the harmonic series, we see that $\sum_{n=1}^{\infty} \frac{1}{x^2+n}$ diverges. Thus, $\sum_{n=1}^{\infty} \frac{(-1)^{n-1}}{x^2+n}$ does not converge absolutely.

The following example shows that a series $\sum_{n=1}^{\infty} a_n(x)$ itself converges absolutely and uniformly, but $\sum_{n=1}^{\infty} |a_n(x)|$ does not converge uniformly.

Example 6.7. *Consider the series $\sum_{n=1}^{\infty} \frac{(-1)^{n-1}x^2}{(1+x^2)^n}$ on $\mathbb{R}$.*

If $x \neq 0$, we find that the alternating series remainder

$$|R_n(x)| = \left| \sum_{k=n+1}^{\infty} \frac{(-1)^{k-1}x^2}{(1+x^2)^k} \right| \le \frac{x^2}{(1+x^2)^{n+1}} = \frac{x^2}{1+(n+1)x^2+\cdots+x^{2(n+1)}} \le \frac{1}{n+1}.$$

So the series converges uniformly on $\mathbb{R}$. Since

$$\sum_{n=1}^{\infty} \left| \frac{(-1)^{n-1}x^2}{(1+x^2)^n} \right| = \sum_{n=1}^{\infty} \frac{x^2}{(1+x^2)^n}$$

is a geometric series with the common ratio $1/(1+x^2) < 1$, it follows that the series converges absolutely. However, the remainder of the series $\sum_{n=1}^{\infty} \frac{x^2}{(1+x^2)^n}$ is given by

$$R_n(x) = \sum_{k=n+1}^{\infty} \frac{x^2}{(1+x^2)^k} = \frac{x^2/(1+x^2)^{n+1}}{1-1/(1+x^2)} = \frac{1}{(1+x^2)^n}.$$

For fixed n, as $x \to 0^+$, $R_n(x) \to 1$. Therefore, $\sum_{n=1}^{\infty} \frac{x^2}{(1+x^2)^n}$ does not converge uniformly.

When the Weierstrass M-test is not applicable, for example, the series is not absolutely convergent, which occurs in Fourier series frequently (see Example 2.14), we often turn to the Able's and Dirichlet's tests for uniform convergence. They are an extension of the corresponding numerical tests Theorems 2.25 and 2.26 by replacing these bounds with the *uniform bounds* (independent of x) and the convergence with uniform convergence. To be specific, in Abel's test (Theorem 2.25), the condition that $|a_{m+1} + \cdots + a_n| < \epsilon$ is replaced by the requirement that

$$|a_{m+1}(x) + \cdots + a_n(x)| < \epsilon \qquad \text{for all } x \in I.$$

This can be asserted by uniform convergence of $\sum_{n=1}^{\infty} a_n(x)$ on I. Thus we have

Theorem 6.4 (Abel's Test for Uniform Convergence). *Assume $\sum_{n=1}^{\infty} a_n(x)$ converges uniformly on I. If the sequence $b_n(x)$ is bounded and monotone, then*

$$\sum_{n=1}^{\infty} a_n(x)b_n(x) = a_1(x)b_1(x) + a_2(x)b_(x) + \cdots + a_n(x)b_n(x) + \cdots$$

converges uniformly on I.

Similarly, we have

Theorem 6.5 (Dirichlet's Test for Uniform Convergence). *Assume the partial sum $\sum_{k=1}^{n} a_n(x)$ is uniform bounded on I. If the sequence $b_n(x)$ is decreasing and converges to 0 uniformly on I, then*

$$\sum_{n=1}^{\infty} a_n(x)b_n(x) = a_1(x)b_1(x) + a_2(x)b_(x) + \cdots + a_n(x)b_n(x) + \cdots$$

converges uniformly on I.

6.2   Importance of Uniform Convergence

Cauchy ultimately recognized the need for uniform convergence in order to assert the continuity of the the sum of a series of continuous functions but even he at that time did not see the error in his use of term-by-term integration of series. — Morris Kline

Equipped with the uniform convergence, we now prove three main theorems which demonstrate the compatibility of uniform convergence with the concepts of analysis.

Theorem 6.6. *If $f_n : I \to \mathbb{R}$ is continuous and f_n converges uniformly to $f(x)$ on I, then $f(x)$ is continuous on I.*

The key step of the proof is displayed in the critical inequality (6.1). We leave the details to the reader. Applying this theorem to the sequence of the partial sums of a series, we have

Corollary 6.1. *If $f_n : I \to \mathbb{R}$ is continuous and $\sum_{n=1}^{\infty} f_n$ converges uniformly to $f(x)$ on I, then $f(x)$ is continuous on I.*

The following consequence provides a simple test for nonuniform convergence.

Corollary 6.2. *Let $f_n : I \to \mathbb{R}$ be continuous, and let $f_n(x) \to f(x)$ pointwise on I. If $f(x)$ is not continuous on I, then $f_n \to f$ is not uniform.*

Let $\{r_1, r_2, \ldots\}$ be ordering rational numbers in $[0, 1]$. Define

$$f_n(x) = \begin{cases} 1, & \text{if } x \in \{r_1, r_2, \ldots, r_n\}, \\ 0, & \text{otherwise.} \end{cases}$$

The set of points of discontinuity of $f_n(x)$ is $\{r_1, r_2, \ldots, r_n\}$ and so f_n is Riemann integrable for every n. Clearly, $f_n(x)$ converges to the Dirichlet function $D(x)$ pointwise. It is well-known that $D(x)$ is not Riemann integrable. Furthermore, Example 6.2 indicates that even if the pointwise limit of a sequence of Riemann integrable functions is Riemann integrable, it may happen that the integral of the limit function is different from the limit of the sequence of integrals. The following theorem shows that uniform convergence is sufficient to resolve this dilemma.

Theorem 6.7. *If $f_n : [a,b] \to \mathbb{R}$ is Riemann integrable and f_n converges uniformly to $f(x)$ on $[a,b]$, then $f(x)$ is Riemann integrable on $[a,b]$, and*

$$\lim_{n\to\infty} \int_a^b f_n(x)\, dx = \int_a^b \lim_{n\to\infty} f_n(x)\, dx = \int_a^b f(x)\, dx. \tag{6.2}$$

Proof. We first show that f is Riemann integrable on $[a,b]$. For every $\epsilon > 0$, the uniform convergence implies that there is an $N \in \mathbb{N}$ such that

$$|f_n(x) - f(x)| < \frac{\epsilon}{3(b-a)} \qquad \text{for all } x \in [a,b] \text{ whenever } n > N. \tag{6.3}$$

Since f_n is integrable, there exists a partition $P = \{x_0, x_1, \ldots, x_n\}$ of $[a,b]$ for which

$$U(P, f_n) - L(P, f_n) < \frac{\epsilon}{3}.$$

Let

$$M_i = \sup_{x\in[x_{i-1},x_i]} f(x), \quad m_i = \inf_{x\in[x_{i-1},x_i]} f(x);$$

$$\overline{M}_i = \sup_{x\in[x_{i-1},x_i]} f_n(x), \quad \overline{m}_i = \inf_{x\in[x_{i-1},x_i]} f_n(x).$$

By (6.3), we have

$$\overline{m}_i - \frac{\epsilon}{3(b-a)} \le m_i \le M_i \le \overline{M}_i + \frac{\epsilon}{3(b-a)}.$$

Hence

$$U(P,f) - L(P,f) = \sum_{i=1}^n (M_i - m_i)\Delta x_i$$

$$\le \sum_{i=1}^n \left(\overline{M}_i + \frac{\epsilon}{3(b-a)}\right)\Delta x_i - \sum_{i=1}^n \left(\overline{m}_i - \frac{\epsilon}{3(b-a)}\right)\Delta x_i$$

$$= \sum_{i=1}^n (\overline{M}_i - \overline{m}_i)\Delta x_i + \frac{\epsilon}{3} + \frac{\epsilon}{3}$$

$$= U(P, f_n) - L(P, f_n) + \frac{2\epsilon}{3} < \epsilon.$$

This implies that $f(x)$ is Riemann integrable on $[a,b]$ by the Darboux Theorem.

Next, we prove (6.2). For every $\epsilon > 0$, there is an $N \in \mathbb{N}$ such that

$$|f_n(x) - f(x)| < \frac{\epsilon}{b-a} \qquad \text{for all } x \in [a,b] \text{ whenever } n > N.$$

Thus, for all $n > N$, we have

$$\left| \int_a^b f_n(x)\, dx - \int_a^b f(x)\, dx \right| = \left| \int_a^b (f_n(x) - f(x))\, dx \right|$$

$$\le \int_a^b |f_n(x) - f(x)|\, dx \le \int_a^b \frac{\epsilon}{b-a}\, dx = \epsilon. \qquad \square$$

The integrability of f can also be asserted by the Lebesgue theorem. Let $D(f)$ and $D(f_n)$ be the set of discontinuities of f and f_n on $[a,b]$, respectively. By the assumption that f_n

is Riemann integrable, Lebesgue theorem implies that $D(f_n)$ has measure zero. Moreover, since $f_n \to f$ uniformly, there is an $N \in \mathbb{N}$ such that

$$|f_N(x) - f(x)| < 1 \quad \text{for all } x \in [a, b].$$

In view of the boundedness of f_N, this implies that $f(x)$ is bounded. If $x_0 \in D(f)$, then x_0 must be a discontinuity point of some f_n. Otherwise, if f_n is continuous at x_0 for each n, by Theorem 6.6, f is also continuous at x_0, which contradicts the assumption that $x_0 \in D(f)$. Therefore, we have

$$D(f) \subseteq \cup_{n=1}^{\infty} D(f_n).$$

Since $D(f_n)$ has measure zero, it follows that $\cup_{n=1}^{\infty} D(f_n)$ has measure zero and so $D(f)$ has measure zero. This concludes that f is Riemann integrable by the Lebesgue theorem.

Applying Theorem 6.7 to the sequence of the partial sums of a series, we have

Corollary 6.3. *If $f_n : [a.b] \to \mathbb{R}$ is Riemann integrable and $\sum_{n=1}^{\infty} f_n$ converges uniformly to $f(x)$ on $[a, b]$, then $f(x)$ is Riemann integrable on $[a, b]$ and*

$$\int_a^b f(x)\, dx = \int_a^b \sum_{n=1}^{\infty} f_n(x)\, dx = \sum_{n=1}^{\infty} \int_a^b f_n(x)\, dx.$$

Similar to Corollary 6.2, we have another simple test for nonuniform convergence as follows.

Corollary 6.4. *Let $f_n : [a, b] \to \mathbb{R}$ be Riemann integrable, and let $f_n(x) \to f(x)$ pointwise on $[a, b]$. If f is Riemann integrable, but*

$$\lim_{n\to\infty} \int_a^b f_n(x)\, dx \neq \int_a^b \lim_{n\to\infty} f_n(x)\, dx,$$

then $f_n \to f$ is not uniform.

Example 6.8. *Consider*

$$f_n(x) = \begin{cases} 4n^2 x, & 0 \le x \le \frac{1}{2n}, \\ 4n(1 - nx), & \frac{1}{2n} < x \le \frac{1}{n}, \\ 0, & \frac{1}{n} < x \le 1. \end{cases}$$

Clearly, $f_n(x)$ is continuous and $f_n \to 0$ pointwise on $[0, 1]$. Since $f(x) = 0$ is continuous, it does not violate Corollary 6.2. However, we have $\lim_{n\to\infty} \int_0^1 f_n(x)\, dx = 1 \neq 0$. Thus Corollary 6.4 concludes that $f_n \to 0$ is not uniform on $[0, 1]$.

Example 6.9. *Consider*

$$f_n(x) = \begin{cases} 2nx, & 0 \le x \le \frac{1}{2n}, \\ 2(1 - nx), & \frac{1}{2n} < x \le \frac{1}{n}, \\ 0, & \frac{1}{n} < x \le 1. \end{cases}$$

Here we have $f_n \to 0$ pointwise. The limit function $f(x) = 0$ is continuous. Moreover

$$\lim_{n\to\infty} \int_a^b f_n(x)\, dx = \int_a^b \lim_{n\to\infty} f_n(x)\, dx,$$

but $f_n \to 0$ is not uniform because $w_n = \sup_{x\in[0,1]} |f_n(x) - f(x)| = f(1/2n) = 1$. This example shows that, for continuity of the limit function, uniform convergence is sufficient but not necessary. However, if f_n is decreasing, the following theorem asserts that uniform convergence is also necessary for the continuity of the limit function.

Theorem 6.8 (Dini's Theorem). *Let $f_n : [a, b] \to \mathbb{R}$ be continuous and decreasing. Suppose that $f_n \to f$ pointwise on $[a, b]$. If f is continuous on $[a, b]$, then $f_n \to f$ uniformly on $[a, b]$.*

Proof. We prove by contradiction. Assume that the convergence is not uniform on $[a, b]$, by Theorem 6.1, we have

$$\omega_n = \sup_{x \in [a,b]} |f_n(x) - f(x)| = \max_{x \in [a,b]} (f_n(x) - f(x)) \not\to 0, \qquad \text{as } n \to \infty.$$

Hence there exists a constant $\alpha > 0$ such that for infinitely many $n \in \mathbb{N}$,

$$\max_{x \in [a,b]} (f_n(x) - f(x)) > \alpha.$$

For each $n \in \mathbb{N}$, since $f_n(x) - f(x)$ is continuous on $[a, b]$, by the extreme value theorem, there is an $x_n \in [a, b]$ such that

$$\max_{x \in [a,b]} (f_n(x) - f(x)) = f_n(x_n) - f(x_n) > \alpha.$$

On the other hand, since $x_n \in [a, b]$ is bounded, by Bolzano-Weierstrass theorem, there exists a subsequence x_{n_k} such that $x_{n_k} \to x_0 \in [a, b]$ with

$$f_{n_k}(x_{n_k}) - f(x_{n_k}) > \alpha.$$

Using $f_{n+1}(x) \le f_n(x)$ on $[a, b]$, for every $n \le n_k$, we have

$$f_n(x_{n_k}) - f(x_{n_k}) > \alpha.$$

Now fixing n and letting $k \to \infty$, by the continuity of f_n and f, we find that

$$f_n(x_0) - f(x_0) \ge \alpha$$

for all $n \in \mathbb{N}$. This contradicts the assumption that $f_n \to f$ pointwise at x_0. $\square$

Recall Example 6.1 where $f_n(x) = x^n$ on $[0, 1]$. Since the limit function is not continuous at $x = 1$, by Corollary 6.2, f_n is not uniformly convergent. However, since f_n is decreasing and $f_n \to 0$ pointwise on $[0, b]$ for any $0 < b < 1$, Dini's theorem implies that f_n converges uniformly on $[0, b]$.

For a series of functions, applying Theorem 6.8 to $S(x) - S_n(x)$ yields

Corollary 6.5. *Let $f_n : [a, b] \to \mathbb{R}$ be continuous and nonnegative. If $S(x) = \sum_{n=1}^{\infty} f_n(x)$ is continuous on $[a, b]$, then the series converges uniformly on $[a, b]$.*

The continuity condition on $S(x)$ is necessary. Otherwise, $f_n(x) = x^{n-1}(1 - x)$ on $[0, 1]$ yields a counterexample.

In contrast to continuity and integrability, proving differentiability is much trickier. We have already seen, in Examples 6.3 and 6.4, that even if $f_n \to f$ uniformly and each f_n is differentiable, nothing promises that $f_n' \to f'$. To guarantee that the limit of the derivative is indeed the derivative of the limit function, we need to impose the uniform convergence on $f_n'(x)$.

Theorem 6.9. *Let $f_n : [a, b] \to \mathbb{R}$ be a sequence of functions such that*

1. *$f_n(x_0)$ converges for some $x_0 \in [a, b]$,*
2. *each f_n is differentiable on $[a, b]$,*
3. *$f_n'(x)$ converges uniformly on $[a, b]$.*

Then f_n converges uniformly on $[a, b]$ to a function $f(x)$, and

$$\frac{d}{dx} \lim_{n \to \infty} f_n(x) = f'(x) = \lim_{n \to \infty} \frac{d}{dx} f_n(x).$$

Proof. By the assumptions, for $\epsilon > 0$, there is an $N \in \mathbb{N}$ such that

$$|f_n(x_0) - f_m(x_0)| < \frac{\epsilon}{2} \quad \text{whenever } n, m > N \tag{6.4}$$

and

$$|f_n'(x) - f_m'(x)| < \frac{\epsilon}{2(b - a)} \quad \text{for all } x \in [a, b] \text{ whenever } n, m > N.$$

This, together with the mean value theorem, yields

$$|(f_n(x) - f_m(x)) - (f_n(t) - f_m(t))| \le \frac{\epsilon}{2(b - a)} |x - t| \le \frac{\epsilon}{2} \tag{6.5}$$

for any $x, t \in [a, b]$ whenever $n, m > N$. Since

$$|f_n(x) - f_m(x)| \le |(f_n(x) - f_m(x)) - (f_n(x_0) - f_m(x_0))| + |f_n(x_0) - f_m(x_0)|,$$

from (6.4) and (6.5), it follows that

$$|f_n(x) - f_m(x)| < \epsilon \quad \text{for all } x \in [a, b] \text{ whenever } n, m > N.$$

By the Cauchy criterion, $f_n(x)$ converges uniformly on $[a, b]$. Let $f(x) = \lim_{n \to \infty} f_n(x)$ for $x \in [a, b]$.

Given $x, t \in [a, b], t \ne x$, define

$$g_n(t) = \frac{f_n(t) - f_n(x)}{t - x}, \quad g(t) = \frac{f(t) - f(x)}{t - x}.$$

By the hypotheses (2), for each $n \in \mathbb{N}$, we have

$$\lim_{t \to x} g_n(t) = f_n'(x). \tag{6.6}$$

The inequality (6.5) implies that

$$|g_n(t) - g_m(t)| \le \frac{\epsilon}{2(b - a)} \quad \text{whenever } n, m > N.$$

So $g_n(x)$ converges uniformly for $t \ne x$. Since f_n converges to f, we have

$$\lim_{n \to \infty} g_n(t) = g(t) \tag{6.7}$$

uniformly for $a \le t \le b, t \ne x$. By (6.6), letting $t \to x$ in (6.7) yields

$$\lim_{t \to x} g(t) = \lim_{n \to \infty} f_n'(x)$$

as desired. $\qquad\qquad\qquad\qquad\qquad\qquad\qquad\qquad\qquad\qquad\qquad\qquad\qquad\qquad\square$

As usual, the following corollary follows from applying Theorem 6.9 to the sequence of the partial sums of a series.

Corollary 6.6. *Let $\sum_{n=1}^{\infty} f_n$ be a series of functions on $[a, b]$ such that*

1. $\sum_{n=1}^{\infty} f_n(x_0)$ *converges for some $x_0 \in [a, b]$,*
2. *each f_n is differentiable on $[a, b]$,*
3. $\sum_{n=1}^{\infty} f_n'(x)$ *converges uniformly on $[a, b]$.*

Then $\sum_{n=1}^{\infty} f_n$ converges uniformly on $[a, b]$ to a function $f(x)$, and

$$\frac{d}{dx} \left(\sum_{n=1}^{\infty} f_n(x) \right) = f'(x) = \sum_{n=1}^{\infty} \left(\frac{d}{dx} f_n(x) \right).$$

6.3 Two Other Convergence Theorems

We know what we are, but know not what we may be. — William Shakespeare

In the previous section, we demonstrated that uniform convergence is sufficient for us to interchange limit and integration operations. In this section we consider two convergence theorems, which allow us to interchange the limit and integration operations without requiring uniform convergence.

We begin with

Example 6.10. *Consider*

$$f_n(x) = \begin{cases} 1 - nx, & 0 \le x \le \frac{1}{n}, \\ 0, & \frac{1}{n} < x \le 1. \end{cases}$$

See Figure 6.5.

We have $f_n \to f(x)$ pointwise on $[0, 1]$, where

$$f(x) = \begin{cases} 1, & x = 0, \\ 0, & 0 < x \le 1. \end{cases}$$

Since $f(x)$ is not continuous on $[0, 1]$, by Corollary 6.2, $f_n \to f$ is not uniform on $[0, 1]$.

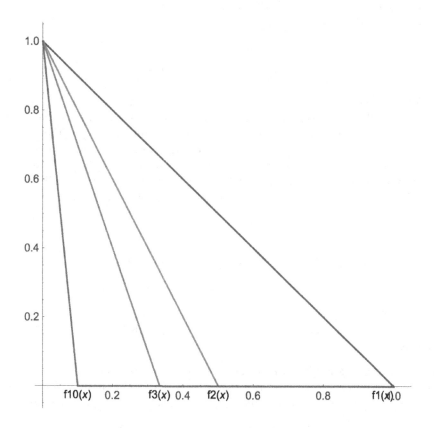

FIGURE 6.5

$f_n(x)$ on $[0, 1]$ for $n = 1, 2, 3, 10$.

However, we do have

$$\lim_{n \to \infty} \int_0^1 f_n(x)\,dx = \lim_{n \to \infty} \frac{1}{2n} = 0 = \int_a^b \lim_{n \to \infty} f_n(x)\,dx.$$

Note that f_n is nonnegative and monotone decreasing with $\lim_{n \to \infty} f_n(x) = 0$ for every $x \in [0,1]$ except at $x = 0$. It turns out that these properties allow us to interchange the limit and integration operations. This is captured in the following theorem.

Theorem 6.10 (Monotone Convergence Theorem). *Let $f_n : [a,b] \to [0,\infty)$ be a sequence of monotone decreasing functions such that*

   1. f_n is Riemann integrable for each $n \in \mathbb{N}$,

   2. $f_n \to 0$ pointwise on (a,b).

Then

$$\lim_{n \to \infty} \int_a^b f_n(x)\,dx = 0.$$

Applying this theorem to the sequence of $f(x) - f_n(x)$, we obtain

Corollary 6.7. *Let $f_n : [a,b] \to \mathbb{R}$ be a sequence of monotone increasing functions such that*

   1. f_n is Riemann integrable for each $n \in \mathbb{N}$,

   2. $f_n(x) \to f(x)$ pointwise on (a,b),

   3. $f(x)$ is Riemann integrable on $[a,b]$.

Then

$$\lim_{n \to \infty} \int_a^b f_n(x)\,dx = \int_a^b \lim_{n \to \infty} f_n(x) = \int_a^b f(x)\,dx.$$

This Monotone convergence theorem is usually presented within the Lebesgue integral. One might think that there is no proof in the context of Riemann integral. Indeed, the deep part of using the Lebesgue integral lies in establishing the integrability of the limit function. If we assume the integrability of the limit function, the proof of the monotone convergence theorem becomes elementary and can be done entirely within the Riemann integral. Here we present a proof based on the Lebesgue criterion of Riemann integrability (Theorem 5.11).

We begin with Cousin's lemma. Recall Definition 5.2. Assume that

$$P = \{a = x_0 < x_1 < \ldots < x_n = b\}$$

is a partition of $[a,b]$ with a set of tags $c_i \in [x_{i-1}, x_i]$, for $i = 1, 2, \ldots, n$. Let $\delta : [a,b] \to (0,\infty)$ be a *gauge function*. We call this tagged partition to be δ-fine if

$$[x_{i-1}, x_i] \subset (c_i - \delta(c_i), c_i + \delta(c_i)), \quad \text{for } i = 1, 2, \ldots, n.$$

See Figure 6.6 for an illustration.

The following Cousin's lemma asserts that every gauge has a δ-fine partition. In particular, we have

Theorem 6.11 (Cousin's Lemma). *Let $\mathcal{C}$ be a collection of closed subintervals of $[a,b]$ with the property that for each $x \in [a,b]$ there exists $\delta = \delta(x) > 0$ such that $\mathcal{C}$ contains all intervals $[c,d] \subset [a,b]$ that contain x and $d - c < \delta$. Then there exists a partition*

$$P = \{a = x_0 < x_1 < \ldots < x_n = b\}$$

of $[a,b]$ such that $[x_{i-1}, x_i] \in \mathcal{C}$ for $i = 1, 2, \ldots, n$.

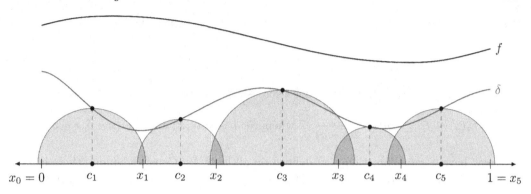

FIGURE 6.6
An example of δ fine partition.

Proof. The proof is by contradiction and is based on the nested interval theorem. Suppose that $\mathcal{C}$ does not contain a δ-fine partition of $[a, b]$. Then, either $[a, (a+b)/2]$ or $[(a+b)/2, b]$ has no δ-fine partition. Let us denote the half of $[a, b]$ without δ-fine partition by $[a_1, b_1]$. We continue this process inductively and obtain a sequence of nested intervals $[a_n, b_n]$ with $b_n - a_n = (b - a)/2^n \to 0$ as $n \to \infty$. By the nested interval theorem, there exists a point ξ which lies in every $[a_n, b_n]$. Since $\delta(\xi) > 0$, there is an $N \in \mathbb{N}$ such that $b_n - a_n < \delta(\xi)$ whenever $n > N$. This indicates that

$$P = \{a_n = x_0 \le \xi \le x_1 = b_n\}$$

is a δ-fine partition of $[a_n, b_n]$. This contradicts the definition of $[a_n, b_n]$. $\qquad\square$

In view of the nature of $\mathcal{C}$, it is indeed a closed covering of $[a, b]$. Similar to the Heine-Borel theorem, Cousin's lemma can be restated that if a closed covering of $[a, b]$ contains all "sufficiently small" closed intervals, then it must contain a partition of $[a, b]$. With this lemma, we are now ready to prove the monotone convergence theorem.

Proof of Theorem 6.10. Since $f_1(x)$ is integrable on $[a, b]$, there is a constant $M > 0$ such that $0 \le f_1(x) \le M$. For every $n \in \mathbb{N}$, let

$$D_n = \{x \in [a, b] \ : \ f_n(x) \text{ is discontinuous at } x\}.$$

By Theorem 5.11, D_n is of measure zero. Let $S = \cup_{n=1}^{\infty} D_n$. Then S is of measure zero as well. Hence for $\epsilon > 0$ there are disjoint open intervals (α_n, β_n) such that

$$S \subset \cup_{n=1}^{\infty}(\alpha_n, \beta_n) \quad \text{and} \quad \sum_{n=1}^{\infty}(\beta_n - \alpha_n) < \frac{\epsilon}{2M}.$$

For each $x \in S$, there is a unique interval (α_n, β_n) such that $x \in (\alpha_n, \beta_n)$, this allows us to choose $\delta(x) > 0$ for which

$$(x - \delta(x), x + \delta(x)) \subset (\alpha_n, \beta_n).$$

On the other hand, for each $x \in [a, b] \setminus S$, since $f_n(x) \to 0$ pointwise, there is an $N(\epsilon, x) \in \mathbb{N}$ such that

$$f_n(x) < \frac{\epsilon}{2(b - a)} \qquad \text{whenever } n \ge N(\epsilon, x)$$

Using the continuity of $f_{N(\epsilon,x)}$ at x, we can find a $\delta(x) > 0$ such that

$$f_{N(\epsilon,x)}(y) < \frac{\epsilon}{2(b-a)} \qquad \text{whenever } y \in (x - \delta, x + \delta).$$

By the assumption that f_n is monotone decreasing, we have

$$f_n(y) \le f_{N(\epsilon,x)}(y) < \frac{\epsilon}{2(b-a)} \qquad \text{whenever } n \ge N(\epsilon,x) \text{ and } y \in (x - \delta, x + \delta).$$

We now have a gauge function $\delta(x)$ defined on all $[a, b]$. By Cousin's lemma, there exists a δ-fine partition P of $[a, b]$ with a set of tags $c_i \in [x_{i-1}, x_i]$, for $i = 1, 2, \ldots, N$. In particular,

$$[x_{i-1}, x_i] \subset (c_i - \delta(c_i), c_i + \delta(c_i)), \qquad \text{for } i = 1, 2, \ldots, N.$$

Let

$$N_\epsilon = \max\{N(\epsilon, c_i) \, : \, i = 1, 2, \ldots, N\}.$$

Then for $n \ge N_\epsilon$ and any tags $d_i \in [x_{i-1}, x_i] \subset (c_i - \delta(c_i), c_i + \delta(c_i))$ for $i = 1, 2, \ldots N$, we have

$$0 \le \sum_{i=1}^{N} f_n(d_i)(x_i - x_{i-1}) = \sum_{i, d_i \in S} f_n(d_i)(x_i - x_{i-1}) + \sum_{i, d_i \in [a,b] \setminus S} f_n(d_i)(x_i - x_{i-1})$$

$$< \sum_{i, d_i \in S} M(x_i - x_{i-1}) + \sum_{i, d_i \in [a,b] \setminus S} \frac{\epsilon}{2(b-a)}(x_i - x_{i-1})$$

$$< M \cdot \frac{\epsilon}{2M} + \frac{\epsilon}{2(b-a)} \cdot (b - a) = \epsilon.$$

Therefore

$$0 \le \int_a^b f_n(x)\,dx \le \sup\left\{ \sum_{i=1}^{N} f_n(d_i)(x_i - x_{i-1}) \right\} \le \epsilon.$$

Since ϵ is arbitrary, we conclude that

$$\lim_{n \to \infty} \int_a^b f_n(x)\,dx = 0. \qquad \square$$

We now turn to the second convergence theorem. Recall Example 6.9 in which $f_n \to 0$ is not uniform on $[0, 1]$. See Figure 6.7.

Clearly, f_n is not monotone neither. But $f_n \le 1$ for all $n \in \mathbb{N}$. Moreover

$$\lim_{n \to \infty} \int_0^1 f_n(x)\,dx = \lim_{n \to \infty} \frac{1}{2n} = 0 = \int_a^b \lim_{n \to \infty} f_n(x)\,dx.$$

The following theorem reveals that the uniform boundedness allows us to interchange limit and integration operations.

Theorem 6.12 (Bounded Convergence Theorem). *Let $f_n : [a, b] \to \mathbb{R}$ be a sequence of Riemann integrable functions such that*

1. *$f_n(x) \to f(x)$ pointwise on (a, b),*

2. *$f(x)$ is Riemann integrable on $[a, b]$,*

3. *$\sup_{x \in [a,b]} |f_n(x)| \le M$ for all n with some positive constant M.*

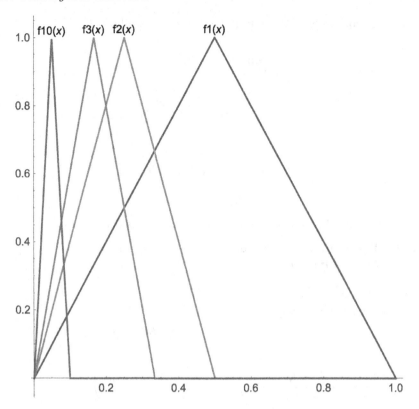

FIGURE 6.7
$f_n(x)$ on $[0, 1]$ for $n = 1, 2, 3, 10$.

Then

$$\lim_{n\to\infty} \int_a^b f_n(x)\, dx = \int_a^b \lim_{n\to\infty} f_n(x)\, dx = \int_a^b f(x)\, dx.$$

Proof. Let $g_n(x) = |f(x) - f_n(x)|$. By the assumptions, we find that $g_n(x)$ is Riemann integrable and $g_n(x) \to 0$ pointwise. Moreover, g_n is uniform bounded since f_n is uniform bounded and f itself is bounded because f is Riemann integrable. Thus, without loss of generality, we assume that

(i) $0 \le f_n \le M$ for all $n \in \mathbb{N}$ and for all $x \in [a, b]$,

(ii) $f_n \to 0$ pointwise on $[a, b]$.

Now for each n and for all $x \in [a, b]$, let

$$F_n(x) = \sup_{k \ge 0}\{f_{n+k}(x)\}.$$

By (i), $F_n(x)$ is well defined. Moreover, we have $0 \le f_n \le F_n$,

$$F_n(x) \ge F_{n+1}(x) \quad \text{for all } x \in [a, b],$$

and for all $x \in [a, b]$, by (ii),

$$\lim_{n\to\infty} F_n(x) = \lim_{n\to\infty} \sup f_n(x) = \lim_{n\to\infty} f_n(x) = 0.$$

By the monotone convergence theorem (Theorem 6.10), we have

$$\lim_{n\to\infty}\int_a^b F_n(x)\,dx = 0.$$

Therefore

$$0 \le \lim_{n\to\infty}\int_a^b f_n(x)\,dx \le \lim_{n\to\infty}\int_a^b F_n(x)\,dx = 0,$$

which implies that

$$\lim_{n\to\infty}\int_a^b f_n(x)\,dx = 0.$$

$\square$

As usual, applying this theorem to the partial sums of a series yields

Theorem 6.13 (Bounded Convergence Theorem for Series). *Let $f_n : [a,b] \to \mathbb{R}$ be a sequence of Riemann integrable functions such that*

1. *$\sum_{n=1}^{\infty} f_n(x) \to f(x)$ pointwise on (a,b),*
2. *$f(x)$ is Riemann integrable on $[a,b]$,*
3. *$|\sum_{i=1}^{n} f_n(x)| \le M$ for all n with some positive constant.*

Then

$$\int_a^b \left(\sum_{n=1}^{\infty} f_n(x)\right) dx = \sum_{n=1}^{\infty}\left(\int_a^b f_n(x)\,dx\right).$$

Example 6.11. *The derivation of Leibniz series*

$$\frac{\pi}{4} = \sum_{n=0}^{\infty}\frac{(-1)^n}{2n+1} = 1 - \frac{1}{3} + \frac{1}{5} - \frac{1}{7} + \cdots$$

offers a perfect application of Theorem 6.13.

Begin with the power series

$$\frac{1}{1+x^2} = \sum_{n=0}^{\infty}(-1)^n x^{2n} = 1 - x^2 + x^4 - x^6 + \cdots, \qquad 0 < x < 1.$$

Let $S_n(x)$ be the nth partial sum of the above series. By Leibniz's test, we have

$$0 < S_n(x) < 1, \qquad \text{for all } n \ge 1 \text{ and every } x \in (0,1).$$

Thus, Theorem 6.13 enables us to integrate term by term to obtain

$$\frac{\pi}{4} = \int_0^1 \frac{1}{1+x^2}\,dx = \int_0^1 \left(\sum_{n=0}^{\infty}(-1)^n x^{2n}\right) dx$$

$$= \int_0^1 \lim_{n\to\infty} s_n\,dx = \lim_{n\to\infty}\int_0^1 s_n(x)\,dx$$

$$= \lim_{n\to\infty}\sum_{n=0}^{n}(-1)^n\int_0^1 x^{2n}\,dx = \sum_{n=0}^{\infty}(-1)^n\int_0^1 x^{2n}\,dx$$

$$= \sum_{n=0}^{\infty}\frac{(-1)^n}{2n+1}.$$

The idea behind the above derivation yields the following generalization. The proof is left to the reader.

Let $f_n : [a, b] \to [0, \infty)$ *be a sequence of Riemann integrable functions. If f_n is monotone decreasing and $\sum_{n=1}^{\infty} (-1)^{n-1} f_n(x)$ is Riemann integrable, then*

$$\int_a^b \left(\sum_{n=1}^{\infty} (-1)^{n-1} f_n(x) \right) dx = \sum_{n=1}^{\infty} (-1)^{n-1} \int_a^b f_n(x) \, dx.$$

6.4   Power Series

In mathematics the art of proposing a question must be held of higher value than solving it. — George Cantor

In this section, we study a special class of series of functions which are represented in the form of a power series:

$$f(x) = \sum_{n=0}^{\infty} a_n x^n = a_0 + a_1 x + a_2 x^2 + a_3 x^3 + \cdots \tag{6.8}$$

or, more generally,

$$f(x) = \sum_{n=0}^{\infty} a_n (x - a)^n = a_0 + a_1(x - a) + a_2(x - a)^2 + a_3(x - a)^3 + \cdots .$$

For simplicity, our discussion will be centered on the case $a = 0$ without any loss of generality.

Historically, power series provided an infinite algorithm to represent general functions. The beauty of this type of series is that all of the relevant information regarding the convergence behavior of the series is encoded in the defining sequence a_n. First, we try to determine the set of points $x \in \mathbb{R}$ for which the resulting series (6.8) converges. Clearly, this set contains $x = 0$. In fact, this set has a very nice structure.

Theorem 6.14. *If a power series $\sum_{n=0}^{\infty} a_n x^n$ converges at some nonzero point $x_0 \in \mathbb{R}$, then it converges absolutely for any x with $|x| < |x_0|$.*

Proof. If the series $\sum_{n=0}^{\infty} a_n x_0^n$ converges, then the sequence $a_n x_0^n$ is bounded. Let $|a_0 x_0^n| < M$ for all $n \in \mathbb{R}$. For every $x \in \mathbb{R}$ satisfying $|x| < |x_0|$, we have

$$|a_n x^n| = |a_n x_0^n| \left| \frac{x}{x_0} \right|^n \le M \left| \frac{x}{x_0} \right|^n .$$

Since $|x/x_0| < 1$, the geometric series

$$\sum_{n=0}^{\infty} M \left| \frac{x}{x_0} \right|^n$$

converges. By the comparison test, $\sum_{n=0}^{\infty} a_n x^n$ converges absolutely. $\square$

Let

$$L = \limsup_{n \to \infty} \sqrt[n]{|a_n|}, \qquad R = \frac{1}{L}. \tag{6.9}$$

Applying the root test, we see that $\sum_{n=0}^{\infty} a_n x^n$ converges absolutely if $|x| < R$ and diverges if $|x| > R$. As a direct consequence of Theorem 6.14, the set for which a power series (6.8) converges must be either $\{0\}$ only or $\mathbb{R}$ entirely or one of the intervals: $(-R, R), [-R, R), (-R, R]$ or $[-R, R]$. The number R is called *the radius of convergence* of the power series. It is customary to assign R the value 0 or ∞ to represent the set $\{0\}$ or $\mathbb{R}$, respectively.

Regarding the uniform convergence of a power series, the Weierstrass M-test leads to an important consequence:

Theorem 6.15. *Let the radius of the power series*

$$a_0 + a_1 x + a_2 x^2 + a_3 x^3 + \cdots$$

be $R > 0$. If $0 < \alpha < R$, then the series converges uniformly on $[-\alpha, \alpha]$. If R is infinite, then the series converges uniformly on $[-M, M]$ for every $M > 0$.

Proof. By the definition (6.9), if R is finite then

$$\limsup_{n \to \infty} \sqrt[n]{|a_n R^n|} = 1.$$

Thus, for every $\epsilon > 0$, there is an $N \in \mathbb{R}$ such that

$$\sqrt[n]{|a_n R^n|} < 1 + \epsilon \quad \text{whenever } n > N.$$

If $0 < |x| < \alpha < R$, let $\epsilon = (\alpha - |x|)/|x|$. Then, for $n > N$, we have

$$\sqrt[n]{|a_n x^n|} = \frac{|x|}{R} \sqrt[n]{|a_n R^n|} < \frac{|x|}{R}(1 + \epsilon) = \frac{\alpha}{R} < 1.$$

Equivalently,

$$|a_n x^n| < \left(\frac{\alpha}{R}\right)^n.$$

Now, the desired conclusion follows from the Weierstrass M-test with $M_n = (\alpha/R)^n$.

If R is infinite, similarly, we have

$$\limsup_{n \to \infty} \sqrt[n]{|a_n|} = 0.$$

So there exists an $N \in \mathbb{N}$ such that $\sqrt[n]{|a_n|} < 1/2M$ for $n > N$. If $0 \leq |x| < M$, then

$$\sqrt[n]{|a_n x^n|} < \frac{|x|}{2M} < \frac{1}{2} \quad \text{for } n > N.$$

Therefore, applying the Weierstrass M-test with $M_n = 1/2^n$, we shows that the series converges uniformly on $[-M, M]$. $\qquad\square$

It should not be too surprising that Theorem 6.15 does not permit us to take uniform convergence all the way to $x = R$ because the power series may not converge at $x = \pm R$. Even if $R = \infty$, i.e., the series converges everywhere, the following example shows that the convergence may not be uniform on $\mathbb{R}$.

Example 6.12. *Prove that $\sum_{n=1}^{\infty} x^n/n!$ does not converge uniformly on $\mathbb{R}$.*

Proof. Let $m = n - 1$. Consider

$$|S_n(x) - S_{n-1}(x)| = \frac{|x|^n}{n!}.$$

Given $0 < \epsilon \leq 1$, we have

$$|S_n(n) - S_{n-1}(n)| = \frac{n^n}{n!} \geq 1 \geq \epsilon.$$

By the Cauchy criterion, $S_n(x)$ does not converge uniformly on $\mathbb{R}$. □

Observe that

$$\limsup_{n \to \infty} \sqrt[n]{|a_n|} = \limsup_{n \to \infty} \sqrt[n]{|a_n| n} = \limsup_{n \to \infty} \sqrt[n]{\frac{|a_n|}{n+1}}.$$

This implies that the radius of convergence of any power series is the same as the radius of convergence of the series of derivatives and the series of integrals as well. Thus, for each $x_0 \in (-R, R)$, there is an interval $x_0 \in [-\alpha, \alpha] \subset (-R, R)$ in which the series, the series of derivatives and the series of integration all converge uniformly. By Corollaries 6.1, 6.3, and 6.5, we conclude that a power series will always be continuous, differentiable, and integrable on $(-R, R)$. Moreover, the differentiation and integration can always be done by differentiating and integrating term-by-term.

But what happens at the endpoints? Does the good behavior of the series on $(-R, R)$ necessarily extend to the endpoint $x = R$? If the series converges absolutely at $x = R$, Theorem 6.15 is still applicable and concludes that the series converges uniformly on $[-R, R]$. So the remaining question is what happens if the series diverges or converges conditionally at $x = R$.

First, we assume that the power series diverges at $x = R$. The following theorem shows that even though the series converges in $(-R, R)$ the convergence is not uniform.

Theorem 6.16. *If $\sum_{n=0}^{\infty} a_n x^n$ diverges at $x = R$, then the convergence of the series on $(-R, R)$ is not uniform.*

Proof. By contradiction, we assume that the convergence is uniform on $(-R, R)$. Thus, for every $\epsilon > 0$, there exists an $N \in \mathbb{N}$ such that

$$|S_n(x) - S_m(x)| = |a_{m+1} x^{m+1} + a_{m+2} x^{m+2} + \cdots + a_n x^n| < \epsilon$$

for $n > m > N$ and every $x \in (-R, R)$. Letting $x \to R^-$ yields

$$|S_n(R) - S_m(R)| = |a_{m+1} R^{m+1} + a_{m+2} R^{m+2} + \cdots + a_n R^n| < \epsilon$$

whenever $n > m > N$. By the Cauchy criterion on the numerical series, $\sum_{n=0}^{\infty} a_n x^n$ converges at $x = R$, which contradicts the assumption. □

Next, we assume that the power series converges conditionally at $x = R$. Notice that if a power series converges conditionally at $x = R$, it may diverge at $x = -R$. For example, the series $\sum_{n=1}^{\infty} \frac{(-1)^n}{n} x^n$ with $R = 1$. To focus our attention on the convergent endpoint, we will prove uniform convergence on $[0, R]$. The proof offers a nice application of Abel's test for uniform convergence (Theorem 6.4).

Theorem 6.17. *If $\sum_{n=0}^{\infty} a_n x^n$ converges at $x = R$, then it converges uniformly on $[0, R]$.*

Proof. Rewrite the series as

$$\sum_{n=0}^{\infty} a_n x^n = \sum_{n=0}^{\infty} a_n R^n \cdot \left(\frac{x}{R} \right)^n.$$

Let

$$A_k = a_k R^k, \quad B_k = \left(\frac{x}{R} \right)^k.$$

By the assumption that $\sum_{n=0}^{\infty} a_n R^n$ converges, it also converges uniformly. Since $0 \leq x \leq R$, it follows that

$$1 \geq \left(\frac{x}{R} \right)^{m+1} \geq \left(\frac{x}{R} \right)^{m+2} \geq \left(\frac{x}{R} \right)^{m+3} \geq \cdots \geq 0.$$

Applying Abel's test to $\sum_{n=0}^{\infty} A_n B_n$ establishes the uniform convergence of the series on $[0, R]$. $\square$

In summary, Theorems 6.15 and 6.17 demonstrate that if a power series converges pointwise on an interval $I \subset \mathbb{R}$, then it converges uniformly on any closed interval containing in I. Combining this with Corollary 6.1, Theorem 6.17 yields a very useful consequence for power series.

Theorem 6.18 (Abel's Continuity Theorem). *Let $R > 0$ be the radius of convergence of $\sum_{n=0}^{\infty} a_n x^n$. If the series converges at $x = R$, then $\sum_{n=0}^{\infty} a_n x^n$ is continuous on $[0, R]$ and*

$$\lim_{x \to R^-} \sum_{n=0}^{\infty} a_n x^n = \sum_{n=0}^{\infty} a_n R^n.$$

Recall that

$$\ln(1 + x) = \sum_{n=1}^{\infty} \frac{(-1)^{n-1}}{n} x^n = x - \frac{1}{2} x + \frac{1}{3} x^3 - \cdots \qquad \text{for } |x| < 1.$$

Here $R = 1$, for $x = 1$ the series converges by the alternating series test. Thus the series represents a function that is continuous at $x = 1$. In particular, we have

$$\ln 2 = 1 - \frac{1}{2} + \frac{1}{3} - \frac{1}{4} + \cdots.$$

The following example exhibits another more sophisticated application of Abel's continuity theorem.

Example 6.13. *Show that*

$$\sum_{n=0}^{\infty} \frac{(-1)^n}{3n + 1} = \frac{1}{3} \left(\ln 2 + \frac{\sqrt{3}}{3} \pi \right).$$

Proof. We consider the power series

$$f(x) := \sum_{n=0}^{\infty} \frac{(-1)^n}{3n + 1} x^{3n+1}.$$

This enables us to convert the series into a geometric series after differentiation. Clearly, the series has the radius of convergence $R = 1$. Moreover, the alternating series test concludes that the series converges at $x = 1$. Thus, apply Abel's continuity theorem, we have

$$\sum_{n=0}^{\infty} \frac{(-1)^n}{3n+1} = \lim_{x \to 1^-} \sum_{n=0}^{\infty} \frac{(-1)^n}{3n+1} x^{3n+1} = \lim_{x \to 1^-} f(x) = \lim_{x \to 1^-} (f(x) - f(0))$$

$$= \lim_{x \to 1^-} \int_0^x f'(t)\, dt = \lim_{x \to 1^-} \int_0^x \sum_{n=0}^{\infty} (-t^3)^n\, dt = \lim_{x \to 1^-} \int_0^x \frac{dt}{1+t^3}$$

$$= \int_0^1 \frac{dt}{1+t^3} = \frac{1}{3} \int_0^1 \left(\frac{1}{1+t} + \frac{2-t}{t^2 - t + 1} \right) dt$$

$$= \frac{1}{3} \left(\ln 2 + \frac{\sqrt{3}}{3} \pi \right).$$

Here the term by term differentiation for $f'(t)$ is justified by Theorem 6.15 with $R = 1$. $\square$

Instead of starting with a power series, if we start with a function f with derivatives of all orders at $x = 0$ and formally define a power series

$$f(x) = \sum_{n=0}^{\infty} a_n x^n,$$

then the coefficients a_n must be given by

$$a_n = \frac{f^{(n)}(0)}{n!}.$$

By Definition 4.5, this power series is the Taylor series for f at $x = 0$. Cauchy's counterexample

$$f(x) = \begin{cases} 0, & \text{if } x = 0, \\ e^{-1/x^2}, & \text{if } x \neq 0 \end{cases}$$

shows that the Taylor series may fail to converge to the original function. This phenomenon shows that not every function with derivatives of all orders has a power series (Taylor series) representation. The following question naturally arises: Is a given power series the Taylor series for some function?

First, we see the answer is yes for a power series with positive radius of convergence. To this end, if

$$f(x) = \sum_{n=0}^{\infty} a_n x^n$$

has radius of convergence $R > 0$, then $f(x)$ has derivatives of all orders in $(-R, R)$ and

$$f^{(k)}(x) = \sum_{n=k}^{\infty} n(n-1) \cdots (n-k+1) a_n x^{n-k}.$$

In particular, we have $f^{(k)}(0) = k! a_k$ or

$$a_k = \frac{f^{(k)}(0)}{k!}.$$

This implies that the series itself is exactly the Taylor series of $f(x)$.

What happens when a power series like $\sum_{n=0}^{\infty} n!x^n$ converges at $x = 0$ only? Gelbaum and Olmsted [44] showed that $\sum_{n=0}^{\infty} n!x^n$ is indeed a Taylor series. As somewhat of a surprise, based on this result, Meyerson [69] asserted that

Theorem 6.19. *Every power series is a Taylor series!*

Proof. We show that given any sequence $a_n \in \mathbb{R}$, there is a function with its Taylor series $\sum_{n=0}^{\infty} a_n x^n$. For $n \in \mathbb{N}$, define

$$g_n(x) = \begin{cases} a_n n!, & \text{if } |x| \leq \frac{1}{(2|a_n|n!+1)}, \\ 0, & \text{if } |x| \geq \frac{2}{(2|a_n|n!+1)} \end{cases}$$

and elsewhere as a "smoothing function" which is monotone in each interval where it is defined and which makes all derivatives of g_n exist. Let $f_0 = g_0$ and

$$f_n(x_n) = \int_0^{x_n} \int_0^{x_{n-1}} \cdots \int_0^{x_2} \int_0^{x_1} g_n(x_0)dx_0 dx_1 \cdots dx_{n-1}.$$

then for $n \geq 1$,

$$|f_n^{(n-1)}(x)| = \left| \int_0^x g_n(t)dt \right| \leq \int_0^{\frac{2}{(2|a_n|n!+1)}} a_n n!\, dx < 1.$$

Thus integrating $n - k - 1$ times yields

$$|f_n^{(k)}(x)| \leq \frac{|x|^{n-k-1}}{(n-k-1)!}, \qquad \text{where } 0 \leq k \leq n-1 \text{ and } f_n^{(0)}(x) = f_n(x).$$

By the Weierstrass M-test, $\sum_{n=0}^{\infty} f_n^{(k)}(x)$ converges uniformly on every bounded interval. Hence $\sum_{n=0}^{\infty} f_n(x)$ converges to some $f(x)$ and $f^{(k)}(x) = \sum_{n=0}^{\infty} f_n^{(k)}(x)$ for all $k \in \mathbb{N}$. Since $f_n^{(k)}(0) = a_n n! \delta_{nk}$, it implies that $f^{(n)}(0) = a_n n!$. $\qquad \square$

Remark. The function with the given Taylor series is not unique. Indeed, for f as above and for every $\alpha \in \mathbb{R}$, let

$$g(x) = f(x) + \alpha\, e^{-1/x^2}$$

if $x \neq 0$ and $g(0) = f(0)$. Then g and f have the same Taylor series at $x = 0$. Therefore there are uncountably many functions with any given Taylor series.

6.5   Weierstrass's Approximation Theorem

Mathematicians do not study objects, but relations among objects; they are indifferent to the replacement of objects by others as long as the relations don't change. Matter is not important, only from interests them. — Henri Poincaré

Observe that if f is defined by a power series with radius of convergence $R > 0$, Theorem 6.15 asserts that f can be approximated uniformly on $[a, b] \subset (-R, R)$ by polynomials—the partial sum of the power series. Since not every function can be represented by a power series, naturally, we want to know what other class of functions can be approximated uniformly by polynomials. With his nowhere differentiable continuous function, Weieistrass (1872) demonstrated how wild a continuous function can be and how far it is from being smooth. It would seem that a continuous function could not be in this class. But, Weieistrass made the case for uniform convergence again. As an affirmative answer to the question above, his following theorem provided a fortuitous connection between continuous functions and polynomials.

Theorem 6.20 (Weierstrass's Approximation Theorem). *Every continuous function on $[a, b]$ can be uniformly approximated by polynomials on $[a, b]$.*

The original proof (1885) by Weierstrass is "existential" in nature—it begins by expressing $f(x)$ as a convolution

$$f(x) = \lim_{h \to 0} \frac{1}{\sqrt{\pi}h} \int_{-\infty}^{\infty} f(y) \exp\left[-\frac{(y-x)^2}{h^2}\right] dy$$

with the Gaussian heat kernel, and relies heavily on analytic limit arguments. Here we present Bernstein's beautiful proof. Philip J. Davis says that "while Bernstein's proof is not the simplest conceptually, it is easily the most elegant." One of the nice features of Bernstein's proof is that the approximating polynomials will be given explicitly. Such polynomials are important theoretically and have been applied in computer-aided design (CAD) and scientific computing [23].

We begin with introducing the Bernstein polynomials and proving some preliminary facts about these polynomials.

Definition 6.5. *Let f be a function defined on $[0, 1]$. For $n \geq 0$, the polynomial*

$$B_n(f) := \sum_{k=0}^{n} f\left(\frac{k}{n}\right) \binom{n}{k} x^k (1-x)^{n-k}$$

is called the nth Bernstein polynomial of f.

This definition was motivated by probability theory. Indeed, consider a series of trials of tossing a coin. In each trial, we assume the coin has probability x of getting heads. In n tosses, the probability of the event that the n trials result in exactly k heads is $\binom{n}{k} x^k (1-x)^{n-k}$. By the binomial theorem, we have

$$\sum_{k=0}^{n} \binom{n}{k} x^k (1-x)^{n-k} = [x + (1-x)]^n = 1. \tag{6.10}$$

Now let X be the number of heads in a series of n trials. As a "random variable," X/n is the ratio of heads to the total number of trials, and it takes the values $\{0, \frac{1}{n}, \ldots, \frac{n}{n}\}$. The expected value of X/n is by definition

$$E[X/n] = \sum_{k=0}^{n} \frac{k}{n} \cdot P(X = k) = \sum_{k=0}^{n} \frac{k}{n} \binom{n}{k} x^k (1-x)^{n-k}.$$

In general, the expected value of $f(X/n)$ is equal to

$$E[f(X/n)] = \sum_{k=0}^{n} f\left(\frac{k}{n}\right) \binom{n}{k} x^k (1-x)^{n-k} = B_n(f).$$

Figure 6.8 shows that the graphs of $f(x) = \sqrt{x}$ and a few Bernstein polynomials of f on $[0, 1]$.

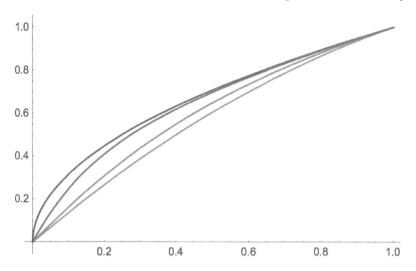

FIGURE 6.8
$f(x) = \sqrt{x}$ and $B_n(f)$ with $n = 2, 3, 10$.

By the definition of $B_n(f)$, (6.10) implies $B_n(1) = 1$. Moreover,

Theorem 6.21. *For $x \in \mathbb{R}$ and $n \geq 0$, we have*

1. $B_n(x) = \sum_{k=0}^{n} \frac{k}{n}\binom{n}{k}x^k(1-x)^{n-k} = x$.

2. $B_n(x^2) = \sum_{k=0}^{n} \left(\frac{k}{n}\right)^2 \binom{n}{k}x^k(1-x)^{n-k} = \frac{(n-1)x^2}{n} + \frac{x}{n}$.

3. $\sum_{k=0}^{n} \left(\frac{k}{n} - x\right)^2 \binom{n}{k}x^k(1-x)^{n-k} = \frac{x(1-x)}{n}$.

Proof. Recall the binomial theorem

$$\sum_{k=0}^{n}\binom{n}{k}x^k y^{n-k} = (x+y)^n.$$

Differentiating this identity with respect to x yields

$$\sum_{k=0}^{n} k\binom{n}{k}x^{k-1}y^{n-k} = n(x+y)^{n-1}.$$

Multiplying x/n on both sides results

$$\sum_{k=0}^{n} \frac{k}{n}\binom{n}{k}x^k y^{n-k} = x(x+y)^{n-1}. \tag{6.11}$$

Letting $y = 1 - x$ in (6.11) proves the identity (1).

Next, differentiating (6.11) with respect to x, then multiplying x/n on both sides yields

$$\sum_{k=0}^{n} \left(\frac{k}{n}\right)^2 \binom{n}{k}x^k y^{n-k} = \frac{(n-1)(x+y)^{n-2}x^2}{n} + \frac{(x+y)^{n-1}x}{n}. \tag{6.12}$$

Letting $y = 1 - x$ in (6.12) proves the identity (2).

Finally, since

$$\sum_{k=0}^{n} \left(\frac{k}{n} - x\right)^2 \binom{n}{k} x^k (1-x)^{n-k} = B_n(x^2) - 2xB_n(x) + x^2 B_n(1),$$

the identity (3) now follows from (6.10) and the identities (1) and (2). $\qquad\square$

We now return to the proof of Theorem 6.20. We may assume that $[a, b] = [0, 1]$. Because if the theorem is proved on $[0, 1]$, we say that there are polynomials $p_n(x) \to f$ uniformly on $[0, 1]$. If $g(x)$ is continuous on $[a, b]$, then

$$f(x) = g(a + (b-a)x)$$

is continuous on $[0, 1]$. Let $q_n(x) = p_n\left(\frac{x-a}{b-a}\right)$. Then $q_n(x) : [a, b] \to \mathbb{R}$ is a polynomial as well. Moreover,

$$q_n(x) - g(x) = p_n\left(\frac{x-a}{b-a}\right) - f\left(\frac{x-a}{b-a}\right) \to 0 \quad \text{uniformly on } [a, b].$$

Bernstain's proof.. Since f is uniformly continuous on $[0, 1]$ by Theorem 3.7, for any $\epsilon > 0$, there is a $\delta > 0$ such that

$$|f(x) - f(y)| < \frac{\epsilon}{2}, \quad \text{whenever } |x - y| < \delta \text{ and } x, y \in [0, 1]. \tag{6.13}$$

Let $B_n(f)$ be the nth Bernstein polynomial of f. Then

$$B_n(f)(x) - f(x) = \sum_{k=0}^{n} f\left(\frac{k}{n}\right) \binom{n}{k} x^k (1-x)^{n-k} - f(x) \sum_{k=0}^{n} \binom{n}{k} x^k (1-x)^{n-k}$$

$$= \sum_{k=0}^{n} \left(f\left(\frac{k}{n}\right) - f(x)\right) \binom{n}{k} x^k (1-x)^{n-k}.$$

This implies that, for each $x \in [0, 1]$,

$$|B_n(f)(x) - f(x)| \leq \sum_{k=0}^{n} \left|\left(f\left(\frac{k}{n}\right) - f(x)\right)\right| \binom{n}{k} x^k (1-x)^{n-k}.$$

To estimate this sum, for the δ given in (6.13), we divide the set $\{0, 1, 2, \ldots, n\}$ into two subsets

- $A = \{k : 0 \leq k \leq n; |x - k/n| < \delta\}$ and
- $B = \{k : 0 \leq k \leq n; |x - k/n| \geq \delta\}$

For $k \in A$, in view of (6.13), we have

$$\sum_{k \in A} \left|\left(f\left(\frac{k}{n}\right) - f(x)\right)\right| \binom{n}{k} x^k (1-x)^{n-k} < \sum_{k \in A} \frac{\epsilon}{2} \binom{n}{k} x^k (1-x)^{n-k}$$

$$\leq \frac{\epsilon}{2} \sum_{k=0}^{n} \binom{n}{k} x^k (1-x)^{n-k} = \frac{\epsilon}{2}.$$

For $k \in B$, let $M = \max_{x \in [0,1]} |f(x)|$. We have $(k/n - x)^2 \geq \delta^2$ and

$$\sum_{k \in B} \left| \left(f\left(\frac{k}{n}\right) - f(x) \right) \right| \binom{n}{k} x^k (1-x)^{n-k} \leq \sum_{k \in B} 2M \binom{n}{k} x^k (1-x)^{n-k}$$

$$\leq \frac{2M}{\delta^2} \sum_{k \in B} \left(\frac{k}{n} - x \right)^2 \binom{n}{k} x^k (1-x)^{n-k}$$

$$\leq \frac{2M}{\delta^2} \sum_{k=0}^{n} \left(\frac{k}{n} - x \right)^2 \binom{n}{k} x^k (1-x)^{n-k}$$

$$= \frac{2M}{n\delta^2} x(1-x) \leq \frac{M}{2n\delta^2},$$

where we have used the identity (3) in Theorem 6.21 and the fact that $x(1-x) \leq 1/4$ for $x \in [0,1]$. Since the δ in (6.13) does not depend on the choice of x, let $N > \frac{M}{\epsilon\delta^2}$. When $n > N$, we have

$$\sum_{k \in B} \left| \left(f\left(\frac{k}{n}\right) - f(x) \right) \right| \binom{n}{k} x^k (1-x)^{n-k} \leq \frac{M}{2n\delta^2} < \frac{\epsilon}{2}.$$

In summary, we have

$$|B_n(f)(x) - f(x)| \leq \sum_{k \in A} \left| \left(f\left(\frac{k}{n}\right) - f(x) \right) \right| \binom{n}{k} x^k (1-x)^{n-k}$$

$$+ \sum_{k \in B} \left| \left(f\left(\frac{k}{n}\right) - f(x) \right) \right| \binom{n}{k} x^k (1-x)^{n-k}$$

$$< \frac{\epsilon}{2} + \frac{\epsilon}{2} = \epsilon.$$

This proves that $B_n(f) \to f$ uniformly. $\square$

Since the proof used the fact that a continuous function on $[a, b]$ is uniformly continuous, Weierstrass's approximation theorem fails on non-closed intervals or on unbounded intervals. For example, $f(x) = 1/x$ can't be uniformly approximated by polynomials on $(0, 1)$ or $(0, 1]$, otherwise, it would be bounded on $(0, 1)$ or $(0, 1]$, which contradicts that $f(x)$ is unbounded on $(0, 1)$ or $(0, 1]$. This function serves as a counterexample on $[1, \infty)$ as well. Suppose that $f(x) = 1/x$ can be uniformly approximated by polynomials. Let $\epsilon = 1$. Then there is a non-constant polynomial $P(x)$ such that

$$\left| \frac{1}{x} - P(x) \right| < 1, \quad \text{for all } x \in [1, \infty).$$

Hence

$$|P(x)| \leq \left| \frac{1}{x} - P(x) \right| + \frac{1}{x} < 1 + 1 = 2, \quad \text{for all } x \in [1, \infty),$$

which is impossible.

As one application of Weierstrass's approximation theorem, we have

Theorem 6.22. *Let f be Riemann integrable on $[a, b]$. Then for any $\epsilon > 0$, there exists a polynomial $p(x)$ such that*

$$\int_a^b |f(x) - p(x)| \, dx < \epsilon.$$

Proof. By the assumption that $f(x)$ is Riemann integrable on $[a, b]$, Theorem 5.4 implies that there is a partition of $[a, b]$

$$P = \{a = x_0 < x_1 < \cdots < x_n = b\}$$

such that

$$\sum_{i=1}^{n} \omega_i(f) \Delta x_i < \frac{\epsilon}{2}.$$

We define a function $g(x)$ on $[a, b]$ as follows

$$g(x) = f(x_{i-1}) + \frac{f(x_i) - f(x_{i-1})}{x_i - x_{i-1}}(x - x_{i-1}), \qquad \text{for } i = 1, 2, \ldots, n.$$

Geometrically, $g(x)$ is the line segment which connects the points $(x_{i-1}, f(x_{i-1}))$ and $(x_i, f(x_i))$ for $i = 1, 2, \ldots, n$. Then

$$\int_a^b |f(x) - g(x)|\, dx = \sum_{i=1}^{n} \int_{x_{i-1}}^{x_i} |f(x) - g(x)|\, dx$$

$$\leq \sum_{i=1}^{n} \int_{x_{i-1}}^{x_i} |\alpha(f(x) - f(x_i)) + (1 - \alpha)(f(x) - f(x_{i-1}))|\, dx \quad \left(\alpha = \frac{x - x_{i-1}}{x_i - x_{i-1}}\right)$$

$$\leq \sum_{i=1}^{n} \int_{x_{i-1}}^{x_i} (\alpha \omega_k + (1 - \alpha)\omega_k)\, dx = \sum_{i=1}^{n} \omega_i(f) \Delta x_i < \frac{\epsilon}{2}.$$

On the other hand, since $g(x)$ is continuous on $[a, b]$, by Weierstrass's approximation theorem, for the ϵ given above, there is a polynomial $p(x)$ such that

$$|g(x) - p(x)| < \frac{\epsilon}{2(b - a)} \qquad \text{for all } x \in [a, b].$$

Therefore

$$\int_a^b |f(x) - p(x)|\, dx \leq \int_a^b |f(x) - g(x)|\, dx + \int_a^b |g(x) - p(x)|\, dx$$

$$\leq \frac{\epsilon}{2} + \frac{\epsilon}{2(b - a)}(b - a) = \epsilon.$$

$\square$

Let $f(x) = \lim_{n \to \infty} B_n(f)(x)$. Must $\lim_{n \to \infty} B'_n(f)(x) = f'(x)$? The answer is no, since there are plenty of continuous functions that are not differentiable. However, we have

Theorem 6.23. *If f is continuously differentiable on $[0, 1]$, then $B'_n(f) \to f'(x)$ uniformly on $[0, 1]$.*

We leave the proof to the reader.

6.6 Worked Examples

Keep in mind that there are millions of theorems but only thousands of proofs, hundreds of proof blocks, and dozens of ideas. Unfortunately, no one has figured out how to transfer the ideas directly yet, so you have to extract them from complicated arguments by yourself. — Fedja Nazarov

1. Determine whether or not $\sum_{n=1}^{\infty} \frac{\sin nx}{n}$ converges uniformly on $[0, 2\pi]$.

   Solution. In Example 2.14, by Dirichlet's test, for every $x \in [0, 2\pi]$, the series converges pointwise. However, the convergence is not uniform. In fact, let $x_n = \pi/4n$. We have

   $$
   \begin{aligned}
   |S_{2n}(x_n) - S_n(x_n)| &= \left| \frac{\sin(n+1)\frac{\pi}{4n}}{n+1} + \cdots + \frac{\sin 2n\frac{\pi}{4n}}{2n} \right| \\
   &\geq n \cdot \frac{\sin(\pi/4)}{2n} = \frac{1}{2\sqrt{2}}.
   \end{aligned}
   $$

   By the Cauchy criterion, this shows that the convergence is not uniform. □

2. Prove that $f(x) = \sum_{n=1}^{\infty} ne^{-nx}$ is continuous on $(0, \infty)$.

   Proof. Let $a_n(x) = ne^{-nx}$. Then

   $$
   \sup_{x \in (0,\infty)} a_n(x) \geq a_n(1/n) = \frac{n}{e} \not\to 0.
   $$

   Thus $a_n(x)$ does not converge to 0 uniformly and so the series does not converge uniformly on $(0, \infty)$. Here we cannot assert the continuity of f on $(0, \infty)$ by Corollary 6.1. To make up for this gap, for any $x_0 \in (0, \infty)$, there is a δ such that $0 < \delta < x_0$. For every $x \geq \delta$, if n is sufficiently large, we have

   $$
   ne^{-nx} \leq ne^{-n\delta} \leq \frac{1}{n^2}.
   $$

   Since $\sum_{n=1}^{\infty} 1/n^2$ converges, by the Weierstrass M-test, it follows that $\sum_{n=1}^{\infty} ne^{-nx}$ converges uniformly on $[\delta, \infty)$. Applying Corollary 6.1 now yields that $f(x)$ is continuous on $[\delta, \infty)$. In particular, $f(x)$ is continuous at x_0. Since $x_0 \in (0, \infty)$ is arbitrary, this concludes that $f(x)$ is continuous on $(0, \infty)$. □

3. Construct an example of a series which converges uniformly but fails by the Weierstrass M-test.

   Solution. For $n \in \mathbb{N}$, let

   $$
   a_n(x) = \begin{cases} \frac{1}{n}, & \text{if } \frac{1}{n+1} \leq x < \frac{1}{n} \\ 0, & \text{elsewhere} \end{cases}
   $$

   For every given $x \in [0, 1]$, there is only one nonzero term in $\sum_{n=1}^{\infty} a_n(x)$, so the series converges pointwise on $[0, 1]$. Since the remainder

   $$
   R_n = \sum_{k=n+1}^{\infty} a_n(x) \leq \frac{1}{n+1}
   $$

   for all $x \in [0, 1]$, it follows that the convergence is uniform on $[0, 1]$. On the other hand, notice

   $$
   \sup_{x \in [0,1]} a_n(x) = \frac{1}{n}
   $$

   and the harmonic series is divergent, thus the Weierstrass M-test is not applicable. □

4. Let $f(x)$ be twice continuously differentiable on $(-\eta, \eta)$ for some $\eta > 0$ with $f(0) = 0, 0 < f'(0) < 1$. Define a sequence $f_n(x)$ by

$$f_1(x) = f(x), f_n(x) = (f \circ f_{n-1})(x), \quad n = 1, 2, \dots.$$

Show that the series $\sum_{n=1}^{\infty} f_n(x)$ converges uniformly on $[-\delta, \delta]$ for some $0 < \delta < \eta$.

Proof. Since f is twice continuously differentiable on $(-\eta, \eta)$, it follows that there is a positive constant M such that $|f''(x)| \leq M$ for $|x| < \eta$. Using the Taylor expansion of f at $x = 0$, we have

$$f(x) = f(0) + f'(0)x + \frac{1}{2!} f''(\xi)x^2,$$

where ξ is between 0 and x. Hence $|\xi| < |x| < \eta$ and

$$|f(x)| \leq |f'(0)||x| + \frac{1}{2}M|x|^2 = \left(f'(0) + \frac{1}{2}M|x| \right)|x|.$$

Let $\delta = \min\{\eta, 2(1 - f'(0))/M\}$. For $|x| \leq \delta$, we have

$$|f(x)| \leq \left(f'(0) + \frac{1}{2}M\delta \right)|x| = q|x|,$$

where $q = f'(0) + \frac{1}{2}M\delta < f'(0) + \frac{1}{2}M \cdot \frac{2}{M}(1 - f'(0)) = 1$. Thus, we have

$$|f_1(x)| \leq q|x|, \quad |f_2(x)| = |f(f(x))| \leq q|f(x)| \leq q^2|x|.$$

By induction, for every $n \in \mathbb{N}$, we have

$$|f_n(x)| \leq q^n|x| \leq q^n\delta.$$

Since $0 < q < 1$, it follows that $\sum_{n=1}^{\infty} q^n\delta$ converges. Therefore, by the Weiertrass M-test, $\sum_{n=1}^{\infty} f_n(x)$ converges uniformly on $[-\delta, \delta]$. $\square$

5. Two more proofs of Dini's theorem.

Proof. Without loss of generality, we assume that

f_n is continuous, decreasing, and $f_n \to 0$ pointwise for all $x \in [a, b]$.

(Proof I) The first proof is based on the Heine-Borel theorem. For every $\epsilon > 0$, given $x \in [a, b]$, there is an $N_x(\epsilon) \in \mathbb{N}$ such that

$$0 \leq f_{N_x}(x) < \epsilon.$$

Since f_{N_x} is continuous at x, there is a δ_x such that

$$0 \leq f_{N_x}(t) < \epsilon, \quad \text{for } t \in (x - \delta_x, x + \delta_x). \tag{6.14}$$

Thus,

$$\cup_{x \in [a,b]} (x - \delta_x, x + \delta_x)$$

constitutes an open cover for $[a, b]$. By the Heine-Borel theorem, there is a finite number of open intervals

$$\{(x_i - \delta_{x_i}, x_i + \delta_{x_i})\}, \quad i = 1, 2, \ldots, m$$

that cover $[a, b]$. Let

$$N = \max_{1 \le i \le m} N_{x_i}.$$

In view of (6.14) and the assumption that f_n is decreasing, if $n > N$, then we have

$$0 \le f_n(x) < \epsilon$$

for all $x \in [a, b]$. This shows that $f_n \to 0$ uniformly.

(Proof II) The second proof is based on the nested interval theorem. We proceed by contradiction. Assume that the convergence is not uniform on $[a, b]$, then there is a $\epsilon_0 > 0$ such that the following statement is true:

For infinitely many n, $f_n(x) < \epsilon$ fails at some $x \in [a, b]$ which may depend on n.

We split the interval into two equal halves and choose $[a_1, b_1]$ as the half which is true:

For infinitely many n, $f_n(x) < \epsilon$ fails at some $x \in [a_1, b_1]$ which may depend on n.

Inductively, we can find a sequence of nested intervals $[a_k, b_k]$

For infinitely many n, $f_n(x) < \epsilon$ fails at some $x \in [a_k, b_k]$ which may depend on n,

where $b_k - a_k \to 0$ as $k \to \infty$. By the nested interval theorem, there is a $\xi \in [a, b]$ contained in every $[a_k, b_k]$. On the other hand, by the assumption that $f_n \to 0$, there is an N such that

$$f_N(\xi) < \epsilon_0.$$

By the continuity of f_N, there is a $\delta > 0$ such that

$$f_N(y) < \epsilon \quad \text{for all } |y - \xi| < \delta.$$

Since $b_k - a_k \to 0$ as $k \to \infty$, there exists a $K \in \mathbb{N}$ such that

$$[a_K, b_K] \subset (\xi - \delta, \xi + \delta).$$

Moreover, for all $y \in [a_K, b_K]$, by assumption that f_n is decreasing, we have

$$f_n(y) \le f_K(y) < \epsilon_0 \quad \text{for all } n > N, K.$$

This contradicts the choice of $[a_K, b_K]$. This completes the proof. □

6. Let the partial sum of $\sum_{n=1}^{\infty} f_n'(x)$ be uniformly bounded on $[a, b]$. Prove that if $\sum_{n=1}^{\infty} f_n(x)$ converges pointwise on $[a, b]$, then it converges uniformly on $[a, b]$.

   Proof. By the assumption, there is a positive constant M such that

   $$\left| \sum_{k=1}^{n} f_n'(x) \right| \le M$$

for every $n \in \mathbb{N}$ and for all $x \in [a, b]$. Given $\epsilon > 0$, let $P = \{x_0, x_1, \cdots, x_l\}$ be a partition of $[a, b]$ such that $\|P\| < \epsilon/4M$. By the Cauchy criterion for pointwise convergence, there exists an $N \in \mathbb{N}$ such that

$$\left| \sum_{k=m+1}^{n} f_k(x_i) \right| \leq \frac{\epsilon}{2}, \quad \text{whenever } n > m > N \text{ and } i = 1, 2, \ldots, l.$$

For any $x \in [a, b]$, without loss of generality, assume that $x \in [x_{i-1}, x_i]$. If $n > m > N$, using the triangle inequality and the mean value theorem, we have

$$\left| \sum_{k=m+1}^{n} f_k(x) \right| \leq \left| \sum_{k=m+1}^{n} f_k(x_i) \right| + \left| \sum_{k=m+1}^{n} (f_k(x) - f_k(x_i)) \right|$$

$$\leq \left| \sum_{k=m+1}^{n} f_k(x_i) \right| + \left| \sum_{k=m+1}^{n} f_k'(\xi_i) \right| |x - x_i| \quad \text{(where } \xi_i \in (x_{i-1}, x_i))$$

$$\leq \frac{\epsilon}{2} + 2M|x - x_i| \leq \frac{\epsilon}{2} + 2M \cdot \frac{\epsilon}{4M} = \epsilon.$$

By the Cauchy criterion, this proves that $\sum_{n=1}^{\infty} f_n(x)$ converges uniformly on $[a, b]$. $\qquad \square$

In view of Abel's continuity theorem (Theorem 6.18 with $R = 1$), if $\sum_{n=0}^{\infty} a_n = s < \infty$, then the power series

$$f(x) = \sum_{n=0}^{\infty} a_n x^n$$

converges for $|x| < 1$ and $f(x) \to s$ as $x \to 1^-$. Consider the power series $\sum_{n=0}^{\infty} (-1)^n x^n$. It has $R = 1$. Moreover, for $|x| < 1$,

$$\sum_{n=0}^{\infty} (-1)^n x^n = \sum_{n=0}^{\infty} (-x)^n = \frac{1}{1+x}.$$

We have

$$\lim_{x \to 1^-} \left(\sum_{n=0}^{\infty} (-1)^n x^n \right) = \lim_{x \to 1^-} \frac{1}{1+x} = \frac{1}{2}.$$

But $\sum_{n=0}^{\infty} (-1)^n$ diverges. This indicates that the converse of Abel's continuity theorem is not true. The following problem shows that the converse of Abel's continuity theorem is valid under the additional hypothesis that $na_n \to 0$.

7. (**Tauber Theorem**). Let $f(x) = \sum_{n=0}^{\infty} a_n x^n$ converge on $(-1, 1)$. If

$$\lim_{x \to 1^-} f(x) = S \quad \text{and} \quad \lim_{n \to \infty} na_n = 0,$$

then $\sum_{n=0}^{\infty} a_n = S$.

Proof. We use the typical "$\epsilon/3$ argument." By the triangle inequality, we have

$$\left| \sum_{k=0}^{n} a_k - S \right| \leq \left| \sum_{k=0}^{n} a_k - \sum_{k=0}^{n} a_k x^k \right| + \left| \sum_{k=n+1}^{\infty} a_k x^k \right| + \left| \sum_{k=0}^{\infty} a_k x^k - S \right|. \quad (6.15)$$

Given any $\epsilon > 0$, we must find an $N \in \mathbb{N}$ so that each term in the right hand side of (6.15) has an upper bound $\epsilon/3$ whenever $n > N$. To this end, in view of the assumption that $\lim_{n \to \infty} na_n = 0$, the Stolz-Cesàro theorem implies that

$$\lim_{n \to \infty} \frac{|a_1| + 2|a_2| + \cdots + n|a_n|}{n} = 0. \tag{6.16}$$

Next, choose $x = x_n = 1 - \frac{1}{n}$. The assumption $\lim_{x \to 1^-} f(x) = S$ yields

$$\lim_{n \to \infty} \left(f\left(1 - \frac{1}{n}\right) - S \right) = 0. \tag{6.17}$$

Now for every $\epsilon > 0$, based on (6.16) and (6.17), there exists an $N \in \mathbb{N}$ such that

$$\frac{|a_1| + 2|a_2| + \cdots + n|a_n|}{n} < \frac{\epsilon}{3}, \quad n|a_n| < \frac{\epsilon}{3}, \quad \text{and} \quad \left| f\left(1 - \frac{1}{n}\right) - S \right| < \frac{\epsilon}{3}$$

whenever $n > N$. For $0 < x < 1$, applying

$$1 - x^k = (1 - x)(1 + x + x^2 + \cdots + x^{k-1}) < k(1 - x)$$

and $x = 1 - 1/n$ yields that

$$\left| \sum_{k=0}^{n} a_k - \sum_{k=0}^{n} a_k x^k \right| < (1 - x) \sum_{k=0}^{n} k|a_k|$$

$$= \frac{|a_1| + 2|a_2| + \cdots + n|a_n|}{n} < \frac{\epsilon}{3}$$

Similarly, for $x = 1 - 1/n$,

$$\left| \sum_{k=n+1}^{\infty} a_k x^k \right| < \frac{1}{n} \sum_{k=n+1}^{\infty} k|a_k|x^k$$

$$\leq \frac{\epsilon}{3n} \sum_{k=n+1}^{\infty} x^k \leq \frac{\epsilon}{3n} \sum_{k=0}^{\infty} x^k$$

$$= \frac{\epsilon}{3n} \frac{1}{1 - x} = \frac{\epsilon}{3}$$

and

$$\left| \sum_{k=0}^{\infty} a_k x^k - S \right| = \left| f\left(1 - \frac{1}{n}\right) - S \right| < \frac{\epsilon}{3}.$$

Finally, putting all three estimates above into (6.15) yields

$$\left| \sum_{k=0}^{n} a_k - S \right| < \epsilon,$$

whenever $n > N$. This proves that $\sum_{n=0}^{\infty} a_n = S$. $\square$

Remark. Inspired by the Tauber theorem, Hardy and Littlewood [53] extended the summation procedures which allow us to assign a series a generalized sum in a natural way. For example, given a series $\sum_{n=0}^{\infty} a_n$, let $f(x) = \sum_{n=0}^{\infty} a_n x^n$. If $\lim_{x\to 1^-} f(x)$ exists, Hardy and Littlewood defined

$$\sum_{n=0}^{\infty} a_n = \lim_{x\to 1^-} f(x)$$

and called the limit the *Abel sum* of the series. Using this approach yields the Abel sums of divergent series such as

$$\sum_{n=0}^{\infty} (-1)^n = \frac{1}{2} \quad \text{and} \quad \sum_{n=0}^{\infty} (-1)^n n = \frac{1}{4}.$$

The Tauber theorem implies that the Abel sum is the same as the regular sum under the condition $na_n \to 0$. Littlewood strengthened the Tauber theorem by weakening the condition $na_n \to 0$ to the boundedness of na_n. A similar technique for summation of a series is based on the averages of the partial sums. Let $S_n = \sum_{k=0}^{n} a_k$ and

$$\sigma_n = \frac{S_0 + S_1 + \cdots + S_n}{n+1}.$$

If $\sigma_n \to \sigma$ as $n \to \infty$, we define

$$\sum_{n=0}^{\infty} a_n = \sigma,$$

which is called the *Cesàro sum* of the series. By the Stolz-Cesàro theorem, $\sigma_n \to s$ whenever $S_n \to s$. But σ_n may still converge when S_n does not. For example, the divergent series $\sum_{n=0}^{\infty} (-1)^n$ has the Cesàro sum $1/2$, which is the same as its Abel sum. In the following, we show that Cesàro summation implies Able summation.

8. If $\sigma_n \to \sigma$ as $n \to \infty$, then $f(x) = \sum_{n=0}^{\infty} a_n x^n$ converge on $(-1, 1)$ and

$$\lim_{x\to 1^-} f(x) = \sigma.$$

This means that if a series is Cesàro summable then it is Able summable, and their sums are identical.

Proof. Applying the Abel summation formula (2.14) twice yields

$$\begin{aligned}
\sum_{n=0}^{\infty} a_n x^n &= a_0 + \sum_{n=1}^{\infty} (S_n - S_{n-1}) x^n = (1-x) \sum_{n=0}^{\infty} S_n x^n \\
&= (1-x)^2 \sum_{n=0}^{\infty} (S_0 + S_1 + \cdots + S_n) x^n \\
&= (1-x)^2 \sum_{n=0}^{\infty} (n+1) \sigma_n x^n.
\end{aligned}$$

Using the identity

$$\sum_{n=0}^{\infty} (n+1) x^n = \frac{1}{(1-x)^2},$$

we have

$$\sum_{n=0}^{\infty} a_n x^n - \sigma = (1-x)^2 \sum_{n=0}^{\infty} (n+1)(\sigma_n - \sigma) x^n.$$

By the assumption $\sigma_n \to \sigma$, for every $\epsilon > 0$, there is an $N \in \mathbb{N}$ such that

$$|\sigma_n - \sigma| < \epsilon \quad \text{for all } n > N.$$

Then for $0 < x < 1$,

$$\left| \sum_{n=0}^{\infty} (n+1)(\sigma_n - \sigma) x^n \right| \leq \sum_{n=0}^{N} (n+1)|\sigma_n - \sigma| x^n + \sum_{n=N+1}^{\infty} (n+1)|\sigma_n - \sigma| x^n$$

$$\leq \sum_{n=0}^{N} (n+1)|\sigma_n - \sigma| x^n + \frac{\epsilon}{(1-x)^2}.$$

When x is sufficiently close to 1, we can make

$$(1-x)^2 \sum_{n=0}^{N} (n+1)|\sigma_n - \sigma| < \epsilon$$

and so

$$\left| \sum_{n=0}^{\infty} (n+1)(\sigma_n - \sigma) x^n \right| < 2\epsilon.$$

This shows that $f(x) \to \sigma$ as $x \to 1^-$. $\square$

9. Let $x \neq n\pi$ for all $n \in \mathbb{Z}$. Show that

$$\frac{1}{\sin^2 x} = \sum_{n \in \mathbb{Z}} \frac{1}{(x+n\pi)^2} = \frac{1}{x^2} + \sum_{n=1}^{\infty} \left(\frac{1}{(x+n\pi)^2} + \frac{1}{(x-n\pi)^2} \right). \qquad (6.18)$$

Proof. First, observe that

$$\frac{1}{\sin^2 t} = \frac{\cos^2(t/2) + \sin^2(t/2)}{4 \sin^2(t/2) \cos^2(t/2)} = \frac{1}{4} \left(\frac{1}{\sin^2(t/2)} + \frac{1}{\sin^2[(\pi + t)/2]} \right).$$

Repeatedly applying this identity yields

$$\frac{1}{\sin^2 x} = \frac{1}{4^2} \left(\frac{1}{\sin^2(x/4)} + \frac{1}{\sin^2[(2\pi + x)/4]} + \frac{1}{\sin^2[(\pi + x)/4]} + \frac{1}{\sin^2[(3\pi + x)/4]} \right)$$

$$= \cdots$$

$$= \frac{1}{2^{2n}} \sum_{k=0}^{2^n - 1} \frac{1}{\sin^2[(k\pi + x)/2^n]}.$$

For $2^{n-1} \leq k \leq 2^n - 1$, applying

$$\sin^2 \frac{k\pi + x}{2^n} = \sin^2 \left(\frac{k\pi + x - 2^n \pi}{2^n} + \pi \right) = \sin^2 \frac{(k - 2^n)\pi + x}{2^n}$$

yields

$$\frac{1}{\sin^2 x} = \frac{1}{2^{2n}} \sum_{k=-2^{n-1}}^{2^{n-1}-1} \frac{1}{\sin^2[(k\pi + x)/2^n]} = R_n + \sum_{k=-2^{n-1}}^{2^{n-1}-1} \frac{1}{(x + k\pi)^2}, \qquad (6.19)$$

where

$$R_n = \frac{1}{2^{2n}} \sum_{k=-2^{n-1}}^{2^{n-1}-1} \left[\frac{1}{\sin^2[(k\pi + x)/2^n]} - \frac{1}{[(x + k\pi)/2^n]^2} \right].$$

Since

$$0 < \frac{1}{\sin^2 t} - \frac{1}{t^2} = 1 + \frac{\cos^2 t}{\sin^2 t} - \frac{1}{t^2} < 1, \qquad \text{for } t \in [-\pi/2, \pi/2],$$

it follows that

$$0 < R_n < \frac{1}{2^{2n}} \sum_{k=-2^{n-1}}^{2^{n-1}-1} 1 = \frac{1}{2^{2n}} \cdot 2^n = \frac{1}{2^n}, \qquad \text{for } x \in [0, \pi/2].$$

Letting $n \to \infty$ in (6.19) concludes (6.18) as desired. $\qquad \square$

Clearly, the series (6.18) converges uniformly on any closed interval which does not contain the points of the form $k\pi$. In particular, we have

$$\frac{1}{3} = \lim_{x \to 0} \left(\frac{1}{\sin^2 x} - \frac{1}{x^2} \right) = 2 \sum_{n=1}^{\infty} \frac{1}{(n\pi)^2}$$

and so

$$\zeta(2) = \sum_{n=1}^{\infty} \frac{1}{n^2} = \frac{\pi^2}{6}.$$

For $x \in (-\pi, \pi)$, by Corollary 6.3, in view of the fact that

$$\int_0^x \left(\frac{1}{\sin^2 t} - \frac{1}{t^2} \right) = \left(\frac{1}{t} - \frac{\cos t}{\sin t} \right) \Big|_0^x = \frac{1}{x} - \frac{\cos x}{\sin x},$$

Integrating (6.18) term by term yields

$$\cot x = \frac{1}{x} + \sum_{n=1}^{\infty} \left(\frac{1}{x + n\pi} + \frac{1}{x - n\pi} \right), \qquad \text{for } x \in (-\pi, \pi). \qquad (6.20)$$

Integrating (6.20) term by term, then taking exponentiation leads to the well-known *Euler infinity product*

$$\frac{\sin x}{x} = \prod_{n=1}^{\infty} \left(1 - \frac{x^2}{(n\pi)^2} \right), \qquad \text{for } x \in (-\pi, \pi). \qquad (6.21)$$

In particular, setting $x = \pi/2$ in (6.21) yields the Wallis's formula again

$$\frac{2}{\pi} = \prod_{n=1}^{\infty} \left(1 - \frac{1}{4n^2} \right).$$

Finally, using the trigonometric identities

$$\tan x = \cot(\pi/2 - x), \quad \sec x = \frac{1}{\sin x} = \cot(x/2) - \cot x,$$

by (6.20), we have

$$\tan x = \sum_{n=1}^{\infty} \left(\frac{1}{(2n-1)\pi/2 - x} - \frac{1}{(2n-1)\pi/2 + x} \right), \qquad (6.22)$$

$$\csc x = \frac{1}{\sin x} = \frac{1}{x} + \sum_{n=1}^{\infty} (-1)^n \left(\frac{1}{x + n\pi} - \frac{1}{x - n\pi} \right). \qquad (6.23)$$

10. Find the power series of $\tan x$ on $(-\pi/2, \pi/2)$.

Solution. For $x \in (-\pi/2, \pi/2)$, by (6.22), we have

$$\tan x \;=\; \sum_{n=1}^{\infty} \frac{2x}{[(2n-1)\pi/2]^2 - x^2}$$

$$\;=\; 2x \sum_{n=1}^{\infty} \sum_{m=0}^{\infty} \frac{1}{[(2n-1)\pi/2]^2} \left(\frac{x}{(2n-1)\pi/2} \right)^{2m}.$$

Interchanging the order of the summations, we find that

$$\tan x = 2x \sum_{m=0}^{\infty} \sum_{n=1}^{\infty} \frac{1}{[(2n-1)\pi/2]^2} \left(\frac{x}{(2n-1)\pi/2} \right)^{2m}.$$

Since

$$\zeta(2k) = \sum_{n=1}^{\infty} \frac{1}{(2n-1)^{2k}} + \sum_{n=1}^{\infty} \frac{1}{(2n)^{2k}} = \sum_{n=1}^{\infty} \frac{1}{(2n-1)^{2k}} + 2^{-2k}\zeta(2k),$$

it follows that

$$\sum_{n=1}^{\infty} \frac{1}{(2n-1)^{2k}} = \frac{2^{2k} - 1}{2^{2k}} \zeta(2k).$$

Therefore, we have

$$\tan x = \sum_{m=0}^{\infty} \frac{2}{\pi^{2(m+1)}} 2(2^{2(m+1)} - 1)\zeta(2(m+1))x^{2m+1} = \sum_{k=1}^{\infty} 2(2^{2k} - 1)\frac{\zeta(2k)}{\pi^{2k}} x^{2k-1}.$$

The justification for interchanging the order without affecting the result is the following statement for the "double sum": Let $s_m = \sum_{n=0}^{\infty} |a_{nm}|$. If $\sum_{m=0}^{\infty} s_m x^m$ converges on $(-R, R)$, then

$$\sum_{n=0}^{\infty} \sum_{m=0}^{\infty} a_{nm} x^m = \sum_{m=0}^{\infty} \left(\sum_{n=0}^{\infty} a_{nm} \right) x^m, \quad \text{for } x \in (-R, R). \qquad \square$$

11. **(Monthly Problem 11848, 2015)**. Prove that

$$\frac{1}{2\pi} \operatorname{Li}_2(e^{-2\pi}) = \ln(2\pi) - 1 - \frac{5\pi}{12} - \sum_{m=1}^{\infty} \frac{(-1)^m \zeta(2m)}{m(2m+1)}.$$

Here, ζ is the Riemann zeta function, and $\operatorname{Li}_2(x) = \sum_{n=1}^{\infty} x^n/n^2$.

Proof. Using the partial fraction

$$\frac{1}{m(2m+1)} = 2\left(\frac{1}{2m} - \frac{1}{2m+1}\right),$$

we have

$$\sum_{m=1}^{\infty} \frac{(-1)^m \zeta(2m)}{m(2m+1)} = 2\left(\sum_{m=1}^{\infty} \frac{(-1)^m \zeta(2m)}{2m} - \sum_{m=1}^{\infty} \frac{(-1)^m \zeta(2m)}{2m+1}\right). \qquad (6.24)$$

In view of the partial fraction decomposition of $\cot x$ (6.20)

$$\pi x \cot \pi x = 1 + \sum_{n=1}^{\infty} \frac{2x^2}{x^2 - n^2}$$

and

$$\frac{x^2}{x^2 - n^2} = -\frac{x^2/n^2}{1 - x^2/n^2} = -\sum_{k=1}^{\infty} \frac{x^{2k}}{n^{2k}},$$

we have

$$\pi x \cot \pi x = 1 - 2\sum_{k=1}^{\infty} \zeta(2k) x^{2k}.$$

Replacing x by ix yields

$$\pi x \coth \pi x = 1 - 2\sum_{k=1}^{\infty} (-1)^k \zeta(2k) x^{2k}.$$

Therefore,

$$\sum_{m=1}^{\infty} \frac{(-1)^m \zeta(2m)}{2m} = \int_0^1 \frac{1}{x}\left(\sum_{m=1}^{\infty} (-1)^m \zeta(2m) x^{2m}\right) dx$$

$$= \int_0^1 \frac{1}{2}\left(\frac{1}{x} - \pi \coth \pi x\right) dx$$

$$= \frac{1}{2}(\ln x - \ln(\sinh \pi x))|_0^1$$

$$= \frac{1}{2}[\ln 2\pi - \pi - \ln(1 - e^{-2\pi})]. \qquad (6.25)$$

Similarly, we have

$$\sum_{m=1}^{\infty} \frac{(-1)^m \zeta(2m)}{2m+1} = \int_0^1 \left(\sum_{m=1}^{\infty} (-1)^m \zeta(2m) x^{2m}\right) dx$$

$$= \int_0^1 \frac{1}{2}(1 - \pi x \coth \pi x) dx$$

$$= \frac{1}{2}\int_0^1 \left(1 - \pi x \left(1 + \frac{2e^{-2\pi x}}{1 - e^{-2\pi x}}\right)\right) dx$$

$$= \frac{1}{2} - \frac{\pi}{4} - \pi \int_0^1 x \sum_{k=1}^{\infty} e^{-2k\pi x} dx$$

$$= \frac{1}{2} - \frac{\pi}{4} - \pi \sum_{k=1}^{\infty} \left(\frac{1}{4n^2\pi^2} - \frac{1}{4n^2\pi^2} e^{-2n\pi} - \frac{1}{2n\pi} e^{-2n\pi}\right).$$

Using the fact that

$$\zeta(2) = \frac{\pi^2}{6} \quad \text{and} \quad -\ln(1-x) = \sum_{n=1}^{\infty} \frac{x^n}{n},$$

we find that

$$\sum_{m=1}^{\infty} \frac{(-1)^m \zeta(2m)}{2m+1} = \frac{1}{2} - \frac{7\pi}{24} + \frac{1}{4\pi} \text{Li}_2(e^{-2\pi}) - \frac{1}{2} \ln(1 - e^{-2\pi}). \qquad (6.26)$$

Plugging (6.25) and (6.26) into (6.24) results in

$$\sum_{m=1}^{\infty} \frac{(-1)^m \zeta(2m)}{m(2m+1)} = \ln 2\pi - 1 - \frac{5\pi}{12} - \frac{1}{2\pi} \text{Li}_2(e^{-2\pi}),$$

which is equivalent to the proposed result.   □

12. Let a_n be a nonnegative and decreasing sequence. Show that $\sum_{n=1}^{\infty} a_n \sin nx$ converges uniformly on $\mathbb{R}$ if and only if $\lim_{n\to\infty} na_n = 0$.

Proof. "$\Longrightarrow$" Assume the series $\sum_{n=1}^{\infty} a_n \sin nx$ converges uniformly on $\mathbb{R}$. Then for every $\epsilon > 0$, there is an $N \in \mathbb{N}$ such that

$$\left| \sum_{k=m}^{n} a_k \sin kx \right| < \epsilon \quad \text{for all } x \in \mathbb{R} \text{ whenever } n > m > N.$$

Let $n > 2N, m = [n/2 + 1]$. Then $m > n/2 > N$. Choosing $x = \pi/2n$, we have

$$\left| \sum_{k=m}^{n} a_k \sin(k\pi/2n) \right| < \epsilon. \qquad (6.27)$$

When $m \le k \le n$, we have

$$\frac{\pi}{4} < \frac{m\pi}{2n} \le \frac{k\pi}{2n} \le \frac{\pi}{2}.$$

Now (6.27) implies that

$$\epsilon > \sum_{k=m}^{n} a_k \sin(k\pi/2n) \ge \sin\frac{\pi}{4} \sum_{k=m}^{n} a_k \ge \frac{\sqrt{2}}{2}(n-m+1)a_n \ge \frac{\sqrt{2}}{4} na_n.$$

This proves that $\lim_{n\to\infty} na_n = 0$.

"$\Longleftarrow$" Assume that $\lim_{n\to\infty} na_n = 0$. Let

$$\alpha_n = \sup_{k \ge n} \{ka_k\}.$$

Then α_n is decreasing and converges to zero as $n \to \infty$. Let

$$S_{m,n}(x) = \sum_{k=m}^{n} a_k \sin kx.$$

we want to show that

$$|S_{m,n}(x)| \leq (\pi + 3)\alpha_m \quad \text{for every } x \in \mathbb{R}. \tag{6.28}$$

Notice that $S_{m,n}(x)$ is an odd and periodic function with period 2π. It suffices to show that (6.28) holds on $[0, \pi]$. To do so, we divide $[0, \pi]$ into $[0, \pi/n], [\pi/n, \pi/m], [\pi/m, \pi]$, and show that (6.28) holds on each such subinterval.

(1) If $0 \leq x \leq \pi/n$, then $kx \in [0, \pi]$ for all $m < k \leq n$. Using $\sin \theta \leq \theta$ yields

$$|S_{m,n}(x)| = \sum_{k=m}^{n} a_k \sin kx \leq x \sum_{k=m}^{n} k a_k$$

$$\leq x \sum_{k=m}^{n} \alpha_k \leq \frac{\pi}{n}(n - m + 1)\alpha_m \leq \pi\alpha_m.$$

(2) If $\pi/n \leq x \leq \pi$, in view of that $\sin \theta \geq \frac{2}{\pi}\theta$ for $\theta \in [0, \pi/2]$, we have

$$\left| \sum_{k=m}^{n} \sin kx \right| = \frac{|\cos(n - 1/2)x - \cos(m + 1/2)x|}{2\sin(x/2)} \leq \frac{1}{\sin(x/2)} \leq \frac{\pi}{x} \leq m.$$

By Abel's lemma, we have

$$|S_{m,n}(x)| = m(a_m + 2a_n) \leq 3ma_m \leq 3\alpha_m.$$

(3) If $\pi/n \leq x \leq \pi/n$, then $m \leq \pi/x \leq n$. Let $L = [\pi/x]$. We have

$$S_{m,n}(x) = \sum_{k=m+1}^{L} a_k \sin kx + \sum_{k=L+1}^{n} a_k \sin kx = S_{m,L}(x) + S_{L+1,n}(x).$$

Since $L \leq \frac{\pi}{x} < L + 1$, it follows that $x \leq \frac{\pi}{L}$. The result of (1) implies that

$$|S_{m,L}(x)| \leq \pi\alpha_m.$$

For $\frac{\pi}{L+1} < x \leq \frac{\pi}{m}$ and $L + 1 > \frac{\pi}{x} \geq m$, by (2) and the decreasing of α_n, we have

$$|S_{L+1,n}(x)| \leq 3\alpha_{L+1}$$

and so

$$|S_{m,n}(x)| \leq (3 + \pi)\alpha_m.$$

In summary, we have proved that (6.28) holds for every $x \in [0, \pi]$. Therefore, the series converges uniformly on $\mathbb{R}$. $\qquad \square$

Remark. This result does not apply to $\sum_{n=1}^{\infty} a_n \cos nx$. Can you give a counterexample?

13. Let $\alpha \in \mathbb{R} \setminus \mathbb{Z}$. Show that, for $x \in (-1, 1)$,

$$\sum_{n=0}^{\infty} \binom{\alpha}{n} x^n = (1 + x)^\alpha, \tag{6.29}$$

where

$$a_n = \binom{\alpha}{n} = \frac{\alpha(\alpha - 1) \cdots (\alpha - n + 1)}{n!}. \quad \text{(Generalized combination numbers)}$$

Solution. Applying the ratio test for absolute convergence yields

$$\left| \frac{a_{n+1}}{a_n} \right| = \left| \frac{\alpha - n}{n+1} \right| \to 1 \quad \text{as } n \to \infty.$$

This shows that the radius of convergence of the series is $R = 1$. Let the series be $f(x)$. For every $x \in (-1, 1)$, term by term differentiating gives

$$
\begin{aligned}
(1+x)f'(x) &= (1+x) \sum_{n=1}^{\infty} n \binom{\alpha}{n} x^{n-1} \\
&= \sum_{n=1}^{\infty} n \binom{\alpha}{n} x^{n-1} + \sum_{n=1}^{\infty} n \binom{\alpha}{n} x^n \\
&= \binom{\alpha}{1} + \sum_{n=1}^{\infty} \left[(n+1) \binom{\alpha}{n+1} + n \binom{\alpha}{n} \right] x^n \\
&= \alpha \left(1 + \sum_{n=1}^{\infty} \binom{\alpha}{n} x^n \right) = \alpha f(x).
\end{aligned}
$$

This implies that

$$[(1+x)^{-\alpha} f(x)]' = -\alpha(1+x)^{-\alpha-1} f(x) + (1+x)^{-\alpha} f'(x) = 0$$

and so $(1+x)^{-\alpha} f(x) = c$ on $(-1, 1)$. In view of that $f(0) = 1$, we have

$$f(x) = (1+x)^{\alpha}.$$

$\square$

Remark. If $\alpha > 0$, since

$$n \left(\frac{|a_n|}{|a_{n+1}|} - 1 \right) = \frac{n(1+\alpha)}{n - \alpha} \to 1 + \alpha > 1 \quad \text{as } n \to \infty,$$

by the Raabe test, the series converges at $x = \pm 1$. Moreover, by Theorem 6.17, the series converges uniformly on $[-1, 1]$.

If $-1 < \alpha < 0$, when $x = -1$, we have

$$(-1)^n a_n = \frac{|\alpha|}{n} \cdot \frac{|\alpha| + 1}{1} \cdots \frac{|\alpha| + n - 1}{n - 1} \geq \frac{|\alpha|}{n}$$

and so the series diverges at $x = -1$. But, at $x = 1$, the series becomes an alternating series. Since $0 < |\alpha| < 1$, we have

$$|a_n| = \left| \frac{\alpha(\alpha - 1) \cdots (\alpha - n + 1)}{n!} \right| \geq \left| \frac{\alpha(\alpha - 1) \cdots (\alpha - n + 1)}{n!} \right| \cdot \left| \frac{\alpha - n}{n + 1} \right| = |a_{n+1}|$$

and $a_n \to 0$ as $n \to \infty$. The alternating series test concludes the series converges at $x = 1$. By Theorem 6.17 again, the series converges to $(1+x)^{\alpha}$ uniformly on $[0, 1]$. In particular, let $\alpha = -1/2$ and replace x by $-x^2$ in (6.29). We find that

$$\frac{1}{\sqrt{1 - x^2}} = \sum_{n=0}^{\infty} \frac{1}{2^{2n}} \binom{2n}{n} x^{2n}, \quad x \in (-1, 1). \tag{6.30}$$

Integrating term by term (6.30) yields

$$\arcsin x = \sum_{n=0}^{\infty} \frac{1}{(2n+1)2^{2n}} \binom{2n}{n} x^{2n+1}, \quad x \in [-1, 1]. \tag{6.31}$$

14. (**Monthly Problem 11897, 2016**). Show for $n \geq 0$ that

$$\sum_{k+l=n, k\geq 0, l\geq 0} \frac{\binom{2k}{k}\binom{2l+2}{l+1}}{k+1} = 2\binom{2n+2}{n}.$$

Proof. We provide two proofs. The first proof is based on the generating functions. Let the left-hand side of the proposed identity be A. Using $\binom{i}{j} = \frac{i}{j}\binom{i-1}{j-1}$ yields

$$A = \sum_{k=0}^{n} \frac{\binom{2k}{k}\binom{2(n-k+1)}{n-k+1}}{k+1} = 2\sum_{k=0}^{n} \frac{\binom{2k}{k}\binom{2(n-k)+1}{n-k}}{k+1}.$$

Replacing x^2 by $4x$ in (6.30) gives

$$\sum_{n=0}^{\infty} \binom{2n}{n} x^n = \frac{1}{\sqrt{1-4x}}.$$

Integrating both sides gives

$$\sum_{n=0}^{\infty} \frac{1}{n+1} \binom{2n}{n} x^n = \frac{1-\sqrt{1-4x}}{2x}.$$

Similarly, we have

$$\sum_{n=0}^{\infty} \binom{2n+1}{n} x^n = \frac{1-\sqrt{1-4x}}{2x\sqrt{1-4x}}.$$

Thus, the Cauchy product gives

$$\sum_{k=0}^{n} \frac{\binom{2k}{k}\binom{2(n-k)+1}{n-k}}{k+1} = [x^n] \frac{1-\sqrt{1-4x}}{2x} \cdot \frac{1-\sqrt{1-4x}}{2x\sqrt{1-4x}}$$

$$= [x^n] \left(-\frac{1}{2x^2} + \frac{1}{2x^2\sqrt{1-4x}} - \frac{1}{x\sqrt{1-4x}} \right)$$

$$= \frac{1}{2}\left(\binom{2(n+2)}{n+2} \right) - \binom{2(n+1)}{n+1}$$

$$= \binom{2n+3}{n+1} - \binom{2n+2}{n+1}.$$

Since

$$\binom{2n+3}{n+1} = \binom{2n+2}{n+1} + \binom{2n+2}{n},$$

it proves that

$$A = 2\left(\binom{2n+3}{n+1} - \binom{2n+2}{n+1} \right) = 2\binom{2n+2}{n}$$

as desired.

The second proof uses the Wilf-Zeilberger algorithm. Let

$$F(n,k) = \frac{1}{k+1}\binom{2k}{k}\binom{2(n-k+1)}{n-k+1},$$

which is hypergeometric in both n and k. Using the Wilf-Zeilberger algorithm—EKHAD (available at `https://sites.math.rutgers.edu/~zeilberg/tokhniot/EKHAD`), we find that

$$G(n,k) = \frac{(-2n+2k-3)(k+1)\binom{2k}{k}\binom{2(n-k+1)}{n-k+1}k!}{(n+2)(k+1)!}$$

and

$$F(n,k) = G(n,k+1) - G(n,k).$$

Adding both sides over k for $0 \le k \le n$, via telescoping, we end up with

$$A = G(n,n+1) - G(n,0) = 2\binom{2n+2}{n}. \qquad \square$$

Remark. The Wilf-Zeilberger algorithm enables us to evaluate sums of binomial coefficients in a mechanical procedure. We refer the interested reader to [71] or [75] for details.

15. Let $f_n(x) : [a,b] \to \mathbb{R}$ be continuous and $f_n(x) \ge 0$ for all $x \in [a,b]$. If $\sum_{n=1}^{\infty} f_n(x) = f(x)$, show that $f(x)$ attains its minimal on $[a,b]$.

Proof. Let $m = \inf_{x \in [a,b]} f(x)$. Then there exists a sequence $x_n \in [a,b]$ such that $\lim_{n \to \infty} f(x_n) = m$. By the Borel-Weierstrass theorem, there is a subsequence $x_{n_k} \to x_0 \in [a,b]$ as $k \to \infty$. Thus, for every $\delta > 0$, we have

$$m = \inf_{x \in [a,b]} f(x) = \inf_{x \in (x_0 - \delta, x_0 + \delta)} f(x). \tag{6.32}$$

Now we show that $f(x_0) = m$. Since clearly $m \le f(x_0)$, it suffices to show that $m \ge f(x_0)$. To this end, in view of the series convergence at x_0, for every $\epsilon > 0$, there exists an $N \in \mathbb{N}$ such that

$$f(x_0) - \sum_{n=1}^{N} f_n(x_0) < \frac{\epsilon}{2}.$$

By the continuity of f_n, there is a $\delta > 0$ such that

$$\left| \sum_{n=1}^{N} f_n(x) - \sum_{n=1}^{N} f_n(x_0) \right| < \frac{\epsilon}{2} \qquad \text{whenever } |x - x_0| < \delta.$$

Hence

$$\left| f(x_0) - \sum_{n=1}^{N} f_n(x) \right| \le f(x_0) - \sum_{n=1}^{N} f_n(x_0) + \left| \sum_{n=1}^{N} f_n(x) - \sum_{n=1}^{N} f_n(x_0) \right| < \epsilon$$

whenever $|x - x_0| < \delta$. In view of the property that $f_n \ge 0$, if $n \ge N$, we have

$$f(x_0) - \epsilon < \sum_{k=1}^{N} f_k(x) \le \sum_{k=1}^{n} f_k(x)$$

whenever $|x - x_0| < \delta$. Letting $n \to \infty$ yields

$$f(x_0) - \epsilon \le f(x) \qquad \text{whenever } |x - x_0| < \delta.$$

(6.32) now implies that

$$m = \inf_{x \in (x_0 - \delta, x_0 + \delta)} f(x) \geq f(x_0) - \epsilon.$$

Since ϵ is arbitrary, it follows that $m \geq f(x_0)$ and so $m = f(x_0)$. □

Is it possible that $f(x)$ attains its maximum on $[a, b]$? We show by example that $f(x)$ need not attain the maximum on $[a, b]$. In view of Dini's Theorem (Problem 4), we need to find a series which converges to f piontwise but not uniformly. Otherwise, by Theorem 6.6, f will be continuous on $[a, b]$. To this end, let $f_n(x) = x^n - x^{n+1}, x \in [0, 1]$. Then

$$f(x) = \sum_{n=1}^{\infty} (x^n - x^{n+1}) = \lim_{n \to \infty} (x - x^{n+2}) = \begin{cases} x, & \text{if } x \neq 1, \\ 0, & \text{if } x = 1. \end{cases}$$

Clearly, $\sup_{x \in [0,1]} f(x) = 1$, but there is no $t \in [0, 1]$ such that $f(t) = 1$.

A similar result is valid for the sequences , we show that if f_n is an increasing and continuous function sequence and $f_n(x) \to f(x)$ pointwise, then $f(x)$ need not attain the maximum. Define $f_n : [0, 1] \to [0, \infty)$ by

$$f_n(x) = \begin{cases} x, & \text{if } x \in [0, 1 - 1/n], \\ L_n(x), & \text{if } x \in [1 - 1/n, 1], \end{cases}$$

where $L_n(x)$ is a linear function which passes through the points $(1 - 1/n, 1 - 1/n)$ and $(1, 0)$. Then $f_n(x)$ satisfies

(a) For each $n \in \mathbb{N}$, $f_n(x)$ is continuous and $\max_{x \in [0,1]} f_n(x) = 1 - 1/n$.
(b) For each $x \in [0, 1]$, $f_n(x) \leq f_{n+1}(x)$ $(n = 1, 2, \ldots)$.
(c) $\lim_{n \to \infty} f_n$ converges to

$$f(x) = \begin{cases} x, & \text{if } x \neq 1, \\ 0, & \text{if } x = 1. \end{cases}$$

Therefore, $\sup_{x \in [0,1]} f(x) = 1$, but, there is no $t \in [0, 1]$ such that $f(t) = 1$.

However, if we replace increasing by decreasing, the conclusion holds.

16. (**Monthly Problem 11765, 2014**). Let C_n be the nth Catalan number, given by $C_n = \frac{1}{n+1} \binom{2n}{n}$. Show that

(a) $\sum_{n=0}^{\infty} \frac{2^n}{C_n} = 5 + \frac{3}{2} \pi$,
(b) $\sum_{n=0}^{\infty} \frac{3^n}{C_n} = 22 + 8\sqrt{3} \pi$.

Proof. First, we establish the generating function for the reciprocal of the central binomial coefficients

$$G(x) := \sum_{n=0}^{\infty} \frac{1}{\binom{2n}{n}} x^n = \frac{4(\sqrt{4 - x} + \sqrt{x} \sin^{-1}(\sqrt{x}/2))}{(4 - x)^{3/2}}. \tag{6.33}$$

To this end, note that

$$\binom{2n}{n}^{-1} = (2n + 1) \int_0^1 t^n (1 - t)^n \, dt.$$

Using

$$\sum_{n=0}^{\infty}(2n+1)x^n = 2\sum_{n=0}^{\infty}nx^n + \sum_{n=0}^{\infty}x^n$$
$$= \frac{2x}{(1-x)^2} + \frac{1}{1-x} = \frac{1+x}{(1-x)^2},$$

we find that

$$G = \sum_{n=0}^{\infty}(2n+1)\int_0^1 x^n t^n (1-t)^n\, dt$$
$$= \int_0^1 \left(\sum_{n=0}^{\infty}(2n+1)[xt(1-t)]^n\right) dt$$
$$= \int_0^1 \frac{1+xt(1-t)}{(1-xt(1-t))^2}\, dt.$$

Now (6.33) follows from the direct calculation of the above integral.

Next, applying $x\frac{d}{dx}$ to both sides of (6.33) yields

$$\sum_{n=0}^{\infty}\frac{1}{C_n}x^n = 2\frac{(8+x)\sqrt{4-x}+12\sqrt{x}\sin^{-1}(\sqrt{x}/2))}{(4-x)^{5/2}}. \tag{6.34}$$

By the ratio test, the radius of convergence of the series in (6.34) equals 4. Thus, letting $x = 2$ and 3 in (6.34) respectively, yields

$$\sum_{n=0}^{\infty}\frac{2^n}{C_n} = 5 + \frac{3}{2}\pi,$$

$$\sum_{n=0}^{\infty}\frac{3^n}{C_n} = 22 + 8\sqrt{3}\,\pi,$$

as claimed. $\qquad\qquad\qquad\qquad\qquad\qquad\qquad\qquad\qquad\qquad\qquad\qquad\qquad\qquad$ □

17. **(Monthly Problem 12297, 2022)**. Prove

$$\int_0^{\pi/2}\left(\frac{\sinh^{-1}(\sin(x))}{\sin(x)}\right)^2 dx = \frac{\pi}{2}\left(\frac{\pi}{2} - \ln 2\right).$$

Proof. Recall that

$$(\sin^{-1}x)^2 = \sum_{n=1}^{\infty}\frac{2^{2n-1}x^{2n}}{n^2\binom{2n}{n}}.$$

Since $\sinh^{-1}x = \frac{1}{i}\sin^{-1}(ix)$ with $i = \sqrt{-1}$, we find

$$(\sinh^{-1}x)^2 = \sum_{n=1}^{\infty}\frac{(-1)^{n-1}2^{2n-1}x^{2n}}{n^2\binom{2n}{n}}.$$

Hence,

$$\int_0^{\pi/2} \left(\frac{\sinh^{-1}(\sin(x))}{\sin(x)} \right)^2 dx = \int_0^{\pi/2} \left(\sum_{n=1}^{\infty} \frac{(-1)^{n-1} 2^{2n-1} \sin^{2(n-1)} x}{n^2 \binom{2n}{n}} \right) dx$$

$$= \sum_{n=1}^{\infty} \frac{(-1)^{n-1} 2^{2n-1}}{n^2 \binom{2n}{n}} \int_0^{\pi/2} \sin^{2(n-1)} x \, dx$$

$$= \sum_{n=1}^{\infty} \frac{(-1)^{n-1} 2^{2n-1}}{n^2 \binom{2n}{n}} \frac{\binom{2(n-1)}{n-1}}{2^{2(n-1)}} \frac{\pi}{2} = \frac{\pi}{2} \sum_{n=1}^{\infty} \frac{(-1)^{n-1}}{n(2n-1)}$$

$$= \frac{\pi}{2} \left(2 \sum_{n=1}^{\infty} \frac{(-1)^{n-1}}{2n-1} - \sum_{n=1}^{\infty} \frac{(-1)^{n-1}}{n} \right)$$

$$= \frac{\pi}{2} \left(2 \cdot \frac{\pi}{4} - \ln 2 \right) = \frac{\pi}{2} \left(\frac{\pi}{2} - \ln 2 \right).$$

Here we have used the well-known Wallis's integral formula:

$$\int_0^{\pi/2} \sin^{2k} x \, dx = \frac{\binom{2k}{k}}{2^{2k}} \frac{\pi}{2}.$$

By Stirling's formula, for large n, we have

$$\binom{2n}{n} \sim \frac{2^{2n}}{\sqrt{n\pi}}.$$

Thus, for large n,

$$\left| \frac{(-1)^{n-1} 2^{2n-1} \sin^{2(n-1)} x}{n^2 \binom{2n}{n}} \right| \leq \frac{2^{2n-1} \sqrt{n\pi}}{n^2 2^{2n}} = \frac{\sqrt{\pi}}{2n^{3/2}}.$$

Since $\sum_{n=1}^{\infty} 1/n^{3/2}$ converges, the Weierstrass M-test justifies the term by term integration. $\square$

18. Find

$$\sum_{n=1}^{\infty} \sum_{m=1}^{\infty} (-1)^{n+m} \frac{H_{n+m}}{n+m},$$

where H_n denotes the nth harmonic number.

Solution. We show the sum is $\frac{\pi^2}{12} - \frac{1}{2} \ln 2 - \frac{1}{2} \ln^2 2$. Denote the double series as H. The substitution $k = n + m$ yields

$$H = \sum_{n=1}^{\infty} \sum_{k=n+1}^{\infty} (-1)^k \frac{H_k}{k}.$$

Using

$$H_n = \int_0^1 \frac{1 - x^n}{1 - x} dx = -n \int_0^1 x^n \ln(1 - x) \, dx,$$

we have

$$
\begin{aligned}
H &= -\sum_{n=1}^{\infty}\sum_{k=n+1}^{\infty}(-1)^k \int_0^1 x^{k-1}\ln(1-x)\,dx \\
&= -\sum_{n=1}^{\infty}\int_0^1 \ln(1-x)\left(\sum_{k=n+1}^{\infty}(-1)^k x^{k-1}\right)dx \\
&= -\sum_{n=1}^{\infty}\int_0^1 \ln(1-x)\,\frac{(-1)^{n+1}x^n}{1+x}\,dx \\
&= -\int_0^1 \frac{x\ln(1-x)}{(1+x)^2}\,dx.
\end{aligned}
$$

Rewriting the integral as

$$
H = -\int_0^1 \frac{\ln(1-x)}{1+x}\,dx + \int_0^1 \frac{\ln(1-x)}{(1+x)^2}\,dx,
$$

in the first integral, using the substitution $t = (1-x)/(1+x)$, we have

$$
\begin{aligned}
\int_0^1 \frac{\ln(1-x)}{1+x}\,dx &= \int_0^1 \frac{1}{1+t}\ln\left(\frac{2t}{1+t}\right)dt \\
&= \int_0^1 \frac{\ln 2}{1+t}\,dt + \int_0^1 \frac{\ln t}{1+t}\,dt - \int_0^1 \frac{\ln(1+t)}{1+t}\,dt \\
&= \ln^2 2 - \frac{\pi^2}{12} - \frac{1}{2}\ln^2 2 = -\frac{\pi^2}{12} + \frac{1}{2}\ln^2 2.
\end{aligned}
$$

Here we have used the fact that

$$
\int_0^1 \frac{\ln t}{1+t}\,dt = \int_0^1 \sum_{n=0}^{\infty}(-t)^n \ln t\,dt = \sum_{n=0}^{\infty}(-1)^{n+1}\frac{1}{(n+1)^2} = -\frac{\pi^2}{12}.
$$

Similarly, applying the same substitution in the second integral yields

$$
\int_0^1 \frac{\ln(1-x)}{(1+x)^2}\,dx = \frac{1}{2}\int_0^1 \ln\left(\frac{2t}{1+t}\right)dt = -\frac{1}{2}\ln 2.
$$

Combining these two integral results, we find that

$$
H = \frac{\pi^2}{12} - \frac{1}{2}\ln 2 - \frac{1}{2}\ln^2 2
$$

as claimed.   □

Remark. In general, for $|x| < 1$, we have

$$
\sum_{n=1}^{\infty}\sum_{m=1}^{\infty}\frac{H_{n+m}}{n+m}x^{n+m} = -\frac{\ln(1-x)}{1-x} - \frac{1}{2}\ln^2(1-x) - \mathrm{Li}_2(x).
$$

19. Evaluate

$$
\sum_{n=1}^{\infty}\frac{(-1)^n}{2n-1}\left(\frac{\pi}{4} - \sum_{k=1}^{n}\frac{(-1)^{k-1}}{2k-1}\right).
$$

Solution. The value of the series is $\pi^2/32$. Recall that

$$\sum_{k=1}^{\infty} \frac{(-1)^{k-1}}{2k-1} = 1 - \frac{1}{3} + \frac{1}{5} - \frac{1}{7} + \cdots = \frac{\pi}{4}.$$

Thus,

$$\frac{\pi}{4} - \sum_{k=1}^{n} \frac{(-1)^{k-1}}{2k-1} = \sum_{k=n+1}^{\infty} \frac{(-1)^{k-1}}{2k-1}.$$

This suggests us to consider

$$f(x) = \sum_{n=1}^{\infty} a_n(x) \sum_{k=n+1}^{\infty} a_k(x),$$

where $a_n(x) = (-1)^{n-1}\frac{x^{2n-1}}{2n-1}$. By symmetry, we get

$$2f(x) = \sum_{n=1}^{\infty} a_n(x) \sum_{k=1}^{\infty} a_k(x) - \sum_{n=1}^{\infty} a_n^2(x) = \arctan^2 x - \sum_{n=1}^{\infty} a_n^2(x).$$

Since

$$\sum_{n=1}^{\infty} a_n^2(1) = \sum_{n=1}^{\infty} \frac{1}{(2n-1)^2} = \frac{\pi^2}{8},$$

we find that

$$f(1) = \frac{1}{2}\left(\frac{\pi^2}{16} - \frac{\pi^2}{8}\right) = -\frac{\pi^2}{32}.$$

This confirms that the value of the series is $-f(1) = \pi^2/32$, as claimed.   □

Remark. The mechanism behind the above technique is based on Abel's summation formula. If both infinite series $S := \sum_{n=1}^{\infty} a_n$ and $T := \sum_{n=1}^{\infty} a_n^2$ converge, let $A_n = \sum_{k=1}^{n} a_k, b_n = \sum_{k=n+1}^{\infty} a_k$. Then, by Abel's summation formula,

$$S_n := \sum_{k=1}^{n} a_k b_k = \frac{1}{2}\left(A_n b_n + SA_n - \sum_{k=1}^{n} a_2^2\right) \to \frac{1}{2}(S^2 - T).$$

20. Let ζ be the Riemann zeta function. Prove that

$$\sum_{n=1}^{\infty} \left(\zeta(2) - \sum_{k=1}^{n} \frac{1}{k^2}\right)^2 = 3\zeta(2) - \zeta^2(2).$$

Proof. We give a proof based on Abel's summation formula. Let $A_n = \sum_{k=1}^{n} a_k$. Recall that

$$\sum_{k=1}^{n} a_k b_k = A_n b_{n+1} + \sum_{k=1}^{n} A_k(b_k - b_{k+1}).$$

The limit version of the above formula leads to

$$\sum_{k=1}^{\infty} a_k b_k = \lim_{n\to\infty} A_n b_{n+1} + \sum_{k=1}^{\infty} A_k(b_k - b_{k+1}).$$

Thus, taking $a_n = 1$ and $b_n = \left(\zeta(2) - \sum_{k=1}^{n} \frac{1}{k^2}\right)^2$ yields

$$\sum_{n=1}^{\infty} \left(\zeta(2) - \sum_{k=1}^{n} \frac{1}{k^2}\right)^2 = \lim_{n \to \infty} n \left(\zeta(2) - \sum_{k=1}^{n+1} \frac{1}{k^2}\right)^2$$

$$+ \sum_{n=1}^{\infty} \frac{n}{(n+1)^2} \left(2\,\zeta(2) - 2\sum_{k=1}^{n} \frac{1}{k^2} - \frac{1}{(n+1)^2}\right)$$

$$= 2\sum_{n=1}^{\infty} \frac{n}{(n+1)^2} \left(\zeta(2) - \sum_{k=1}^{n} \frac{1}{k^2}\right) - \sum_{n=1}^{\infty} \frac{n}{(n+1)^4}$$

$$= 2\sum_{n=1}^{\infty} \left(\frac{1}{n+1} - \frac{1}{(n+1)^2}\right)\left(\zeta(2) - \sum_{k=1}^{n} \frac{1}{k^2}\right) - \zeta(3) + \zeta(4).$$

Since

$$\zeta(2) - \sum_{k=1}^{n} \frac{1}{k^2} = \sum_{k=n+1}^{\infty} \frac{1}{k^2} = -\int_0^1 \frac{x^n \ln x}{1-x}\, dx,$$

it follows that

$$\sum_{n=1}^{\infty} \frac{1}{n+1}\left(\zeta(2) - \sum_{k=1}^{n} \frac{1}{k^2}\right) = -\int_0^1 \frac{\ln x}{1-x} \sum_{n=1}^{\infty} \frac{x^n}{n+1}\, dx$$

$$= \int_0^1 \frac{\ln x}{1-x}\left(1 + \frac{\ln(1-x)}{x}\right) dx$$

$$= -\zeta(2) + 2\zeta(3).$$

Similarly, we have

$$\sum_{n=1}^{\infty} \frac{1}{(n+1)^2}\left(\zeta(2) - \sum_{k=1}^{n} \frac{1}{k^2}\right) = -\zeta(2) + \frac{7}{4}\zeta(4).$$

Hence,

$$\sum_{n=1}^{\infty} \left(\zeta(2) - \sum_{k=1}^{n} \frac{1}{k^2}\right)^2 = 3\zeta(2) - \frac{5}{2}\zeta(4) = 3\zeta(2) - \zeta^2(2),$$

where we have used $\zeta(4) = \pi^4/90 = \frac{2}{5}(\pi^2/6)^2 = \frac{2}{5}\zeta^2(2)$.   □

Remark. Another way to evaluate this series is in terms of the *multiple zeta function* (see [18, Chapter 3], which is defined by

$$\zeta(a, b) = \sum_{n=1}^{\infty} \sum_{m=1}^{n-1} \frac{1}{n^a m^b}.$$

Indeed, rewrite

$$\sum_{n=1}^{\infty} \left(\zeta(2) - \sum_{k=1}^{n} \frac{1}{k^2}\right)^2 = \sum_{n=1}^{\infty} \sum_{i=n+1}^{\infty} \sum_{j=n+1}^{\infty} \frac{1}{i^2 j^2}.$$

Rearranging the sums by the cases: $i = j, i < j$ and $i > j$, respectively, gives

$$\sum_{n=1}^{\infty}\sum_{i=n+1}^{\infty}\sum_{j=n+1}^{\infty}\frac{1}{i^2 j^2} = \left(\sum_{i=2}^{\infty}\sum_{j=2}^{\infty}+\sum_{i=3}^{\infty}\sum_{j=3}^{\infty}+\sum_{i=4}^{\infty}\sum_{j=4}^{\infty}+\cdots\right)\frac{1}{i^2 j^2}$$

$$= \left(\sum_{j=2}^{\infty}\frac{j-1}{j^4}+\sum_{i=2}^{\infty}\frac{i-1}{i^2}\sum_{j=i+1}^{\infty}\frac{1}{j^2}+\sum_{j=2}^{\infty}\frac{j-1}{j^2}\sum_{i=j+1}^{\infty}\frac{1}{i^2}\right)$$

$$= \left(\sum_{j=1}^{\infty}\frac{j-1}{j^4}+\sum_{i=1}^{\infty}\sum_{j>i}^{\infty}\frac{i-1}{i^2 j^2}+\sum_{j=1}^{\infty}\sum_{i>j}^{\infty}\frac{j-1}{i^2 j^2}\right).$$

In terms of the Riemann zeta function ζ and the multiple zeta function $\zeta(a,b)$, we have

$$\sum_{n=1}^{\infty}\left(\zeta(2)-\sum_{k=1}^{n}\frac{1}{k^2}\right)^2 = \zeta(3)-\zeta(4)+\zeta(2,1)-\zeta(2,2)+\zeta(2,1)-\zeta(2,2)$$

$$= \zeta(3)+2\zeta(2,1)-\zeta(4)-2\zeta(2,2).$$

Since

$$\zeta(2,1)=\zeta(3), \quad \zeta(a,b)+\zeta(b,a)+\zeta(a+b)=\zeta(a)\zeta(b),$$

we obtain

$$\sum_{n=1}^{\infty}\left(\zeta(2)-\sum_{k=1}^{n}\frac{1}{k^2}\right)^2 = 3\zeta(3)-\zeta^2(2)$$

again. Along the same lines, in general, for $p, q \geq 2$, we can establish

$$\sum_{n=1}^{\infty}\left(\zeta(p)-\sum_{k=1}^{n}\frac{1}{k^p}\right)\left(\zeta(q)-\sum_{k=1}^{n}\frac{1}{k^q}\right) = \zeta(p,q-1)+\zeta(q,p-1)+\zeta(p+q-1)-\zeta(p)\zeta(q).$$

In particular, since

$$\zeta(2,2)=\frac{3}{4}\zeta(4), \; \zeta(3,1)=\frac{1}{4}\zeta(4), \; \zeta(3,2)=3\zeta(2)\zeta(3)-\frac{11}{2}\zeta(5),$$

$$\zeta(3,3)=\frac{1}{2}\left(\zeta^2(3)-\zeta(6)\right), \; \zeta(4,2)=\zeta^2(3)-\frac{4}{3}\zeta(6),$$

we find that

$$\sum_{n=1}^{\infty}\left(\zeta(2)-\sum_{k=1}^{n}\frac{1}{k^2}\right)\left(\zeta(3)-\sum_{k=1}^{n}\frac{1}{k^3}\right) = 2\zeta(4)-\zeta(2)\zeta(3),$$

$$\sum_{n=1}^{\infty}\left(\zeta(3)-\sum_{k=1}^{n}\frac{1}{k^3}\right)\left(\zeta(3)-\sum_{k=1}^{n}\frac{1}{k^3}\right) = -10\zeta(5)+6\zeta(2)\zeta(3)-\zeta^2(3),$$

$$\sum_{n=1}^{\infty}\left(\zeta(3)-\sum_{k=1}^{n}\frac{1}{k^3}\right)\left(\zeta(4)-\sum_{k=1}^{n}\frac{1}{k^4}\right) = \frac{3}{2}\zeta^2(3)-\frac{5}{6}\zeta(6)-\zeta(3)\zeta(4).$$

We now end this section with one more example. It demonstrates a new approach to prove inequalities via power series.

21. (**Nesbitt's inequality**). For all positive real numbers a, b, and c, show that

$$\frac{a}{b+c} + \frac{b}{c+a} + \frac{c}{a+b} \geq \frac{3}{2}.$$

Proof. This inequality, and along with several natural variations, has appeared in various mathematical competitions. Typically, it was proved by using either AM-GM inequality or Jensen's inequality. Here we reprove it by using power series. By symmetry, we assume, without loss of generality, that $a + b + c = 1$. So the original inequality becomes

$$\frac{a}{1-a} + \frac{b}{1-b} + \frac{c}{1-c} \geq \frac{3}{2} \quad \text{for } a, b, c \in (0,1).$$

Note that the fractions in the left-hand side of the inequality above are the sums of some convergent geometric series. Thus,

$$\frac{a}{1-a} + \frac{b}{1-b} + \frac{c}{1-c} = \sum_{n=1}^{\infty} (a^n + b^n + c^n) = 3 \sum_{n=1}^{\infty} \frac{a^n + b^n + c^n}{3}$$

$$\geq 3 \sum_{n=1}^{\infty} \left(\frac{a+b+c}{3}\right)^n = 3 \cdot \sum_{n=1}^{\infty} \frac{1}{3^n} = \frac{3}{2},$$

where the mean inequality is used. This proves the claimed inequality. $\qquad\square$

6.7 Exercises

The world of ideas is not revealed to us in one stroke; we must both permanently and unceasingly recreate it in our consciousness. — René Thom

1. Let $f_n(x) = xn^\alpha e^{-nx}$, $(n = 1, 2, \ldots)$. Determine the value of α such that
   (a) $f_n(x)$ pointwise converges on $[0, 1]$,
   (b) $f_n(x)$ uniformly converges on $[0, 1]$,
   (c) $\lim_{n\to\infty} \int_0^1 f_n(x)\,dx = \int_0^1 \lim_{n\to\infty} f_n(x)\,dx$.

2. Let $k \geq 0$ be an integer and define a sequence $f_n : \mathbb{R} \to \mathbb{R}$ by

$$f_n(x) = \frac{x^k}{x^2 + n}, \quad n \in \mathbb{N}.$$

   For which values of k does the sequence converge uniformly on $\mathbb{R}$? On every bounded subset of $\mathbb{R}$?

3. Prove that $\sum_{n=1}^{\infty} (-1)^{n+1} x^n (1-x)$ converges absolutely as well as uniformly on $[0, 1]$. But the corresponding absolute series does not converge uniformly on $[0, 1]$.

4. Let

$$S(x) = \sum_{n=1}^{\infty} \frac{1}{n^2 \ln(n+1)} x^n.$$

Prove that

(a) $S(x)$ is continuous on $[-1, 1]$.

(b) $S'_+(-1)$ is finite but $S'_-(1) = \infty$.

5. Determine whether or not that following series converges uniformly on $(0, \infty)$:

$$(a) \sum_{n=1}^{\infty} \frac{x}{1 + n^4 x^2}; \qquad (b) \sum_{n=1}^{\infty} x^2 e^{-nx}; \qquad (c) \sum_{n=1}^{\infty} \frac{1}{n} \left[e^x - \left(1 + \frac{x}{n}\right)^n \right].$$

6. Let $P_n(x)$ be a sequence of real polynomials with degree not to exceed N , a fixed integer. Assume that $P_n(x) \to 0$ pointwisely for $0 \le x \le 1$. Show that $P_n(x) \to 0$ uniformly on $[0, 1]$.

7. Let $f, f_n : \mathbb{R} \to \mathbb{R}$ be functions such that $f_n(x_n) \to f(x)$ as $n \to \infty$ whenever $x_n \to x$. Prove that $f(x)$ is continuous.

8. Construct a sequence of continuous functions f_n which is not uniformly convergent, but $f_n(x) \to f(x)$ pointwisely and $f(x)$ is continuous.

9. Construct a sequence of continuous functions f_n such that $f_n(x) \to f(x)$ pointwisely but nowhere uniformly.

10. Show that there exists a positive sequence $a_1 < a_2 < \cdots$ and a compact set C such that the sequence $\sin(a_n x)$ converges pointwisely on C but not uniformly.

11. Let $f_n : \mathbb{R} \to \mathbb{R}$ be differentiable for each $n \in \mathbb{N}$ with $|f'_n(x)| \le 1$ for all n and x . Assume

$$\lim_{n \to \infty} f_n(x) = f(x)$$

for all $x \in \mathbb{R}$. Show that $f(x)$ is continuous on $\mathbb{R}$

12. Let $f : [0, 1] \to \mathbb{R}$ be a continuous function. Define recurrently the sequence of functions on $[0, 1]$ by

$$f_1(x) = \int_0^x f(t) \, dt, \quad f_{n+1}(x) = \int_0^x f_n(t) \, dt, \quad \text{for } n \ge 1.$$

Show that, for $n \in \mathbb{N}$,

$$f_n(x) = \frac{1}{(n-1)!} \int_0^x (x - t)^{n-1} f(t) \, dt.$$

13. Let $f_0(x)$ be integrable on $[a, b]$. Define the sequence

$$f_n(x) = \int_a^x f_{n-1}(t) dt, \quad (n = 1, 2, \dots).$$

(a) Show that $f_n(x)$ converges to 0 uniformly on $[a, b]$.

(b) Show that $f(x) = \sum_{n=1}^{\infty} f_n(x) \, dx$ is continuous on $[a, b]$ and find a simple expression for $f(x)$.

14. Let f_n $(n \in \mathbb{N})$ be continuously differentiable functions on $[0, 1)$ such that

$$f_1(x) = 1, \quad f'_{n+1}(x) = f_n(x) f_{n+1}(x), \quad f_{n+1}(0) = 1.$$

Show that $\lim_{n \to \infty} f_n(x)$ exists for every $x \in [0, 1)$ and find its limit.

15. Define the polynomial sequence by

$$P_0(x) = 0, \quad P_n(x) = P_{n-1}(x) + \frac{1}{2}\left(x - P_{n-1}^2(x)\right), \quad n \in \mathbb{N}.$$

Show that $P_n(x)$ converges to $\sqrt{x}$ uniformly on $\mathbb{R}$.

16. Let $p_1(x) = 0$ and

$$p_{n+1}(x) = p_n(x) + \frac{x^2 - p_n^2(x)}{2}, \quad n = 1, 2, \ldots.$$

Show that $p_n(x) \to |x|$ on $[-1, 1]$ uniformly. Notice that each $p_n(x)$ is a polynomial, this gives another explicit example of the Weierstrass approximation theorem.

17. Let $p_n(x)$ be a sequence of polynomials. If $p_n(x) \to p(x)$ on $\mathbb{R}$ uniformly, then $p(x)$ is a polynomial. This result indicates that the Weierstrass approximation theorem holds only on finite closed intervals.

18. Let $B_n(f)$ be the nth Bernstein polynomial of f. If f is Lipschitz continuous on $[0, 1]$ with the Lipschitz constant L, show that

$$|B_n(f) - f(x)| \leq \frac{1.2L}{n^{1/3}}.$$

19. Let $B_n(f)$ be the nth Bernstein polynomial of f. If $|f''(x)| \leq M$ on $[0, 1]$, show that

$$|B_n(f) - f(x)| \leq \frac{1}{2n}Mx(1 - x).$$

20. Let $B_n(f)$ be the nth Bernstein polynomial of f. Show that

$$\lim_{n \to \infty} n\left(B_n(x^3) - x^3\right) = 3(1 - x)x^2.$$

21. Let $B_n(f)$ be the nth Bernstein polynomial of f. Prove that

(a) If f is increasing on $[0, 1]$, then so is $B_n(f)$. *Hint:* Let $y_k = f(k/n)$. Show that $\sum_{k=m}^{n}\binom{n}{k}x^k(1 - x)^{n-k}$ is increasing on $[0, 1]$ for $0 < m < n$. Then use Abel's summation formula.

(b) If f is convex on $[0, 1]$, then so is $B_n(f)$.

22. (**Kelisky and Rivlin**). Let f^k be the kth iterate of the function f. For any function $f(x)$ which is defined on $[0, 1]$, show that

$$\lim_{k \to \infty} (B_n(f))^k = f(0) + (f(1) - f(0))x.$$

This indicates that the only functions invariant under the map B_n are the linear function $f(x) = ax + b$.

23. Let a_n, b_n be positive sequences. Suppose that the radius of convergence for $\sum_{n=0}^{\infty} a_n x^n$ is 1 and $\sum_{n=0}^{\infty} a_n$ diverges. If $\lim_{n \to \infty} b_n/a_n = A$, show that

$$\lim_{x \to 1^-} \frac{\sum_{n=0}^{\infty} b_n x^n}{\sum_{n=0}^{\infty} a_n x^n} = A.$$

24. Let f be continuous on $\mathbb{R}$. Define

$$f_n(x) = \frac{1}{n} \sum_{k=0}^{n-1} f\left(x + \frac{k}{n}\right), \quad n \in \mathbb{N}.$$

Show that $f_n(x)$ converges uniformly on every finite interval $[a, b]$.

25. Let $f(x) : \mathbb{R} \to \mathbb{R}$ be continuous and $|f(x)| < |x|$ for each $x \neq 0$. Define a sequence $f_n(x)$ by

$$f_1(x) = f(x), f_n(x) = (f \circ f_{n-1})(x), \quad n = 1, 2, \dots.$$

Show that the sequence $f_n(x)$ converges uniformly on $[-A, A]$ for any positive constant A.

26. Let $f_n : [0, 1] \to [0, \infty)$ be continuous for each $n \in \mathbb{N}$ and

$$f_1(x) \geq f_2(x) \geq f_3(x) \geq \cdots \quad \text{for all } x \in [0, 1].$$

If $f_n(x) \to f(x)$ pointwisely, show that there exists $t \in [0, 1]$ such that

$$f(t) = \sup_{x \in [0,1]} f(x).$$

27. Let $f(x) = 2x(1 - x), x \in \mathbb{R}$. Define a sequence $f_n(x)$ by

$$f_1(x) = f(x), f_n(x) = (f \circ f_{n-1})(x), \quad n = 1, 2, \dots.$$

Evaluate $\int_0^1 f_n(x)\, dx$.

28. Let $f(x) : [0, 1] \to \mathbb{R}$ be continuous. Define the sequence $f_n : [0, 1] \to \mathbb{R}$ by

$$f_0(x) = f(x), f_n(x) = \frac{1}{x} \int_0^x f_{n-1}(t)dt, \quad (n = 1, 2, \dots.).$$

Determine $\lim_{n\to\infty} f_n(x)$ for every $x \in [0, 1]$.

29. Show that the infinite series

$$\sum_{n=1}^{\infty} \frac{(2 + \sin n)^n}{n\, 3^n}$$

converges.

30. Let

$$f(x) = \sum_{n=0}^{\infty} \frac{\sin(2^n x)}{n!}.$$

Show that this series diverges everywhere except $x = 0$.

31. Let

$$\phi(x) = \begin{cases} x, & 0 \leq x \leq 1/2, \\ 1 - x, & 1/2 < x \leq 1. \end{cases}$$

Then extend $\phi(x)$ on $\mathbb{R}$ such that $\phi(x + 1) = \phi(x)$. Define

$$f(x) = \sum_{n=o}^{\infty} \frac{\phi(4^n x)}{4^n}.$$

Show that $f(x)$ is nowhere differentiable on $\mathbb{R}$.

32. (**Peano Curve**). Let

$$\phi(x) = \begin{cases} 0, & 0 \le x \le 1/3, \\ 3x - 1, & 1/3 < x \le 2/3, \\ 1, & 2/3 < x \le 1. \end{cases}$$

Then extend $\phi(x)$ on $\mathbb{R}$ such that $\phi(x+2) = \phi(x), \phi(-x) = \phi(x)$. Define

$$x(t) = \sum_{n=0}^{\infty} \frac{\phi(3^{2n}t)}{2^{n+1}},$$

$$y(t) = \sum_{n=0}^{\infty} \frac{\phi(3^{2n+1}t)}{2^{n+1}}.$$

Prove that $x(t)$ and $y(t)$ are nowhere differentiable and determine the range of $(x(t), y(t)) = [0,1] \times [0,1]$ for $t \in [0,1]$.

33. Let

$$f_n(x) = \frac{x^2}{x^2 + (1-nx)^2}, \quad x \in [0,1].$$

Show that the sequence f_n has no uniformly convergent subsequence.

34. Let f be continuous on $[1/2, 1]$. Show that the sequence $x^n f(x)$ converges uniformly on $[1/2, 1]$ if and only if $f(1) = 0$.

35. Determine if the series

$$\sum_{n=1}^{\infty} \frac{x^n}{1 + x^n}$$

converges uniformly on $(-1, 1)$.

36. Prove that for every odd continuous function f on $[-1, 1]$ and for every $\epsilon > 0$ there is an $n \in \mathbb{N}$ and $c_1, c_2, \ldots, c_n \in \mathbb{R}$ such that

$$\max_{x \in [-1,1]} \left| f(x) - \sum_{k=1}^{n} c_k x^{2k+1} \right| < \epsilon.$$

37. Let $f : [a, b] \to \mathbb{R}$ be continuously differentiable. Define

$$F_n(x) = \frac{n}{2} \left[f\left(x + \frac{1}{n}\right) - f\left(x - \frac{1}{n}\right) \right].$$

Show that $F_n(x)$ converges uniformly on $[\alpha, \beta]$ $(a < \alpha < \beta < b)$ and

$$\lim_{n \to \infty} \int_{\alpha}^{\beta} F_n(x)dx = f(\beta) - f(\alpha).$$

38. Prove that there is a unique series in the form of

$$\sin(\pi x) = \sum_{n=1}^{\infty} a_n x^n (1-x)^n$$

that converges on $\mathbb{R}$ and $|a_n| \le M \frac{\pi^{2n}}{(2n)!}$ for some constant M and for all $n \in \mathbb{N}$.

39. Let
$$f_n(x) = \frac{(-1)^{n-1}}{x^2 + n}, \quad x \in \mathbb{R}, \, n \in \mathbb{N}.$$

Show that the series $\sum_{n=1}^{\infty} f_n(x)$ converges uniformly on $\mathbb{R}$ and nowhere converges absolutely.

40. Show that $\sum_{n=2}^{\infty} \frac{1}{n \ln n} \cos nx$ does not converge uniformly on $(0, \pi]$.

41. Show that
$$\sum_{n=1}^{\infty} \frac{x^n}{1 + x + x^2 + \cdots + x^{2n-1}} \cos nx$$

converges uniformly on $(0, 1]$.

42. Let $\{r_1, r_2, \cdots, r_n, \cdots\}$ be all rational number in $(0, 1)$. For each $x \in (0, 1)$, define
$$f(x) = \sum_{x > r_n} \frac{1}{2^n}.$$

Evaluate $\int_0^1 f(x)\, dx$.

43. Let $\{r_1, r_2, \cdots, r_n, \cdots\}$ be all rational number in $[0, 1]$. For each $x \in [0, 1]$, define
$$f(x) = \sum_{n=1}^{\infty} \frac{|x - r_n|}{3^n}.$$

Show that $f(x)$ is continuous on $[0, 1]$ and differentiable at all irrationals in $[0, 1]$.

44. Let the function $f : (0, 1) \to [0, \infty)$ be zero except at the distinct points $\{a_1, a_2, \ldots\}$. Let $b_n = f(a_n)$.

   (a) Prove that if $\sum_{n=1}^{\infty} b_n$ converges, then f is differentiable for least one point $x \in (0, 1)$.

   (b) Prove that for any nonnegative sequence b_n with $\sum_{n=1}^{\infty} b_n$ diverges, there exists a sequence a_n such that the function f defined as above is nowhere differentiable.

45. Show that
$$\int_0^1 x^{-x}\, dx = \sum_{n=1}^{\infty} \frac{1}{n^n} \quad \text{and} \quad \int_0^1 x^x\, dx = \sum_{n=1}^{\infty} \frac{(-1)^{n-1}}{n^n}.$$

46. Let $k \in \mathbb{N}$ and $x > 0$. Show that
$$\sum_{n=1}^{\infty} \frac{1}{(x+p)(2x+p)\cdots(nx+p)} x^{n-1} = e \int_0^1 t^{p-1+x} e^{-t^x}\, dt.$$

47. Let $k \in \mathbb{N}$ and $|x| < 1$. Show that
$$\sum_{n=1}^{\infty} \frac{1}{n}\left(\frac{1}{1 - x^k} - \sum_{i=0}^{n} x^{ik}\right) = -\frac{\ln(1 - x^k)}{1 - x^k}.$$

48. (**Monthly Problem 11597, 2011**). Let $f(x) = x/\ln(1 - x)$. Prove that for $0 < x < 1$,
$$\sum_{n=1}^{\infty} \frac{x^n(1 - x)^n}{n!} f^{(n)}(x) = -\frac{1}{2} x f(x).$$

49. Suppose the power series of $f(x)$ converges in $(-R, R)$. Show that, for $|x| < R$,

$$\sum_{n=0}^{\infty} f^{(n)}(0) \left(e^x - \sum_{k=0}^{n} \frac{1}{k!} x^k \right) = \int_0^x e^{x-t} f(t) \, dt.$$

50. Suppose the power series of $f(x)$ converges in $(-R, R)$. Let $T_n(x)$ be its nth degree Taylor polynomial.

   (a) For $|x| < R$, find $\sum_{n=1}^{\infty} (f(x) - T_n(x))$.
   (b) For $|x| < R$, find $\sum_{n=1}^{\infty} (-1)^{n-1} (f(x) - T_n(x))$.

51. Let $K_n(x)$ be the *Fejér kernel* that is defined by

$$K_n(x) = \frac{1}{2(n+1)} \left(\frac{\sin[(n+1)x/2]}{\sin(x/2)} \right)^2.$$

   Prove that

   (a) for $n \in \mathbb{N}$ and $0 < |x| < \pi$,

$$0 \le K_n(x) \le \frac{\pi}{(n+1)x^2}.$$

   (b) for $n \in \mathbb{N}$,

$$\frac{1}{\pi} \int_{-\pi}^{\pi} K_n(x) \, dx = 1.$$

   (c) if f is continuous on $[-\pi, \pi]$, then

$$\lim_{n \to \infty} \frac{1}{\pi} \int_{-\pi}^{\pi} K_n(x) f(x) \, dx = f(0).$$

52. Let the sequence $\phi_n(x) : [-1, 1] \to \mathbb{R}$ satisfy

   (a) $\phi_n(x)$ is nonnegative and continuous on $[-1, 1]$;
   (b) $\lim_{n \to \infty} \int_{-1}^{1} \phi_n(x) \, dx = 1$;
   (c) For any $0 < c < 1$, $\phi_n(x)$ uniformly converges to 0 on $[-1, -c]$ and $[c, 1]$.
   If $f(x)$ is continuous on $[-1, 1]$, prove that

$$\lim_{n \to \infty} \int_{-1}^{1} \phi_n(x) f(x) \, dx = f(0).$$

53. Show that Riemann zeta function $\zeta(x) = \sum_{n=1}^{\infty} \frac{1}{n^x}$ is continuous and differentiable on $(1, \infty)$, but it does not converge uniformly on $(1, \infty)$.

54. Let $f_k : [a, b] \to \mathbb{R}$ be a sequence of functions. $f_k(x)$ is *uniformly integrable* on $[a, b]$ if for every $\epsilon > 0$, there exists a $\delta > 0$, which is independent of k, such that

$$S_n(P, f_k) = \sum_{i=1}^{n} w_{ik} \Delta x_i < \epsilon,$$

   for all $k \in \mathbb{N}$ and for all partitions P of $[a, b]$ with $\|P\| < \delta$. Show that if $f_n(x)$ is uniformly integrable on $[a, b]$ and $f_n \to f$ pointwise, then f is integrable on $[a, b]$ and

$$\lim_{n \to \infty} \int_a^b f_n(x) \, dx = \int_a^b \lim_{n \to \infty} f_n(x) \, dx.$$

55. (**Monthly Problem 12207, 2020**). Let $f : [0,1] \to \mathbb{R}$ be a continuous function satisfying $\int_0^1 f(x)\,dx = 1$. Evaluate

$$\lim_{n\to\infty} \frac{n}{\ln n} \int_0^1 x^n f(x^n) \ln(1-x)\,dx.$$

56. Let f and g be positive continuous functions on $[a,b]$. Let $M = \max_{x\in[a,b]} f(x)$. Prove that

(a) $\lim_{n\to\infty} \sqrt[n]{\int_a^b g(x) f^n(x)\,dx} = M$.

(b) $\lim_{n\to\infty} \dfrac{\sqrt[n]{\int_a^b g(x) f^{n+1}(x)\,dx}}{\sqrt[n]{\int_a^b g(x) f^n(x)\,dx}} = M$.

(c) $\lim_{n\to\infty} n \left(\sqrt[n]{\int_a^b g(x) f^{n+1}(x)\,dx} - \sqrt[n]{\int_a^b g(x) f^n(x)\,dx} \right) = M \ln M$.

57. Let $f : [0,1] \to (0,1]$ and $g : [0,1] \to \mathbb{R}$ be continuous functions with $\int_0^1 g(x)\,dx = 1$. Prove that

$$\lim_{n\to\infty} \left(\int_0^1 \sqrt[n]{f(x)}\, g(x)\,dx \right)^n = \exp\left(\int_0^1 g(x) \ln f(x)\,dx \right).$$

58. (**Monthly Problem 12242, 2021**). For $n \geq 1$, let

$$I_n = \int_0^1 \frac{(\sum_{k=0}^n x^k/(2k+1))^{2022}}{(\sum_{k=0}^{n+1} x^k/(2k+1))^{2021}}\,dx.$$

Let $L = \lim_{n\to\infty} I_n$. Compute L and $\lim_{n\to\infty} n(I_n - L)$.

59. Let a_n be a positive sequence, and let $f_n(x) = \sum_{k=1}^n a_k x^k$. Define

$$I(n,m) = \int_0^1 \frac{(f_n(x))^{m+1}}{(f_{n+1}(x))^m}\,dx.$$

If $\lim_{n\to\infty} n a_n = A$, show that, for any $m \in \mathbb{N}$,

$$\lim_{n\to\infty} I(n,m) = L := \sum_{k=1}^\infty \frac{a_k}{k+1},$$

which is independent of m, and

$$\lim_{n\to\infty} n(I(n,m) - L) = -A.$$

60. Let $\sum_{n=1}^\infty a_n = A$. Show that

$$\int_0^\infty e^{-x} \left(\sum_{n=1}^\infty a_n \frac{x^n}{n!} \right) dx = A.$$

Using your result calculate

$$\int_0^\infty e^{-x} \left(\int_0^x \frac{\sin t}{t}\,dt \right) dx.$$

61. Evaluate
$$\lim_{x \to 1^-} \sum_{n=1}^{\infty} \frac{(1-x)x^n}{1+x^n}.$$

62. Evaluate
$$\lim_{x \to 0^+} \sum_{n=1}^{\infty} \frac{(-1)^n}{n^x}.$$

63. Let $f(x) = \sum_{n=2}^{\infty} \frac{1}{\ln n} \cos(nx)$. Show that
$$\int_0^{\infty} \frac{f(x)}{1+x^2}\, dx = \frac{\pi}{2} \sum_{n=2}^{\infty} \frac{1}{e^n \ln n}.$$

64. Let $\alpha > 0$ and let a_n be a positive sequence such that
$$\lim_{n \to \infty} \frac{1}{n^{\alpha}} \sum_{k=1}^{n} a_k = L.$$

For every continuous function f on $[0,1]$, show that
$$\lim_{n \to \infty} \frac{1}{n^{\alpha}} \sum_{k=1}^{n} f\left(\frac{k}{n}\right) a_k = \alpha L \int_0^1 x^{\alpha-1} f(x)\, dx.$$

65. Let $f(x) = \sum_{n=0}^{\infty} e^{-n} \cos n^3 x$. Show that the Taylor series of $f(x)$ converges at $x = 0$ only.

66. Show that, for $x \in [-1, 1]$,
$$(\arcsin x)^2 = \frac{1}{2} \sum_{n=1}^{\infty} \frac{1}{n^2 \binom{2n}{n}} (2x)^{2n},$$
$$(\arcsin x)^4 = \frac{3}{2} \sum_{n=1}^{\infty} \left\{ \sum_{k=1}^{n-1} \frac{1}{k^2} \right\} \frac{1}{n^2 \binom{2n}{n}} (2x)^{2n}.$$

67. Let C_n be the Catalan numbers. Prove that
$$\sum_{n=0}^{\infty} \frac{1}{C_n} x^n = \begin{cases} 1 + \frac{x(4-x)^{3/2}+6x\sqrt{4-x}+24\sqrt{x}\arcsin(\sqrt{x}/2)}{(4-x)^{5/2}}, & x \in [0, 4); \\ 1 - \frac{|x|(4-x)^{3/2}+6\sqrt{|x|(4-x)}+24\sqrt{|x|}\ln[(\sqrt{|x|}+\sqrt{4-x})/2]}{(4-x)^{5/2}}, & x \in (-4, 0]. \end{cases}$$

68. Let C_n be the Catalan numbers. Prove that
    (a) $\sum_{n=0}^{\infty} \frac{(-1)^n}{(2n+1)^3} = \frac{\pi^3}{32}$.
    (b) $\sum_{n=0}^{\infty} \frac{n+1}{4^{2n}(2n+1)^3} C_n = \frac{7\pi^3}{216}$.
    (c) $\sum_{n=1}^{\infty} \frac{2^n(n+1)}{nC_n} \left(1 + \frac{1}{2^2} + \cdots + \frac{1}{(n-1)^2}\right) = \frac{\pi^3}{48}$.

69. Let C_n be the Catalan numbers. Prove that
    (a) $\sum_{n=0}^{\infty} \frac{(n+1)}{4^n(2n+1)^2} C_n = \frac{\pi}{2} \ln 2$.
    (b) $\sum_{n=0}^{\infty} \frac{(n+1)(2n)!!}{4^n(2n+1)^2(2n+1)!!} C_n = \frac{7}{8}\zeta(3)$.

70. (**Monthly Problem 12051, 2018**). Prove that

$$\sum_{n=0}^{\infty} \binom{2n}{n} \frac{1}{4^n (2n+1)^3} = \frac{\pi^3}{48} + \frac{\pi}{4} \ln^2 2.$$

71. Show that

$$\int_0^1 \frac{(\arcsin x^2)^2}{\sqrt{1-x^2}} \, dx = \frac{\pi^3}{4} - \frac{3}{4} \pi \ln^2 2 - 2\pi \sum_{n=1}^{\infty} \frac{1}{2^{n/2} \, n^2}.$$

72. Let $k \in \mathbb{N}$, and let $(\arctan x)^k = \sum_{n=1}^{\infty} a(n,k) x^n$. Prove that

$$\sum_{n=1}^{\infty} a(n,k)\beta\left(\frac{n+1}{2}\right) = \frac{1}{2^{2k+1}(k+1)} \pi^{k+1},$$

where $\beta(x)$ is the *incomplete beta function* and

$$\beta(n+1) = (-1)^n \left(\ln 2 - \left(1 - \frac{1}{2} + \cdots + \frac{(-1)^{n-1}}{n} \right) \right).$$

73. Evaluate

$$\int_0^{2\pi} e^{\cos x} \cos(\sin x) \cos nx \, dx.$$

74. Evaluate the following infinite series in closed form.
    (a) $1 - \frac{1}{4} + \frac{1}{6} - \frac{1}{9} + \frac{1}{11} - \frac{1}{14} + \cdots$
    (b) $\sum_{n=0}^{\infty} \frac{1}{n^2+n+1}$.
    (c) $\sum_{n=1}^{\infty} \frac{4^n}{\binom{2n}{n}(4n^2-1)}$.

75. For $x \in (-1, 1]$, calculate

$$\sum_{n=0}^{\infty} (-1)^n \left(\sum_{k=n+1}^{\infty} (-1)^{k-n-1} \frac{x^k}{k} \right).$$

76. Determine the sum of

$$\frac{x}{x+1} + \frac{x^2}{(x+1)(x^2+1)} + \frac{x^4}{(x+1)(x^2+1)(x^4+1)} + \cdots.$$

77. Show that

$$\lim_{x \to 1^-} \sum_{n=1}^{\infty} \frac{(-1)^{n-1}(1-x)x^n}{1-x^{2n}} = \frac{1}{2} \ln 2.$$

78. Show that

$$\sum_{n=2}^{\infty} \frac{(-1)^n \zeta(n)}{n(n+1)} = \frac{1}{2} (\ln(2\pi) + \gamma - 2).$$

79. Show that

$$\sum_{n=1}^{\infty} \frac{1}{n!(n+1)!} = \frac{1}{\pi} \int_{-1}^{1} \frac{xe^{2x}}{\sqrt{1-x^2}} \, dx.$$

80. Show that

(a) $\frac{13}{8} + \sum_{n=0}^{\infty} \frac{(-1)^{n+1}(n+2)\binom{2n+1}{n}}{4^{2n+3}} = \phi,$

(b) $\prod_{n=1}^{\infty} \left(1 + \frac{(-1)^n}{F_n^2}\right) = \phi,$

where $\phi = (1 + \sqrt{5})/2$ is the golden ratio and F_n is the nth Fibonacci number.

81. Let $x \in \mathbb{R}$. Define the sequence x_n recursively by $x_1 = 1$ and $x_{n+1} = x^n + nx_n$ for $n \geq 1$. Prove that

$$\prod_{n=1}^{\infty} \left(1 - \frac{x^n}{x_{n+1}}\right) = e^{-x}.$$

82. (**Monthly Problem 11299, 2007**). Show that

$$\prod_{n=2}^{\infty} \left(\frac{1}{e}\left(\frac{n^2}{n^2 - 1}\right)^{n^2-1}\right) = \frac{e^{3/2}}{2\pi}.$$

Hint: Express the logarithm of the infinite product as the combinations of $\sum_{n=1}^{\infty} \frac{\zeta(2n)-1}{n}$ and $\sum_{n=1}^{\infty} \frac{\zeta(2n)-1}{n+1}$.

83. Show that

$$\sum_{n=0}^{\infty} \frac{(-1)^n}{2n+1} \sum_{k=0}^{2n} \frac{1}{2n+4k+3} = \frac{3}{8}\pi \ln\phi - \frac{1}{16}\pi \ln 5,$$

where $\phi = (1 + \sqrt{5})/2$ is the golden ratio.

84. Let $\Phi_n(x)$ be the nth cyclotomic polynomial which is define by

$$\Phi_n(x) = \prod_{1 \leq k \leq n,\, \gcd(k,n)=1} (x - \xi^k), \quad n \in \mathbb{N}$$

where $\xi = e^{2\pi i/n}$. For example, $\Phi_1(x) = x - 1$, $\Phi_2(x) = x + 1$, $\Phi_3(x) = x^2 + x + 1$. Show that

$$\sum_{n=1}^{\infty} \frac{1}{n^4} \frac{\Phi_n'(e^{2\pi})}{\Phi_n(e^{2\pi})} = \frac{1}{\pi e^{2\pi}}\left(\frac{7}{4} + \frac{45}{\pi^3}\zeta(3)\right).$$

85. (**Putnam Problem 2016-B6**). Evaluate

$$\sum_{k=1}^{\infty} \frac{(-1)^{k-1}}{k} \sum_{n=0}^{\infty} \frac{1}{k2^n + 1}.$$

86. Let $\{x\} = x - [x]$ be the fractional part of x. Prove that for $n \in \mathbb{N}$,

$$\int_0^1 x^n \left\{\frac{1}{x}\right\} dx = 1 - \frac{1}{n+1}\sum_{k=2}^{n+1} \zeta(k).$$

87. Let $\{x\} = x - [x]$ be the fractional part of x. Prove that

(a) $\int_0^1 \left\{\frac{1}{x}\right\}\left\{\frac{1}{1-x}\right\} dx = 2\gamma - 1.$

(b) $\int_0^1 \left\{\frac{1}{x}\right\}^2 \left\{\frac{1}{1-x}\right\}^2 dx = 4\ln(2\pi) - 4\gamma - 5.$

88. Let $\{x\} = x - [x]$ be the fractional part of x. For $n \in \mathbb{N}$, prove that

$$\int_0^1 \int_0^1 (x+y)^n \left\{ \frac{1}{x+y} \right\}^n \, dx\,dy = 1 - \frac{1}{(n+1)(n+2)} \sum_{k=2}^{n+1} k\zeta(k+1).$$

Test your result by

(a) $\int_0^1 \int_0^1 (x+y) \left\{ \frac{1}{x+y} \right\} dx\,dy = 1 - \frac{1}{3}\zeta(3)$.

(b) $\int_0^1 \int_0^1 (x+y)^2 \left\{ \frac{1}{x+y} \right\}^2 dx\,dy = 1 - \frac{1}{6}\zeta(3) - \frac{1}{4}\zeta(4)$.

89. Let $|\alpha| < 1$. Evaluate

$$\int_0^\pi \frac{\ln(1 + \alpha \cos x)}{\cos x} \, dx.$$

90. Evaluate

$$\int_0^1 \frac{\ln(1 - 2x \cos\theta + x^2)}{x} \, dx.$$

91. Let

$$f(x) := \int_0^1 \frac{\ln(1-t)}{x-t} \, dt$$

Show that $f(x)$ satisfies that

$$f(x) + f(1/x) + \ln(1-x) \cdot \ln(1 - 1/x) = \frac{\pi^2}{6} = \zeta(2).$$

92. Let a_n be a numerical sequence. Show that series

$$\sum_{n=1}^\infty \frac{a_n}{n!} \int_0^x t^n e^{-t} \, dt$$

uniformly converges on $(0, \infty)$ if and only if $\sum_{n=1}^\infty a_n$ converges.

93. Evaluate

$$\lim_{x \to 0^+} x^2 \sum_{n=1}^\infty n^2 \int_{(n-1/2)x}^{(n+1/2)x} e^{-t^2/2} \, dt.$$

94. Show that

$$\lim_{x \to 0^+} \sum_{n=1}^\infty \left(\frac{1}{n^{1+x}} - \frac{1}{x} \right) = \gamma.$$

95. Show that

$$\sum_{n=2}^\infty (-1)^n \frac{\ln n}{n} = \gamma \ln 2 - \frac{1}{2} \ln^2 2.$$

96. Show that

$$\lim_{x \to \infty} \sum_{n=1}^\infty \frac{nx}{(n^2 + x^2)^2} = \frac{1}{2}.$$

97. Let H_n be the nth harmonic number. Define

$$H(k) = \sum_{n=1}^{\infty} \frac{(-1)^{n+1}}{n+k} H_n, \quad (k = 0, 1, 2, \ldots).$$

Evaluate $H(k)$ in a closed form. Test your results by

$$H(0) = \frac{\pi^2}{12} - \frac{1}{2} \ln^2 2, \quad H(1) = \frac{1}{2} \ln^2 2.$$

98. (**Monthly Problem 11916, 2016**). Show that if n, r, and s are positive integers, then

$$\binom{n+r}{n} \sum_{k=0}^{s-1} \binom{r+k}{r-1} \binom{n+k}{n} = \binom{n+s}{n} \sum_{k=0}^{r-1} \binom{s+k}{s-1} \binom{n+k}{n}.$$

Hint: Use the Wilf-Zeilberger algorithm.

99. (**Monthly Problem 12060, 2018**). Show that

$$\sum_{n=2}^{\infty} \frac{H_n H_{n+1}}{n(n^2-1)} = \frac{5}{2} - \frac{1}{4}\zeta(2) - \zeta(3).$$

100. (**Monthly Problem 12026, 2018**). For $n \in \mathbb{N}$, let

$$S_n = \sum_{k=1}^{n} \frac{(-1)^{n-k}}{k} \sum_{j=1}^{k} H_j.$$

Find $\lim_{n\to\infty} S_n / \ln n$ and $\lim_{n\to\infty} (S_{2n} - S_{2n-1})$.

101. Define $h_n = \sum_{k=1}^{n} \frac{1}{2k-1}$. Prove that

(a) $\sum_{n=1}^{\infty} h_n x^{2n-1} = -\frac{\ln\left(\frac{1-x}{1+x}\right)}{2(1-x^2)}$.

(b) $\sum_{n=1}^{\infty} \frac{h_n}{n} x^{2n} = \frac{1}{4} \ln^2\left(\frac{1-x}{1+x}\right)$.

102. Let

$$h_n = \sum_{k=1}^{n} \frac{1}{2k-1}, \quad h_n^* = \sum_{k=1}^{n} \frac{(-1)^{k-1}}{2k-1}.$$

Prove that

(a) For $|x| \le 1$,

$$\sum_{n=1}^{\infty} \frac{(-1)^{n-1} h_n}{n^2} x^{2n} = 2 \int_0^1 \frac{(\arctan(xt))^2}{t} \, dt.$$

(b) For $|x| \le 1$,

$$\sum_{n=1}^{\infty} \left(h_n^* - \frac{\pi}{4}\right) \frac{h_n}{n} x^{2n} = \int_0^1 \frac{(\arctan(xt))^2}{1+t^2} \, dt.$$

(c) $\sum_{n=1}^{\infty} \left(h_n^* - \frac{\pi}{4}\right) \frac{h_n}{n} = \frac{\pi^3}{192}$.

103. Let $S_n = \sum_{k=0}^{n} 1/\binom{n}{k}$. For $|x| < 1$, show that

$$\sum_{n=0}^{\infty} \frac{S_n}{n+1} x^{n+1} = -\frac{2\ln(1-x)}{2-x}.$$

104. Let $k \geq 2$ be positive integer. Evaluate

   (a) $S(k) := \sum_{n=1}^{\infty} \frac{1}{n} \left(\zeta(k) - \sum_{i=1}^{n} \frac{1}{i^k} \right).$
   (b) $t(k) := \sum_{n=1}^{\infty} \frac{1}{(n+1)^k} \left(\sum_{i=1}^{n} \frac{1}{i^k} \right).$
   (c) $T(k) := \sum_{n=1}^{\infty} \frac{1}{(n+1)^2} \left(\zeta(k) - \sum_{i=1}^{n} \frac{1}{i^k} \right).$

105. (**Monthly Problem 3815, 1937**). Let

$$f(x) = \sum_{n=1}^{\infty} \frac{1}{n} \sin \frac{x}{4^n}.$$

   Show that there is a positive constant c_1, independent of x, such that

$$|f(x)| < c_1 \ln(\ln x), \qquad x > e.$$

   Show also that there is a sequence $x_1 < x_2 < \cdots \to \infty$ and a positive constant c_2, independent of x, such that

$$|f(x_k)| > c_2 \ln(\ln x), \qquad k = 1, 2, \ldots.$$

106. (**Monthly Problem 11410, 2009**). For $0 < \phi < \pi/2$, find

$$\lim_{x \to 0} x^{-2} \left(\frac{1}{2} \ln \cos \phi + \sum_{n=1}^{\infty} \frac{(-1)^{n-1}}{n} \frac{\sin^2(nx)}{(nx)^2} \sin^2(n\phi) \right).$$

107. (**Monthly Problem 11515, 2010**). Find a closed form expression for

$$\sum_{n=1}^{\infty} 4^n \sin^4(2^{-n}x).$$

108. (**Monthly Problem 11659, 2012**). Let x be real with $0 < x < 1$, and consider the sequence a_n given by $a_0 = 0, a_1 = 1$, and for $n > 1$,

$$a_n = \frac{a_{n-1}^2}{xa_{n-2} + (1-x)a_{n-1}}.$$

   Show that

$$\lim_{n \to \infty} \frac{1}{a_n} = \sum_{k=-\infty}^{\infty} (-1)^k x^{k(3k-1)/2}.$$

109. (**Monthly Problem 11885, 2016**). Prove that

$$\sum_{p=1}^{\infty} \sum_{n=1}^{\infty} \sum_{m=1}^{\infty} \frac{1}{(m+n)^4 + ((m+n)(m+p))^2} = \frac{3}{2} \zeta(3) - \frac{5}{4} \zeta(4),$$

   where ζ is the Riemann zeta function.

110. (**Monthly Problem 11973, 2017**). Let G be the Catalan constant. Prove

$$G = \frac{\pi}{2} \sum_{n=0}^{\infty} \frac{\zeta(2n)}{(2n+1)4^n} \left(1 - \frac{2}{4^n}\right)$$

with $\zeta(0) = -1/2$.

111. (**Monthly Problem 11982, 2017**). Calculate

$$\lim_{x \to \infty} \left(\sum_{n=1}^{\infty} \left(\frac{x}{n}\right)^n\right)^{1/x}.$$

112. (**Monthly Problem 11890, 2016**). Find all x in $(1, \infty)$ such that

$$\left(\frac{1}{x} + \frac{x-1}{x+1}\right) + \frac{1}{3}\left(\frac{1}{x^3} + \left(\frac{x-1}{x+1}\right)^3\right) + \cdots = \frac{1}{2}\int_0^x \frac{dt}{\sqrt{1+t^2}}.$$

113. (**Putnam Problem 2014-A1**). Prove the every nonzero coefficient of the Taylor series of

$$(1 - x + x^2)e^x$$

about $x = 0$ is a rational number whose numerator (in lowest terms) is either 1 or a prime number.

114. (**Putnam Problem 2015-B6**). For $k \in \mathbb{N}$, let $A(k)$ be the number of odd divisors of k in the inteval $[1, \sqrt{2k})$. Evaluate

$$\sum_{k=1}^{\infty} (-1)^{k-1} \frac{A(k)}{k}.$$

115. Find

$$\lim_{n \to \infty} \sum_{k=0}^{n} (-1)^k \sqrt{\binom{n}{k}}.$$

116. Evaluate

$$S_0 := \sum_{n=2}^{\infty} n \coth^{-1} n(4n^2 - 3).$$

Hint: Using $\coth^{-1} x = \frac{1}{2}\ln\frac{x+1}{x-1}$ and

$$e^{2S_0} = \prod_{n=2}^{\infty} \left(\frac{2n-1}{2n+1}\right)^{2n} \left(\frac{n+1}{n-1}\right)^n.$$

117. Show that

$$n! = \left(\frac{n}{e}\right)^n \sqrt{2\pi n} \prod_{k=n}^{\infty} \{(1 + 1/k)^{k+1/2}/e\}.$$

118. For $x > 0$, show that

$$\sum_{n=0}^{\infty} \frac{1}{\cosh(2n+1)x} = \sum_{n=0}^{\infty} \frac{(-1)^n}{\sinh(2n+1)x}.$$

119. Evaluate the double sum

$$\sum_{i=1}^{\infty} \sum_{j=1,(i,j)=1}^{\infty} \frac{x^{i-1}y^{j-1}}{1-x^iy^j}.$$

120. Let $n, m \in \mathbb{N}$ and $(n, m) = 1$. Find the closed form of

$$S(n,m) = \sum_{i=1}^{\infty} \sum_{j=1,nj\neq mi}^{\infty} \frac{1}{n^2j^2 - m^2i^2}$$

and show that $S(n, m) + S(m, n) = \pi^2/4nm$.

121. Show that

$$\frac{1}{\sqrt{1}} + \frac{1}{\sqrt{2}} + \cdots + \frac{1}{\sqrt{n}} = 2\sqrt{n} + C + \frac{1}{2\sqrt{n}} + O\left(\frac{1}{n^{3/2}}\right),$$

where

$$C = -(1+\sqrt{2})\left(1 - \frac{1}{\sqrt{2}} + \frac{1}{\sqrt{3}} - \frac{1}{\sqrt{4}} + \cdots\right).$$

122. Show that

$$\sum_{n=1}^{\infty} \frac{1}{\sqrt{n}(\sqrt{n} + \sqrt{n-1})^2} = (1+\sqrt{2})\sum_{n=1}^{\infty} \frac{(-1)^{n+1}}{\sqrt{n}}.$$

123. (**An Elementary Approximation for the Stirling Formula**). Let $n \in \mathbb{N}$. For $n \geq 2$, show that

$$\frac{7}{8} < \ln(n!) - \left(n + \frac{1}{2}\right)\ln n + n < 1.$$

124. For $0 < x < 1$, show that

$$x < \frac{1}{2}\ln\left(\frac{1-x}{1+x}\right) < x + \frac{x^3}{3(1-x^2)}.$$

125. For positive $x, y, z \in \mathbb{R}$, show that

(a) $x + y + z \leq 2\left(\frac{x^2}{y+z} + \frac{y^2}{x+z} + \frac{z^2}{x+y}\right)$,

(b) $\left(\frac{x+y}{x+y+z}\right)^{1/2} + \left(\frac{x+z}{x+y+z}\right)^{1/2} + \left(\frac{y+z}{x+y+z}\right)^{1/2} \leq \sqrt{6}$.

(c) Let $k \geq 0$ be an integer and let $\alpha > 0$ be a real number. For $x, y, z \in (-1, 1)$, prove that

$$\frac{x^{2k}}{(1-x^2)^\alpha} + \frac{y^{2k}}{(1-y^2)^\alpha} + \frac{z^{2k}}{(1-z^2)^\alpha} \geq \frac{x^k y^k}{(1-xy)^\alpha} + \frac{y^k z^k}{(1-yz)^\alpha} + \frac{x^k z^k}{(1-xz)^\alpha}.$$

7

Improper and Parametric Integration

Nature laughs at the difficulties of integration.

— Pierre-Simon de Laplace

I could never resist an integral.

— Godfrey Harold Hardy

The Riemann integration in Chapter 5 has been defined only for functions that are bounded on a finite closed interval. In this chapter, with an extra limiting process, we develop rules which enable us to integrate some functions that are unbounded or are defined on an unbounded interval. When the integrand involves a parameter, a study of uniform convergence of the integral is needed in order to determine whether or not the differentiation and integration on the parameter are allowable under the integral sign. We establish the required theorems in Section 7.2. As an application, we introduce the basic theory of the gamma function and its relation to the beta function. The gamma function is often the first of the so-called *special functions* that the reader meets beyond the level of calculus, and frequently appears in formulas of analysis. Lots of analysis techniques have been invoked in the process of establishing various properties of the gamma function.

7.1 Improper Integrals

Mathematics is not the rigid and rigidity-producing schema that the layman thinks it is; rather, in it we find ourselves at that meeting point of constraint and freedom that is the very essence of human nature. — Hermann Weyl

First, we extend the definition of the Riemann integral $\int_a^b f(x)\,dx$ to integrands which are unbounded at $x = b$. Let's begin with $[a, b] = [0, 1]$. Let $f(x) = (1 - x)^{-1}$ and $g(x) = 1/\sqrt{1 - x}$, respectively. For each $0 < c < 1$, we have

$$\int_0^c f(x)\,dx = -\ln(1 - c), \quad \int_0^c g(x)\,dx = 2(1 - \sqrt{1 - c}).$$

As $c \to 1$, the value of the first integral tends to ∞, but the second integral has the finite limit 2. With these examples in mind we define an improper integral as follows.

Definition 7.1. *Let f be Riemann integrable on $[a, c]$ for each $c < b$. We define*

$$\int_a^b f(x)\,dx = \lim_{c \to b^-} \int_a^c f(x)\,dx.$$

If this limit exists then the improper integral converges, otherwise it diverges.

Recall that if $p \neq 1$, for any $a < c < b$, we have

$$\int_a^c \frac{1}{(b - x)^p}\,dx = \frac{1}{1 - p}\left((b - a)^{1-p} - (b - c)^{1-p}\right).$$

DOI: 10.1201/9781003304135-7

Based on Definition 7.1, we find that

Example 7.1. *The improper integral $\int_a^b (b-x)^{-p}\,dx$ converges if and only if $0 < p < 1$.*

In analogy with convergence of infinite sequences and series, it is important to establish criteria that tests when improper integrals converge. The following result offers a basic tool in determining the convergence.

Theorem 7.1 (Comparison Test). *Let f be continuous on $[a,b)$ and $|f(x)| \le g(x)$ for all $x \in [a,b)$. If $\int_a^b g(x)\,dx$ converges, then $\int_a^b f(x)\,dx$ converges and*

$$\left| \int_a^b f(x)\,dx \right| \le \int_a^b g(x)\,dx.$$

As a corollary, we have the following comparison test for divergence.

Corollary 7.1 *Let f and g be continuous on $[a,b)$ and $0 \le g(x) \le f(x)$ for all $x \in [a,b)$. If $\int_a^b g(x)\,dx$ diverges, then $\int_a^b f(x)\,dx$ diverges.*

We now turn to the convergence of integrals in which the integrand is bounded but where the interval of integration is unbounded.

Definition 7.2. *Let f be Riemann integrable on $[a,R]$ for each $R > a$. Then*

$$\int_a^\infty f(x)\,dx = \lim_{R \to \infty} \int_a^R f(x)\,dx.$$

If this limit exists then the improper integral converges, otherwise it diverges.

In contrast to Example 7.1, we have

Example 7.2. *The improper integral $\int_a^\infty x^{-p}\,dx\,(a>0)$ converges if and only if $p > 1$.*

As with convergence of sequences and series, there is a Cauchy criterion for improper integrals.

Theorem 7.2 (Cauchy Criterion). *$\int_a^\infty f(x)\,dx$ converges if and only if for every $\epsilon > 0$ there is an $R \in \mathbb{R}$ such that*

$$\left| \int_A^B f(x)\,dx \right| < \epsilon \qquad \text{whenever } B > A > R.$$

The corresponding comparison test becomes

Theorem 7.3 (Comparison Test). *Let $f(x)$ and $g(x)$ be continuous functions on $[a,\infty)$ with $0 \le f(x) \le g(x)$ for all $x \in [a,\infty)$.*

  1. If $\int_a^\infty g(x)\,dx$ converges, then $\int_a^\infty f(x)\,dx$ converges.

  2. If $\int_a^\infty f(x)\,dx$ diverges, then $\int_a^\infty g(x)\,dx$ diverges.

The convergence of integrals with unbounded integrands while the interval is infinite may be treated by combining Theorems 7.1 and 7.2. The following example illustrates the method.

Example 7.3. *Test convergence for*

$$\int_0^\infty \frac{e^{-x}}{\sqrt{x}}\,dx$$

In view of the improper nature of the integral, we decompose it into two parts:

$$\int_0^1 \frac{e^{-x}}{\sqrt{x}}\,dx \quad \text{and} \quad \int_1^\infty \frac{e^{-x}}{\sqrt{x}}\,dx.$$

Since

$$e^{-x}/\sqrt{x} \le 1/\sqrt{x} \quad \text{for } x \in [0,1]$$

and

$$e^{-x}/\sqrt{x} \le e^{-x} \quad \text{for } x \in [1,\infty),$$

the comparison test yields that both integrals above converge. Therefore, the original integral converges.

The integral test for infinite series (Theorem 2.22) suggests that much of the theory of convergence for improper integrals is similar to that of infinite series. To be specific, we have

Theorem 7.4. *The improper $\int_a^\infty f(x)\,dx$ converges if and only if for every sequence $a = a_0 < a_1 < a_2 < \cdots$ with $a_n \to \infty$ as $n \to \infty$, the series*

$$\sum_{n=0}^\infty \int_{a_n}^{a_{n+1}} f(x)\,dx$$

converges and

$$\int_a^\infty f(x)\,dx = \sum_{n=0}^\infty \int_{a_n}^{a_{n+1}} f(x)\,dx.$$

Proof. "$\Longrightarrow$" Suppose that $\int_a^\infty f(x)\,dx$ converges. By the Cauchy criterion, for every $\epsilon > 0$ there is an $R \in \mathbb{R}$ such that

$$\left| \int_A^B f(x)\,dx \right| < \epsilon \quad \text{whenever } B > A > R.$$

Since $a_n \to \infty$ as $n \to \infty$, there exists some $N \in \mathbb{N}$ such that $a_m > a_n > R$ whenever $m > n > N$. Hence

$$\left| \int_{a_n}^{a_{n+1}} f(x)\,dx + \int_{a_{n+1}}^{a_{n+2}} f(x)\,dx + \cdots \int_{a_{m-1}}^{a_m} f(x)\,dx \right| = \left| \int_{a_n}^{a_m} f(x)\,dx \right| < \epsilon.$$

This implies that the series $\sum_{n=1}^\infty \int_{a_n}^{a_{n+1}} f(x)\,dx$ converges by the Cauchy criterion for the series.

"$\Longleftarrow$" We proceed by contradiction. Suppose that the series $\sum_{n=1}^\infty \int_{a_n}^{a_{n+1}} f(x)\,dx$ converges for every sequence a_n, but $\int_a^\infty f(x)\,dx$ diverges. Thus, there is an $\epsilon > 0$ such that for any $R > a$

$$\left| \int_A^\infty f(x)\,dx \right| \ge \epsilon \quad \text{whenever } A > R.$$

In particular, let $a_0 = a$, $R_0 = \max\{1, a\}$. There is $a_1 > R_0$ such that

$$\left| \int_{a_1}^\infty f(x)\,dx \right| \ge \epsilon \quad \text{whenever } a_1 > R_0.$$

Next, let $R_1 = \max\{2, a_1\}$. Then there is $a_2 > R_1$ such that

$$\left| \int_{a_2}^\infty f(x)\,dx \right| \ge \epsilon \quad \text{whenever } a_2 > R_1.$$

In general, let $R_n = \max\{n + 1, a_n\}$. Then there is $a_{n+1} > R_n$ such that

$$\left| \int_{a_{n+1}}^\infty f(x)\, dx \right| \geq \epsilon \quad \text{whenever } a_n > R_{n-1}.$$

Clearly, a_n is increasing and $a_n \to \infty$ as $n \to \infty$. Moreover, we have

$$\left| \sum_{k=n}^\infty \int_{a_k}^{a_{k+1}} f(x)\, dx \right| = \left| \int_{a_n}^\infty f(x)\, dx \right| \geq \epsilon.$$

This contradicts that the series $\sum_{n=1}^\infty \int_{a_{n-1}}^{a_n} f(x)\, dx$ converges for every sequence a_n. $\quad\square$

This following example demonstrates explicitly how to use the series convergence tests in determining the convergence of improper integrals.

Example 7.4. *Test convergence for*

$$I = \int_0^\infty \frac{\sin x}{x}\, dx$$

Let $a_n = n\pi$ for $n = 0, 1, \ldots$. We first show the corresponding series

$$\sum_{n=0}^\infty \int_{n\pi}^{(n+1)\pi} \frac{\sin x}{x}\, dx$$

converges. Indeed, let $x = n\pi + t$. Then

$$\sum_{n=0}^\infty \int_{n\pi}^{(n+1)\pi} \frac{\sin x}{x}\, dx = \sum_{n=0}^\infty (-1)^n \int_0^\pi \frac{\sin t}{n\pi + t}\, dt.$$

Since $\int_0^\pi \frac{\sin t}{n\pi + t}\, dt$ is a decreasing sequence and

$$\int_0^\pi \frac{\sin t}{n\pi + t}\, dt \leq \int_0^\pi \frac{1}{n\pi}\, dt = \frac{1}{n} \to 0 \quad (\text{as } n \to \infty),$$

by the alternating series test, it follows that the series converges.

Let $S = \sum_{n=0}^\infty \int_{n\pi}^{(n+1)\pi} \frac{\sin x}{x}\, dx$. Then for every $\epsilon > 0$, there is an $N \in \mathbb{N}$ such that

$$\left| \int_0^{n\pi} \frac{\sin x}{x}\, dx - S \right| = \left| \sum_{k=0}^{n-1} \int_{k\pi}^{(k+1)\pi} \frac{\sin x}{x}\, dx - S \right| < \epsilon \quad \text{whenever } n > N. \tag{7.1}$$

To prove that I converges, assume $R > N\pi$. Then there is an $n_0 \in \mathbb{N}$ such that $n_0\pi \leq R < (n_0 + 1)\pi$. Since $\sin x$ does not change sign on $(n_0\pi, (n_0 + 1)\pi)$, it follows that the integral $\int_0^R \frac{\sin x}{x}\, dx$ is between $\int_0^{n_0\pi} \frac{\sin x}{x}\, dx$ and $\int_0^{(n_0+1)\pi} \frac{\sin x}{x}\, dx$. In view of (7.1), it implies that both $\int_0^{n_0\pi} \frac{\sin x}{x}\, dx$ and $\int_0^{(n_0+1)\pi} \frac{\sin x}{x}\, dx$ are between $S - \epsilon$ and $S + \epsilon$. Therefore,

$$\left| \int_0^R \frac{\sin x}{x}\, dx - S \right| < \epsilon \quad \text{whenever } R > N\pi$$

and so

$$I = \lim_{R \to \infty} \int_0^R \frac{\sin x}{x}\, dx = S.$$

Consequently, as an application of the expansion of $\csc x$ (6.23), we find that

$$
\begin{aligned}
S &= \sum_{n=0}^{\infty} \int_{n\pi}^{n\pi+\pi/2} \frac{\sin x}{x}\, dx + \sum_{n=0}^{\infty} \int_{n\pi+\pi/2}^{(n+1)\pi} \frac{\sin x}{x}\, dx \\
&= \sum_{n=0}^{\infty} \int_{0}^{\pi/2} (-1)^n \frac{\sin t}{n\pi + t}\, dt + \sum_{n=0}^{\infty} \int_{0}^{\pi/2} (-1)^n \frac{\sin t}{(n+1)\pi - t}\, dt \\
&= \int_{0}^{\pi/2} \frac{\sin t}{t}\, dt + \sum_{n=1}^{\infty} \int_{0}^{\pi/2} (-1)^n \left(\frac{1}{t+n\pi} + \frac{1}{t-n\pi} \right) \sin t\, dt \\
&= \int_{0}^{\pi/2} \left[\frac{1}{t} + \sum_{n=1}^{\infty} (-1)^n \left(\frac{1}{t+n\pi} + \frac{1}{t-n\pi} \right) \right] \sin t\, dt \\
&= \int_{0}^{\pi/2} \csc t \sin t\, dt = \frac{\pi}{2}.
\end{aligned}
$$

Here the term by term integration is justified by showing that

$$
\sum_{n=1}^{\infty} (-1)^n \left(\frac{1}{t+n\pi} + \frac{1}{t-n\pi} \right) \sin t
$$

uniformly converges on $[0, \pi/2]$. This follows from the Weierstrass M-test with $M_n = \frac{1}{\pi(n^2 - 1/4)}$.

It is interesting to notice that I does not converge absolutely, i.e., the integral $\int_0^\infty \frac{|\sin x|}{x}\, dx$ diverges. Indeed, we may obtain as before that

$$
\int_0^\infty \frac{|\sin x|}{x}\, dx = \sum_{n=0}^{\infty} \int_{n\pi}^{(n+1)\pi} \frac{|\sin x|}{x}\, dx = \sum_{n=0}^{\infty} \int_0^\pi \frac{\sin t}{n\pi + t}\, dt.
$$

Since $n\pi + t \le (n+1)\pi$ for $t \in [0, \pi]$, it follows that

$$
\int_0^\pi \frac{\sin t}{n\pi + t}\, dt \ge \frac{1}{(n+1)\pi} \int_0^\pi \sin t\, dt = \frac{2}{(n+1)\pi}.
$$

But the series $\frac{2}{\pi} \sum_{n=1}^{\infty} \frac{1}{n+1}$ diverges, this implies that $\int_0^\infty \frac{|\sin x|}{x}\, dx$ diverges.

When the improper integral does not necessarily absolutely converge, the comparison test is not applicable. Combining the Cauchy criterion and the second mean value theorem for integrals, we establish two more robust tests which improve the existing Dirichlet's and Abel's tests. We begin with an example.

Example 7.5. *Let f be Riemann integrable on $[a, R]$ $(a > 1)$ for any $R > a$. Prove that if $\int_a^\infty x f(x)\, dx$ converges, so does $\int_0^\infty f(x)\, dx$ itself.*

Proof. For every $B > A > a > 1$, we have

$$
\left| \int_A^B f(x)\, dx \right| = \left| \int_A^B x f(x) \cdot \frac{1}{x}\, dx \right|.
$$

In view of the fact that $1/x$ is decreasing and nonnegative, $xf(x)$ is integrable on $[A, B]$, applying the second mean value theorem for integral yields

$$\left| \int_A^B f(x)\, dx \right| = \left| \frac{1}{A} \int_A^\xi x f(x)\, dx \right|$$

for some $\xi \in (A, B)$. By the assumption, if we take A and B sufficiently large, for every given $\epsilon > 0$, we will have

$$\left| \int_A^B f(x)\, dx \right| < \epsilon.$$

This proves that $\int_0^\infty f(x)\, dx$ converges by the Cauchy criterion.   □

Using the ingredients contained in the proof of this example, we first establish Dirichlet's Test. Unlike the existing literature, we show that Dirichlet's test is not only necessary but also sufficient.

Theorem 7.5 (Dirichlet's Test). *Let $f : [a, \infty) \to \mathbb{R}$ be Riemann integrable on $[a, R]$ for every $R > a$. The improper integral $\int_a^\infty f(x)\, dx$ converges if and only if $f(x) = u(x)v(x)$ such that*

1. $u(x)$ is monotone and $\lim_{x \to \infty} u(x) = 0$,

2. For every $R > a$, $\int_a^R v(x)\, dx$ exists and bounded.

Proof. "$\Longrightarrow$" Suppose that $\int_a^\infty f(x)\, dx$ converges. By the Cauchy criterion, there is $A_1 > a$ such that

$$\left| \int_A^B f(x)\, dx \right| < 1 \quad \text{whenever } B > A \geq A_1.$$

By induction, for $n \geq 2$, there is $A_n \geq A_{n-1} + 1$ such that

$$\left| \int_A^B f(x)\, dx \right| < \frac{1}{n^3} \quad \text{whenever } B > A \geq A_n.$$

Clearly, the sequence $A_n \to \infty$ monotonically. For each $n \in \mathbb{N}$, let

$$u(x) = \begin{cases} 1, & x \in [a, A_1] \\ \frac{1}{n}, & x \in (A_n, A_{n+1}] \end{cases}$$

and

$$v(x) = \frac{f(x)}{u(x)}, \quad x \in [a, \infty).$$

Thus, $f(x) = u(x)v(x)$ and $u(x)$ satisfies (1) as expected. We now show that $v(x)$ satisfies (2). Since both $f(x)$ and $1/u(x)$ are integrable on $[a, R]$ for any $R > a$, it follows that $\int_a^R v(x)\, dx$ exists, and so it suffices to show that $\int_a^R v(x)\, dx$ is bounded. If $a \leq R \leq A_1$, here $v(x) = f(x)$. By assumption, there is a positive constant M such that

$$\left| \int_a^R v(x)\, dx \right| < M.$$

If $R > A_1$, there is $m \in \mathbb{N}$ such that $A_m < R \leq A_{m+1}$. Moreover,

$$
\begin{aligned}
\left| \int_a^R v(x)\,dx \right| &= \left| \int_a^{A_1} v(x)\,dx + \sum_{n=1}^{m-1} \int_{A_n}^{A_{n+1}} v(x)\,dx + \int_{A_m}^R v(x)\,dx \right| \\
&= \left| \int_a^{A_1} f(x)\,dx + \sum_{n=1}^{m-1} \int_{A_n}^{A_{n+1}} nf(x)\,dx + \int_{A_m}^R mf(x)\,dx \right| \\
&\leq \left| \int_a^{A_1} f(x)\,dx \right| + \sum_{n=1}^{m-1} \left| \int_{A_n}^{A_{n+1}} nf(x)\,dx \right| + m \left| \int_{A_m}^R f(x)\,dx \right| \\
&\leq M + 1 + 2 \cdot \frac{1}{2^3} + \cdots + (m-1) \cdot \frac{1}{(m-1)^3} + m \cdot \frac{1}{m^3} \\
&= M + \sum_{n=1}^m \frac{1}{n^2} < M + \pi^2/6.
\end{aligned}
$$

"$\Longleftarrow$" Let $V(R) = \int_a^R v(x)\,dx$. By assumption, there is a positive constant K such that $|V(R)| \leq K$ for every $R > a$. Applying the second mean value theorem for integrals yields

$$
\int_A^B u(x)v(x)\,dx = u(A) \int_A^\xi v(x)\,dx + u(B) \int_\xi^B v(x)\,dx \quad \text{whenever } B > A > a,
$$

where $\xi \in (A, B)$. Also,

$$
\left| \int_A^\xi v(x)\,dx \right| = |V(\xi) - V(A)| \leq 2K, \quad \left| \int_\xi^B v(x)\,dx \right| \leq 2K.
$$

On the other hand, for every $\epsilon > 0$, there is a $X > 0$ such that $|u(x)| < \frac{\epsilon}{4K}$ whenever $x > X$. Therefore, for $B > A > X$,

$$
\begin{aligned}
\left| \int_A^B u(x)v(x)\,dx \right| &\leq |u(A)| \left| \int_A^\xi v(x)\,dx \right| + |u(B)| \left| \int_\xi^B v(x)\,dx \right| \\
&< \frac{\epsilon}{4K} \cdot 2K + \frac{\epsilon}{4K} \cdot 2K = \epsilon.
\end{aligned}
$$

This shows that $\int_a^\infty f(x)\,dx$ converges by the Cauchy criterion. $\square$

As a direct consequence of this enhanced Dirichlet's test, the regular Abel's test can be improved through the following theorem.

Theorem 7.6 (Abel's Test). *Let $f : [a, \infty) \to \mathbb{R}$ be Riemann integrable on $[a, R]$ for every $R > a$. The improper integral $\int_a^\infty f(x)\,dx$ converges if and only if $f(x) = u(x)v(x)$ such that*

1. *$u(x)$ is monotone and bounded,*
2. *$\int_a^\infty v(x)\,dx$ converges.*

These two tests can be used to determine the convergence of series such as

$$
\int_0^\infty \frac{x \sin ax}{b^2 + x^2}\,dx, \quad \int_0^\infty e^{\sin x} \frac{\sin 2x}{x^\alpha}\,dx, \quad \int_0^\infty \frac{\sin(x + x^2)}{x^\alpha}\,dx,
$$

where $a, b, \alpha > 0$.

7.2   Integrals with Parameters

"I had learned to do integrals by various methods shown in a book that my high school physics teacher Mr. Bader had given me. [It] showed how to differentiate parameters under the integral sign—it's a certain operation. It turns out that's not taught very much in the universities; they don't emphasize it. But I caught on how to use that method, and I used that one damn tool again and again. [If] guys at MIT or Princeton had trouble doing a certain integral, [then] I come along and try differentiating under the integral sign, and often it worked. So I got a great reputation for doing integrals, only because my box of tools was different from everybody elses, and they had tried all their tools on it before giving the problem to me." — Richard Feynman

The solutions of practical problems, especially in differential equations, are frequently given in terms of integrals with parameters. For example, the arc length of the ellipse $x^2/a^2 + y^2/b^2 = 1$ is given by

$$L = 4 \int_0^{\pi/2} \sqrt{a^2 \sin^2 t + b^2 \cos^2 t}\, dt = 4b \int_0^{\pi/2} \sqrt{1 - k^2 \sin^2 t}\, dt,$$

where $k = \sqrt{b^2 - a^2}/b$ is the *eccentricity*. This leads to the elliptic integral

$$E(k) = \int_0^{\pi/2} \sqrt{1 - k^2 \sin^2 t}\, dt$$

with the parameter k. In general,

Definition 7.3. *Let $D \subseteq \mathbb{R}$ and $f(x,t) : [a,b] \times D \to \mathbb{R}$. For every $t \in D$, $f(x,t)$ is Riemann integrable on $[a,b]$. Then*

$$I(t) = \int_a^b f(x,t)\, dx \qquad (7.2)$$

is called an integral with parameter t.

Observe that the integral of a sequence of functions

$$I_n = \int_a^b f_n(x)\, dx$$

is a special case of (7.2) with $D = \mathbb{N}$ and the parameter n.

For the functions define by (7.2), we are interested in when it is possible to interchange the calculus operations (limit, differentiation and integration) with the integral—commonly known as calculus operations under the integral sign.

First, similar to Definition 6.2, we introduce

Definition 7.4. *$f(x,t) : [a,b] \times D \to \mathbb{R}$ converges to $g(x)$ uniformly as $t \to t_0$ if*

1. $\lim_{t \to t_0} f(x,t) = g(x)$ for each $x \in [a,b]$.

2. For every $\epsilon > 0$ there exists a $\delta > 0$, which is independent on x, such that

$$|f(x,t) - g(x)| < \epsilon \quad \text{whenever } |t - t_0| < \delta \text{ and for all } x \in [a,b].$$

Equipped with this definition, if $D = [c, d]$ is bounded and f is Riemann integrable on $[a, b]$ for every $t \in D$, under very mild conditions, we are allowed to take all the calculus operations under the integral sign. To be specific, we have

Theorem 7.7. *Let f be integrable on $[a, b]$ for every $t \in D$. If $f(x, t)$ converges to $g(x)$ uniformly as $t \to t_0$, then $g(x)$ is integrable on $[a, b]$ and*

$$\lim_{t \to t_0} \int_a^b f(x, t)\, dx = \int_a^b \lim_{t \to t_0} f(x, t)\, dx = \int_a^b g(x)\, dx.$$

Theorem 7.8. *If $f(x, t)$ is continuous on $[a, b] \times [c, d]$, then*

$$I(t) = \int_a^b f(x, t)\, dx$$

is continuous on $[c, d]$.

The hypotheses on the continuity of f on $[a, b] \times [c, d]$ is necessary. The function

$$f(x, t) = \begin{cases} \frac{t}{x^2 + t^2}, & (x, t) \neq (0, 0) \\ 0, & (x, t) = (0, 0) \end{cases}$$

provides a counterexample on $[0, 1] \times [0, 1]$. Indeed, we have

$$I(t) = \int_0^1 f(x, t)\, dx = \arctan\left(\frac{1}{t}\right) \to \frac{\pi}{2} \text{ as } t \to 0^+.$$

This implies $I(t)$ is not continuous at $t = 0$ since $I(0) = \int_0^1 f(x, 0)\, dx = 0$.

Theorem 7.9 (Fubini's Theorem). *If $f(x, t)$ is continuous on $[a, b] \times [c, d]$, then*

$$\int_c^d \int_a^b f(x, t)\, dx dt = \int_a^b \int_c^d f(x, t)\, dt dx.$$

Theorem 7.10 (Leibniz's Rule). *If $f(x, t), f_t(x, t)$ are continuous on $[a, b] \times [c, d]$, then*

$$I(t) = \int_a^b f(x, t)\, dx$$

is differentiable on $[c, d]$ and

$$I'(t) = \int_a^b f_t(x, t)\, dx.$$

Here we give a proof of Leibniz's rule only. The proofs of Theorems 7.7–7.9 are pretty straight forward and are left to the reader.

Proof of Leibniz's Rule. Let

$$g(t) = \int_a^b f_t(x, t)\, dx, \quad t \in [c, d].$$

For any $t \in [c, d]$, since $f_t(x, t)$ is continuous on $[a, b] \times [c, d]$, by Fubini's theorem, we have

$$\begin{aligned} \int_c^t g(s)\, ds &= \int_c^t \left(\int_a^b f_t(x, s)\, dx \right) ds \\ &= \int_a^b \left(\int_c^t f_t(x, s)\, ds \right) dx \\ &= \int_a^b (f(x, t) - f(x, c))\, dx \\ &= I(t) - I(c). \end{aligned}$$

Since $g(t)$ is continuous on $[c, d]$ by Theorem 7.8, using the Second fundamental theorem of calculus, we have

$$I'(t) = \frac{d}{dt} \left(\int_c^t g(s)\, ds \right) = g(t) = \int_a^b f_t(x, t)\, dx,$$

as desired. □

We illustrate some applications of these theorems with three examples.

Example 7.6. *For $0 < a < b$, find*

$$I = \int_0^1 \frac{x^b - x^a}{\ln x}\, dx.$$

Solution. Observe that

$$\frac{x^b - x^a}{\ln x}\, dx = \int_a^b x^t\, dt.$$

Clearly, $f(x, t) = x^t$ is continuous on $[0, 1] \times [a, b]$. By Fubini's theorem, we find that

$$I = \int_0^1 \left(\int_a^b x^t\, dt \right) dx = \int_a^b \left(\int_0^1 x^t\, dx \right) dt = \int_a^b \frac{dt}{1 + t} = \ln \frac{1 + b}{1 + a}.$$

□

Example 7.7. *Evaluate*

$$I(t) = \int_0^{\pi/2} \ln \left(\frac{1 + t \cos x}{1 - t \cos x} \right) \frac{dx}{\cos x}, \qquad (\text{for } |t| < 1).$$

Solution. We present two solutions based on Fubini's theorem and Leibniz's rule, respectively. First, let

$$f(x, t) = \frac{1}{\cos x} \ln \left(\frac{1 + t \cos x}{1 - t \cos x} \right).$$

Since

$$\lim_{x \to \pi/2} f(x, t) = \lim_{x \to \pi/2} \left(\frac{\ln(1 + t \cos x)}{\cos x} - \frac{\ln(1 - t \cos x)}{\cos x} \right) = 2t,$$

it follows that $f(x, t)$ is continuous on $[0, \pi/2] \times [-T, T]$ for any $0 < T < 1$. In view of that

$$f(x, t) = \frac{\ln(1 + t \cos x)}{\cos x} - \frac{\ln(1 - t \cos x)}{\cos x} = \int_{-t}^t \frac{dy}{1 + y \cos x},$$

by Fubini's theorem, we have

$$I(t) = \int_{-t}^t \int_0^{\pi/2} \frac{1}{1 + y \cos x}\, dx dy$$

$$= \int_{-t}^t \int_0^{\pi/2} \frac{1 - y \cos x}{1 - y^2 \cos^2 x}\, dx dy$$

$$= 2 \int_0^t \int_0^{\pi/2} \frac{1}{1 - y^2 \cos^2 x}\, dx dy$$

$$= 2 \int_0^t \frac{\pi}{2} \frac{1}{\sqrt{1 - y^2}}\, dy = \pi \arcsin t,$$

where we have used the fact that

$$\int_{-t}^{t} \frac{y \cos x}{1 - y^2 \cos^2 x} \, dy = 0$$

because the integrand is odd in y.

Next, notice that

$$f_t(x, t) = \frac{2}{1 - t^2 \cos^2 x}$$

is continuous on $[0, \pi/2] \times [-T, T]$ for any $0 < T < 1$. Applying Leibniz's rule, we have

$$\begin{aligned}
I'(t) &= \int_0^{\pi/2} \frac{2}{1 - t^2 \cos^2 x} \, dx \\
&= 2 \int_0^{\pi/2} \frac{\sec^2 x}{\sec^2 x - t^2} \, dx \quad (\text{use } u = \tan x) \\
&= 2 \int_0^{\pi/2} \frac{1}{1 - t^2 + u^2} \, du \\
&= \frac{\pi}{\sqrt{1 - t^2}}.
\end{aligned}$$

This again leads to

$$I(t) = I(0) + \int_0^t I'(s) \, ds = \int_0^t \frac{\pi}{\sqrt{1 - s^2}} \, ds = \pi \arcsin t.$$

$\square$

Example 7.8. *Prove that, for every $t \in \mathbb{R}$,*

$$F(t) = \int_0^{2\pi} e^{t \cos x} \cos(t \sin x) \, dx = 2\pi.$$

Proof. Notice that $F(0) = 2\pi$. We now show that $F(t)$ is a constant. To this end, applying Leibniz's rule yields

$$\begin{aligned}
F'(t) &= \int_0^{2\pi} e^{t \cos x} \cos(t \sin x) \cos x \, dx - \int_0^{2\pi} e^{t \cos x} \sin(t \sin x) \sin x \, dx \\
&= \int_0^{2\pi} e^{t \cos x} \cos(t \sin x + x) \, dx.
\end{aligned}$$

By induction, for each $n \in \mathbb{N}$, we have

$$F^{(n)}(t) = \int_0^{2\pi} e^{t \cos x} \cos(t \sin x + nx) \, dx.$$

Hence,

$$F^{(n)}(0) = \int_0^{2\pi} \cos nx \, dx = 0, \quad n = 1, 2, \ldots.$$

Applying Taylor's theorem (Theorem 4.21) yields

$$F(t) = \sum_{k=0}^{n} \frac{F^{(k)}(0)}{k!} x^k + R_n(x) = 2\pi + \frac{t^n}{n!} F^{(n)}(\xi),$$

where ξ is between 0 and t. Since

$$|F^{(n)}(\xi)| = \left| \int_0^{2\pi} e^{\xi \cos x} \cos(\xi \sin x + nx) \, dx \right| \le \int_0^{2\pi} e^{|\xi|} \, dx = 2\pi e^{|\xi|},$$

for each fixed $t \in \mathbb{R}$, it follows that

$$\lim_{n \to \infty} R_n(x) = \lim_{n \to \infty} \frac{t^n}{n!} F^{(n)}(\xi) = 0.$$

This proves that $F(t) \equiv F(0) = 2\pi$. $\qquad\qquad\qquad\qquad\qquad\qquad\qquad$ $\square$

We now take up the improper integrals with parameter in the form of

$$I(t) = \int_a^\infty f(x, t) \, dx. \qquad\qquad (7.3)$$

To ensure integrability we assume that $f(x, t)$ is continuous on $[a, \infty) \times I$ for some interval I.

The following example shows the difference between (7.2) and (7.3). Specifically, we show that Theorem 7.7 does not hold for improper integral (7.3) even if $f(x, t)$ converges to $g(x)$ uniformly at t_0.

Example 7.9. *Let*

$$f(x, t) = \begin{cases} \frac{t}{x^3} e^{-t/2x^2}, & x > 0, \, t \ge 1, \\ 0, & x = 0, \, t \ge 1. \end{cases}$$

Since

$$\max_{x \in [0, \infty)} f(x, t) = f(\sqrt{t/3}, t) = \frac{3\sqrt{3}}{\sqrt{t}} e^{-3/2},$$

it follows that $f(x, t) \to 0$ uniformly on $[0, \infty)$ as $t \to \infty$. But, for any $t > 0$,

$$\int_0^\infty f(x, t) \, dx = \lim_{R \to \infty} \int_0^R f(x, t) \, dx = \lim_{R \to \infty} e^{-t/2x^2} \Big|_0^R = 1.$$

To eliminate this kind of drawback, just as we did for infinite series of functions, we introduce the uniform convergence for integrals, which determines the nature of convergence of the integral for all $t \in I$.

Definition 7.5. *The integral $\int_a^\infty f(x, t) \, dx$ converges uniformly on I if for every $\epsilon > 0$ there is $R > a$, which is independent of t, such that*

$$\left| \int_A^\infty f(x, t) \, dx \right| < \epsilon \quad \text{whenever } A > R \text{ and for all } t \in I.$$

This definition is seldom a practical method for deciding whether or not a specific integral converges uniformly. Analogous to the uniform convergence tests for the series of functions, we have the following uniform convergence tests for integrals. Their proofs are similar to the series version and are left to the reader.

Theorem 7.11. *Let*

$$\omega(A) = \sup_{t \in I} \left| \int_A^\infty f(x, t) \, dx \right|.$$

Then $\int_a^\infty f(x, t) \, dx$ converges uniform on I if and only if

$$\lim_{A \to \infty} \omega(A) = 0.$$

Example 7.10. *Show that $\int_0^\infty te^{-tx}\, dx$ converges uniformly on $[c,\infty)$ for any $c > 0$, but not uniform on $(0,\infty)$.*

Proof. For any $c > 0$, we have

$$\omega(A) = \sup_{t \in [c,\infty)} \left| \int_A^\infty te^{-tx}\, dx \right| = e^{-Ac} \to 0 \quad \text{as } A \to \infty.$$

By Theorem 7.11, the integral converges uniformly on $[c,\infty)$. On the other hand, since

$$\omega(A) = \sup_{t \in (0,\infty)} \left| \int_A^\infty te^{-tx}\, dx \right| = \sup_{t \in (0,\infty)} \{e^{-tA}\} = 1 \nrightarrow 0,$$

by Theorem 7.11 again, the integral does not converge uniformly. $\qquad \square$

Theorem 7.12 (Cauchy Criterion). $\int_a^\infty f(x,t)\, dx$ *converges uniformly on I if and only if for every $\epsilon > 0$ there exists a $R > a$, which is independent on t, such that*

$$\left| \int_A^B f(x,t)\, dx \right| < \epsilon \quad \text{whenever } B > A > R \text{ and for all } t \in I.$$

Theorem 7.13 (Weierstrass M-Test). *Let f be continuous on $[a,\infty) \times I$. If there exists a continuous function $g(x)$ such that*

1. $|f(x,t)| \le g(x)$ *for all $x \in [a,\infty)$ and for all $t \in I$,*
2. $\int_a^\infty g(x)\, dx$ *converges,*

then $\int_a^\infty f(x,t)\, dx$ converges uniformly on I.

Theorem 7.14 (Dirichlet's Test). *Assume that*

1. $g(x,t)$ *is monotone in x and $g(x,t) \to 0$ uniformly on I,*
2. $\int_a^A f(x,t)\, dx$ *is bounded for every $A > a$ and for all $t \in I$,*

then $\int_a^\infty f(x,t)g(x,t)\, dx$ converges uniformly on I.

Theorem 7.15 (Abel's Test). *Assume that*

1. $g(x,t)$ *is monotone in x and bounded on $[a,\infty) \times I$,*
2. $\int_a^\infty f(x,t)\, dx$ *converges uniformly on I,*

then $\int_a^\infty f(x,t)g(x,t)\, dx$ converges uniformly on I.

Remark. As the consequence of either Dirichlet's test or Abel's test, if $\int_a^\infty f(x)\, dx$ converges, then integrals such as

$$\int_a^\infty e^{-xt} f(x)\, dx, \quad \int_a^\infty e^{-x^2 t} f(x)\, dx, \quad (a \ge 0)$$

converge uniformly on $t \in [0,\infty)$.

Example 7.11. *Consider*

$$I(t) = \int_0^\infty \frac{\sin x^2}{1 + x^t}\, dx.$$

Test whether or not the convergence is uniform on $[0,\infty)$.

Since

$$\int_0^\infty \sin x^2 \, dx = \int_0^\infty \frac{\sin u}{2\sqrt{u}} \, du$$

converges (for example, by Dirichlet's test) and is independent of t, it follows that $\int_0^\infty \sin x^2 \, dx$ converges uniformly on $[0, \infty)$. For every fixed $t \in [0, \infty)$, $1/(1 + x^t)$ is monotone on x and

$$\left| \frac{1}{1 + x^t} \right| \le 1.$$

Thus, by Abel's test, $I(t)$ converges uniformly on $[0, \infty)$.

Here is an alternate proof using Dirichlet's test: Rewrite $I(t)$ as

$$I(t) = \int_0^\infty x \sin x^2 \, \frac{1}{x(1 + x^t)} \, dx.$$

For any $A > 0$, we have

$$\left| \int_0^A x \sin x^2 \, dx \right| = \left| -\frac{1}{2} \cos x^2 \Big|_0^A \right| \le 1.$$

For every fixed $t \in [0, \infty)$, $\frac{1}{x(1 + x^t)}$ is monotone in x and

$$\left| \frac{1}{x(1 + x^t)} \right| \le \frac{1}{x} \to 0 \quad \text{as } x \to \infty.$$

Thus, by Dirichlet's test, $I(t)$ converges uniformly on $[0, \infty)$.

Similar to Theorem 7.4, we have the following uniform convergence test in terms of series. The proof is very similar to the proof of Theorem 7.4. We leave that proof to the reader.

Theorem 7.16. *The integral $\int_a^\infty f(x, t) \, dx$ converges uniformly on I if and only if for every sequence $a = a_0 < a_1 < a_2 < \cdots$ with $a_n \to \infty$ as $n \to \infty$, the series*

$$\sum_{n=0}^\infty \int_{a_n}^{a_{n+1}} f(x, t) \, dx$$

converges uniformly on I.

The next three theorems illustrate the importance of uniform convergence. We leave the proofs to the reader. The first says that the uniformly convergent integral of a continuous function is continuous.

Theorem 7.17. *Let $f(x, t)$ be continuous on $[a, \infty) \times [c, d]$. If $\int_a^\infty f(x, t) \, dx \to F(t)$ uniformly on $[c, d]$, then $F(t)$ is continuous on $[c, d]$ and for any $t_0 \in [c, d]$,*

$$F(t_0) = \lim_{t \to t_0} \int_a^\infty f(x, t) \, dx = \int_a^\infty \lim_{t \to t_0} f(x, t) \, dx = \int_a^\infty f(x, t_0) \, dx.$$

Using this result on continuity, the next theorem allows us to interchange the order of a Riemann integral and an improper integral.

Theorem 7.18. *Let $f(x, t)$ be continuous on $[a, \infty) \times [c, d]$. If $\int_a^\infty f(x, t) \, dx \to F(t)$ uniformly on $[c, d]$, then $F(t)$ is integrable on $[c, d]$ and*

$$\int_c^d F(t) \, dt = \int_c^d \int_a^\infty f(x, t) \, dx dt = \int_a^\infty \int_c^d f(x, t) \, dt dx.$$

The next theorem tells us when we can differentiate under the integral sign.

Theorem 7.19. *Let $f(x,t)$ and $f_t(x,t)$ be continuous on $[a,\infty) \times [c,d]$. If $\int_a^\infty f(x,t)\,dx$ converges to $F(t)$ on $[c,d]$ and $\int_a^\infty f_t(x,t)\,dx$ converges uniformly on $[c,d]$, then $F(t)$ is differentiable and for each $t \in (c,d)$*

$$F'(t) = \frac{d}{dt} \int_a^\infty f(x,t)\,dx = \int_a^\infty f_t(x,t)\,dtdx.$$

Example 7.12. *For $\alpha \neq 0$, show that*

$$F(t) = \int_0^\infty \frac{1}{x}\left(1 - e^{-xt}\right)\cos\alpha x\,dx$$

is continuous on $[0,\infty)$ and differentiable on $(0,\infty)$.

Proof. Let

$$f(x,t) = \begin{cases} \frac{1}{x}\left(1 - e^{-xt}\right)\cos\alpha x, & x > 0, t \geq 0; \\ t, & x = 0, t \geq 0. \end{cases}$$

Then $f(x,t)$ is continuous on $[0,\infty) \times [0,\infty)$. For $\alpha \neq 0$, we see that $\int_0^\infty \frac{\cos\alpha x}{x}\,dx$, which is independent of t, converges uniformly on $[0,\infty)$. Since $1 - e^{-xt}$ is monotone on t and $|1 - e^{-xt}| \leq 2$, Abel's test concludes that $F(t)$ converges uniformly on $[0,\infty)$ and so $F(t)$ is continuous on $[0,\infty)$ by Theorem 7.17.

Next, in view of the fact that

$$f_t(x,t) = e^{-at}\cos\alpha x, \qquad x \geq 0,\ t \geq 0,$$

we see that both $f(x,t)$ and $f_t(x,t)$ are continuous on $[0,\infty) \times [0,\infty)$. Furthermore, For every $\epsilon > 0$, when $t \geq \epsilon$, we have

$$|f_t(x,t)| \leq e^{-xt} \leq e^{-\epsilon x}, \quad x \in [0,\infty).$$

Since $\int_0^\infty e^{-\epsilon x}\,dx$ converges, the Weierstrass M-test implies that $\int_0^\infty f_t(x,t)\,dx$ converges uniformly on $[\epsilon,\infty)$. By Theorem 7.19, $F(t)$ is differentiable on $[\epsilon,\infty)$. Since ϵ is arbitrary, it follows that $F(t)$ is differentiable on $(0,\infty)$ and

$$F'(t) = \int_0^\infty e^{-xt}\cos\alpha x\,dx. \tag{7.4}$$

$\square$

Calculating the integral in (7.4) gives

$$F'(t) = \frac{e^{-xt}}{t^2 + \alpha^2}\left(\alpha\sin\alpha x - t\cos\alpha x\right)\Big|_0^\infty = \frac{t}{t^2 + \alpha^2}.$$

Integrating this result yields

$$F(t) = \frac{1}{2}\ln(t^2 + \alpha^2) + c,$$

where c is an arbitrary constant. Since $F(t)$ is continuous and $F(0) = 0$, we find that

$$F(t) = \frac{1}{2}\ln\frac{t^2 + \alpha^2}{\alpha^2}.$$

This echos the fact that F is continuous on $[0,\infty)$ and differentiable on $(0,\infty)$.

When the range of parameter t also becomes infinite, in order to switch the order of integration, Fubini's theorem (Theorem 7.18) requires further conditions.

Theorem 7.20 (Fubini's Theorem). *If* $f(x,t) : [a, \infty) \times [c, \infty) \to \mathbb{R}$ *satisfies*

  1. $f(x,t)$ *is continuous on* $[a, \infty) \times [c, \infty)$,

  2. The improper integrals

$$\int_a^\infty f(x,t)\, dx \quad and \quad \int_c^\infty f(x,t)\, dt$$

  exist and converge uniformly for x *and* t *restricted to every finite interval, respectively,*

  3. One of the integrals

$$\int_c^\infty \int_a^\infty |f(x,t)|\, dx dt \quad and \quad \int_a^\infty \int_c^\infty |f(x,t)|\, dt dx$$

converges,

then both integrals

$$\int_c^\infty \int_a^\infty f(x,t)\, dx dt \quad and \quad \int_a^\infty \int_c^\infty f(x,t)\, dt dx$$

converge and

$$\int_c^\infty \int_a^\infty f(x,t)\, dx dt = \int_a^\infty \int_c^\infty f(x,t)\, dt dx.$$

Proof. Without loss of generality, we assume that $\int_a^\infty \int_c^\infty |f(x,t)|\, dt dx$ converges. The Cauchy criterion implies that $\int_a^\infty \int_c^\infty f(x,t)\, dt dx$ converges as well. Thus,

$$\int_c^\infty \int_a^\infty f(x,t)\, dx dt \;=\; \lim_{R \to \infty} \int_c^R \int_a^\infty f(x,t)\, dx dt$$
$$=\; \lim_{R \to \infty} \int_a^\infty \int_c^R f(x,t)\, dt dx.$$

Notice that

$$\int_a^\infty \int_c^\infty f(x,t)\, dt dx = \int_a^\infty \int_c^R f(x,t)\, dt dx + \int_a^\infty \int_R^\infty f(x,t)\, dt dx.$$

It suffices to show that

$$\lim_{R \to \infty} \int_a^\infty \int_R^\infty f(x,t)\, dt dx = 0.$$

To see this, for every $\epsilon > 0$, in view of that $\int_a^\infty \int_c^\infty |f(x,t)|\, dt dx$ converges, there is $A > a$ such that

$$0 \le \int_A^\infty \int_c^\infty |f(x,t)|\, dt dx < \frac{\epsilon}{2}.$$

For fixed A, since $\int_c^\infty f(x,t)\, dt$ uniformly converges on $x \in [a, A]$, it follows that there is $R' > c$ such that

$$\left| \int_\beta^\infty f(x,t)\, dt \right| < \frac{\epsilon}{2(A-a)}, \quad \text{for every } x \in [a, A] \text{ and } \beta > R'.$$

Hence, for $R > R'$,

$$\left| \int_a^\infty \int_R^\infty f(x,t)\,dt\,dx \right| = \left| \int_a^A \int_R^\infty f(x,t)\,dt\,dx + \int_A^\infty \int_R^\infty f(x,t)\,dt\,dx \right|$$

$$\leq \int_a^A \left| \int_R^\infty f(x,t)\,dt \right|\,dx + \int_A^\infty \int_R^\infty |f(x,t)|\,dt\,dx$$

$$< \frac{\epsilon}{2(A-a)}(A-a) + \int_A^\infty \int_c^\infty |f(x,t)|\,dt\,dx$$

$$< \frac{\epsilon}{2} + \frac{\epsilon}{2} = \epsilon.$$

This completes the proof. $\qquad\qquad\qquad\qquad\qquad\qquad\qquad\qquad\qquad\qquad\square$

The following example indicates that Condition (3) in Theorem 7.20 is necessary.

Example 7.13. *Consider*

$$f(x,t) = \frac{t^2 - x^2}{(x^2 + t^2)^2}, \qquad (x,t) \in [1,\infty) \times [1,\infty).$$

Since

$$\left| \int_A^\infty \frac{t^2 - x^2}{(x^2 + t^2)^2}\,dx \right| = \frac{A}{A^2 + t^2} \leq \frac{1}{A},$$

by the Cauchy criterion, the improper integral $\int_1^\infty \frac{t^2 - x^2}{(x^2 + t^2)^2}\,dx$ converges uniformly on $t \in [1,\infty)$. Similarly, the improper integral $\int_1^\infty \frac{t^2 - x^2}{(x^2 + t^2)^2}\,dt$ also converges uniformly on $x \in [1,\infty)$. Thus, $f(x,t)$ satisfies the conditions (1) and (2) in Theorem 7.20. But, we have

$$\int_1^\infty \left(\int_1^\infty \frac{t^2 - x^2}{(x^2 + t^2)^2}\,dx \right) dt = -\frac{\pi}{4} \neq \int_1^\infty \left(\int_1^\infty \frac{t^2 - x^2}{(x^2 + t^2)^2}\,dt \right) dx = \frac{\pi}{4}.$$

Direct calculations show that both integrals

$$\int_1^\infty \left(\int_1^\infty \frac{|t^2 - x^2|}{(x^2 + t^2)^2}\,dx \right) dt \text{ and } \int_1^\infty \left(\int_1^\infty \frac{|t^2 - x^2|}{(x^2 + t^2)^2}\,dt \right) dx$$

diverge!

If, in addition, that $f(x,t) \geq 0$ on $[a,\infty) \times [c,d]$, we show that the uniform convergence of $\int_0^\infty f(x,t)\,dx$ on $[c,d]$ becomes a necessary condition for the continuity of $I(t)$.

Theorem 7.21. *Let $f : [a,\infty) \times [c,d] \to [0,\infty)$ be continuous. If $I(t) = \int_0^\infty f(x,t)\,dx$ is continuous on $[c,d]$, then $\int_0^\infty f(x,t)\,dx$ converges uniformly on $[c,d]$.*

Proof. Rewrite

$$I(t) = \int_a^\infty f(x,t)\,dx = \sum_{n=1}^\infty \int_{a+n-1}^{a+n} f(x,t)\,dx = \sum_{n=1}^\infty a_n(t),$$

where

$$a_n(t) = \int_{a+n-1}^{a+n} f(x,t)\,dx, \qquad n = 1, 2, \dots.$$

By the assumptions of f, we see that $a_n(t)$ is continuous and nonnegative on $[c, d]$, and so the series $\sum_{n=1}^{\infty} a_n(t)$ converges uniformly on $[c, d]$ by Dini's theorem (see Example 5 in Section 6.5). Thus, for any $\epsilon > 0$, there is some N such that

$$\sum_{n=N+1}^{\infty} a_n(t) < \epsilon \quad \text{for any } t \in [c, d].$$

We now choose $R = a + N$. For $A > R$, in view of the fact that $f \geq 0$, we have

$$\int_A^{\infty} f(x, t)\, dx \leq \int_{N+a}^{\infty} f(x, t)\, dx = \sum_{n=N+1}^{\infty} a_n(t) < \epsilon, \quad \text{for any } t \in [c, d].$$

This proves that $\int_a^{\infty} f(x, t)\, dx$ converges uniformly on $[c, d]$. $\qquad\square$

This theorem, together with Theorem 7.20, suggests the following theorem in which that the continuity of the integrals $\int_a^{\infty} f(x, t)\, dx$ and $\int_c^{\infty} f(x, t)\, dt$ justify us to switch the order of the integration.

Theorem 7.22 (Dini's Theorem). *If $f(x, t) : [a, \infty) \times [c, \infty) \to \mathbb{R}$ satisfies*

1. *$f(x, t)$ is continuous and nonnegative on $[a, \infty) \times [c, \infty)$,*

2. *The functions $I(t) = \int_a^{\infty} f(x, t)\, dx$ is continuous on $[c, \infty)$ and $J(x) = \int_c^{\infty} f(x, t)\, dt$ is continuous on $[a, \infty)$,*

3. *One of integrals*

$$\int_c^{\infty} I(t)\, dt \quad and \quad \int_a^{\infty} J(x)\, dx$$

   converges,

then

$$\int_c^{\infty} I(t)\, dt = \int_c^{\infty} \int_a^{\infty} f(x, t)\, dx dt = \int_a^{\infty} J(x)\, dx = \int_a^{\infty} \int_c^{\infty} f(x, t)\, dt dx.$$

Combining this with the Weierstrass M-test, we have the following theorem which will be needed to justify the desired steps in most of our subsequent discussion.

Theorem 7.23 (Extended Fubini's Theorem). *If $f(x, t) : [a, \infty) \times [c, \infty) \to \mathbb{R}$ satisfies*

1. *$f(x, t)$ is continuous on $[a, \infty) \times [c, \infty)$,*

2. *The improper integrals*

$$\int_a^{\infty} f(x, t)\, dx \quad and \quad \int_c^{\infty} f(x, t)\, dt$$

   both exist and converge uniformly for t and x restricted to every finite interval, respectively,

3. *For $B > A > a$*

$$\left| \int_A^B f(x, t)\, dx \right| \leq M(t)$$

   and $\int_b^{\infty} M(t)\, dt$ converges,

then

$$\int_b^\infty \int_a^\infty f(x,t)\,dx dt = \int_a^\infty \int_b^\infty f(x,t)\,dt dx.$$

As a powerful application of this theorem, in view of

$$\frac{1}{x} = \int_0^\infty e^{-xt}\,dt$$

after verifying that $f(x,t) = e^{-xt}\sin x$ satisfies all conditions in Theorem 7.23, in one line, we recover the result that $\int_0^\infty \frac{\sin x}{x}\,dx = \frac{\pi}{2}$ (see Example 7.4).

$$\int_0^\infty \left(\int_0^\infty e^{-xt}\sin x\,dt \right) dx = \int_0^\infty \left(\int_0^\infty e^{-xt}\sin x\,dx \right) dt = \int_0^\infty \frac{dt}{1+t^2} = \frac{\pi}{2}.$$

We end this section by establishing another classical result:

$$I = \int_0^\infty e^{-x^2}\,dx = \frac{\sqrt{\pi}}{2}. \tag{7.5}$$

This formula frequently appears in the formulas of the gamma function in the next section and plays a central role in probability theory. There are many ways to calculate this integral. The method used here is based only on a substitution and Dini's Theorem (Theorem 7.22). For $u > 0$, the substitution $x = ut$ leads to

$$I = \int_0^\infty ue^{-u^2 t^2}\,dt.$$

Hence

$$I^2 = I \int_0^\infty e^{-u^2}\,du = \int_0^\infty I e^{-u^2}\,du$$

$$= \int_0^\infty \int_0^\infty ue^{-u^2(1+t^2)}\,dt du$$

$$= \int_0^\infty \left(\int_0^\infty ue^{-u^2(1+t^2)}\,du \right) dt \quad \text{(using Theorem 7.22)}$$

$$= \int_0^\infty -\frac{e^{-u^2(1+t^2)}}{2(1+t^2)} \Big|_0^\infty dt = \frac{1}{2}\int_0^\infty \frac{dt}{1+t^2} = \frac{\pi}{4}$$

and so $I = \sqrt{\pi}/2$, which confirms the formula (7.5). To justify the interchange the order of integration, we verify the three conditions in Theorem 7.22. Clearly, $f(t,u) = ue^{-u^2(1+t^2)}$ is continuous and nonnegative on $[0,\infty) \times [0,\infty)$. Moreover,

$$\phi(u) = \int_0^\infty f(t,u)\,dt = e^{-u^2}\int_0^\infty e^{-(ut)^2}\,d(ut) = I\,e^{-u^2}$$

is continuous on $[0,\infty)$ and

$$\psi(t) = \int_0^\infty f(t,u)\,dt = \frac{1}{2(1+t^2)}$$

is also continuous on $[0,\infty)$. Moreover, $\int_0^\infty \psi(t)\,dt$ converges. Thus, Theorem 7.22 is applicable.

7.3 The Gamma Function

Yet the record is this: each generation has found something new of interest to say about the gamma function. — Paul Nahin

The gamma function was introduced into analysis in the year 1729 by Euler while seeking a generalization of the factorial $n!$ for non-integral values of x. Initially, he expressed his solution in integral form

$$x! = \int_0^1 (-\ln t)^x \, dt.$$

Later, Legendre proposed the notation $\Gamma(x)$ and the now standard definition of the gamma function.

Definition 7.6. *For $x > 0$, the gamma function is defined by*

$$\Gamma(x) = \int_0^\infty t^{x-1} e^{-t} \, dt.$$

A change of variable show that $\Gamma(x+1)$ is the same as Euler's definition $x!$. Moreover, using integration by parts, we find that

$$\Gamma(x+1) = \int_0^\infty t^x e^{-t} \, dt = x \int_0^\infty t^{x-1} e^{-t} \, dt = x\Gamma(x). \tag{7.6}$$

Since $\Gamma(1) = 1$, for $n \in \mathbb{N}$, applying (7.6) repeatedly yields that $\Gamma(n+1) = n!$. A graph of the gamma function is shown in Figure 7.1.

The graph suggests that $\Gamma(x)$ is a convex function. In fact, we have the following stronger property.

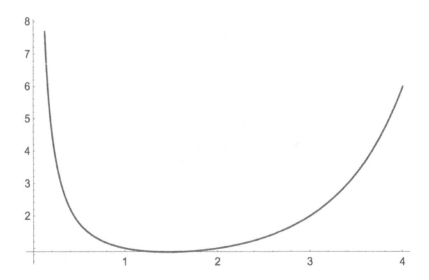

FIGURE 7.1
Graph of $y = \Gamma(x)$.

Theorem 7.24. $\ln \Gamma(x)$ *is convex on* $(0, \infty)$.

Proof. Recall the definition of convex (Definition 4.3), we assume that $0 < a < b$ and $0 < \alpha < 1$. Then

$$
\begin{aligned}
\Gamma(\alpha a + (1-\alpha)b) &= \int_0^\infty t^{\alpha a + (1-\alpha)b - 1} e^{-t}\, dt \\
&= \int_0^\infty \left(t^{a-1}e^{-t}\right)^\alpha \cdot \left(t^{b-1}e^{-t}\right)^{1-\alpha}\, dt \\
&\le \left(\int_0^\infty t^{a-1}e^{-t}\, dt\right)^\alpha \cdot \left(\int_0^\infty t^{b-1}e^{-t}\, dt\right)^{1-\alpha} \\
&= \Gamma(a)^\alpha \cdot \Gamma(b)^{1-\alpha}.
\end{aligned}
$$

where Hölder's inequality has been applied with $p = 1/\alpha, q = 1/(1-\alpha)$. Taking logarithms of both sides yields

$$
\ln \Gamma(\alpha a + (1-\alpha)b) \le \alpha \ln \Gamma(a) + (1-\alpha)\Gamma(b),
$$

which shows that $\ln \Gamma(x)$ is convex by definition. □

The graph of $\ln \Gamma(x)$ is displayed in Figure 7.2.

Remark. Given $f : I \to \mathbb{R}$, if $\ln f(x)$ is convex on I, i.e., for any $0 < \alpha < 1$, $a, b \in I$, we have

$$
\ln f(\alpha a + (1-\alpha)b) \le \alpha \ln f(a) + (1-\alpha) \ln f(b).
$$

Taking exponents on both sides, then using the weighted AM-GM inequality (4.3), we have

$$
f(\alpha a + (1-\alpha)b) \le f(a)^\alpha f(b)^{1-\alpha} \le \alpha f(a) + (1-\alpha)f(b).
$$

This implies that f itself is convex. Thus log-convex is a stronger property than convexity.

Since the time of Euler, mathematicians have wondered if the log-convex of $\Gamma(x)$ can be of any use. Surprisingly, Bohr and Mollerup discovered that the gamma function is

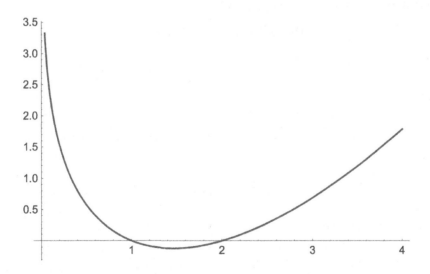

FIGURE 7.2
Graph of $y = \ln \Gamma(x)$.

actually characterized by this property—having fixed the integer values of the function, the logarithmic convexity will restrict the growth of the function in such a way as it must be the gamma function. For the following theorem, Davis [35] once said: "The proof: one page. The discovery: 193 years."

Theorem 7.25 (Bohr-Mollerup Theorem). *If a positive function $f(x)$ on $(0, \infty)$ satisfies*

> 1. $f(1) = 1$,
> 2. $f(x+1) = xf(x)$,
> 3. $\ln f(x)$ *is convex,*

then $f(x) = \Gamma(x)$. i.e., the gamma function is the only function satisfying the three properties.

Proof. The hypotheses (1) and (2) imply that $f(n+1) = n!$ for all $n \in \mathbb{N}$. By (2), without loss of generality, we consider $0 < x < 1$ only. For any positive integer n and for $0 < x < 1$, we write

$$n + x = (1 - x)n + x(n + 1).$$

Then by the convexity hypothesis (3),

$$\ln f(n + x) \le (1 - x)\ln f(n) + x \ln f(n + 1),$$

or equivaletly

$$f(n + x) \le f(n)^{1-x} f(n+1)^x = ((n-1)!)^{1-x}(n!)^x = n! \, n^{x-1}. \tag{7.7}$$

Along the same lines, the convex combination

$$n + 1 = x(n + x) + (1 - x)(n + x + 1),$$

together with $f(n + x + 1) = (n + x)f(n + x)$, gives the inequality

$$n! = f(n + 1) \le f(n + x)^x f(n + x + 1)^{1-x} = f(n + x)(n + x)^{1-x}. \tag{7.8}$$

Combining inequalities (7.7) and (7.8) yields

$$n! \, (n + x)^{x-1} \le f(n + x) \le n! \, n^{x-1}. \tag{7.9}$$

Applying

$$f(n + x) = (n + x - 1)(n + x - 2) \cdots (x + 1)xf(x)$$

to (7.9) leads to

$$\frac{n! \, (n + x)^x}{x(x + 1) \cdots (x + n)} \le f(x) \le \frac{n! \, n^x}{x(x + 1) \cdots (x + n - 1)n}.$$

Rewriting this as

$$\frac{n! \, n^x}{x(x + 1) \cdots (x + n)} \cdot \frac{(n + x)^x}{n^x} \le f(x) \le \frac{n! \, n^x}{x(x + 1) \cdots (x + n)} \cdot \frac{x + n}{n}$$

then letting $n \to \infty$, by the squeeze theorem, we find that

$$f(x) = \lim_{n \to \infty} \frac{n! \, n^x}{x(x + 1) \cdots (x + n)}.$$

Hence $f(x)$ is uniquely determined by (1)–(3). Since $\Gamma(x)$ satisfies (1)–(3) as well, it follows that $f(x) = \Gamma(x)$. $\qquad\square$

As a by-product we recapture the *Gauss product formula*:

$$\Gamma(x) = \lim_{n\to\infty} \frac{n!\,n^x}{x(x+1)\cdots(x+n)}. \tag{7.10}$$

In fact, the formula (7.10) is Euler's original construction for $x!$. But Gauss rediscovered this formula and recognized its importance.

Implicit in the above calculations is the following generalization of Stirling's formula $n! \sim \sqrt{2\pi n}(n/e)^n$. In fact, for any $0 < a \le 1$, (7.10) can be rewritten as

$$\lim_{n\to\infty} \frac{\Gamma(n+a+1)}{n!\,n^a} = 1.$$

Applying Stirling's formula for $n!$ in this limit gives

$$\lim_{n\to\infty} \frac{\Gamma(n+a)}{n^{n+a-1/2}\,e^{-n}} = \sqrt{2\pi}$$

or

$$\lim_{n\to\infty} \frac{(n+a)\Gamma(n+a)}{(n+a)^{n+a+1/2}\,e^{-n-a}} = \sqrt{2\pi},$$

where the fact that $(1+a/n)^n \to e^a$ uniformly on $(0,1]$ has been used. This establishes the following Stirling's formula for the gamma function.

Theorem 7.26 (Stirling's Formula for $\Gamma(x)$). *As $x \to \infty$,*

$$\Gamma(x+1) \sim \sqrt{2\pi x}\left(\frac{x}{e}\right)^x. \tag{7.11}$$

A more accurate approximation is given in Exercise 78 at the end of this chapter. As another application of (7.10), the following theorem shows that the Gauss product formula (7.10) is essentially the same as the infinite product representation of $1/\Gamma(x)$, which Weierstrass used as a definition of $\Gamma(x)$.

Theorem 7.27. *For all $x \in \mathbb{R}$ with $x \ne 0, -1, -2, \ldots$,*

$$\frac{1}{\Gamma(x)} = xe^{\gamma x} \prod_{n=1}^{\infty} \left(1+\frac{x}{n}\right) e^{-x/n}, \tag{7.12}$$

where γ is the Euler-Mascheroni constant.

Proof. In view of Theorem 2.31, the convergence of the infinite product in (7.12) follows from

$$\left(1+\frac{x}{n}\right) e^{-x/n} = \left(1+\frac{x}{n}\right)\left(1-\frac{x}{n}+\frac{x^2}{2n^2}-\cdots\right) = 1 - \frac{x^2}{2n^2} + O\left(\frac{1}{n^3}\right)$$

and the series $\sum_{n=1}^{\infty} 1/n^2$ converges. By (7.10), we have

$$\frac{1}{\Gamma(x)} = \lim_{n\to\infty} \frac{x(x+1)\cdots(x+n)}{n!\,n^x}$$

$$= \lim_{n\to\infty} x\left(1+\frac{x}{1}\right)\left(1+\frac{x}{2}\right)\cdots\left(1+\frac{x}{n}\right) e^{-x\ln n}$$

$$= \lim_{n\to\infty} xe^{x(1+1/2+\cdots+1/n-\ln n)} \prod_{k=1}^{n} \left(1+\frac{x}{k}\right) e^{-x/k}$$

$$= xe^{\gamma x} \prod_{k=1}^{\infty} \left(1+\frac{x}{k}\right) e^{-x/k}.$$

This proves (7.12). $\qquad\square$

We now return to the Bohr-Mollerup theorem. Aside from its aesthetic appeal, this theorem can be used to verify formulas involving the gamma function.

Theorem 7.28 (Legendre's Duplication Formula). *For $x > 0$,*

$$\Gamma(2x) = \frac{2^{2x-1}}{\sqrt{\pi}} \Gamma(x)\Gamma\left(x + \frac{1}{2}\right). \tag{7.13}$$

Proof. Let

$$f(x) = \frac{2^{x-1}}{\sqrt{\pi}} \Gamma\left(\frac{x}{2}\right)\Gamma\left(\frac{x+1}{2}\right).$$

To show that $f(x) = \Gamma(x)$, it suffices to check that $f(x)$ satisfies the three conditions in the Bohr-Mollerup Theorem. By (7.5), we have

$$\Gamma(1/2) = \int_0^\infty t^{-1/2}e^{-t}\,dt \overset{(t=x^2)}{=} 2\int_0^\infty e^{-x^2}\,dx = \sqrt{\pi},$$

which implies that

$$f(1) = \frac{1}{\sqrt{\pi}}\Gamma(1/2)\Gamma(1) = 1.$$

Also,

$$f(x+1) = \frac{2^x}{\sqrt{\pi}}\Gamma\left(\frac{x+1}{2}\right)\Gamma\left(\frac{x}{2}+1\right) = \frac{x}{2}\frac{2^x}{\sqrt{\pi}}\Gamma\left(\frac{x+1}{2}\right)\Gamma\left(\frac{x}{2}\right) = xf(x).$$

The logarithmic convexity of f follows at once from that of the gamma function. Hence it follows from the Bohr-Mollerup theorem that $f(x) = \Gamma(x)$. Replacing x by $2x$ yields the formula (7.13). $\qquad\square$

Legendre's duplication formula can be generalized to the more general case. Indeed, let

$$f(x) = \frac{m^{x-1/2}}{(2\pi)^{(n-1)/2}}\Gamma\left(\frac{x}{m}\right)\Gamma\left(\frac{x+1}{m}\right)\cdots\Gamma\left(\frac{x+m-1}{m}\right). \tag{7.14}$$

Clearly, $\ln f$ is convex on $(0,\infty)$. Moreover,

$$
\begin{aligned}
f(x+1) &= \frac{m^{x+1/2}}{(2\pi)^{(n-1)/2}}\Gamma\left(\frac{x+1}{m}\right)\Gamma\left(\frac{x+2}{m}\right)\cdots\Gamma\left(\frac{x+m-1}{m}\right)\Gamma\left(\frac{x+m}{m}\right)\\
&= \frac{m^{x+1/2}}{(2\pi)^{(n-1)/2}}\Gamma\left(\frac{x+1}{m}\right)\Gamma\left(\frac{x+2}{m}\right)\cdots\Gamma\left(\frac{x+m-1}{m}\right)\Gamma\left(\frac{x}{m}+1\right)\\
&= \frac{m^{x+1/2}}{(2\pi)^{(n-1)/2}}\Gamma\left(\frac{x+1}{m}\right)\Gamma\left(\frac{x+2}{m}\right)\cdots\Gamma\left(\frac{x+m-1}{m}\right)\frac{x}{m}\Gamma\left(\frac{x}{m}\right)\\
&= x\frac{m^{x-1/2}}{(2\pi)^{(n-1)/2}}\Gamma\left(\frac{x}{m}\right)\Gamma\left(\frac{x+1}{m}\right)\cdots\Gamma\left(\frac{x+m-1}{m}\right)\\
&= xf(x).
\end{aligned}
$$

We now show that $f(1) = 1$, i.e.,

$$\frac{m^{1/2}}{(2\pi)^{(m-1)/2}}\Gamma\left(\frac{1}{m}\right)\Gamma\left(\frac{2}{m}\right)\cdots\Gamma\left(\frac{m}{m}\right) = 1.$$

For this purpose, applying the Gauss product formula (7.10) yields

$$\Gamma\left(\frac{k}{m}\right) = \lim_{n\to\infty} \frac{n!\, n^{k/m}\, m^{n+1}}{k(k+m)(k+2m)\cdots(k+nm)}, \quad (k = 1, 2, \ldots, m).$$

Hence

$$\Gamma\left(\frac{1}{m}\right)\Gamma\left(\frac{2}{m}\right)\cdots\Gamma\left(\frac{m}{m}\right) = \lim_{n\to\infty} \frac{(n!)^m\, n^{(m+1)/2} m^{(n+1)m}}{(m+nm)!}. \tag{7.15}$$

Since

$$\frac{(m+nm)!}{(nm)!\,(nm)^m} = \left(1+\frac{1}{nm}\right)\left(1+\frac{2}{nm}\right)\cdots\left(1+\frac{m}{nm}\right) \to 1, \quad \text{as } n \to \infty,$$

this transforms (7.15) into

$$\Gamma\left(\frac{1}{m}\right)\Gamma\left(\frac{2}{m}\right)\cdots\Gamma\left(\frac{m}{m}\right) = \lim_{n\to\infty} \frac{(n!)^m\, m^{nm}}{(nm)!\, n^{(m-1)/2}}.$$

Appealing to the Stirling's formula for $n!$, we have

$$\frac{(n!)^m}{(nm)!} \sim \frac{(2\pi)^{(m-1)/2} n^{(m-1)/2}}{m^{nm+1/2}}.$$

Therefore,

$$\Gamma\left(\frac{1}{m}\right)\Gamma\left(\frac{2}{m}\right)\cdots\Gamma\left(\frac{m}{m}\right) = \frac{(2\pi)^{(m-1)/2}}{m^{1/2}}.$$

This shows that $f(1) = 1$ and so $f(x) = \Gamma(x)$ by the Bohr-Mollerup Theorem. Moreover, replacing x by mx in (7.14) results in

Theorem 7.29 (Gauss Multiplication Formula). *For $m \in \mathbb{N}, x > 0$,*

$$\Gamma(mx) = \frac{m^{mx-1/2}}{(2\pi)^{(m-1)/2}}\, \Gamma(x)\Gamma\left(x+\frac{1}{m}\right)\cdots\Gamma\left(x+\frac{m-1}{m}\right). \tag{7.16}$$

Another significant consequence of the Bohr-Mollerup theorem is that it enables us to associate the Euler beta function with the gamma function.

Definition 7.7. *For $x, y > 0$, the Euler beta function is defined by*

$$B(x, y) = \int_0^1 t^{x-1}(1-t)^{y-1}\, dt. \tag{7.17}$$

During his search for the definition of $x!$ and the systematic building of the formulas of integrals, Euler introduced (7.17) based on Wallis's formula. The following theorem reveals a striking relation between the beta function and the gamma function. They are commonly used to evaluate a number of integrals.

Theorem 7.30. *For all $x, y > 0$,*

$$B(x, y) = \frac{\Gamma(x)\,\Gamma(y)}{\Gamma(x+y)}. \tag{7.18}$$

Proof. For each fixed $y \in (0, \infty)$, let

$$f(x) = \frac{\Gamma(x+y)}{\Gamma(y)} B(x, y), \quad x \in (0, \infty).$$

We now verify that $f(x)$ satisfies all three conditions in the Bohr-Mollerup theorem. First,

$$f(1) = \frac{\Gamma(1+y)}{\Gamma(y)} B(1, y) = yB(1, y).$$

Since

$$B(1, y) = \int_0^1 (1 - t)^{y-1} dt = \int_0^1 u^{y-1} du = \frac{1}{y},$$

it follows that $f(1) = 1$. Next, rewrite

$$B(x+1, y) = \int_0^1 t^x (1-t)^{y-1} dy = \int_0^1 \left(\frac{t}{1-t}\right)^x (1-t)^{x+y-1} dt.$$

Integrating by parts yields

$$B(x+1, y) = \left[-\frac{t^x(1-t)^y}{x+y}\right]_0^1 + \int_0^1 \frac{(1-t)^{x+y}}{x+y} \cdot \frac{xt^{x-1}}{(1-t)^{x+1}} dt = \frac{x}{x+y} B(x, y).$$

Thus

$$f(x+1) = \frac{\Gamma(x+y+1)}{\Gamma(y)} B(x+1, y) = \frac{(x+y)\Gamma(x+y)}{\Gamma(y)} \frac{x}{x+y} B(x, y) = xf(x).$$

It remains to show that $\ln f(x)$ is convex. Notice that

$$\ln f(x) = \ln \Gamma(x+y) + \ln B(x, y) - \ln \Gamma(y).$$

Since $\ln \Gamma(x+y)$ is convex for x (Theorem 7.24), it suffices to show that $\ln B(x, y)$ is convex in x. The proof follows the same lines as that of Theorem 7.24. Let $0 < \alpha < 1$. Then

$$B(\alpha a + (1-\alpha)b, y) = \int_0^1 t^{\alpha a + (1-\alpha)b - 1}(1-t)^{y-1} dt$$

$$= \int_0^1 \left(t^{a-1}(1-t)^{y-1}\right)^\alpha \left(t^{b-1}(1-t)^{y-1}\right)^{1-\alpha} dt$$

$$\leq \left(\int_0^1 t^{a-1}(1-t)^{y-1} dt\right)^\alpha \left(\int_0^1 t^{b-1}(1-t)^{y-1} dt\right)^{1-\alpha}$$

$$= B(a, y)^\alpha B(b, y)^{1-\alpha}.$$

Taking logarithms yields

$$\ln B(\alpha a + (1-\alpha)b, y) \leq \alpha B(a, y) + (1-\alpha) B(b, y).$$

Hence $\ln B(x, y)$ is convex and so $\ln f(x)$ is convex. By the Bohr-Wollerup theorem, $f(x) = \Gamma(x)$. This demonstrates (7.18) as stated. $\quad\square$

As a consequence of (7.18), together with the properties of the gamma function, we obtain

Theorem 7.31. *For* $p, q > 0$,

(1). $B(p,q) = B(q,p)$.

(2). $B(p+1, q+1) = \frac{pq}{(p+q+1)(p+q)} B(p,q)$.

(3). $B(p,q)$ is continuous and has continuous derivatives of all order on $(0, \infty) \times (0, \infty)$.

By the recurrence relation of the gamma function, for $x \in (0,3) \setminus \{1, 2\}$, we define

$$\Gamma(1-x) := \frac{\Gamma(3-x)}{(2-x)(1-x)}.$$

Then, for any $x \in (0,1)$,

$$\Gamma(x+2)\Gamma(1-(x+2)) = \Gamma(x)\Gamma(1-x).$$

This indicates that the product $\Gamma(x)\Gamma(1-x)$ is a periodic function with period 2. It is well-known that $\sin \pi x$ is also a periodic function with period 2. However, who could anticipate that they can be expressed in terms of each other? Davis [35] poetically describes the following relation (7.19) as "a fine example of the delicate patterns which make the mathematics of the period so magical."

Theorem 7.32 (Euler's Reflection Formula). *For* $0 < x < 1$,

$$\Gamma(x)\Gamma(1-x) = \frac{\pi}{\sin \pi x}. \tag{7.19}$$

Proof. There are many different proofs (real and complex variables) of this formula. Here we present two real number proofs. The first one is based on the beta function and partial fraction decomposition for $\csc x$. The second is to apply the Sine infinite product (6.21) and the infinite product formula (7.12).

Proof I. By (7.18), we have

$$\Gamma(x)\Gamma(1-x) = B(x, 1-x) = \int_0^1 t^{x-1}(1-t)^{-x} \, dt.$$

Applying the substitution $u = t/(1-t)$ yields

$$\Gamma(x)\Gamma(1-x) = \int_0^\infty \frac{u^{x-1}}{1+u} \, du = \int_0^1 \frac{u^{x-1}}{1+u} \, du + \int_1^\infty \frac{u^{x-1}}{1+u} \, du.$$

Notice that the substitution $y = 1/u$ gives

$$\int_1^\infty \frac{u^{x-1}}{1+u} \, du = \int_0^1 \frac{y^{-x}}{1+y} \, dy.$$

Hence

$$\Gamma(x)\Gamma(1-x) = \int_0^1 \frac{u^{x-1} + u^{-x}}{1+u} \, du$$

$$= \int_0^1 u^{x-1} \, du + \int_0^1 \frac{u^{-x} - u^x}{1+u} \, du \quad \left(\text{use } \tfrac{1}{1+u} = 1 - \tfrac{u}{1+u}\right)$$

$$= \frac{1}{x} + \int_0^1 (u^{-x} - u^x) \sum_{n=0}^\infty (-1)^n u^n \, du$$

$$= \frac{1}{x} + \int_0^1 \sum_{n=0}^\infty (-1)^n (u^{-x+n} - u^{x+n}) \, du$$

$$= \frac{1}{x} + \sum_{n=1}^\infty (-1)^n \left(\frac{1}{n+x} - \frac{1}{n-x} \right).$$

In view of (6.23), we arrive at (7.19) as desired. The justification of the term by term integration is as follows: Let $S_n(u)$ and $R_n(u)$ be the nth partial sum and remainder of the series $\sum_{k=0}^{\infty} (-1)^k u^k (u^{-x} - u^x)$. We show that $\int_0^1 R_n(u)\, du \to 0$ as $n \to \infty$. In fact,

$$\int_0^1 |R_n(u)|\, du = \int_0^1 \frac{u^{n+1}(u^{-x} - u^x)}{1+u}\, du = \int_0^1 u^n \left(\frac{u^{1-x} - u^{1+x}}{1+u} \right) du.$$

Since $0 < x < 1$, the function $(u^{1-x} - u^{1+x})/(1+u)$ is continuous on $[0,1]$, and so there is a $M > 0$ such that $(u^{1-x} - u^{1+x})/(1+u) \le M$ for all $x \in (0,1)$. Thus,

$$\int_0^1 |R_n(u)|\, du \le M \int_0^1 u^n\, du = \frac{M}{n+1} \to 0 \ \text{as}\ n \to \infty.$$

This completes the proof.

Proof II. By the recurrence relation and the infinite product representation (7.12) of $\Gamma(x)$, we have

$$\frac{1}{\Gamma(x)\Gamma(1-x)} = \frac{1}{\Gamma(x)(-x)\Gamma(-x)}$$

$$= xe^{\gamma x} e^{-\gamma x} \prod_{n=1}^{\infty} \left(1 + \frac{x}{n}\right) e^{-x/n} \cdot \left(1 - \frac{x}{n}\right) e^{x/n}$$

$$= x \prod_{n=1}^{\infty} \left(1 - \frac{x^2}{n^2}\right).$$

In view of the sine infinite product (6.21), we conclude that

$$\frac{1}{\Gamma(x)\Gamma(1-x)} = \frac{\sin \pi x}{\pi},$$

which is equivalent to (7.19). □

As a powerful application of (7.19), the expression $P := \prod_{k=1}^{m-1} \Gamma(k/m)$ can be recaptured as follows:

$$P^2 = \prod_{k=1}^{m-1} \Gamma\left(\frac{k}{m}\right) \cdot \prod_{k=1}^{m} \Gamma\left(1 - \frac{k}{m}\right)$$

$$= \prod_{k=1}^{m-1} \Gamma\left(\frac{k}{m}\right) \Gamma\left(1 - \frac{k}{m}\right)$$

$$= \prod_{k=1}^{m-1} \frac{\pi}{\sin\left(\frac{k\pi}{m}\right)} = \frac{\pi^{m-1}}{\prod_{k=1}^{m-1} \sin\left(\frac{k\pi}{m}\right)} = \frac{(2\pi)^{m-1}}{m}.$$

Again, we obtain

$$\prod_{k=1}^{m} \Gamma\left(\frac{k}{m}\right) = \frac{(2\pi)^{(m-1)/2}}{\sqrt{m}}.$$

Finally, without utilizing the complex contour integration, we calculate a special integral based on only the standard properties of the gamma function. This integral will be used to derive a functional equation for the Riemann zeta function.

Example 7.14. *For $0 < \alpha < 2$, show that*

$$\int_0^\infty x^{-\alpha} \sin x \, dx = \frac{\pi}{2\Gamma(\alpha)\sin(\alpha\pi/2)}. \tag{7.20}$$

Proof. First, for $x > 0$, replacing t by xt yields

$$\Gamma(\alpha) = \int_0^\infty t^{\alpha-1} e^{-t} \, dt = x^\alpha \int_0^\infty t^{\alpha-1} e^{-xt} \, dt.$$

Next, choose $\epsilon > 0$, let

$$I(\epsilon) = \int_0^\infty e^{-\epsilon x} x^{-\alpha} \sin x \, dx.$$

Then

$$
\begin{aligned}
I(\epsilon) &= \frac{1}{\Gamma(\alpha)} \int_0^\infty e^{-\epsilon x} \sin x \int_0^\infty t^{\alpha-1} e^{-xt} \, dt dx \\
&= \frac{1}{\Gamma(\alpha)} \int_0^\infty t^{\alpha-1} \int_0^\infty e^{-x(\epsilon+t)} \sin x \, dx dt \\
&= \frac{1}{\Gamma(\alpha)} \int_0^\infty \frac{t^{\alpha-1}}{1 + (\epsilon+t)^2} \, dt.
\end{aligned}
$$

Here the factor e^{-xt} makes the integrand absolutely integrable on $[0, \infty) \times [0, \infty)$. By Fubini's theorem, the interchange of the order of the integration is justified. Moreover, recall the remark after the Dirichlet's test (Theorem 7.14). It implies that $\int_0^\infty e^{-\epsilon x} x^{-\alpha} \sin x \, dx$ is uniformly converges on $[0, \infty)$, so $I(\epsilon)$ is continuous and

$$\int_0^\infty x^{-\alpha} \sin x \, dx = \lim_{\epsilon \to 0^+} I(\epsilon) = \frac{1}{\Gamma(\alpha)} \int_0^\infty \frac{t^{\alpha-1}}{1+t^2} \, dt.$$

To compute the last integral, by the substitution $u = t^2$, we have

$$\int_0^\infty \frac{t^{\alpha-1}}{1+t^2} \, dt = \frac{1}{2} \int_0^\infty \frac{u^{\alpha/2-1}}{1+u} \, du = \frac{1}{2} B\left(\frac{\alpha}{2}, 1 - \frac{\alpha}{2}\right).$$

Applying (7.18) and (7.19) gives

$$B\left(\frac{\alpha}{2}, 1 - \frac{\alpha}{2}\right) = \Gamma\left(\frac{\alpha}{2}\right) \Gamma\left(1 - \frac{\alpha}{2}\right) = \frac{\pi}{\sin(\alpha\pi/2)},$$

which proves (7.20). □

7.4 Worked Examples

There is no subject so old that something new cannot be said about it.
— Fyodor Dostoevsky
You are never sure whether or not a problem is good unless you actually solve it.
— Mikhail Gromov

In this section, we collect some problems to demonstrate the calculations and applications of the improper integration.

1. Test the convergence of

$$I = \int_0^\infty \frac{dx}{1 + x^\alpha |\sin x|^\beta} \qquad (\alpha, \beta > 0).$$

Solution. We consider four cases.
(1) Assume $\alpha \leq 1$. Since

$$\frac{1}{1 + x^\alpha |\sin x|^\beta} \geq \frac{1}{1 + x^\alpha},$$

by the comparison test, the integral diverges.
(2) Assume that $\alpha \leq \beta$. we have

$$
\begin{aligned}
I &= \sum_{n=0}^\infty \int_{n\pi}^{(n+1)\pi} \frac{dx}{1 + x^\alpha |\sin x|^\beta} \\
&= \sum_{n=0}^\infty \int_0^\pi \frac{dt}{1 + (n\pi + t)^\alpha |\sin t|^\beta} \\
&\geq \sum_{n=0}^\infty \int_0^{1/(n+1)\pi} \frac{dt}{1 + (n\pi + t)^\alpha |\sin t|^\beta} \\
&\geq \frac{1}{2\pi} \sum_{n=0}^\infty \frac{1}{n+1}.
\end{aligned}
$$

In the last inequality, we have used the inequality: If $0 < t < 1/(n+1)\pi$,

$$(n\pi + t)^\alpha |\sin t|^\beta < (n+1)^\alpha \pi^\alpha t^\beta < (n+1)^\beta \pi^\beta \left(\frac{1}{(n+1)\pi} \right)^\beta = 1.$$

Here the divergence of the harmonic series implies the integral diverges.
(3) Assume $\alpha > \beta > 1$. Observe that

$$I = \sum_{n=0}^\infty \int_0^{\pi/2} \frac{dt}{1 + (n\pi + t)^\alpha \sin t^\beta} + \sum_{n=0}^\infty \int_0^{\pi/2} \frac{dt}{1 + (n\pi - t)^\alpha \sin t^\beta} = I_1 + I_2.$$

For $t \in (0, \pi/2)$ and $n \geq 1$,

$$(n\pi + t)^\alpha \sin t^\beta \geq (n\pi)^\alpha \left(\frac{2}{\pi} t \right)^\beta = n^\alpha c^\beta t^\beta, \quad \text{(with } c = 2\pi^{\alpha/\beta - 1}),$$

in which the well-known inequality $\sin x > \frac{2}{\pi} x$ has been used. Thus, the general term in I_1 does not exceed

$$\int_0^{\pi/2} \frac{dt}{1 + n^\alpha c^\beta t^\beta} = \frac{1}{n^{\alpha/\beta} c} \int_0^{n^{\alpha/\beta} c\pi/2} \frac{dt}{1 + t^\beta} \leq \frac{C_n}{n^{\alpha/\beta}},$$

where $C_n = \frac{1}{c} \int_0^\infty \frac{dt}{1+t^\beta}$. The p-series test implies that I_1 converges. By a similar approach, we see that I_2 converges as well.
(4) The leftover case: $\alpha > 1$ and $\alpha > \beta$. Taking a number $\eta \geq \beta$ such that $\alpha > \eta > 1$ converts this case into case (3). So the corresponding integral converges. In summary, we have that the integral I converges if $\alpha > \max\{1, \beta\}$ and diverges if $\alpha \leq \max\{1, \beta\}$. $\qquad \square$

2. If $f(x) > 0$ is decreasing, show that $\int_a^\infty f(x)\, dx$ and $\int_a^\infty f(x) \sin^2 x\, dx$ both converge or both diverge.

Proof. Since $\sin x \leq 1$, by the comparison test, it follows that $\int_a^\infty f(x) \sin^2 x\, dx$ converges if $\int_a^\infty f(x)\, dx$ converges. We now prove the converse by contradiction. Assume $\int_a^\infty f(x)\, dx$ diverges. In view of

$$\int_a^\infty f(x)\, dx = \sum_{n=0}^\infty \int_{a+n\pi}^{a+(n+1)\pi} f(x)\, dx$$

and the assumption that f is decreasing, we have

$$\pi \sum_{n=0}^\infty f(a + n\pi) \geq \sum_{n=0}^\infty \int_{a+n\pi}^{a+(n+1)\pi} f(x)\, dx = \infty. \tag{7.21}$$

On the other hand, we have

$$\int_a^\infty f(x) \sin^2 x\, dx = \sum_{n=0}^\infty \int_{a+n\pi}^{a+(n+1)\pi} f(x) \sin^2 x\, dx$$

$$\geq \sum_{n=0}^\infty \int_{a+n\pi}^{a+(n+1)\pi} f(a + (n+1)\pi) \sin^2 x\, dx$$

$$= \sum_{n=0}^\infty f(a + (n+1)\pi) \int_{a+n\pi}^{a+(n+1)\pi} \sin^2 x\, dx$$

$$= \frac{\pi}{2} \sum_{n=0}^\infty f(a + (n+1)\pi).$$

This implies that $\int_a^\infty f(x) \sin^2 x\, dx$ diverges from (7.21). □

3. Let $f_n(x) : [a, \infty) \to \mathbb{R}$ converge to $f(x)$. If

(1). For any $A > a$, f_n converges uniformly on $[a, A]$,

(2). $\int_a^\infty f_n(x)\, dx$ converges uniformly on n,

prove that $\int_a^\infty f(x)\, dx$ converges and

$$\int_a^\infty f(x)\, dx = \lim_{n \to \infty} \int_a^\infty f_n(x)\, dx.$$

Proof. By the assumption (2), for any $\epsilon > 0$, there is $R > a$ such that

$$\left| \int_A^B f_n(x)\, dx \right| < \epsilon \quad \text{whenever } B > A > R \text{ and for all } n \in \mathbb{N}.$$

Since f_n converges uniformly to f on $[A, B]$, by Theorem 7.7, letting $n \to \infty$ leads to

$$\left| \int_A^B f(x)\, dx \right| \leq \epsilon.$$

This implies that $\int_a^\infty f(x)\, dx$ converges. Now we choose $A' > a$ such that, for every $n \in \mathbb{N}$,

$$\left| \int_{A'}^\infty f(x)\, dx \right| < \frac{\epsilon}{3} \quad \text{and} \quad \left| \int_{A'}^\infty f_n(x)\, dx \right| < \frac{\epsilon}{3}.$$

By (1), there is an N such that

$$|f_n(x) - f(x)| < \frac{\epsilon}{3(A' - a)} \qquad \text{whenever } x \in [a, A'].$$

Thus

$$\left| \int_a^\infty f_n(x)\, dx - \int_a^\infty f(x)\, dx \right| \leq \int_a^{A'} |f_n(x) - f(x)|\, dx + \left| \int_{A'}^\infty f_n(x)\, dx \right|$$

$$+ \left| \int_{A'}^\infty f(x)\, dx \right|$$

$$< \frac{\epsilon}{3(A' - a)}(A' - a) + \frac{\epsilon}{3} + \frac{\epsilon}{3} = \epsilon.$$

This proves the stated limit. $\qquad\qquad\qquad\qquad\qquad\qquad\qquad\qquad\qquad\qquad\square$

Remark. Recall from Theorem 6.7 that the uniform convergence on $[a, b]$ is the only requirement necessary to switch the limit and integral. For the improper integrals, the following example shows that the additional condition (2) is necessary. Let

$$f_n(x) = \begin{cases} \frac{1}{n^2}, & 0 \leq x \leq n^2, \\ 0, & x > n^2. \end{cases}$$

Since $0 \leq f_n(x) \leq 1/n^2$ for all $x \in [0, \infty)$, f_n converges to 0 uniformly. But

$$\lim_{n \to \infty} \int_0^\infty f_n(x)\, dx = \lim_{n \to \infty} \int_0^{n^2} \frac{1}{n}\, dx = \infty \neq 0.$$

4. Another real number proof of (7.18).

Proof. Begin with

$$B(x, y + 1) = \int_0^1 t^{x-1}(1 - t)^y\, dt$$

$$= \int_0^1 t^{x-1}(1 - t)(1 - t)^{y-1}\, dt$$

$$= B(x, y) - B(x + 1, y).$$

On the other hand, integration by parts gives

$$B(x, y + 1) = \int_0^1 \frac{1}{x}(1 - t)^y\, d(t^x)$$

$$= \frac{1}{x} t^x (1 - t)^y \Big|_0^1 + \frac{y}{x} \int_0^1 t^x (1 - t)^{y-1}\, dt$$

$$= \frac{y}{x} B(x + 1, y).$$

Combining these two equations yields

$$B(x,y) = \frac{x+y}{y} B(x, y+1).$$

Repeatedly using this identity gives

$$B(x,y) = \frac{(x+y)(x+y+1)}{y(y+1)} B(x, y+2) = \cdots$$
$$= \frac{(x+y)(x+y+1)\cdots(x+y+n-1)}{y(y+1)\cdots(y+n-1)} B(x, y+n).$$

Keeping the Gauss product formula (7.10) in mind, we rewrite $B(x,y)$ as

$$B(x,y) = \frac{(x+y)(x+y+1)\cdots(x+y+n)n!n^y}{y(y+1)\cdots(y+n)n!n^{x+y}} \frac{y+n}{x+y+n} n^x B(x, y+n)$$
$$= \frac{y+n}{x+y+n} \frac{(x+y)(x+y+1)\cdots(x+y+n)}{n!n^{x+y}} \frac{n!n^y}{y(y+1)\cdots(y+n)} n^x$$
$$\int_0^1 t^{x-1}(1-t)^{y+n-1}\, dt$$
$$= \frac{y+n}{x+y+n} \frac{(x+y)(x+y+1)\cdots(x+y+n)}{n!n^{x+y}} \frac{n!n^y}{y(y+1)\cdots(y+n)}$$
$$\int_0^n u^{x-1}\left(1 - \frac{u}{n}\right)^{y+n-1}\, du.$$

Letting $n \to \infty$, then invoking the Gauss product formula (7.10), we find that

$$B(x,y) = \frac{\Gamma(y)}{\Gamma(x+y)} \lim_{n\to\infty} \int_0^n u^{x-1}\left(1 - \frac{u}{n}\right)^{y+n-1}\, du$$
$$= \frac{\Gamma(y)}{\Gamma(x+y)} \int_0^\infty u^{x-1}e^{-u}\, du = \frac{\Gamma(x)\Gamma(y)}{\Gamma(x+y)},$$

where the limit process is justified by the Dominated convergence theorem. □

Remark. Here we have taken advantage of the integral definition for $\Gamma(x)$. Since we have the powerful tools of integration such as integration by parts and change of variables, we can deal with related properties of the gamma function more efficiently. As an example, we give another beautiful derivation of (7.18) due to Jacobi. Rewrite

$$\Gamma(x)\Gamma(y) = \int_0^\infty t^{x-1}e^{-t}\, dt \int_0^\infty s^{y-1}e^{-s}\, ds = \int_0^\infty \int_0^\infty t^{x-1}s^{y-1}e^{-(s+t)}\, ds dt.$$

The equality of the iterated integral and the double integral is justified by the absolute convergence of the double integral. On the other hand, making the change of variables:

$$s = uv, \quad t = u(1-v),$$

which is initially suggested by $u = s + t$, we find that

$$(u,v) \in (0,\infty) \times (0,1) \quad \text{if} \quad (s,t) \in (0,\infty) \times (0,\infty).$$

Since

$$J = \frac{D(s,t)}{D(u,v)} = \begin{vmatrix} v & u \\ 1-v & -u \end{vmatrix} = -u,$$

we find that

$$\Gamma(x)\Gamma(y) = \int_0^\infty \int_0^1 u^{x-1}(1-v)^{x-1}u^{y-1}v^{y-1}e^{-u}u\,du\,dv$$

$$= \int_0^\infty u^{x+y-1}e^{-u}\,du \int_0^1 v^y(1-v)^{x-1}\,dv$$

$$= \Gamma(x+y)B(y,x).$$

(7.18) now follows from $B(x,y) = B(y,x)$.

5. Evaluate $\int_0^\infty \frac{\sin^2 x}{x^2}\,dx$.

Solution. Appealing to the identity

$$\sin(2n-1)x - \sin(2n-3)x = 2\sin x\cos 2(n-1)x,$$

for each $n \in \mathbb{N}$, we have

$$\int_0^{\pi/2} \frac{\sin(2n-1)x}{\sin x}\,dx = \int_0^{\pi/2} \frac{\sin(2n-3)x}{\sin x}\,dx = \cdots = \int_0^{\pi/2} \frac{\sin x}{\sin x}\,dx = \frac{\pi}{2}.$$

Next, in view of the identity $\sin^2 nx - \sin^2(n-1)x = \sin(2n-1)x\sin x$, we have

$$\frac{\sin^2 nx}{\sin^2 x} = \frac{\sin^2(n-1)x}{\sin^2 x} + \frac{\sin(2n-1)x}{\sin x} = \sum_{k=1}^n \frac{\sin(2k-1)x}{\sin x}\,dx.$$

Hence

$$\int_0^{\pi/2} \frac{\sin^2 nx}{\sin^2 x}\,dx = \sum_{k=1}^n \int_0^{\pi/2} \frac{\sin(2k-1)x}{\sin x}\,dx = \frac{\pi}{2}n.$$

Applying the well-known inequality $x - x^3/3! \leq \sin x \leq x$ for $x \in [0,\pi/2]$ yields

$$\frac{1}{n}\int_0^{\pi/2} \frac{\sin^2 nx}{x^2}\,dx \leq \frac{\pi}{2} = \frac{1}{n}\int_0^{\pi/2} \frac{\sin^2 nx}{\sin^2 x}\,dx$$

$$\leq \frac{1}{n}\int_0^\delta \frac{\sin^2 nx}{x^2}\left(1 - \frac{\delta^2}{6}\right)^{-2}\,dx$$

$$+ \frac{1}{n}\int_\delta^{\pi/2} \frac{1}{x^2}\left(1 - \frac{\pi^2}{24}\right)^{-2}\,dx.$$

Using the substitution $x = t/n$, then letting $n \to \infty$ leads to

$$\int_0^\infty \frac{\sin^2 x}{x^2}\,dx \leq \frac{\pi}{2} \leq \left(1 - \frac{\delta^2}{6}\right)^{-2}\int_0^\infty \frac{\sin^2 x}{x^2}\,dx.$$

Since δ is arbitrary, it follows that

$$\int_0^\infty \frac{\sin^2 x}{x^2}\,dx = \frac{\pi}{2}. \qquad \square$$

Remark. If we know that

$$\int_0^\infty \frac{\sin x}{x}\,dx = \frac{\pi}{2},$$

using integration by parts, we simply have

$$\int_0^\infty \frac{\sin^2 x}{x^2}\,dx = \int_0^\infty \frac{2\sin x\cos x}{x}\,dx = \int_0^\infty \frac{\sin 2x}{x}\,dx = \frac{\pi}{2}.$$

6. (**Monthly Problem 12288, 2021**). Prove

$$\int_0^\infty \left(1 - x^2 \sin^2(1/x)\right)^2 dx = \frac{\pi}{5}.$$

Proof. Let the proposed integral be I. The substitution $x = 1/t$ leads to

$$I = \int_0^\infty \frac{(t^2 - \sin^2 t)^2}{t^6} dt.$$

Rewrite I as

$$I = 2\int_0^\infty \frac{t^2 - \sin^2 t}{t^4} dt - \int_0^\infty \frac{t^4 - \sin^4 t}{t^6} dt.$$

In general, this suggests us to consider

$$I_n := \int_0^\infty \frac{t^n - \sin^n t}{t^{n+2}} dt$$

where $n \geq 2$ is a positive even number. We now evaluate I_n via the Laplace transform. Recall that

$$\mathcal{L}[t^k] = \int_0^\infty t^k e^{-st} dt = \frac{k!}{s^{k+1}}.$$

We find

$$I_n = \frac{1}{(n+1)!} \int_0^\infty (t^n - \sin^n t) \mathcal{L}[s^{n+1}] dt$$

$$= \frac{1}{(n+1)!} \int_0^\infty s^{n+1} \int_0^\infty (t^n - \sin^n t) e^{-st} dt ds$$

$$= \frac{1}{(n+1)!} \int_0^\infty s^{n+1} \left(\mathcal{L}[t^n](s) - \mathcal{L}[\sin^n t](s)\right) ds$$

$$= \frac{1}{(n+1)!} \int_0^\infty s^{n+1} \left(\frac{n!}{s^{n+1}} - \frac{n!}{s(s^2 + 2^2) \cdots (s^2 + n^2)}\right) ds$$

$$= \frac{1}{n+1} \int_0^\infty \left(1 - \frac{s^n}{(s^2 + 2^2) \cdots (s^2 + n^2)}\right) ds,$$

where we have used the fact that

$$\mathcal{L}[\sin^n t](s) = \frac{n(n-1)}{s^2 + n^2} \mathcal{L}[\sin^{n-2} t](s),$$

which results from

$$s^2 \mathcal{L}[\sin^n t](s) = \mathcal{L}[(\sin^n t)''](s) = n(n-1)\mathcal{L}[\sin^{n-2} t](s) - n^2 \mathcal{L}[\sin^n t](s).$$

In particular, we have

$$I_2 = \frac{1}{3} \int_0^\infty \left(1 - \frac{s^2}{s^2 + 2^2}\right) ds = \frac{4}{3} \int_0^\infty \frac{ds}{s^2 + 2^2} = \frac{\pi}{3}$$

and

$$I_4 = \frac{1}{5} \int_0^\infty \left(1 - \frac{s^4}{(s^2 + 2^2)(s^2 + 4^2)}\right) ds$$

$$= \frac{1}{5} \int_0^\infty \left(\frac{64}{3(s^2 + 4^2)} - \frac{4}{3(s^2 + 2^2)}\right) ds = \frac{7\pi}{15}.$$

Hence

$$I = 2I_2 - I_4 = 2 \cdot \frac{\pi}{3} - \frac{7\pi}{15} = \frac{\pi}{5}.$$

□

In general, we challenge the reader to prove that

$$I_n = \frac{\pi}{2^n(n+1)!} \sum_{k=0}^{[(n-1)/2]} (-1)^k \binom{n}{k} (n-2k)^{n+1}$$

and that for any $n \in \mathbb{N}$,

$$\int_0^\infty \left(1 - x^2 \sin^2(1/x)\right)^n \, dx$$

is a rational multiple of π.

7. (**Monthly Problem 12281, 2021**). Evaluate

$$\int_0^\infty \left(\frac{\cosh x}{\sinh^2 x} - \frac{1}{x^2}\right) (\ln x)^2 \, dx.$$

Solution. Let $u = (\ln x)^2$ and $v = 1/x - 1/\sinh x$. Since

$$\frac{1}{x} - \frac{1}{\sinh x} = \frac{1}{6}x - \frac{7}{360}x^3 + o(x^3) \quad \text{(for } x \text{ nearby } 0),$$

integrating by parts yields

$$I := \int_0^\infty \left(\frac{\cosh x}{\sinh^2 x} - \frac{1}{x^2}\right) (\ln x)^2 \, dx = 2 \int_0^\infty \left(\frac{1}{\sinh x} - \frac{1}{x}\right) \frac{\ln x}{x} \, dx.$$

Recall the partial fraction expansion:

$$\frac{1}{\sin x} = \frac{1}{x} + 2x \sum_{n=1}^\infty \frac{(-1)^n}{x^2 - n^2\pi^2}.$$

This implies, together with $\sinh x = -i \sin(ix)$,

$$\frac{1}{\sinh x} = \frac{1}{x} + 2x \sum_{n=1}^\infty \frac{(-1)^n}{x^2 + n^2\pi^2}.$$

Hence,

$$\begin{aligned}
I &= 4 \int_0^\infty \left(\sum_{n=1}^\infty \frac{(-1)^n}{x^2 + n^2\pi^2}\right) \ln x \, dx \\
&= 4 \sum_{n=1}^\infty (-1)^n \int_0^\infty \frac{\ln x}{x^2 + n^2\pi^2} \, dx.
\end{aligned}$$

For $a > 0$, we have

$$\begin{aligned}
\int_0^\infty \frac{\ln x}{x^2 + a^2} \, dx &= \frac{1}{a} \int_0^\infty \frac{\ln(at)}{t^2 + 1} \, dt \quad \text{(use } x = at) \\
&= \frac{1}{a} \left(\int_0^\infty \frac{\ln a}{t^2 + 1} \, dt + \int_0^\infty \frac{\ln t}{t^2 + 1} \, dt\right) = \frac{\pi}{2a} \ln a,
\end{aligned}$$

because

$$\int_0^\infty \frac{\ln t}{t^2+1}\, dt = \int_0^1 \frac{\ln t}{t^2+1}\, dt + \int_1^\infty \frac{\ln t}{t^2+1}\, dt$$

$$= \int_0^\infty \frac{\ln t}{t^2+1}\, dt - \int_0^1 \frac{\ln u}{u^2+1}\, du = 0, \ \text{(use } u = 1/t\text{)}.$$

Let $a = n\pi$. Finally we find that

$$I = 2\sum_{n=1}^\infty \frac{(-1)^n \ln(n\pi)}{n} = 2\left(\sum_{n=1}^\infty \frac{(-1)^n \ln n}{n} + \sum_{n=1}^\infty \frac{(-1)^n \ln \pi}{n}\right)$$

$$= 2\left(\frac{1}{2}(2\gamma \ln 2 - \ln^2 2) - \ln 2 \ln \pi\right) = \ln 2(2\gamma - \ln 2 - 2\ln \pi),$$

where γ is the Euler-Mascheroni constant. □

8. Let $f(x)$ be monotone on $[0, \infty)$. If $\int_0^\infty f(x)\, dx$ converges, prove that

$$\lim_{h\to 0+} h \sum_{n=1}^\infty f(nh) = \int_0^\infty f(x)\, dx. \tag{7.22}$$

Proof. Without loss of generality, we assume that f is decreasing. In this case, f is nonnegative on $[0, \infty)$. Indeed, if there exists some x_0 such that $f(x_0) < 0$, then

$$\int_{x_0}^A f(x)\, dx \le f(x_0)(A - x_0) \to -\infty \quad (\text{as } A \to \infty),$$

which contradicts the convergence of the integral. Therefore, for every $n \in \mathbb{N}$, we have

$$\int_h^{(n+1)h} f(x)\, dx \le h \sum_{k=1}^n f(kh) \le \int_0^{nh} f(x)\, dx.$$

This implies that $S_n := h \sum_{k=1}^n f(kh)$ is monotone increasing and bounded from above. Letting $n \to \infty$ yields

$$\int_h^\infty f(x)\, dx \le h \sum_{k=1}^\infty f(kh) \le \int_0^\infty f(x)\, dx.$$

This proves (7.22) as desired by letting $h \to 0^+$. □

Remark. Applying (7.22) with $f(x) = \sin x/x$ recaptures the result of Example 7.4 as follows

$$\int_0^\infty \frac{\sin x}{x}\, dx = \lim_{h\to 0} \sum_{n=1}^\infty \frac{\sin nh}{n} = \lim_{h\to 0} \frac{\pi - h}{2} = \frac{\pi}{2}.$$

9. **(Frullani Integral).** Let $f : [0, \infty) \to \mathbb{R}$ be continuous and $a, b > 0$. If $f(\infty) = \lim_{x\to\infty} f(x)$ exists, show that

$$\int_0^\infty \frac{f(ax) - f(bx)}{x}\, dx = (f(0) - f(\infty)) \ln \frac{b}{a}. \tag{7.23}$$

Proof. We establish the convergence of the integral in the process of the computation. For $0 < r < R < \infty$, using substitution yields

$$\int_r^R \frac{f(ax) - f(bx)}{x}\,dx = \int_r^R \frac{f(ax)}{x}\,dx - \int_r^R \frac{f(bx)}{x}\,dx$$

$$= \int_{ar}^{aR} \frac{f(x)}{x}\,dx - \int_{br}^{bR} \frac{f(x)}{x}\,dx$$

$$= \int_{ar}^{br} \frac{f(x)}{x}\,dx - \int_{aR}^{bR} \frac{f(x)}{x}\,dx. \qquad (7.24)$$

Applying the first mean value theorem for integrals to the two integrals in (7.24), respectively, gives

$$\int_{ar}^{br} \frac{f(x)}{x}\,dx = f(\xi) \int_{ar}^{br} \frac{dx}{x} = f(\xi)\ln\frac{b}{a}, \quad (ar < \xi < br);$$

$$\int_{aR}^{bR} \frac{f(x)}{x}\,dx = f(\eta) \int_{aR}^{bR} \frac{dx}{x} = f(\eta)\ln\frac{b}{a}, \quad (ar < \eta < br).$$

Let $r \to 0^+$ and $R \to \infty$, respectively. Then $\xi \to 0$ and $\eta \to \infty$. In view of the continuity of f,

$$\lim_{\xi \to 0} f(\xi) = f(0) \quad \text{and} \quad \lim_{\eta \to \infty} f(\eta) = f(\infty),$$

we conclude that the proposed integral exists and

$$\int_0^\infty \frac{f(ax) - f(bx)}{x}\,dx = (f(0) - f(\infty))\ln\frac{b}{a}.$$

$\square$

Remark. In view of (7.24), we obtain two variants of Frullani integral:

(a) If $f(\infty) = \lim_{x \to \infty} f(x)$ does not exist, but for some $A > 0$, the integral

$$\int_A^\infty \frac{f(x)}{x}\,dx$$

converges, then

$$\int_0^\infty \frac{f(ax) - f(bx)}{x}\,dx = f(0)\ln\frac{b}{a}.$$

(b) If $\lim_{x \to 0^+} f(x)$ does not exist, but for some $A > 0$, the integral

$$\int_0^A \frac{f(x)}{x}\,dx$$

converges, then

$$\int_0^\infty \frac{f(ax) - f(bx)}{x}\,dx = -f(\infty)\ln\frac{b}{a} = f(\infty)\ln\frac{a}{b}.$$

10. **(Monthly Problem 11709, 2013).** Find

$$\int_0^\infty \frac{dx}{x} \int_0^x \frac{dy}{y}\{\cos(x - y) - \cos x\}.$$

Solution. The integral value is $\zeta(2) = \pi^2/6$. Let $y = xt$ in the inner integral. Then

$$\int_0^x \frac{dy}{y}\{\cos(x-y) - \cos x\} = \int_0^1 \frac{\cos(1-t)x - \cos x}{t}\, dt.$$

Interchanging the order of the integration yields

$$\int_0^\infty \frac{dx}{x}\int_0^1 \frac{\cos(1-t)x - \cos x}{t}\, dt = \int_0^1 \frac{dt}{t}\int_0^\infty \frac{\cos(1-t)x - \cos x}{x}\, dx.$$

Appealing to the variant (a) of (7.23), we have

$$\int_0^\infty \frac{\cos px - \cos qx}{x}\, dx = \ln\left(\frac{q}{p}\right).$$

Hence

$$\int_0^1 \frac{dt}{t}\int_0^\infty \frac{\cos(1-t)x - \cos x}{x}\, dx \;=\; \int_0^1 \frac{-\ln(1-t)}{t}\, dt.$$

Since

$$\frac{-\ln(1-t)}{t} = \sum_{n=1}^\infty \frac{t^{n-1}}{n}$$

is a positive series for $t \in (0,1)$, integrating term by term gives

$$\int_0^1 \frac{-\ln(1-t)}{t}\, dt = \sum_{n=1}^\infty \frac{1}{n}\int_0^1 t^{n-1}\, dt = \sum_{n=1}^\infty \frac{1}{n^2} = \zeta(2) = \frac{\pi^2}{6}$$

as claimed. $\qquad\square$

Remark. A careful reader may find a missing justification in our interchange of the order of integration. To overcome this defect, for $R > 0$ and $y = xt$, observe that

$$\int_0^R \frac{dx}{x}\int_0^x \frac{dy}{y}\{\cos(x-y) - \cos x\} = \int_0^R \int_0^1 \int_{1-t}^t \frac{\sin(sx)}{t}\, ds\,dt\,dx.$$

Since $|\sin(sx)| \leq 1$, the triple integral is absolutely convergent. Fubini's theorem now justifies an interchange of the order of integration. Therefore,

$$\int_0^R \frac{dx}{x}\int_0^x \frac{dy}{y}\{\cos(x-y) - \cos x\} = \int_0^1 \int_{1-t}^1 \frac{1}{t}\int_0^R \sin(sx)\, dx\,ds\,dt$$

$$= \int_0^1 \frac{1 - \cos(Rs)}{s}\int_{1-s}^1 \frac{1}{t}\, dt\,ds$$

$$= -\int_0^1 \frac{\ln(1-s)}{s}\, ds + \int_0^1 \frac{\ln(1-s)}{s}\cos(Rs)\, ds.$$

Since $\ln(1-s)/s$ is integrable on $[0,1]$, by the Riemann Lemma (see Example 8 in Section 5.5) we have

$$\lim_{R\to\infty} \int_0^1 \frac{\ln(1-s)}{s}\cos(Rs)\, ds = 0.$$

Hence we find

$$\lim_{R\to\infty} \int_0^R \frac{dx}{x}\int_0^x \frac{dy}{y}\{\cos(x-y) - \cos x\} = -\int_0^1 \frac{\ln(1-s)}{s}\, ds = \frac{\pi^2}{6}.$$

11. If $\int_0^\infty f(x^2)\,dx$ converges, show that

$$\int_0^\infty f\left[\left(Ax - \frac{B}{x}\right)^2\right]dx = \frac{1}{A}\int_0^\infty f(u^2)\,du, \quad (A, B > 0). \tag{7.25}$$

Proof. By the substitution $u = Ax - B/x$, we have

$$\int_{-\infty}^\infty f(u^2)\,du = \int_0^\infty f\left[\left(Ax - \frac{B}{x}\right)^2\right]\left(A + \frac{B}{x^2}\right)dx$$

$$= A\int_0^\infty f\left[\left(Ax - \frac{B}{x}\right)^2\right]dx + B\int_0^\infty f\left[\left(Ax - \frac{B}{x}\right)^2\right]\frac{dx}{x^2}.$$

Making the substitution $x = -B/At$ in the second integral above yields

$$A\int_{-\infty}^0 f\left[\left(At - \frac{B}{t}\right)^2\right]dt.$$

Hence

$$\int_{-\infty}^\infty f(u^2)\,du = A\int_{-\infty}^\infty f\left[\left(Ax - \frac{B}{x}\right)^2\right]dx.$$

This is equivalent to the claimed equality because the integrands on both sides are even functions. $\square$

Let $f(x) = e^{-x}$. By (7.25), for $A, B > 0$, we find that

$$\int_0^\infty e^{-(Ax - B/x)^2}\,dx = \frac{1}{A}\int_0^\infty e^{-u^2}\,du = \frac{\sqrt{\pi}}{2A}.$$

12. Show that, for $s > 1$,

$$\zeta(s) = \sum_{n=1}^\infty \frac{1}{n^s} = \frac{1}{\Gamma(s)}\int_0^\infty \frac{x^{s-1}}{e^x - 1}\,dx.$$

This gives an integral representation for the Riemann zeta function.

Proof. In view of

$$\frac{1}{e^x - 1} = \frac{e^{-x}}{1 - e^{-x}} = \sum_{n=1}^\infty e^{-nx},$$

for every $A > 0$, we have

$$\int_0^A \frac{x^{s-1}}{e^x - 1}\,dx = \int_0^A x^{s-1}\left(\sum_{n=1}^\infty e^{-nx}\right)dx.$$

For fixed $s > 1$, the series $\sum_{n=1}^\infty x^{s-1}e^{-nx}$ converges uniformly on $(0, \infty)$. Thus,

$$\int_0^A \frac{x^{s-1}}{e^x - 1}\,dx = \sum_{n=1}^\infty \int_0^A x^{s-1}e^{-nx}\,dx$$

$$= \sum_{n=1}^\infty \int_0^{nA} \frac{1}{n}\left(\frac{u}{n}\right)^{s-1}e^{-u}\,du$$

$$= \sum_{n=1}^\infty \frac{1}{n^s}\int_0^{nA} u^{s-1}e^{-u}\,du.$$

The last series on A converges uniformly. Letting $A \to \infty$ yields

$$\int_0^\infty \frac{x^{s-1}}{e^x - 1} \, dx = \Gamma(s) \sum_{n=1}^\infty \frac{1}{n^s} = \Gamma(s)\zeta(s),$$

which is equivalent to the proposed equality. $\qquad\qquad\qquad\qquad\qquad$ $\square$

13. Define the *Hurwitz zeta function*

$$\zeta(s, a) := \sum_{n=0}^\infty \frac{1}{(n+a)^s}.$$

Show that

(a) For $s > 1$, we have

$$\zeta(s, a) = \frac{1}{\Gamma(s)} \int_0^\infty \frac{x^{s-1} e^{-ax}}{1 - e^{-x}} \, dx.$$

(b) For $0 < a < 1, k \in \mathbb{N}$, we have

$$\zeta(k, a) - (-1)^{n-1}\zeta(k, 1 - a) = \frac{(-1)^{k-1}\pi}{(k-1)!}(\cot(a\pi))^{(k-1)},$$

where $f^{(k)}$ denotes the kth derivative.

Proof. The proof of (a) is similar to Problem 12 above. We turn to proving (b). Notice that

$$\zeta(k, a) - (-1)^{k-1}\zeta(k, 1 - a) = \frac{1}{(k-1)!} \int_0^\infty \frac{x^{k-1}(e^{-ax} - (-1)^{k-1}e^{-(1-a)x})}{1 - e^{-x}} \, dx.$$

We show that, via the parametric differentiation, this integral can be related to

$$f(a) = \int_0^\infty \frac{e^{-ax} - e^{-(1-a)x}}{1 - e^{-x}} \, dx.$$

In fact, in view of

$$\frac{1}{1 - e^{-x}} = \sum_{n=0}^\infty e^{-nx},$$

we have

$$f(a) = \int_0^\infty \frac{e^{-ax}}{1 - e^{-x}} \, dx - \int_0^\infty \frac{e^{-(1-a)x}}{1 - e^{-x}} \, dx$$

$$= \sum_{n=0}^\infty \frac{1}{n+a} - \sum_{n=0}^\infty \frac{1}{n+1-a}$$

$$= \sum_{n=0}^\infty \frac{1}{n+a} + \sum_{n=0}^\infty \frac{1}{-(n+1)+a}$$

$$= \sum_{n=-\infty}^\infty \frac{1}{n+a}.$$

Using Euler's elegant partial fraction decomposition (6.20):

$$\cot x = \sum_{n=-\infty}^{\infty} \frac{1}{n\pi + x},$$

we find that

$$f(a) = \pi \cot(a\pi).$$

Taking the $(k-1)$th derivative of $f(a)$ gives the desired result (b).   □

Remark. The formula (b) is often referred to the *reflection formula* for the Hurwitz zeta function. As an application, note that

$$\sum_{n=0}^{\infty} \frac{1}{(3n+1)^3} - \sum_{n=0}^{\infty} \frac{1}{(3n+2)^3} = \frac{1}{3^3} \left(\zeta(3,1/3) - \zeta(3,2/3) \right).$$

Hence

$$\sum_{n=0}^{\infty} \frac{1}{(3n+1)^3} - \sum_{k=0}^{\infty} \frac{1}{(3n+2)^3} = \frac{\pi}{2! \cdot 3^3} \left(\cot(a\pi) \right)'' \big|_{a=1/3} = \frac{4}{3^5} \sqrt{3}\,\pi^3.$$

In conjunction with

$$\sum_{n=0}^{\infty} \frac{1}{(3n+1)^3} + \sum_{k=0}^{\infty} \frac{1}{(3n+2)^3} = (1 - 3^{-3})\zeta(3),$$

we find that

$$\sum_{k=0}^{\infty} \frac{1}{(3k+1)^3} = \frac{1}{2}(1 - 3^{-3})\xi(3) + \frac{2}{3^5}\sqrt{3}\,\pi^3,$$

$$\sum_{k=0}^{\infty} \frac{1}{(3k+2)^3} = \frac{1}{2}(1 - 3^{-3})\xi(3) - \frac{2}{3^5}\sqrt{3}\,\pi^3.$$

14. **Evaluate**

$$R_0 = \int_0^1 \ln \Gamma(x)\, dx.$$

Solution. Replacing x by $1 - x$ yields

$$R_0 = \int_0^1 \ln \Gamma(1 - x)\, dx.$$

Thus, by Euler's reflection formula, we have

$$\begin{aligned}
2R_0 &= \int_0^1 \ln\left(\Gamma(x)\Gamma(1-x) \right)\, dx \\
&= \int_0^1 \ln\left(\frac{\pi}{\sin \pi x} \right)\, dx \\
&= \ln \pi - \frac{1}{\pi} \int_0^\pi \ln \sin x\, dx \\
&= \ln(2\pi).
\end{aligned}$$

Hence $R_0 = \ln \sqrt{2\pi}$. In general, for $x > 0$, we have

$$\int_x^{x+1} \ln \Gamma(t)\, dt = x(\ln x - 1) + \ln \sqrt{2\pi}.$$ □

15. Evaluate

$$\int_0^\infty \frac{e^{-ax^2} - e^{-bx^2}}{x^2} \, dx, \quad (a, b > 0).$$

Solution. We calculate this integral in two different ways: (i) parametric differentiation (based on Theorem 7.17) and (ii) parametric integration (based on Theorem 7.16).

I. *Use parametric differentiation.* Take a as the parameter and let

$$I(a) = \int_0^\infty \frac{e^{-ax^2} - e^{-bx^2}}{x^2} \, dx.$$

In view of (7.5), applying Theorem 7.19 and differentiating under the integral sign yields

$$I'(a) = -\int_0^\infty e^{-ax^2} \, dx = -\frac{1}{\sqrt{a}} \int_0^\infty e^{-t^2} \, dt = -\frac{\sqrt{\pi}}{2\sqrt{a}}.$$

Since $I(b) = 0$, we find that

$$I(a) = I(b) + \int_b^a \left(-\frac{\sqrt{\pi}}{2\sqrt{t}} \right) dt = \sqrt{\pi}(\sqrt{b} - \sqrt{a}).$$

It remains to verify that $f(x, a) = (e^{-ax^2} - e^{-bx^2})/x^2$ satisfies the conditions in Theorem 7.19. In fact, if we define $f(0, a) = b - a$, we see that $f(x, a)$ and $f_a(x, a)$ are continuous on $[0, \infty) \times [0, \infty)$. For any $\delta > 0$, we then have

$$|f_a(x, a)| = e^{-ax^2} \le e^{-\delta x^2} \quad \text{(for all } x \in [0, \infty) \text{ and } a \in [\delta, \infty)).$$

Since $\int_0^\infty e^{-\delta x^2} \, dx$ converges, by Weierstrass M-test, $\int_0^\infty f_a(x, a) \, dx$ converges uniformly on $[\delta, \infty)$. Thus, by Theorem 7.19, we have

$$\frac{d}{da} \int_0^\infty f(x, a) \, dx = \int_0^\infty f_a(x, a) \, dx, \quad \text{for every } a \in [\delta, \infty).$$

Since δ is arbitrary, this proves that differentiating under the integral sign holds for all $a \in (0, \infty)$.

II. *Ues parametric integration.* Observe that

$$\int_a^b e^{-x^2 t} \, dt = \frac{e^{-ax^2} - e^{-bx^2}}{x^2}.$$

This yields

$$\int_0^\infty \frac{e^{-ax^2} - e^{-bx^2}}{x^2} \, dx = \int_0^\infty \int_a^b e^{-x^2 t} \, dt dx.$$

Interchanging the order of the integration (Theorem 7.18) gives

$$\int_a^b \int_0^\infty e^{-x^2 t} \, dx dt = \int_a^b \left(\int_0^\infty \frac{1}{\sqrt{t}} e^{-u^2} \, du \right) dt = \int_a^b \frac{\sqrt{\pi}}{2\sqrt{t}} \, dt = \sqrt{\pi}(\sqrt{b} - \sqrt{a}).$$

$\square$

16. **(Fresnel Integral).** Show that

$$\int_0^\infty \sin x^2 \, dx = \int_0^\infty \cos x^2 \, dx = \frac{1}{2}\sqrt{\frac{\pi}{2}}.$$

Proof. Let $x^2 = t$. We have

$$\int_0^\infty \sin x^2 \, dx = \frac{1}{2} \int_0^\infty \frac{\sin t}{\sqrt{t}} \, dt; \quad \int_0^\infty \cos x^2 \, dx = \frac{1}{2} \int_0^\infty \frac{\cos t}{\sqrt{t}} \, dt.$$

By Dirichlet's test, both integrals converge. We show the result for the first integral only. Recall that

$$\frac{1}{\sqrt{t}} = \frac{2}{\sqrt{\pi}} \int_0^\infty e^{-tu^2} \, du.$$

Formally, we have

$$\begin{aligned}
\int_0^\infty \sin x^2 \, dx &= \frac{1}{2} \int_0^\infty \sin t \left(\int_0^\infty \frac{2}{\sqrt{\pi}} \int_0^\infty e^{-tu^2} \, du \right) dt \\
&= \frac{1}{\sqrt{\pi}} \int_0^\infty \left(\int_0^\infty e^{-tu^2} \sin t \, dt \right) du = \frac{1}{\sqrt{\pi}} \int_0^\infty \frac{du}{1 + u^4} \\
&= \frac{1}{\sqrt{\pi}} \frac{\pi}{2\sqrt{2}} = \frac{1}{2} \sqrt{\frac{\pi}{2}}.
\end{aligned}$$

Here, however justification of interchange of the order of integration is not easy. To avoid this difficult, we introduce the convergent factor $e^{-\alpha t}$ and consider

$$I(\alpha) = \int_0^\infty e^{-\alpha t} \frac{\sin t}{\sqrt{t}} \, dt.$$

Thus, for $\alpha > 0$, we have

$$\begin{aligned}
I(\alpha) &= \frac{2}{\sqrt{\pi}} \int_0^\infty \left(\int_0^\infty e^{-t(u^2 + \alpha)} \sin t \, du \right) dt \\
&= \frac{2}{\sqrt{\pi}} \int_0^\infty \left(\int_0^\infty e^{-t(u^2 + \alpha)} \sin t \, dt \right) du \\
&= \frac{2}{\sqrt{\pi}} \int_0^\infty \frac{du}{1 + (u^2 + \alpha)^2}.
\end{aligned}$$

Since $|e^{-t(u^2+\alpha)} \sin t| \leq e^{-\alpha t}$ and $\int_0^\infty e^{-\alpha t} \, dt$ converges, by Weierstrass M-test, it follows that $\int_0^\infty e^{-t(u^2+\alpha)} \sin t \, dt$ converges uniformly on $u \in [0, \infty)$. Next, we show that $\int_0^\infty e^{-t(u^2+\alpha)} \sin t \, du$ converges uniformly on $t \in [0, \infty)$. Indeed, in view of the following limit

$$\lim_{t \to 0^+} \frac{e^{\alpha t} \sin t}{\sqrt{t}} = 0,$$

for every $\epsilon > 0$, there is a $\delta > 0$ such that

$$\left| \frac{e^{\alpha t} \sin t}{\sqrt{t}} \right| < \frac{2\epsilon}{\sqrt{\pi}}, \quad \text{for any } t \in (0, \delta).$$

Thus, for any $A > 0$ and $t \in [0, \delta)$,

$$\begin{aligned}
\left| \int_A^\infty e^{-t(u^2+\alpha)} \sin t \, du \right| &= \left| e^{-\alpha t} \sin t \int_A^\infty e^{-u^2 t} \, du \right| \\
&= \left| \frac{e^{-\alpha t} \sin t}{\sqrt{t}} \int_{\sqrt{t}A}^\infty e^{-y^2} \, dy \right| \\
&< \frac{2\epsilon}{\sqrt{\pi}} \int_0^\infty e^{-y^2} \, dy = \epsilon.
\end{aligned}$$

On the other hand, when $t \in [\delta, \infty)$, we have

$$|e^{-t(u^2+\alpha)} \sin t| \le e^{-\delta(\alpha+u^2)} \le e^{-\delta u^2}.$$

Since $\int_0^\infty e^{-\delta u^2} \, du$ converges, it follows that $\int_0^\infty e^{-t(u^2+\alpha)} \sin t \, du$ converges uniformly on $t \in [\delta, \infty)$ from the Weierstrass M-test. Combining the arguments above confirms that $\int_0^\infty e^{-t(u^2+\alpha)} \sin t \, du$ converges uniformly on $t \in [0, \infty)$. Furthermore, since the integral

$$\int_0^\infty \left(\int_0^\infty e^{-t(u^2+\alpha)} \, dt \right) du = \int_0^\infty \frac{du}{u^2 + \alpha}$$

converges, it follows that

$$\int_0^\infty \left(\int_0^\infty e^{-t(u^2+\alpha)} |\sin t| dt \right) du$$

also converges. This justifies the interchange of the order of integration. Finally, in view of

$$0 < \frac{1}{1 + (u^2 + \alpha)^2} < \frac{1}{1 + u^4},$$

again, the Weierstruss M-test implies that $\int_0^\infty du/(1 + (u^2 + \alpha)^2)$ uniformly converges on $\alpha \in [0, \infty)$. By Abel's test, the integral $I(\alpha)$ also uniformly converges on $\alpha \in [0, \infty)$. Thus,

$$\int_0^\infty \sin x^2 \, dx = \lim_{\alpha \to 0} \frac{1}{2} I(\alpha) = \frac{1}{2} \int_0^\infty \frac{du}{1 + u^2} = \frac{1}{2} \sqrt{\frac{\pi}{2}}. \qquad \square$$

17. **The Fresnel Integral Revisited.**

Many Monthly articles [42, 66, 93] have been devoted to determining the values of the Fresnel integrals:

$$F_c = \int_0^\infty \cos x^2 \, dx \quad \text{and} \quad F_s = \int_0^\infty \sin x^2 \, dx.$$

For example, Flanders [42] considers

$$F_c(t) = \int_0^\infty e^{-tx^2} \cos x^2 \, dx \quad \text{and} \quad F_s(t) = \int_0^\infty e^{-tx^2} \sin x^2 \, dx \qquad (7.26)$$

and shows that they satisfy the functional equations

$$F_c^2(t) - F_s^2(t) = \frac{\pi t}{4(1 + t^2)}, \quad 2F_c(t)F_s(t) = \frac{\pi}{4(1 + t^2)}.$$

He then solves these simultaneous quadratic equations to find the values of $F_c(t)$ and $F_s(t)$. The cleverness of his method resides in the introduction of polar coordinates, as is usually done to evaluate $\int_0^\infty e^{-x^2} \, dx$. The following solution offers a simpler method which does not use the double integral, nor the system of quadratic equations. The main ingredients of the method are the consideration of some related derivatives and linear differential equations. The method applies to a number of pairs of integrals.

In analogy of the function pairs (7.26), for $t > 0, p \in \mathbb{R}$, we introduce

$$F_c(p) = \int_0^\infty e^{-tx^2} \cos(px^2)\, dx, \quad F_s(p) = \int_0^\infty e^{-tx^2} \sin(px^2)\, dx.$$

We will prove that

$$F_c'(p) = -\frac{pF_c(p) + tF_s(p)}{2(t^2 + p^2)}, \quad F_s'(p) = \frac{tF_c(p) - pF_s(p)}{2(t^2 + p^2)}, \tag{7.27}$$

and if $Z(p) = F_c(p) + i\, F_s(p)$, that

$$Z(p) = \frac{\sqrt{\pi}}{2}\, \frac{\sqrt{t + pi}}{\sqrt{t^2 + p^2}}. \tag{7.28}$$

Since

$$\sqrt{t + pi} = \sqrt{\frac{t + \sqrt{t^2 + p^2}}{2}} + \sqrt{\frac{-t + \sqrt{t^2 + p^2}}{2}}\, i,$$

upon establishing (7.28), as an immediate consequence we find

$$\begin{aligned} F_c(p) &= \frac{1}{2}\sqrt{\frac{\pi}{2}}\sqrt{\frac{t + \sqrt{t^2 + p^2}}{t^2 + p^2}}, \\[2mm] F_s(p) &= \frac{1}{2}\sqrt{\frac{\pi}{2}}\sqrt{\frac{-t + \sqrt{t^2 + p^2}}{t^2 + p^2}}. \end{aligned}$$

Theorem 7.15 asserts that $F_c(p)$ and $F_s(p)$ both converge uniformly for all $t > 0$. Hence F_c and F_s are continuous functions of t. In particular, setting $p = 1$ and letting $t \to 0^+$, we obtain the Fresnel integrals

$$\int_0^\infty \cos x^2\, dx = \int_0^\infty \sin x^2\, dx = \frac{1}{2}\sqrt{\frac{\pi}{2}}$$

as desired.

To prove (7.28), for any $p \in \mathbb{R}$ and $t \geq t_0 > 0$, since

$$|x^2 e^{-tx^2} \cos(px^2)| \leq x^2 e^{-t_0 x^2}, \quad |x^2 e^{-tx^2} \sin(px^2)| \leq x^2 e^{-t_0 x^2}$$

and $\int_0^\infty x^2 e^{-t_0 x^2}\, dx$ converges, the integrals

$$\int_0^\infty x^2 e^{-tx^2} \cos(px^2)\, dx \quad \text{and} \quad \int_0^\infty x^2 e^{-tx^2} \sin(px^2)\, dx$$

converge uniformly for any p. By using the Leibniz's rule, differentiating under the integral sign gives

$$F_c'(p) = -\int_0^\infty x^2 e^{-tx^2} \sin(px^2)\, dx, \quad F_s'(p) = \int_0^\infty x^2 e^{-tx^2} \cos(px^2)\, dx.$$

Next, integrating by parts leads to

$$\begin{aligned} F_c'(p) &= \frac{1}{2t} \int_0^\infty x \sin(px^2)\, d(e^{-tx^2}) \\[2mm] &= \frac{1}{2t}\left(-\int_0^\infty e^{-tx^2} \sin(px^2)\, dx - 2p \int_0^\infty x^2 e^{-tx^2} \cos(px^2)\, dx\right) \\[2mm] &= -\frac{1}{2t} F_s(p) - \frac{p}{t} F_s'(p). \end{aligned}$$

Similarly,

$$F'_s(p) = \frac{1}{2t} F_c(p) + \frac{p}{t} F'_c(p).$$

Finally, solving for $F'_c(p)$ and $F'_s(p)$ yields (7.27) as expected.

We prove (7.28) by observing:

$$\frac{d}{dp} Z(p) = \frac{-p + ti}{2(t^2 + p^2)} Z(p) = \frac{i}{2} \frac{Z(p)}{t - pi}. \qquad (7.29)$$

To solve this equation, instead of using separation of variables, which will involve the complex logarithm, direct calculation shows that (7.29) is equivalent to

$$\frac{d}{dp} \left(Z(p) \cdot \sqrt{t - pi} \right) = 0.$$

Thus,

$$Z(p) \cdot \sqrt{t - pi} = \text{const.}$$

Notice that

$$Z(0) = F_c(0) = \int_0^\infty e^{-tx^2} dx = \frac{1}{2} \sqrt{\frac{\pi}{t}}.$$

It follows that

$$Z(p) = \frac{\sqrt{\pi}}{2} \frac{1}{\sqrt{t - pi}} = \frac{\sqrt{\pi}}{2} \frac{\sqrt{t + pi}}{\sqrt{t^2 + p^2}}$$

as claimed.

Remarks. It is interesting to see that this technique is also strong enough to capture the parameter integrals:

$$G_c(p) = \int_0^\infty e^{-x^2} \cos\left(\frac{p^2}{x^2}\right) dx, \quad G_s(p) = \int_0^\infty e^{-x^2} \sin\left(\frac{p^2}{x^2}\right) dx.$$

Neither of these can be evaluated as the double integral via polar coordinates. By pursuing the same line of reasoning as used above, we have

$$G''_c(p) = 4 G_s(p), \quad G''_s(p) = -4G_c(p).$$

Furthermore, let $W(p) = G_c(p) + G_s(p) i$. Then $W''(p) = -4i W(p)$. This yields, subject to the boundedness of $W(p)$ and $W(0) = \sqrt{\pi}/2$,

$$W(p) = \frac{\sqrt{\pi}}{2} e^{-\sqrt{2}p} \cos(\sqrt{2}p) + i \frac{\sqrt{\pi}}{2} e^{-\sqrt{2}p} \sin(\sqrt{2}p).$$

Thus,

$$G_c(p) = \frac{\sqrt{\pi}}{2} e^{-\sqrt{2}p} \cos(\sqrt{2}p), \quad G_s(p) = \frac{\sqrt{\pi}}{2} e^{-\sqrt{2}p} \sin(\sqrt{2}p).$$

For interested reader, establishing the following formulas shows the wider applicability of our method.

$$\int_0^\infty \sin(a^2 x^2) \cos\left(\frac{p^2}{x^2}\right) dx = \frac{1}{4a} \sqrt{\frac{\pi}{2}} \left(\sin 2ap + \cos 2ap + e^{-2ap}\right),$$

$$\int_0^\infty \cos(a^2 x^2) \cos\left(\frac{p^2}{x^2}\right) dx = \frac{1}{4a} \sqrt{\frac{\pi}{2}} \left(\cos 2ap - \sin 2ap + e^{-2ap}\right),$$

$$\int_0^\infty e^{-ax^2} x^{2p-1} \cos(bx^2) dx = \frac{\Gamma(p)}{2(a^2 + b^2)^{p/2}} \cos(p\theta_0),$$

$$\int_0^\infty e^{-ax^2} x^{2p-1} \sin(bx^2) dx = \frac{\Gamma(p)}{2(a^2 + b^2)^{p/2}} \sin(p\theta_0),$$

where $a > 0, p > 0$ and $\theta_0 = \arctan(b/a)$.

18. Show that

$$\sum_{n=1}^{\infty} \frac{1}{n \binom{2n}{n}} = \frac{\pi}{3\sqrt{3}}. \tag{7.30}$$

Proof. The following proof is based on converting the series into a definite integral.

$$\sum_{n=1}^{\infty} \frac{1}{n \binom{2n}{n}} = \sum_{n=1}^{\infty} \frac{(n-1)! \, n!}{(2n)!} = \sum_{n=1}^{\infty} \frac{\Gamma(n)\Gamma(n+1)}{\Gamma(2n+1)}$$

$$= \sum_{n=1}^{\infty} B(n, n+1) = \sum_{n=1}^{\infty} B(n+1, n)$$

$$= \sum_{n+1}^{\infty} \int_0^1 t^n (1-t)^{n-1} \, dx = \int_0^1 t \sum_{n=1}^{\infty} [t(1-t)]^{n-1} \, dt$$

$$= \int_0^1 \frac{dt}{t^2 - t + 1} = \frac{1}{2} \int_0^1 \frac{(2t-1)+1}{t^2 - t + 1} \, dt = \frac{\pi}{3\sqrt{3}}.$$

The interchange the integration and summation is justified by the positivity of the integrands.   □

Remark. Based on Example 12 in 6.4, we have the following *generating function*

$$\sum_{n=1}^{\infty} \frac{1}{n \binom{2n}{n}} (2x)^{2n} = \frac{2x \arcsin x}{\sqrt{1-x^2}}.$$

It enables us to find the exact values for a large class of interesting series including (7.30). For more details, see [25, Chapter 18].

19. **(Monthly Problem 12317, 2022)**. Prove

$$\int_0^{\pi/2} \frac{\sin(4x)}{\ln(\tan x)} \, dx = -14 \frac{\zeta(3)}{\pi^2},$$

where $\zeta(3)$ is Apéry's constant $\sum_{n=1}^{\infty} 1/n^3$.

Proof. Since $\sin(4x) = 2\sin(2x)\cos(2x) = 4\sin x \cos x(1 - 2\sin^2 x)$, by the substitution $u = \tan x$, we have

$$I := \int_0^{\pi/2} \frac{\sin(4x)}{\ln(\tan x)} \, dx = \int_0^{\infty} \frac{4u(1-u^2)}{\ln u (1+u^2)^3} \, du$$

$$= -4 \int_0^{\infty} \frac{t-1}{\ln t (1+t)^3} \, dt \quad (\text{use } t = u^2)$$

We now use the parametric integration to compute the last integral. To this end, observe that

$$\int_0^1 t^p \, dp = \frac{1}{\ln t} t^p \Big|_0^1 = \frac{t-1}{\ln t}.$$

We have

$$I = -4 \int_0^{\infty} \int_0^1 \frac{t^p}{(1+t)^3} \, dp \, dt = -4 \int_0^1 \left(\int_0^{\infty} \frac{t^p}{(1+t)^3} \, dt \right) dp. \tag{7.31}$$

For $p \in (0,1)$, let $s = 1/(1+t)$. Then

$$\int_0^\infty \frac{t^p}{(1+t)^3}\,dt = \int_0^1 s^{1-p}(1-s)^p\,ds$$

$$= B(2-p, p+1) = \frac{\Gamma(2-p)\Gamma(p+1)}{\Gamma(3)}$$

$$= \frac{1}{2}p(1-p)\Gamma(1-p)\Gamma(p) \quad (\text{use } \Gamma(1+x) = x\Gamma(x))$$

$$= \frac{1}{2}p(1-p)\frac{\pi}{\sin(p\pi)} \quad (\text{use } \Gamma(p)\Gamma(1-p) = \tfrac{\pi}{\sin p\pi}).$$

By (7.31) we obtain

$$I = -2\pi \int_0^1 \frac{p(1-p)}{\sin(p\pi)}\,dp = -\frac{2}{\pi^2}\int_0^\pi \frac{x(\pi-x)}{\sin x}\,dx \quad (\text{use } x = p\pi).$$

Recall the Fourier sine series of $x(\pi - x)$ on $[0, \pi]$:

$$x(\pi - x) = \frac{8}{\pi}\sum_{n=0}^\infty \frac{1}{(2n+1)^3}\sin((2n+1)x).$$

We finally find that

$$I = -\frac{16}{\pi^3}\sum_{n=0}^\infty \frac{1}{(2n+1)^3}\int_0^\pi \frac{\sin((2n+1)x)}{\sin x}\,dx$$

$$= -\frac{16}{\pi^3}\sum_{n=0}^\infty \frac{\pi}{(2n+1)^3} = -\frac{16}{\pi^2}\left(1 - \frac{1}{2^3}\right)\sum_{n=1}^\infty \frac{1}{n^3}$$

$$= -\frac{16}{\pi^2}\frac{7}{8}\zeta(3) = -14\frac{\zeta(3)}{\pi^2},$$

as claimed. Here we have used the fact that, for $n \in \mathbb{N}$,

$$\int_0^\pi \frac{\sin nx}{\sin x}\,dx = \begin{cases} 0, & \text{if } n \text{ is even,} \\ \pi, & \text{if } n \text{ is odd.} \end{cases}$$

This can be derived as follows: For $n > 2$,

$$\int_0^\pi \frac{\sin nx}{\sin x}\,dx - \int_0^\pi \frac{\sin(n-2)x}{\sin x}\,dx = \int_0^\pi \frac{\sin nx - \sin(n-2)}{\sin x}\,dx$$

$$= 2\int_0^\pi \cos(n-1)x\,dx = 0. \qquad \square$$

20. (**Monthly Problem 11564, 2011**). Prove that

$$\int_0^\infty \frac{e^{-x}(1-e^{-6x})}{x(1+e^{-2x}+e^{-4x}+e^{-6x}+e^{-8x})}\,dx = \ln\left(\frac{3+\sqrt{5}}{2}\right).$$

Proof. We prove it using parametric differentiation. Let

$$I(p) := \int_0^\infty \frac{e^{-px}(1-e^{-6x})}{x(1+e^{-2x}+e^{-4x}+e^{-6x}+e^{-8x})}\,dx.$$

In view of

$$1 + e^{-2x} + e^{-4x} + e^{-6x} + e^{-8x} = \frac{1 - e^{-10x}}{1 - e^{-2x}},$$

we find that

$$I(p) = \int_0^\infty \frac{e^{-px}}{x}(1 - e^{-6x})(1 - e^{-2x}) \sum_{n=0}^\infty e^{-10nx}\, dx,$$

and so

$$
\begin{aligned}
I'(p) &= -\int_0^\infty e^{-px}(1 - e^{-6x})(1 - e^{-2x}) \sum_{n=0}^\infty e^{-10nx}\, dx \\
&= -\sum_{n=0}^\infty \int_0^\infty e^{-px}(1 - e^{-6x})(1 - e^{-2x})e^{-10nx}\, dx \\
&= -\sum_{n=0}^\infty \left(\frac{1}{p + 10n} - \frac{1}{p + 2 + 10n} - \frac{1}{p + 6 + 10n} + \frac{1}{p + 8 + 10n} \right).
\end{aligned}
$$

Here interchange of the summation and integration is justified by the monotone converges theorem since the integrands are positive. By the digamma function identity

$$\psi(\alpha) - \psi(\beta) = -\sum_{n=0}^\infty \left(\frac{1}{n + \alpha} - \frac{1}{n + \beta} \right),$$

appealing to $\psi(x) = d(\ln \Gamma(x))/dx$, we get

$$I(p) = \ln \frac{\Gamma(p/10)\Gamma((p+8)/10)}{\Gamma((p+2)/10)\Gamma((p+6)/10)}.$$

Now, the reflection formula

$$\Gamma(\alpha)\Gamma(1 - \alpha) = \frac{\pi}{\sin \alpha\pi}$$

yields

$$I(1) = \ln \frac{\Gamma(1/10)\Gamma(9/10)}{\Gamma(3/10)\Gamma(7/10)} = \ln \frac{\sin(3\pi/10)}{\sin(\pi/10)} = \ln \left(\frac{3 + \sqrt{5}}{2} \right)$$

as desired.   □

21. Let $n \in \mathbb{N}$. Evaluate

$$\int_0^1 \frac{1 + x + \cdots + x^{n-1}}{1 + x + \cdots + x^n}\, dx.$$

Solution. Rewrite the integral as

$$I := \int_0^1 \frac{1 - x^n}{1 - x^{n+1}}\, dx.$$

Along the same lines as in Example 7.12, choosing $\epsilon \in (0,1)$, define

$$I(\epsilon) := \int_0^1 \frac{1 - x^n}{(1 - x^{n+1})^{1-\epsilon}}\, dx.$$

Then

$$
\begin{aligned}
I(\epsilon) &= \int_0^1 (1-x^{n+1})^{\epsilon-1}\,dx - \int_0^1 x^n(1-x^{n+1})^{\epsilon-1}\,dx \quad (\text{set } y = x^{n+1}) \\
&= \frac{1}{n+1}\left(\int_0^1 y^{1/(n+1)-1}(1-y)^{\epsilon-1}\,dy - \int_0^1 (1-y)^{\epsilon-1}\,dy\right) \\
&= \frac{1}{n+1}\left(B\left(\frac{1}{n+1},\epsilon\right) - B(1,\epsilon)\right) \quad (\text{use Formula (7.18)}) \\
&= \frac{\Gamma(\epsilon)}{n+1}\left(\frac{\Gamma(1/(n+1))}{\Gamma(\epsilon+1/(n+1))} - \frac{1}{\Gamma(\epsilon+1)}\right).
\end{aligned}
$$

Since, for $\epsilon \ll 1$,

$$
\Gamma(x+\epsilon) = \Gamma(x) + \Gamma'(x)\epsilon + O(\epsilon^2),
$$

It follows that

$$
\frac{1}{\Gamma(x+\epsilon)} = \frac{1}{\Gamma(x)} \cdot \frac{1}{1+\psi(x)\epsilon+O(\epsilon^2)} = \frac{1}{\Gamma(x)} - \frac{\psi(x)}{\Gamma(x)}\epsilon + O(\epsilon^2).
$$

In view of $\psi(1) = -\gamma$ and $\Gamma(\epsilon) = 1/\epsilon + O(1)$, this yields

$$
I(\epsilon) = -\frac{1}{n+1}\left(\psi(1/(n+1)) + \gamma\right) + O(\epsilon),
$$

and so

$$
I = \lim_{\epsilon \to 0^+} I(\epsilon) = -\frac{1}{n+1}\left(\psi(1/(n+1)) + \gamma\right).
$$

By the Gauss formula (see Exercise 81), explicitly,

$$
I = \frac{\pi}{2(n+1)}\cot\frac{\pi}{n+1} + \frac{\ln(n+1)}{n+1} - \frac{1}{n+1}\sum_{k=1}^{n}\cos\frac{2k\pi}{n+1}\ln\left(2\sin\frac{k\pi}{n+1}\right).
$$

The following problem presents a formula for the digamma function in terms of a double integral.

22. (**Monthly Problem 11937, 2016**). Let s be a complex number not a zero of the gamma function $\Gamma(s)$. Prove

$$
\int_0^1 \int_0^1 \frac{(xy)^{s-1} - y}{(1-xy)\ln(xy)}\,dx\,dy = \frac{\Gamma'(s)}{\Gamma(s)} = \psi(s).
$$

Proof. Denote the double integral by S. Expanding $1/(1-xy)$ into a geometric series yields

$$
\begin{aligned}
S &= \int_0^1 \int_0^1 \frac{(xy)^{s-1}-y}{\ln(xy)}\sum_{n=0}^{\infty}(xy)^n\,dx\,dy \\
&= \sum_{n=0}^{\infty}\int_0^1 \int_0^1 \frac{(xy)^{s-1}-y}{\ln(xy)}(xy)^n\,dx\,dy.
\end{aligned}
$$

Here the justification of the interchange of summation and integration is ensured by the terms having the same sign. In view of the fact that, for $0 < xy < 1$,

$$
\frac{(xy)^k}{\ln(xy)} = -\int_k^{\infty}(xy)^t\,dt,
$$

we find that

$$
\begin{aligned}
S &= -\sum_{n=0}^{\infty} \int_0^1 \int_0^1 \left(\int_{n+s-1}^{\infty} (xy)^t \, dt - \int_n^{\infty} (xy)^t \, dt \right) dx \, dy \\
&= -\sum_{n=0}^{\infty} \left(\int_{n+s-1}^{\infty} \int_0^1 \int_0^1 (xy)^t \, dx \, dy \, dt - \int_n^{\infty} \int_0^1 \int_0^1 y (xy)^t \, dx \, dy \, dt \right) \\
&= -\sum_{n=0}^{\infty} \left(\int_{n+s-1}^{\infty} \frac{dt}{(t+1)^2} - \int_n^{\infty} \frac{dt}{(t+1)(t+2)} \right) \\
&= -\sum_{n=0}^{\infty} \left(\frac{1}{n+s} - \ln \left(\frac{n+2}{n+1} \right) \right) \\
&= -\sum_{n=1}^{\infty} \left(\frac{1}{n+s-1} - \ln \left(\frac{n+1}{n} \right) \right).
\end{aligned}
$$

Since the integrands in the first equality are nonnegative, it allows us to reverse the order of integrations. By the definition

$$
\gamma = \lim_{n \to \infty} \left(1 + \frac{1}{2} + \cdots + \frac{1}{n} - \ln n \right),
$$

we have

$$
\gamma = \sum_{n=1}^{\infty} \left(\frac{1}{n} - \ln \left(\frac{n+1}{n} \right) \right).
$$

Thus, we can rewrite S as

$$
S = -\gamma - \sum_{n=1}^{\infty} \left(\frac{1}{s+n-1} - \frac{1}{n} \right).
$$

On the other hand, taking the logarithmic derivative of (7.12) gives

$$
-\frac{\Gamma'(s)}{\Gamma(s)} = \gamma + \sum_{n=1}^{\infty} \left(\frac{1}{s+n-1} - \frac{1}{n} \right).
$$

This proves that

$$
S = \frac{\Gamma'(s)}{\Gamma(s)}
$$

as desired.   □

23. (**Fontana Formula**). Let $H_n = \sum_{i=1}^n 1/i$. Show that

$$
H_n = \gamma + \ln n + \frac{1}{2n} - \sum_{i=2}^{\infty} \frac{(i-1)! \, G_i}{n(n+1) \cdots (n+i-1)},
$$

where G_i is the *Gregory coefficient* defined by

$$
\frac{x}{\ln(1-x)} = 1 + \sum_{i=1}^{\infty} G_i x^i, \qquad \text{for } |x| < 1.
$$

Proof. We give a proof based on the Euler beta function and Frullani formula. Since $G_1 = \frac{1}{2}$, denote

$$S_n = \sum_{i=1}^{\infty} \frac{(i-1)!\,G_i}{n(n+1)\cdots(n+i-1)}.$$

By using the Euler beta function, we have

$$\frac{(i-1)!}{n(n+1)\cdots(n+i-1)} = \int_0^1 x^{n-1}(1-x)^{i-1}\,dx.$$

Appealing to the generating function of G_i yields

$$S_n = \int_0^1 x^{n-1} \sum_{i=1}^{\infty} G_k(1-x)^{i-1}\,dx = \int_0^1 \left(\frac{1}{1-x} + \frac{1}{\ln x}\right) x^{n-1}\,dx.$$

Here the interchange of the integral and summation is justified since the summands are nonnegative. Thus

$$
\begin{aligned}
S_n - S_{n+1} &= \int_0^1 \left(x^{n-1} + \frac{x^{n-1} - x^n}{\ln x}\right) dx \\
&= \frac{1}{n} - \int_0^{\infty} \frac{e^{-ny} - e^{-(n+1)y}}{y}\,dy. \qquad \text{(Let } y = -\ln x\text{)}
\end{aligned}
$$

Using the Frullani formula, for $0 < a < b$,

$$\int_0^{\infty} \frac{e^{-ay} - e^{-by}}{y}\,dy = \ln\frac{b}{a},$$

we find that

$$S_n - S_{n+1} = \frac{1}{n} - \ln\frac{n+1}{n}.$$

Notice that

$$S_1 = \int_0^1 \left(\frac{1}{1-x} + \frac{1}{\ln x}\right) dx = \gamma.$$

Telescoping yields

$$\gamma - S_n = H_{n-1} - \ln n,$$

which is equivalent to the desired formula. $\qquad\square$

The following two problems offer a method for proving a class of inequality by using improper integral.

24. Let $a_i, b_i\ (1 \le i \le n)$ be positive real numbers. Show that

$$\left(\sum_{j=1}^n \frac{a_i}{b_j}\right)^2 - 2\sum_{j,k=1}^n \frac{a_j a_k}{(b_j + b_k)^2} \le 2\sqrt{2}\left(\sum_{j,k=1}^n \frac{a_j a_k}{(b_j + b_k)}\sum_{j,k=1}^n \frac{a_j a_k}{(b_j + b_k)^3}\right)^{1/2}.$$

Proof. Define $f : (0, \infty) \to (0, \infty)$ by

$$f(x) = \sum_{j=1}^n a_j e^{-b_j x}.$$

Then, for $m \in \{0, 1, 2\}$,

$$C := \int_0^\infty f(x)\,dx = \sum_{j=1}^n \frac{a_i}{b_j}, \quad D_m := \frac{1}{m!} \int_0^\infty x^m f^2(x)\,dx = \frac{1}{m!} \sum_{j,k=1}^n \frac{a_j a_k}{(b_j + b_k)^{m+1}}.$$

By the Cauchy-Schwarz inequality, for $t > 0$

$$\begin{aligned}
\left(\int_0^\infty f(x)\,dx \right)^2 &= \left(\int_0^\infty \frac{1}{x+t}(x+t)f(x)\,dx \right)^2 \\
&\le \left(\int_0^\infty \frac{dx}{(x+t)^2} \right) \left(\int_0^\infty (x+t)^2 f^2(x)\,dx \right) \\
&= t \int_0^\infty f^2(x)\,dx + 2 \int_0^\infty x f^2(x)\,dx + \frac{1}{t} \int_0^\infty x^2 f^2(x)\,dx.
\end{aligned}$$

It follows that for any $t > 0$,

$$C^2 - 2D_1 \le tD_0 + \frac{2}{t} D_2.$$

In particular, letting $t = (2D_2/D_0)^{1/2}$ gives the minimum value over all t of $tD_0 + 2D_2/t$

$$C^2 - 2D_1 \le 2\sqrt{2}(D_0 D_2)^{1/2},$$

which is equivalent to the proposed inequality. $\qquad\square$

25. (**Monthly Problem 11680, 2012**). Let $x_1, x_2, \ldots, x_n$ be nonnegative numbers. Show that

$$\left(\sum_{i=1}^n \frac{x_i}{i} \right)^4 \le 2\pi^2 \sum_{i,j=1}^n \frac{x_i x_j}{i+j} \sum_{i,j=1}^n \frac{x_i x_j}{(i+j)^3}.$$

Proof. We show the proposed inequality is the consequence of Carlson's inequality: if $f \ge 0$ on $[0, \infty)$ such that $f, xf \in L^2([0, \infty))$, then

$$\left(\int_0^\infty f(x)\,dx \right)^4 \le \pi^2 \int_0^\infty f^2(x)\,dx \int_0^\infty x^2 f^2(x)\,dx.$$

To this end, let $f(x) = \sum_{i=1}^n x_i e^{-ix}$. Direct calculations give

$$\begin{aligned}
\int_0^\infty f(x)\,dx &= \sum_{i=1}^n x_i \int_0^\infty e^{-ix}\,dx = \sum_{i=1}^n \frac{x_i}{i}, \\
\int_0^\infty f^2(x)\,dx &= \sum_{i,j=1}^n x_i x_j \int_0^\infty e^{-(i+j)x}\,dx = \sum_{i,j=1}^n \frac{x_i x_j}{i+j}, \\
\int_0^\infty x^2 f^2(x)\,dx &= \sum_{i,j=1}^n x_i x_j \int_0^\infty x^2 e^{-(i+j)x}\,dx = 2 \sum_{i,j=1}^n \frac{x_i x_j}{(i+j)^3}.
\end{aligned}$$

Thus, the proposed inequality follows from Carlson's inequality. $\qquad\square$

Remark. The discrete version of Carlson's inequality is:

$$\left(\sum_{n=1}^\infty a_n \right)^4 \le \pi^2 \sum_{n=1}^\infty a_n^2 \sum_{n=1}^\infty n^2 a_n^2.$$

Here we present Hardy's elegant elementary proof which only used the Cauchy-Schwarz inequality. Let

$$S := \sum_{n=1}^{\infty} a_n^2, \quad T = \sum_{n=1}^{\infty} n^2 a_n^2.$$

If either S or T are unbounded, there is nothing to prove. So we may assume that both S and T are finite. Using the Cauchy-Schwarz inequality, for any $\alpha, \beta > 0$, we find that

$$\left(\sum_{n=0}^{\infty} a_n \right)^2 = \left(\sum_{n=0}^{\infty} a_n \sqrt{\alpha + \beta n^2} \frac{1}{\sqrt{\alpha + \beta n^2}} \right)^2$$
$$\leq \sum_{n=0}^{\infty} a_n^2 (\alpha + \beta n^2) \sum_{n=1}^{\infty} \frac{1}{\alpha + \beta n^2}$$
$$\leq (\alpha S + \beta T) \int_0^{\infty} \frac{dx}{\alpha + \beta x^2}$$
$$= \frac{\pi}{2} \left(S \sqrt{\frac{\alpha}{\beta}} + T \sqrt{\frac{\beta}{\alpha}} \right)$$
$$= \frac{\pi}{2} \left(tS + \frac{T}{t} \right),$$

where $t = \sqrt{\frac{\alpha}{\beta}}$. Thus, for fixed S and T, the right-hand side can be minimized with respect to t. Now Carlson's inequality follows if we just take $t = \sqrt{\frac{T}{S}}$.

7.5   Exercises

The problem is there, you must solve it. — David Hilbert

1. Let $f : (0, 1] \to \mathbb{R}$ be continuous and $\lim_{x \to 0+} f(x) = \infty$. For each $n \in \mathbb{N}$, define $f_n(x) = \min\{f(x), n\}$ on $(0, 1]$. Show that the improper integral $\int_0^1 f(x)\, dx$ converges if and only if $\lim_{n \to \infty} \int_0^1 f_n(x)\, dx$ exists.

2. Prove the following statements, which are concerned with the behavior of integrands at infinity.

   (a) If $\int_a^{\infty} f(x)\, dx$ converges and $\lim_{n \to \infty} f(x)\, dx$ exists, then $\lim_{n \to \infty} f(x) = 0$.

   (b) If $\int_a^{\infty} f(x)\, dx$ converges and $f(x)$ is monotone, then $\lim_{n \to \infty} x f(x) = 0$.

   (c) If $\int_a^{\infty} f(x)\, dx$ converges and $f(x)$ is uniformly continuous on $[a, \infty)$, then $\lim_{n \to \infty} f(x) = 0$.

   (d) If $\int_a^{\infty} f(x)\, dx$ converges and $f(x)$ is continuous on $[a, \infty)$, then there exists a sequence $x_n \in [a, \infty)$ such that
   $$\lim_{n \to \infty} x_n = \infty, \quad \lim_{n \to \infty} f(x_n) = 0.$$

   (e) If $\int_a^{\infty} |f(x)|\, dx$ converges and $f(x)$ is continuous on $[a, \infty)$, then there exists a sequence $x_n \in [a, \infty)$ such that
   $$\lim_{n \to \infty} x_n = \infty, \quad \lim_{n \to \infty} x_n f(x_n) = 0.$$

Give an example to show that the conclusion need not hold if absolute convergence is replaced by conditional convergence.

3. Consider the improper integral

$$I(p, q) = \int_0^{\pi/2} \frac{dx}{\sin^p x \cos^q x}.$$

Determine the range of (p, q) for which the integral diverges, converges conditionally, and converges absolutely, respectively.

4. Show that

$$\int_0^\infty \frac{x\, dx}{1 + x^6 \sin^2 x}$$

converges. This indicates that $\limsup f(x) = \infty$ even if the improper integral converges,

5. Let $p > 0$. Prove that

$$\int_0^\infty \frac{\sin x}{x^p + \sin x}\, dx$$

diverges for $p \leq 1/2$, converges conditionally for $1/2 < p \leq 1$ and converges absolutely for $p > 1$.

6. Let $m \geq 1$ and $n \geq m + 2$. Evaluate

$$\int_0^\infty \frac{1 + x + \cdots + x^{m-1}}{1 + x + \cdots + x^{n-1}}\, dx.$$

7. Let the rational function $f(x) = P(x)/Q(x)$ be integrable on $\mathbb{R}$. If $Q(x)$ has no real zero and the partial fraction decomposition is given by

$$\frac{P(x)}{Q(x)} = \sum_k \left[\frac{A_k}{x - x_k} + \frac{A'_k}{(x - x_k)^2} + \cdots \right].$$

Show that

$$\int_{-\infty}^\infty \frac{P(x)}{Q(x)}\, dx = 2\pi i \sum_k A_k,$$

In particular, if x_k is a simple zero, then $A_k = P(x_k)/Q'(x_k)$.

8. Applying the above result establish

(a) Let $m, n \in \mathbb{N}$ and $m < n$. Then $\int_{-\infty}^\infty \frac{x^{2m}}{1+x^{2n}}\, dx = \frac{\pi}{n} \csc\left(\frac{2m+1}{2n}\pi\right).$

(b) Let $p, q, n \in \mathbb{N}$ and $p, q < n$. Then

$$\int_{-\infty}^\infty \frac{x^{2p} - x^{2q}}{1 - x^{2n}}\, dx = \frac{\pi}{n}\left[\cot\left(\frac{2p+1}{2n}\pi\right) - \cot\left(\frac{2q+1}{2n}\pi\right)\right].$$

(c) Let $m, n \in \mathbb{N}$ with $m < n$ and $\theta \in (-\pi, \pi)$. Then

$$\int_{-\infty}^\infty \frac{x^{2m}}{x^{4n} + 2x^{2n}\cos\theta + 1}\, dx = \frac{\pi}{n}\frac{\sin\left(1 - \frac{2m+1}{2n}\pi\right)\theta}{\sin\theta \sin\left(\frac{2m+1}{2n}\pi\right)}.$$

9. Let $x > 1$. Show that

$$\int_1^\infty \frac{dt}{1 + t^x} = \sum_{n=1}^\infty \frac{1 - 2^{1-n}}{x^n}\zeta(n).$$

10. Let $\alpha, \alpha_k > 0$ for $k = 1, 2, \ldots, n$ and $\alpha > \sum_{k=1}^{n} \alpha_k$. Show that

$$\int_0^\infty \frac{\sin \alpha x}{x} \cdot \left(\prod_{k=1}^{n} \frac{\sin \alpha_k x}{x} \right) dx = \frac{\pi}{2} \left(\prod_{k=1}^{n} \alpha_k \right).$$

11. (**Monthly Problem 12260, 2021**). Prove

$$\int_0^\infty \frac{\sin^2 x - x \sin x}{x^3} \, dx = \frac{1}{2} - \ln 2.$$

12. Show that

$$\int_0^\infty f(\sin^2 x) \frac{\sin x}{x} \, dx = \int_0^{\pi/2} f(\sin^2 x) \, dx,$$

provided the integral on the left hand side converges.

13. Assume $\int_1^\infty f(x) \, dx$ converges. Prove that there is a $\xi \in (1, \infty)$ such that

$$\int_1^\infty \frac{f(x)}{x} \, dx = \int_1^\xi f(x) \, dx.$$

14. Let $f(x)$ be nonnegative on $\mathbb{R}$. Furthermore assume the following

$$\int_{-\infty}^\infty f(x) \, dx = 1, \quad \int_{-\infty}^\infty x f(x) \, dx = 0, \quad \int_{-\infty}^\infty x^2 f(x) \, dx = 1.$$

Prove that

(a) $\int_{-\infty}^x f(t) \, dt \geq \frac{x^2}{1+x^2}$ for $x > 0$,

(b) $\int_{-\infty}^x f(t) \, dt \leq \frac{x^2}{1+x^2}$ for $x < 0$.

15. (**Monthly Problem 12149, 2020**). Let Γ be the gamma function. Prove

$$x^x y^y \left(\Gamma\left(\frac{x+y}{2} \right) \right)^2 \leq \left(\frac{x+y}{2} \right)^{x+y} \Gamma(x)\Gamma(y),$$

for all positive real numbers x and y.

16. (**Landau's inequality**). Let $a_n \geq 0$ for all $n \in \mathbb{N}$. Show that

$$\left(\sum_{n=1}^\infty a_n \right)^4 \leq \pi^2 \sum_{n=1}^\infty a_n^2 \sum_{n=1}^\infty (n - 1/2)^2 a_n^2.$$

Show the analogous integral inequality

$$\left(\int_0^\infty f(x) \, dx \right)^4 \leq C \int_0^\infty f^2(x) \, dx \int_0^\infty (x - 1/2)^2 f^2(x) \, dx$$

does not hold for $C = \pi^2$, but it holds for $C = 4\pi^2$.

17. Let $\lfloor x \rfloor$ be the integer part of x. Evaluate

$$\int_1^\infty \left(\frac{1}{\lfloor x \rfloor} - \frac{1}{x} \right).$$

18. Evaluate
$$\int_0^1 \frac{x \ln x}{\sqrt{1-x^2}}\, dx.$$

19. Evaluate
$$\int_0^1 \frac{\ln^2 x}{x^2 - \sqrt{2}x + 1}\, dx.$$

20. Evaluate
$$\int_0^\infty \frac{\sin x}{x} \ln x\, dx.$$

21. Evaluate
$$\int_0^1 \sin \pi x \ln \Gamma(x)\, dx.$$

22. Evaluate
$$\int_0^\infty \left(\frac{x}{e^x - e^{-x}} - \frac{1}{2} \right) \frac{dx}{x^2}.$$

23. Let $\{x\}$ be the fraction part of x. Show that
$$\int_0^\infty \frac{\{x\}^{n+1}}{(1+x)^{n+2}}\, dx = 1 - \gamma - \sum_{i=2}^{n-1} \frac{1}{i} (\zeta(i) - 1),$$

where γ is the Euler-Mascheroni constant.

24. Show that
$$\int_0^{\pi/2} x \ln(1 - \cos x)\, dx = \frac{35}{16}\zeta(3) - \frac{3}{4}\ln 2\zeta(2) - \pi G,$$

where G is the Catalan constant.

25. Show that
$$\int_0^{\pi/4} \ln^2(\tan x)\, dx = \frac{\pi^3}{16}.$$

26. (**Monthly Problem 12054, 2018**). Prove
$$\int_0^1 \frac{\arctan x}{x} \ln\left(\frac{1+x^2}{(1-x)^2} \right) dx = \frac{\pi^3}{16}.$$

27. (**Monthly Problem 12184, 2020**). Prove
$$\int_1^\infty \frac{\ln(x^4 - 2x^2 + 1)}{x\sqrt{x^2 - 1}}\, dx = \pi \ln(2 + \sqrt{2}).$$

28. (**Monthly Problem 12221, 2020**). Prove
$$\int_0^1 \frac{\ln(x^6 + 1)}{x^2 + 1}\, dx = \frac{\pi \ln(6)}{2} - 3G,$$

where G is the Catalan constant $\sum_{n=0}^\infty (-1)^n/(2n+1)^2$.

29. (**Monthly Problem 12228, 2021**). Prove

$$\int_0^1 \frac{(\ln x)^2 \ln(2\sqrt{x}/(x^2+1))}{x^2-1} \, dx = 2G^2,$$

where G is the Catalan constant $\sum_{n=0}^{\infty}(-1)^n/(2n+1)^2$.

30. (**Monthly Problem 12274, 2021**). Evaluate

$$\int_0^1 \frac{\arctan x}{1+x^2} \ln^2\left(\frac{2x}{1-x^2}\right) dx.$$

31. Show that

(a) $\int_0^{\pi/2} x \ln(\sin x) \, dx = \frac{7}{16}\zeta(3) - \frac{\pi^2}{8}\ln 2$. (Euler)

(b) $\int_0^{\pi/2} x^2 \ln(\sin x) \, dx = \frac{3\pi}{16}\zeta(3) - \frac{\pi^3}{24}\ln 2$.

Remark. It is interesting to find the closed form of

$$\int_0^{\pi/2} x^n \ln(\sin x) \, dx, \quad (n \in \mathbb{N}).$$

32. Show that

$$\int_0^{\pi/2} x^2 \ln^2(\cos x) \, dx = \frac{\pi}{16}\left(4\ln^2\zeta(2) + 8\ln 2\zeta(3) + 11\zeta(4)\right).$$

33. Show that

$$\int_0^1 \frac{x\ln x + 1 - x}{x\ln^2 x} \ln(1+x) \, dx = \ln\left(\frac{4}{\pi}\right).$$

34. For $0 < p < 1$, evaluate

$$\int_0^{\infty} \frac{x^{p-1}\ln x}{1+x} \, dx.$$

35. Let $a, \alpha > 0$. Evaluate

$$\int_0^{\infty} \frac{e^{-ax}\cos bx - e^{-\alpha x}\cos \beta x}{x} \, dx.$$

Hint: Differentiate the integral with respect to a and b, respectively.

36. Let $n, m \in \mathbb{N}$. Evaluate

(a) $\int_0^{\infty} \frac{\ln^n x\, dx}{1+x^2}$.

(b) $\int_0^{\infty} \frac{\ln x\, dx}{(1+x^2)^m}$.

(c) $\int_0^{\infty} \frac{\ln^n x\, dx}{(1+x^2)^m}$.

37. Let $a, b > 0$. Show that

(a) $\int_0^{\infty} \frac{\ln(1+a^2x^2)}{b^2+x^2} \, dx = \frac{\pi}{b}\ln(ab+1)$.

(b) $\int_0^{\infty} \frac{\arctan ax}{x(1+x^2)} \, dx = \frac{\pi}{2}\ln(1+a)$.

(c) $\int_0^{\infty} \frac{\arctan ax \cdot \arctan bx}{x^2} \, dx = \frac{\pi}{2}\ln\frac{(a+b)^{a+b}}{a^a \cdot b^b}$.

38. Let $n \in \mathbb{N}$. Show that

$$\int_0^\infty \frac{\ln(1+x^{4n})}{1+x^2}\,dx = \pi \ln\left\{ 2^n \prod_{k=0}^{n-1}\left(1+\sin\left(\frac{(2k+1)\pi}{4n}\right)\right)\right\}.$$

39. Let $n > 1$ be positive integer and $\alpha > 0$. Prove that

(a) $\int_0^\infty \frac{n^2 x^n \ln x}{1+x^{2n}}\,dx = \frac{\pi^2}{4}\frac{\sin(\pi/2n)}{\cos^2(\pi/2n)}$.

(b) $\lim_{n\to\infty}\int_0^\infty \frac{n^2 x^n \ln \alpha x}{1+x^{2n}}\,dx = \frac{\pi}{2}\ln\alpha$.

40. Let $\alpha, \beta, k > 0$. Evaluate

(a) $\int_0^\infty \frac{1-\cos\alpha x}{x}e^{-kx}\,dx$.

(b) $\int_0^\infty \frac{\sin\alpha x}{x}\cdot\frac{\sin\beta x}{x}e^{-kx}\,dx$.

41. Show that

$$\int_0^1 \int_0^1 \ln(1-x)\ln(1-xy)\,dx\,dy = 3 - 2\zeta(3),$$

where ζ is the Riemann zeta function.

42. Show that

(a) $\int_0^1 \int_0^1 \frac{\ln(1+xy)}{(1+x)(1+y)}\,dx\,dy = \frac{\pi^2}{12}\ln 2 + \frac{1}{3}\ln^3 2 - \frac{1}{2}\zeta(3)$.

(b) $\int_0^1 \int_0^1 \frac{(xy)^2}{(1+x)(1+y)(1+x^2 y^2)}\,dx\,dy = \frac{7}{8}\ln^2 2 + \frac{\pi}{8}\ln 2 - \frac{\pi^2}{48} - \frac{1}{2}G$.

(c) $\int_0^1 \int_0^1 \frac{x\ln(x)\ln(y)}{(1+x^2)(1+xy)}\,dx\,dy = \frac{1}{48}\pi^2 G$, (**Pi Mu Epsilon Journal Problem 1376, 2021**)

where $G = \sum_{n=0}^\infty (-1)^n/(2n+1)^2$ is the Catalan constant.

43. Show that

$$\int_0^{\pi/2} \int_0^{\pi/2} \frac{\ln(1+\cos x) - \ln(1+\cos y)}{\cos x - \cos y}\,dx\,dy = 2\pi G - \frac{7}{2}\zeta(3),$$

where $G = \sum_{n=0}^\infty (-1)^n/(2n+1)^2$ is the Catalan constant.
Hint: Note that

$$\frac{\ln(1+\cos x) - \ln(1+\cos y)}{\cos x - \cos y} = \int_0^1 \frac{dt}{(1+t\cos x)(1+t\cos y)}.$$

44. Let $\alpha > 0$. Show that

$$\int_0^\infty \left(\frac{\sin x}{x}\right)^2 e^{-2\alpha x}\,dx = \text{arccot}(\alpha) + \alpha \ln\left(\frac{\alpha}{\sqrt{1+\alpha^2}}\right).$$

45. For $\alpha \in (0,1)$, show that

$$\int_0^\infty \cos(x^\alpha)\,dx = \frac{\pi\csc(\pi/2\alpha)}{2\alpha\Gamma(1-\alpha)}.$$

46. Show that

$$\int_0^{\pi/2} \cos^{\alpha+\beta}\theta \cos(\alpha-\beta)\theta\,d\theta = \frac{\pi\Gamma(\alpha+\beta+1)}{2^{\alpha+\beta+1}\Gamma(\alpha+1)\Gamma(\beta+1)}.$$

47. Show that

$$\sum_{n=1}^{\infty} (-1)^{n-1} \frac{1}{2n} B(1/2, n/2) = \frac{\pi^2}{4}.$$

48. Evaluate

$$\int_0^{\infty} \frac{\ln x}{1 + e^x} \, dx.$$

Using this result proves that

$$\sum_{n=1}^{\infty} \frac{(-1)^n \ln n}{n} = \gamma \ln 2 - \frac{1}{2} \ln^2 2,$$

where γ is the Euler-Mascheroni constant.

49. Let $k \in \mathbb{N}$. Prove that

$$\int_0^1 \ln x \ln^k (1 - x) \, dx = (-1)^{k-1} k! \left(k + 1 - \zeta(2) - \zeta(3) - \cdots - \zeta(k+1) \right).$$

50. Let $k \in \mathbb{N}, k \geq 2$. Prove that

$$\int_0^1 \frac{\ln x \ln^k (1 - x)}{x} \, dx = (-1)^{k-1} k! \left(\frac{k+1}{2} \zeta(k+2) - \frac{1}{2} \sum_{i=1}^{k-1} \zeta(i+1) \zeta(k+1-i) \right).$$

Test your result with

$$\int_0^1 \frac{\ln x \ln^2 (1 - x)}{x} \, dx = -\frac{1}{2} \zeta(4).$$

$$\int_0^1 \frac{\ln x \ln^3 (1 - x)}{x} \, dx = 12 \zeta(5) - \pi^2 \zeta(3).$$

51. Study the following extension: For $m, n \in \mathbb{N}, n \geq 2$, evaluate

$$\int_0^1 \frac{\ln^m x \ln^n (1 - x)}{x} \, dx.$$

One possible way is based on Euler integral

$$\int_0^1 (1 - t)^p t^{q-1} \, dt = \frac{\Gamma(p+1)\Gamma(q+1)}{\Gamma(p+q+1)}.$$

Differentiating with respect to p and q yields

$$\int_0^1 \frac{\ln^m x \ln^n (1 - x)}{x} \, dx = D_p^n D_q^m \left[\frac{\Gamma(p+1)\Gamma(q+1)}{q\Gamma(p+q+1)} \right] \Big|_{p=q=0}.$$

In particular, show that

$$\int_0^1 \frac{\ln^m x \ln^2 (1 - x)}{x} \, dx = (-1)^m m! \left((m+2)\zeta(m+3) - \sum_{i=1}^{m} \zeta(i+1)\zeta(m+2-i) \right).$$

52. Show that

$$\int_0^1 \frac{\ln(1-x)\ln(1+x)}{1+x}\,dx = \frac{1}{3}\ln^3 2 - \frac{1}{2}\ln 2\zeta(2) + \frac{1}{8}\zeta(3).$$

Let H_n be the nth harmonic number. Use this result to evaluate

$$\sum_{n=1}^{\infty} \frac{(-1)^{n+1}H_n}{(n+1)^2} \quad \text{and} \quad \sum_{n=1}^{\infty} \frac{(-1)^{n+1}H_n^2}{n+1}.$$

Remark. In general, it is interesting to evaluate

$$I(k) = \int_0^1 \frac{\ln^k(1+x)\ln(1-x)}{1+x}\,dx$$

in closed form. With Mathematica experiments on $k = 1, 2, 3$, it is expected that the final answer should involve the values of $Li_n(1/2)$, which is the polylogarithm defined by $Li_n(x) = \sum_{k=1}^{\infty} x^k/k^n$.

53. **(Monthly Problem 11993, 2017).** Prove

$$\int_0^1 \frac{\ln(1-x)\ln^2(1+x)}{x}\,dx = -\frac{\pi^4}{240} = -\frac{3}{8}\zeta(4).$$

54. Let H_n be the nth harmonic number. Show that

$$\sum_{n=1}^{\infty} \frac{(-1)^{n+1}}{(n+1)^2}H_nH_{n+1} = \frac{3}{16}\zeta(4).$$

55. Prove

(a) $\int_0^1 \frac{\ln x \ln(1-x)\ln^2(1+x)}{x}\,dx = \frac{7}{8}\zeta(2)\zeta(3) - \frac{25}{16}\zeta(5).$

(b) $\int_0^1 \frac{\ln^2 x \ln(1-x)\ln(1+x)}{x}\,dx = \frac{3}{4}\zeta(2)\zeta(3) - \frac{27}{16}\zeta(5).$

(c) $\int_0^1 \frac{\ln^2 x \ln(1+x)}{1-x}\,dx = \frac{7}{2}\ln^2 2\zeta(3) - \frac{19}{8}\zeta(4).$

(d) $\int_0^1 \frac{\ln x \ln(1-x)\ln^2(1+x)}{1-x}\,dx = -\ln^3 2\zeta(2) + \frac{7}{2}\ln^2 2\zeta(3) - 3\zeta(2)\zeta(3) - \frac{15}{8}\zeta(4) + 6\zeta(5).$

Hint: Use the identity $6ab^2 = (a+b)^3 + (a-b)^3 - 3a^3$ to reduce the number of logarithms with distinct arguments.

56. Prove
$$\int_0^1 \frac{\ln^2(\sin x)\ln^2(\cos x)}{\sin x \cos x}\,dx = \frac{1}{2}\zeta(5) - \frac{1}{4}\zeta(2)\zeta(3)).$$

57. Show that
$$\int_0^1 \frac{\ln\Gamma(x+1)}{x}\,dx = -\gamma + \sum_{n=2}^{\infty} \frac{(-1)^n}{n^2}\zeta(n),$$

where γ is the Euler-Mascheroni constant.

58. Let $\alpha, \beta \geq 0$ and $\gamma, p, q > 0$. Show that

$$\int_0^1 \frac{x^{p-1}(1-x)^{q-1}}{[\alpha x + \beta(1-x) + \gamma]^{p+q}}\,dx = \frac{B(p,q)}{(\alpha+\gamma)^p(\beta+\gamma)^q}.$$

59. For $\alpha, \beta, p > 0$, express the following integrals in terms of the beta function.

$$\int_0^1 \frac{x^{\alpha-1}(1-x)^{\beta-1}}{(x+p)^{\alpha+\beta}}\, dx, \quad \int_{-1}^1 \frac{(1+x)^{2\alpha-1}(1-x)^{2\beta-1}}{(1+x^2)^{\alpha+\beta}}\, dx.$$

Hint: Use the substitutions $u = \frac{(1+p)x}{x+p}$ and $u = \frac{(1+x)^2}{2(1+x^2)}$, respectively.

60. (**Ramanujan**). For $n \in \mathbb{N}, |x| < 2\pi$, show that

$$\int_0^x t^n \cot(t/2)\, dt = 2\cos(n\pi/2)n!\zeta(n+1) - 2\sum_{k=0}^n (-1)^{k(k+1)/2}\frac{n!}{(n-k)!} x^{n-k}\mathrm{Cl}_{k+1}(x),$$

where $\mathrm{Cl}_n(x)$ is the *Clausen functions* defined by

$$\mathrm{Cl}_{2n}(x) = \sum_{k=1}^\infty \frac{\sin(kx)}{k^{2n}}, \quad \mathrm{Cl}_{2n+1}(x) = \sum_{k=1}^\infty \frac{\cos(kx)}{k^{2n+1}}.$$

61. Show that

$$\int_0^{\pi/2} t^n \cot t\, dt = \frac{1}{n+1}\int_0^{\pi/2} t^{n+1} \csc^2 t\, dt.$$

Then use the above Ramanujan result to evaluate

$$I(p,q) := \int_0^{\pi/2} x^p \cot^{2q} x\, dx, \quad (p, q \in \mathbb{N}).$$

62. Let $0 < p, q < 1$. Show that

$$I(p,q) = \int_0^\infty \frac{x^{p-1} - x^{q-1}}{(1+x)\ln x}\, dx = \ln\frac{\tan(p\pi/2)}{\tan(q\pi/2)}$$

by using the parametric differentiation and integration, respectively.
Hint: Notice that $\int_q^p x^{t-1}\, dt = \frac{x^{p-1} - x^{q-1}}{\ln x}$

63. Let p be prime and $n = p^m$. Show that

$$\prod_{(k,n)=1} \Gamma\left(\frac{k}{n}\right) = \frac{(2\pi)^{\phi(n)/2}}{\sqrt{p}},$$

where $\phi(n)$ is the Euler-phi function, which denotes the number of positive integers not exceeding n that are relative prime to n. For example, $\phi(9) = 6$ since that $1, 2, 4, 5, 7, 8$ are relative prime to 9.

64. Let $a_1, a_2, \ldots, a_n \in \mathbb{R}$. Show that

$$\left(\sum_{k=1}^n \frac{a_k}{k}\right)^2 \le \sum_{i=1}^n \sum_{j=1}^n \frac{a_i a_j}{i+j-1}.$$

65. Let $n, m \in \mathbb{N}$. Show that

$$\sum_{k=0}^n (-1)^k \binom{n}{k}\frac{1}{k+m+1} = \sum_{k=0}^m (-1)^k \binom{n}{k}\frac{1}{k+n+1}.$$

66. Let $p, q \in \mathbb{N}$. Show that

$$(2\pi)^{-q/2} q^{x + \frac{pq-p-q}{2}} \prod_{j=0}^{q-1} \Gamma\left(\frac{x + pj}{q}\right) = (2\pi)^{-p/2} p^{x + \frac{pq-p-q}{2}} \prod_{j=0}^{p-1} \Gamma\left(\frac{x + qj}{p}\right).$$

67. Establish the *Dirichlet Formula*: for $\alpha, \beta, p, q > 0$,

$$\Gamma(q) \int_0^\infty \frac{e^{-\alpha x} x^{p-1}}{(\beta + x)^q} \, dx = \Gamma(p) \int_0^\infty \frac{e^{-\beta x} x^{q-1}}{(\alpha + x)^p} \, dx.$$

Hint: Using

$$\frac{\Gamma(q)}{(\beta + x)^q} = \int_0^\infty e^{-(\beta + x)y} y^{q-1} \, dy.$$

68. Let $a > 0$. Show that

$$\int_0^\infty \frac{x e^{-x^2}}{\sqrt{x^2 + a^2}} \, dx = \frac{a}{\sqrt{\pi}} \int_0^\infty \frac{e^{-x^2}}{x^2 + a^2} \, dx = e^{a^2} \int_a^\infty e^{-t^2} \, dt.$$

69. If a positive function $f(x)$ is twice continuously differentiable on $(0, \infty)$ and satisfies

    (a) $f(x + 1) = x f(x)$,
    (b) $f(2x) = \frac{2^{2x-1}}{\sqrt{\pi}} f(x) f(x + 1/2)$,

    prove that $f(x) = \Gamma(x)$.

70. The *Gauss hypergeometric series* is defined by

$$F(\alpha, \beta, \gamma, x) := 1 + \sum_{n=1}^\infty \frac{\alpha(\alpha + 1) \cdots (\alpha + n - 1)\beta(\beta + 1) \cdots (\beta + n - 1)}{n! \, \gamma(\gamma + 1) \cdots (\gamma + n - 1)} x^n.$$

Show that

$$F(\alpha, \beta, \gamma, 1) = \frac{\Gamma(\gamma)\Gamma(\gamma - \alpha - \beta)}{\Gamma(\gamma - \alpha)\Gamma(\gamma - \beta)}.$$

71. For $n \in \mathbb{N}$, show that

$$\int_0^{\pi/2} \sin^{2n} \theta \ln \sin \theta \, d\theta = -\frac{\sqrt{\pi}}{2} \frac{\Gamma(n + 1/2)}{\Gamma(n + 1)} \int_0^1 \frac{u^{2n}}{1 + u} \, du$$

$$= \frac{\pi}{2} \frac{(2n - 1)!!}{(2n)!!} \left(1 - \frac{1}{2} + \frac{1}{3} - \cdots - \frac{1}{2n} - \ln 2\right).$$

72. Let $\alpha, \beta > -1, \alpha + \beta > -1$. Show that

$$\int_0^1 \frac{(1 - x^\alpha)(1 - x^\beta)}{(1 - x) \ln x} \, dx = \ln \frac{\Gamma(\alpha + 1)\Gamma(\beta + 1)}{\Gamma(\alpha + \beta + 1)}.$$

Extend your method to evaluate

$$\int_0^1 \frac{x^\alpha(1 - x^\beta)(1 - x^\gamma)}{(1 - x) \ln x} \, dx \quad \text{and} \quad \int_0^1 \frac{(1 - x^\alpha)(1 - x^\beta)(1 - x^\gamma)}{(1 - x) \ln x} \, dx.$$

73. Define the *digamma function*

$$\psi(x) = \frac{\Gamma'(x)}{\Gamma(x)} = \frac{d}{dx}\Gamma(x).$$

Show that

(a) For $n \in \mathbb{N}$,

$$\psi(x+n) = \psi(x) + \sum_{k=0}^{n-1} \frac{1}{x+k}.$$

(b) For $0 < x < 1$,

$$\psi(x) - \psi(1-x) = -\frac{\pi}{\tan \pi x}.$$

74. For $x > 0$, show that

$$\psi(x) = -\gamma + \int_0^1 \left(\frac{1-u^{x-1}}{1-u}\right) du$$

and so

$$\gamma = \int_0^1 \left(\frac{1}{1-u} + \frac{1}{\ln u}\right) du = \int_0^\infty e^{-t} \ln t \, dt.$$

75. For $x > 0$, show that

$$\psi(x) = -\gamma + (x-1) - \frac{(x-1)(x-2)}{2 \cdot 2!} + \frac{(x-1)(x-2)(x-3)}{3 \cdot 3!} + \cdots .$$

Hint: First establish the absolute and uniform convergence of the series on $(0, \infty)$.

76. Let B_n be the nth Bernoulli polynomial. For $x \to \infty$, prove that

$$\psi(x) \sim \ln(x - 1/2) - \sum_{n=1}^\infty \frac{B_{2n}(1/2)}{2n} (x - 1/2)^{-2n}.$$

More precisely, if one truncates the series at N, show that the resulting sum is a lower bound for even N and $x > 1/2$. If N is odd, the sum is an upper bound.

77. Prove the duplication formula for the digamma function:

$$\psi(2x) = \frac{1}{2}\left(\psi(x) + \psi\left(x + \frac{1}{2}\right)\right) + \ln 2.$$

78. Prove the following refinement of (7.11): As $x \to \infty$,

$$\Gamma(x+1) \sim \sqrt{2\pi x} \left(\frac{x}{e}\right)^x \left\{1 + \frac{1}{12x} + \frac{1}{288x^2} - \frac{139}{51840x^3} - \cdots \right\}.$$

79. Let $p, q > 0, q < p + 1$. Define

$$I(p, q) = \int_0^{\pi/2} x \cos^{p-1} x \sin(qx) \, dx$$

and $p_1 = (p + q + 1)/2, q_1 = (p - q + 1)/2$.

(a) Show that

$$I(p, q) = \frac{\psi(p_1) - \psi(q_1)}{2^{p+1}\Gamma(p_1)\Gamma(q_1)}\Gamma(p)\pi.$$

(b) Show that

$$\int_0^{\pi/2} x^2 \cos^{p-1} x \, dx = \frac{\psi'((p+1)/2)}{2^{p+1}\Gamma^2((p+1)/2)}\Gamma(p)\pi.$$

80. Let $f(x) = x(\ln x - \psi(x))$. Show that f is strictly decreasing and convex on $(0, \infty)$ and

$$\lim_{x \to 0} f(x) = 1 \quad \text{and} \quad \lim_{x \to \infty} f(x) = \frac{1}{2}.$$

81. Let $f(x) = x(\ln x - \psi(x))$. Show that, for all $x, y \geq 0$,

$$1 \leq \frac{f(x)f(y)}{f(x+y)} < 2.$$

Prove that both bounds are the best possible by the stress testing method.

82. A function f is called *completely monotonic* on I if f has derivatives of all orders on I and

$$(-1)^n f^{(n)}(x) \geq 0 \quad \text{for all } x \in I \text{ and for all } n \geq 0.$$

Show that $f(\alpha, x) = x^\alpha(\ln x - \psi(x))$ is strictly completely monotonic on $(0, \infty)$ if and only if $\alpha \leq 1$.

83. Let $f(x) = \ln\Gamma(x)/x$. For $x > -1, n \in \mathbb{N}$, show that

$$(-1)^n f^{(n+1)}(x) = (n+1)! \int_0^1 t^{n+1}\zeta(n+2, xt+1) \, dt$$

and $f'(x)$ is completely monotonic. Here $\zeta(s, a)$ is the Hurwitz zeta function (see Problem 13 in Section 7.4).

84. (**Simpson's Dissection Formula**). Let $f(x) = \sum_{n=0}^{\infty} a_n x^n$. Show that

$$\sum_{n=0}^{\infty} a_{kn+m}x^{kn+m} = \frac{1}{k}\sum_{j=0}^{k-1} \omega^{-jm} f(\omega^j x),$$

where $\omega = e^{2\pi i/k}$ is a primitive kth root of unity.

85. Use Simpson's dissection formula to show that, for $p, q \in \mathbb{N}, 0 < p < q$,

$$\psi\left(\frac{p}{q}\right) = -\gamma - \frac{\pi}{2}\cot\frac{p\pi}{q} - \ln q + 2\sum_{n=1}^{[q/2]} \cos\frac{2np\pi}{q} \ln\left(2\sin\frac{n\pi}{q}\right).$$

86. A polygonal number is a type of figurate number that is a generalization of triangular, square, etc., to an n-gon for n an arbitrary positive integer. Let $P(n, r)$ be the nth r-gon number. Show that

$$P(n, r) = \frac{1}{2}n[(r-2)n - (r-4)]$$

and for $r \geq 5$ find

$$\sum_{n=1}^{\infty} \frac{1}{P(n, r)}$$

in closed form.

87. (**The Herglotz trick**). Here is an outline of the proof of the Euler's reflection formula due to Herglotz. Let

$$g(x) = \frac{\pi}{\tan \pi x} - \lim_{N \to \infty} \sum_{n=-N}^{N} \frac{1}{n+x}.$$

Verify that

(a) $g'(x) = -\frac{\pi^2}{\sin^2 \pi x} + \sum_{n=-\infty}^{\infty} \frac{1}{(n+x)^2}$.

(b) $g'(x)$ is continuous for $0 \le x \le 1$ if $g'(0) = g'(1) = 0$.

(c) $g'(x/2) + g'((x+1)/2) = 4g'(x)$.

(d) Let $M = \max_{0 \le x \le 1} |g'(x)|$. Then $M \le M/2$ which implies that $M = 0$.

(e) $g(x/2) - g((x+1)/2) = \frac{2\pi}{\sin \pi x} - 2 \sum_{n=-\infty}^{\infty} \frac{(-1)^n}{n+x}$.

(f) $g(x+1) = g(x)$.

(g) $g(x)$ is a constant.

(h) $\int_0^\infty \frac{t^{x-1}}{1+t} dt = \sum_{n=-\infty}^{\infty} \frac{(-1)^n}{n+x}$.

88. For $x, \lambda > 0$ and $\alpha \in (-\pi/2, \pi/2)$, prove that

$$\int_0^\infty t^{x-1} e^{-\lambda t \cos \alpha} \cos(\lambda t \sin \alpha) \, dt = \lambda^{-x} \Gamma(x) \cos \alpha x;$$

$$\int_0^\infty t^{x-1} e^{-\lambda t \cos \alpha} \sin(\lambda t \sin \alpha) \, dt = \lambda^{-x} \Gamma(x) \sin \alpha x.$$

89. Establish the functional equation

$$\zeta(x, s) = \frac{2\Gamma(1-s)}{(2\pi)^{1-s}} \left\{ \sin(\pi s/2) \sum_{m=1}^{\infty} \frac{\cos 2m\pi x}{m^{1-s}} + \cos(\pi s/2) \sum_{m=1}^{\infty} \frac{\sin 2m\pi x}{m^{1-s}} \right\}.$$

90. Let $\zeta(s, a)$ be the Hurwitz zeta function (see Problem 13 in Section 7.4). Show that

$$\zeta(s, a) = \frac{s^{-a}}{2} + \frac{s^{1-a}}{a-1} + 2 \int_0^\infty \frac{(s^2+x^2)^{-a/2} \sin(a \arctan x/s)}{e^{2\pi x} - 1} \, dx.$$

Using this result concludes that

$$\zeta(s, 2) = \frac{1}{2s^2} + \frac{1}{s} + \int_0^\infty \frac{4sx \, dx}{(s^2 + x^2)^2 (e^{2\pi x} - 1)}.$$

91. Let $\psi(x)$ be the digamma function. Show that

$$\psi(x) = \ln x - \frac{1}{2x} - \int_0^\infty \frac{2t \, dt}{(x^2 + t^2)(e^{2\pi t} - 1)}.$$

Hint: Note that $\psi'(x) = \zeta(x, s)$.

92. Prove *Binet's formula*:

$$\ln \Gamma(x) = \left(x - \frac{1}{2} \right) \ln x - x + \frac{1}{2} \ln(2\pi) + 2 \int_0^\infty \frac{\arctan(t/x)}{e^{2\pi t} - 1} \, dt.$$

93. (**Kummer's Fourier Expansion for** $\ln \Gamma(x)$). For $0 < x < 1$, show that

$$\ln \frac{\Gamma(x)}{\sqrt{2\pi}} = -\frac{1}{2}\ln(2\sin \pi x) + \frac{1}{2}(\gamma + \ln(2\pi))(1 - 2x) + \frac{1}{\pi}\sum_{k=1}^{\infty}\frac{\ln k}{k}\sin(2\pi kx).$$

94. For $\alpha \in (0, 1)$, prove that

$$\int_0^1 \frac{\ln \ln(1/x)}{1 - 2\cos(2\alpha\pi)x + x^2}\,dx = \frac{\pi}{2\sin(2\alpha\pi)}\left((1 - 2\alpha)\ln(2\pi) + \ln \frac{\Gamma(1 - \alpha)}{\Gamma(\alpha)}\right).$$

95. (**Monthly Problem 11806, 2015**) Prove that

$$\int_0^{2\pi}\log \Gamma\left(\frac{x}{2\pi}\right)e^{\cos x}\sin(x + \sin x)\,dx = (e - 1)(\log(2\pi) + \gamma) + \sum_{n=2}^{\infty}\frac{\log n}{n!}.$$

96. Refine Carlson's inequality by replacing the multiplier $\sqrt{\alpha + \beta x^2}$, which is used in Hardy's proof, by $\sqrt{\alpha + \beta x^2 + \gamma x^4}$.

97. (**Monthly Problem 11746, 2013**). Let f be a continuous function from $[0, \infty)$ to $\mathbb{R}$ such that the following integrals converge: $S = \int_0^{\infty} f^2(x)\,dx$, $T = \int_0^{\infty} x^2 f^2(x)\,dx$, and $U = \int_0^{\infty} x^4 f^2(x)\,dx$. Let $V = T + \sqrt{T^2 + 3SU}$. Given that f is not identical to 0, show that

$$\left(\int_0^{\infty}|f(x)|\,dx\right)^4 \leq \pi^2 \frac{S(T + V)^2}{9V}.$$

Remark. This result is an improvement of Carlson's bound $\pi^2 ST$ only if $SU/T < (11 + 5\sqrt{5})/2$. But, based on the refined Carlson's inequality above, a better bound is given by $\pi^2 ST - \pi^2 T^3/U$.

98. Given $f : [1, \infty) \to (0, \infty)$ and a constant $c > 0$, show that if

$$\int_1^t f(x)\,dx \leq ct^2 \quad \text{for all } t \geq 1 \text{ then} \quad \int_1^{\infty}\frac{dx}{f(x)} = \infty.$$

If the bound ct^2 is replaced by $ct^2 \ln t$, does the reciprocal integral still diverge?

99. (**Monthly Problem 11697, 2013**). Let n and q be integers, with $2n > q \geq 1$. Let

$$f(t) = \int_{\mathbb{R}^q}\frac{e^{-t(x_1^{2n} + \cdots + x_q^{2n})}}{1 + x_1^{2n} + \cdots + x_q^{2n}}\,dx_1 \cdots dx_q.$$

Prove that

$$\lim_{n\to\infty} t^{q/2n} f(t) = \left(\frac{1}{n}\Gamma\left(\frac{1}{2n}\right)\right)^q.$$

100. (**Monthly Problem 11793, 2014**). Prove that

$$\sum_{n=1}^{\infty}\frac{\log(n + 1)}{n^2} = -\zeta'(2) + \sum_{n=3}^{\infty}(-1)^{n+1}\frac{\zeta(n)}{n - 2},$$

where ζ' denotes its derivative.

101. (**Monthly Problem 11796, 2014**). Find

$$\int_0^\infty \frac{\sin((2n+1)x)}{\sin x} e^{-\alpha x} x^{m-1}\, dx$$

in terms of α, m, and n, when $\alpha > 0, m \geq 1$, and n is a nonnegative integer.

102. (**Monthly Problem 11802, 2014**). Let $H_{n,2} = \sum_{k=1}^n k^{-2}$, and let $D_n = n! \sum_{k=0}^n (-1)^k/k!$. (This is the derangement number of n, that is, the number of permutations of $\{1, 2, \ldots, n\}$ without fixed pionts.) Prove that

$$\sum_{n=1}^\infty H_{n,2} \frac{(-1)^n}{n!} = \frac{\pi^2}{6e} - \sum_{n=0}^\infty \frac{D_n}{n!(n+1)^2}.$$

103. (**Monthly Problem 11842, 2015**). Let ψ be the Digamma function, that is, $\psi(x) = (\ln \Gamma(x))'$. Let $\phi = (1 + \sqrt{5})/2$. Prove that

$$\sum_{n=1}^\infty \frac{\psi(n+\phi) - \psi(n-1/\phi)}{n^2 + n - 1} = \frac{\pi^2}{2\sqrt{5}} + \frac{\pi^2 \tan^2(\sqrt{5}\pi/2)}{\sqrt{5}} + \frac{4}{5}\pi \tan(\sqrt{5}\pi/2).$$

104. (**Monthly Problem 12285, 2021**). Prove that, for $a > 0$,

$$\sum_{n=1}^\infty \int_0^\infty \frac{t \cos t}{t^2 + n^2 a^2}\, dt = \int_0^\infty \left(-\frac{\pi}{2a} \cos t + \sum_{n=1}^\infty \frac{t \cos t}{t^2 + n^2 a^2} \right) dt.$$

105. Prove that there exists a unique function $\Gamma, \Gamma(x) > 0$ for $x \geq 1$, such that $L(x) = \ln \Gamma(x+1)$ in which $L(x)$ has the following properties

    (a) $L(0) = 0$ (Initial condition, $1! = 1$),
    (b) $L(x+1) = \ln(x+1) + L(x)$ (functional equation),
    (c) $L(x+n) = L(n) + x \ln(n+1) + r_n(x)$, where $\lim_{n\to\infty} r_n(x) = 0$.

    This problem gives another characterization of the gamma function.

106. Show that

$$\int_0^1 \frac{\ln x \tanh^{-1} x}{x(1-x^2)}\, dx = -\frac{\pi^2}{8} \ln 2 - \frac{7}{16}\zeta(3).$$

107. Prove that

$$\ln \frac{\Gamma(x+1/2)}{\sqrt{x}\,\Gamma(x)} = -\sum_{k=1}^n \frac{(1-2^{-2k})B_{2k}}{k(2k-1)} \frac{1}{x^{2k-1}} + O\left(\frac{1}{x^{2n-1/2}} \right),$$

    where B_{2k} are Bernoulli numbers.

108. Let $y = x/(1 - e^{-x})$. Evaluate

$$I = \int_0^\infty e^{-y/2} \sqrt{y}\, dx.$$

109. Let $J_0(x)$ be the Bessel function of zero order. For $n \in \mathbb{N}$, evaluate

$$I_n = \int_0^\infty \frac{\sin nx}{x} J_0^n(x)\, dx.$$

110. Let

$$f(x) = \int_0^{\pi/2} (1 - e^{x(1-\csc\theta)}) \sec^2\theta \, d\theta.$$

For $\alpha > 3$, show that

$$\int_0^\infty x^{\alpha-3} e^{-2x} f^2(x) \, dx = 2^{\alpha-3} \frac{\Gamma^4(\alpha/2)}{\Gamma(\alpha)}.$$

111. Evaluate

$$M_n(x) = n(n-1) \int_0^1 \int_0^u (u-v)^2 \left(\frac{uv}{(1-u)(1-v)} \right)^{x/2} dv\, du$$

in terms of the gamma function. This integral is called the *moment generating function* for the mid-range of a random n-sample from the logistic distribution.

112. Evaluate

$$I_n = \int_{-\infty}^\infty \int_0^\infty \frac{e^{-kx} \cos xy}{\sqrt{x^2+a^2}\,(\mu x + \sqrt{x^2+a^2})^n} \, dx\, dy.$$

Hint: Use

$$\int_0^\infty e^{-\eta x} \cos xy \, dx = \frac{\eta}{y^2+\eta^2}.$$

113. Find all the continuous functions $f : [1, \infty) \to \mathbb{R}$ such that

$$\int_x^{x^n} f(t) \, dt = \int_1^x (t^{n-1} + t^{n-2} + \cdots + t) f(t) \, dt$$

for every $x \in [1, \infty)$ and every $n \in \mathbb{N}$ with $n \geq 2$.

114. Find a continuous function $F(t)$ for $t \geq 0$ satisfying

$$tF^2(t) = \int_0^1 \frac{F(ts) - F(t-ts)}{1-2s} \, ds$$

and $F(0) = C > 0$.

115. Prove that

(a) $\sum_{k=0}^n \frac{1}{\binom{n}{k}} = \frac{n+1}{2^{n+1}} \sum_{k=1}^{n+1} \frac{2^k}{k}$.

(b) $\sum_{k=0}^n \frac{1}{\binom{2n}{2k}} = \frac{2n+1}{2^{2n+1}} \sum_{k=0}^{2n+1} \frac{2^k}{k+1}$.

(c) $\sum_{k=0}^n \frac{1}{\binom{pn}{pk}} = \frac{pn+1}{p} \sum_{k=0}^{(n+1)p-1} \frac{1}{[(n+1)p-k]2^{k-p}} \left(\frac{1}{2} + \sum_{i=1}^{[(p-1)/2]} \frac{(-1)^i \cos(\pi i k/p)}{(\cos(i\pi/p))^{k+2-p}} \right)$.

Hint: Use

$$\binom{n}{k}^{-1} = (n+1) \int_0^1 x^k (1-x)^{n-k} \, dx.$$

116. Let $f : [0,1] \to \mathbb{R}$ be integrable. Show that

$$\int_0^1 \int_0^1 f(xy) \, dx\, dy = -\int_0^1 f(x) \ln x \, dx.$$

Using this result derive

$$\int_0^1 \int_0^1 \frac{\ln(1+xy)}{1-xy} \, dx\, dy = \frac{1}{4}\pi^2 \ln 2 - \zeta(3).$$

117. Let p and q be nonnegative integers. Evaluate

$$\int_0^1 \int_0^1 -\frac{\ln(xy)}{1-xy} x^p y^q \, dx dy.$$

Test your result with

$$\int_0^1 \int_0^1 -\frac{\ln(xy)}{1-xy} x^n y^n \, dx dy = 2\left(\zeta(3) - \sum_{k=1}^n \frac{1}{k^3}\right).$$

118. (**Hadjicostas-Chapman formula**). Let $s > -2$. Prove that

$$\int_0^1 \int_0^1 \frac{1-x}{(1-xy)}(-\ln xy)^s \, dx dy = \Gamma(s+2)\left(\zeta(s+2) - \frac{1}{s+1}\right).$$

119. (**J. Sondow**). Let γ be the Euler-Mascheroni constant. Prove that

(a) $\gamma = \sum_{n=1}^\infty \left(\frac{1}{n} - \ln\frac{n+1}{n}\right) = \int_0^1 \int_0^1 \frac{1-x}{(1-xy)(-\ln xy)} \, dx dy.$

(b) $\ln\frac{4}{\pi} = \sum_{n=1}^\infty (-1)^{n-1}\left(\frac{1}{n} - \ln\frac{n+1}{n}\right) = \int_0^1 \int_0^1 \frac{1-x}{(1+xy)(-\ln xy)} \, dx dy.$

120. Let $f(x)$ be a continuous extension of H_n. Show that

$$f(x) = \sum_{k=1}^\infty \left(\frac{1}{k} - \frac{1}{x+k}\right) = x \sum_{k=1}^\infty \frac{1}{x(x+k)}.$$

Verify that $f(x) = f(x-1) + 1/x$ and

$$\int_0^1 f(x) \, dx = \gamma, \qquad \int_0^n f(x) \, dx = n\gamma + \ln n!.$$

121. (**George Stoica**). It is well-known that $\int_{-\infty}^x e^{-t^2/2} \, dt$ cannot be integrated in finite terms. Let P be a polynomial. Prove that

$$\int_{-\infty}^x P(t)e^{-t^2/2} \, dt$$

can be integrated in finite terms if and only if $\int_{-\infty}^\infty P(t)e^{-t^2/2} \, dt = 0$.
Hint: Let H_n be *Hermite polynomial* which is defined by

$$H_n(t) = (-1)^n e^{t^2/2}\left(e^{-t^2/2}\right)^{(n)}, \qquad n = 0, 1, 2, \ldots.$$

Show that

$$\int_{-\infty}^x H_n(t)e^{-t^2/2} \, dt = -H_{n-1}(x)e^{-x^2/2}.$$

122. (**Gregory summation formula**)—an analog of Euler-Maclaurin summation formula. Show that

$$\lim_{n\to\infty}\left\{\sum_{i=a}^n f(i) - \frac{1}{2}f(n) - \int_a^n f(t) \, dt\right\} = \sum_{k=1}^\infty (-1)^{k-1} G_k \Delta^{k-1} f(a),$$

where G_i is the Gregory coefficient defined by

$$\frac{x}{\ln(1-x)} = 1 + \sum_{i=1}^\infty G_i x^i, \qquad \text{for } |x| < 1.$$

123. Show that
$$\sum_{k=1}^{\infty}\left(H_k - \ln k - \gamma - \frac{1}{2k}\right) = \frac{1}{2}\left(1 - \ln(2\pi) + \gamma\right).$$

124. Let $p_n(x)$ be the Legendre polynomial on $(0,1)$:
$$p_n(x) = \frac{1}{n!}\frac{d^n}{dx^n}\{x^n(1-x)^n\}.$$

Show that there exists integers A_n and B_n such that
$$0 \neq \int_0^1\int_0^1 -\frac{\ln(xy)}{1-xy}p_n(x)p_n(y)\,dxdy = (A_n + B_n\zeta(3))d_n^{-3},$$

where $d_n = \operatorname{lcm}(1, 2, \ldots, n)$.

125. For A_n and B_n as in the previous problem, show that
$$0 < |A_n + B_n\zeta(3)|d_n^{-3} < 2(\sqrt{2}-1)^{4n}\zeta(3).$$

Remark. This result implies that $\zeta(3)$ is irrational.

A

List of Problems from MAA

Problems from *The American Mathematical Monthly* (*AMM*), *Putnam Competition*, *Mathematics Magazine* (*MM*), and *The College Mathematics Journal* (*CMJ*) ©Mathematical Association of America, 2022. All rights reserved.

Chapter 1

AMM

12143, 2019	E1557, 1963	11604, 2011	12270, 2021
12220, 2020	12153, 2020	11528, 2010	11786, 2014
11935, 2016	11771, 2014	11875, 2015	12120, 2019
E2860, 1980	11376, 2008	11559, 2011	11811, 2014
11821, 2015	11973, 2002	12129, 2019	11367, 2008
12210, 2020	11153, 2005	11851, 2015	11852, 2015
11976, 2017	11995, 2017	12166, 2020	

Putnam

1970-A4	1966-A3	1953-A6	1966-A6
2012-B4	2001-B6	2002-A5	2018-B4
2016-A2	1969-A6	1949-B5	1963-A4
1950-A3	2008-B2	1947-A1	2006-B6

CMJ	965, 2011	1211, 2021
MM	2136, 2022	2087, 2020

Chapter 2

AMM

12101, 2019	12084, 2019	11829, 2015	11954, 2017
11400, 2008	11409, 2009	12241, 2021	12287, 2021
12004, 2017	11649, 2012	11910, 2016	11999, 2017
11384, 2008	11910, 2016	11930, 2016	11932, 2016
11952, 2017	11810, 2015	12134, 2019	12194, 2020
12102, 2019	12206, 2020	12215, 2020	12262, 2021
11068, 2004	12026, 2018	11809, 2015	11260, 2006
11982, 2017	11515, 2010	11853, 2015	11333, 2007

11299, 2007	12029, 2018	12110, 2019	11739, 2013
11685, 2013	11438, 2009	11499, 2010	12289, 2021

Putnam

2020-A3	2008-A4	2016-B1	2016-B6
2021-B2	2002-A6	2011-A2	2001-B3
1948-A3	1949-B2	1966-B3	1994-A1
2004-B5	2014-A3		

MM   2097, 2020

Chapter 3

AMM

11555, 2011	10818, 2000	11872, 2015

Putnam

2021-A2	1998-B1	1986-B4	1964-B3
2012-A3	2013-B3	2021-A6	1971-B2
1988-A5	1990-B5	1996-A6	2014-B6

Chapter 4

AMM

11369, 2008	10261, 1992	10604, 1997	11957, 2017
10739, 1999	11892, 2016	6585, 1988	11641, 2012

Putnam

2003-A2	2010-A2	2002-B3	2014-A1
1986-A1	1999-B4	2005-B3	2010-B5
	2009-B5		

Chapter 5

AMM

11819, 2015	11780, 2014	11812, 2015	12308, 2022
11961, 2017	12046, 2018	11872, 2015	11933, 2015
12205, 2020	12318, 2022	11152, 2005	11924, 2016
11457, 2009	11941, 2016	11535, 2016	11548, 2011
11517, 2010	11417, 2009	11946, 2016	12229, 2021
11133, 2005	11981, 2017	12193, 2020	

Putnam

1968-B4	2007-B2	2006-B5	1980-A3

2016-A3	1940-A3	1968-A1	1987-B1
1991-A5	2014-B2		

Chapter 6

AMM

11848, 2015	11897, 2016	11765, 2014	12297, 2022
11597, 2011	12207, 2020	12242, 2021	12051, 2018
11299, 2007	11916, 2016	12060, 2018	12026, 2018
3815, 1937	11410, 2009	11515, 2010	11659, 2012
11885, 2016	11973, 2017	11982, 2017	11890, 2016

Putnam

2014-A1	2015-B6	2016-B6

Chapter 7

AMM

12288, 2021	12281, 2021	11709, 2013	12317, 2022
11564, 2011	11937, 2016	11680, 2012	12260, 2021
12149, 2020	12054, 2018	12184, 2020	12221, 2020
12228, 2021	12274, 2021	11993, 2017	11806, 2015
11746, 2013	11697, 2013	11793, 2014	11796, 2014
11802, 2014	11842, 2015	12285, 2021	

Bibliography

[1] A. Abian, An ultimate proof Rolle's theorem, *Amer. Math. Monthly*, **86**(1979), 484–485.

[2] M. Aigner and G. M. Ziegler, *Proofs from the Book*, 4th edition, Springer-Verlag, New York, 2010.

[3] G. L. Alexanderson, L. Klosinski and L. C. Larson (Eds), *The William Lowell Putnam Mathematical Competition Problems and Solutions:1965-1984*, MAA Problem Books, The Mathematical Association of America, Washington, D.C., 1985.

[4] G. Almkvist, C. Krattenthaler and J. Petersson, Some new formulas for π, *Experimental Math.* **12** (2000), 441–456.

[5] H. Alzer, J. L. Brenner and O. G. Ruehr, On Mathieu's inequality, *J. Math. Anal. Appl.*, **218**(1998), 607–610.

[6] G. Andrews, R. Askey and R. Roy, *Special Functions*, Encyclopedia of Mathematics and Its Applications 71, Cambridge University Press, Cambridge, 1999.

[7] R. Apéry, Irrationalité de $\zeta(2)$ et $\zeta(3)$, *Astérisque*, **61**(1979), 11–13.

[8] T. Apostol, *Mathematical Analysis*, 2nd edition, Addison Wesley, Reading, MA, 1974.

[9] T. M. Apostol, A proof that Euler missed: Evaluating $\zeta(2)$ the easy way, *Math. Intelligence*, **5**(1983), 59–60.

[10] D. H. Bailey, A compendium of BBP-type formulas for mathematical constants. Available at
http://www.davidhbailey.com/dhbpapers/bbp-formulas.pdf

[11] D. H. Bailey, J. M. Borwein P. B. Borwein and S. Plouffe, The Quest for Pi, *Mathematical Intelligencer*, **1**(1997), 50–56.

[12] R. G. Bartle, *The Elementary of Real Analysis*, 2nd Edition, John Wiley, New York, 1964.

[13] E. Beckenbach and R. Bellman, *Inequalities*, Springer-Verlag, Berlin, 1961.

[14] B. Berndt, *Ramanujan's Notebooks, Part I*, Springer-Verlag, New York, 1985.

[15] E. Bloch, *The Real Numbers and Real Analysis*, Springer-Verlag, New York, 2010.

[16] D. D. Bonar and M. J. Khoury, *Real Infinite Series*, The Mathematical Association of America, Washington D.C., 2006.

[17] G. M. Boros and V. H. Moll, *Irresistible Integrals*, Cambridge University Press, Cambridge, 2004.

[18] J. Borwein, D. Bailey and R. Girgensohn, *Experimentation in Mathematics*, A. K. Peters, Massachusetts, 2004.

[19] J. Borwein and P. Borwein, *Pi and the AGM*, John Wiley, New York, 1987.

[20] D. Borwein and J. M. Borwein, On an intriguing integral and some series related to $\zeta(4)$, *Proc. Amer. Math. Soc.* **123**(1995), 1191–1198.

[21] D. M. Bressoud, *A Radical Approach to Real Analysis*, The Mathematical Association of America, Washington D.C., 1994.

[22] D. M. Bressoud, *Proofs and Confirmations, The Story of the Alternating Sign Matrix Conjecture*, The Mathematical Association of America, Washington D.C., 1999.

[23] B. Cassellman, From Bézier to Bernstein, Feature Column, American Mathematical Society, 2008.
Available at `http://www.ams.org/publicoutreach/feature-column/fcarc-bezier`

[24] A. L. Cauchy, *Analyse Algebrique* (Note 2; Oeuves Completes), Ser. 2, Vol. 3, Paris, (1897), 375–377.

[25] H. Chen, *Excursions in Classical Analysis*, The Mathematical Association of America, Washington D.C., 2010.

[26] H. Chen, *Monthly Problem Gems*, A. K. Peters/CRC Press, Boca Raton, 2021.

[27] H. Chen, Evaluations of some variant Euler sums, *J. Integer Seq.* **9**(2006), Article 06.2.3. Available at
`https://cs.uwaterloo.ca/journals/JIS/VOL9/Chen/chen78.html`

[28] H. Chen, The Fresnel Integrals Revisited, *Coll. Math. J.*, **40**(2009), 259–262.

[29] H. Chen, Means generated by an integral, *Math. Magazine*, **78**(2005), 397–399.

[30] B. R. Choe, An elementary proof of $\sum_{k=1}^{\infty} 1/n^2 = \pi^2/6$, *Amer. Math. Monthly*, **94**(1987), 662–663.

[31] S. Chowla and A. Selberg, On Epstein's zeta function, *J. Reine Agnew. Math.*, **227**(1967), 86–110.

[32] W. Chu, Summations on trigonometric functions, *Appl. Math. Comp.*, **141**(2003), 161–176.

[33] W. Chu and L. D. Donno, Hypergeometric series and harmonic number identities, *Adv. Appl. Math.* **34**(2005), 123–137.

[34] R. Courant and F. John, *Introduction to calculus and analysis*, Springer-Verlag, New York, 1989.

[35] P. J. Davis, Leonhard Euler's integral: a historical profile of the gamma function, *Amer. Math. Monthly*, **66**(1959), 849–869.

[36] T. P. Dence and J. B. Dence, A survey of Euler's constant, *Math. Mag.* **84**(2009), 255–265.

[37] R. Devaney, *An Introduction to Chaotic Dynamical Systems*, 2nd edition, Addison-Wesley, Reading, MA, 1989.

[38] J. Dieudonne, *Mathematics – The Music of Reason*, Springer-Verlag, New York, 1998.

[39] W. Dunham, *Euler: The Master of Us All*, The Mathematical Association of America, 1990.

[40] W. Dunham, *Journey Through Genius: The Great Theorems of Mathematics*, John Wiley, New York, 1990.

[41] J. A. Ewell, A new series representation for $\zeta(3)$, *Amer. Math. Monthly*, **97**(1990), 219–220.

[42] H. Flanders, On the Fresnel integrals, *Amer. Math. Monthly*, **89**(1982), 264–266.

[43] O. Furdui, *Limits, Series, and Fractional Part Integrals*, Springer-Verlag, New York, 2013.

[44] B. R. Gelbaum and J. H. Olmsted, *Counterexamples in Analysis*, Dover Books on Mathematics, Courier Corporation, 2012.

[45] R. Gelca and T. Andreescu, *Putnam and Beyond*, Springer-Verlag, Berlin Heidelberg, 2017.

[46] R. Graham, D. Knuth and O. Patashnik, *Concrete Mathematics*, 2nd edition, Addison-Wesley, Reading, MA, 1994.

[47] H. Gould, *Combinatorial Identities*, published by the author, revised edition, 1972.

[48] I. Gradshteyn and I. Ryzhik, *Table of Integrals, Series, and Product*, 6th edition, edited by A. Jeffrey and D. Zwillinger, Academic Press, New York, 2000.

[49] M. Gromov and M. Taylor, Finite propagation speed, kernel estimates for functions of the Laplace operator, and the geometry of complete Riemannian manifolds, *J. Diff. Geom.*, **17** (1982), 15–53.

[50] E. Hairer and G. Wanner, *Analysis by its History*, Undergraduate Texts in Mathematics, Springer-Verlag, New York, 1996.

[51] J. Hale and H. Kocak, *Dynamics and Bifurcations*, Texts in Applied Mathematics, No. 3, Springer-Verlag, New York, 1991.

[52] E. Hansen, *A Table of Series and Products*, Prentice Hall, New Jersey, 1975.

[53] G. H. Hardy, *Divergent Series*, Cambridge University Press, Cambridge, 1949.

[54] G. H. Hardy, J. E. Littlewood and G. Polya, *Inequalities*, Cambridge University Press, Cambridge, 1967.

[55] J. Havil, *Gamma: Exploring Euler's Constant*, Princeton University Press, Princeton, NJ, 2003.

[56] K. Hardy and K. S. Williams, *The Green Book—100 practice problems for undergraduate mathematics competitions*, Integer Press, Ottawa, 1985.

[57] K. Hardy and K. S. Williams, *The Red Book of Mathematics Problems*, Dover, 1988.

[58] P. Halmos, *Problems for Mathematicians Young and Old*, The Mathematical Association of America, Washington D. C., 1991.

[59] K. S. Kedlaya, B. Poonen and R. Vakil, *The William Lowell Putnam Mathematical Competition 1985-2000*, MAA Problem Books, The Mathematical Association of America, Washington, D.C., 2002.

[60] K. Knopp, Theory and Application of Infinite Series, 2nd edition, Dover, 1990.

[61] D. Knuth, Problem 11369, *Amer. Math. Monthly*, **115**(2008), 567.

[62] D. Knuth, Problem 11832, *Amer. Math. Monthly* **122**(2015), 390.

[63] L. C. Larson, *Problem-Solving through Problems*, Problem Books in Mathematics, Springer-Verlag, New York, 1983.

[64] J. Lagarias, An elementary problem equivalent to the Riemann Hypothesis, *Amer. Math. Monthly*, **109**(2002), 534–543.

[65] D. H. Lehmer, Interesting series involving the central binomial coefficient, *Amer. Math. Monthly*, **92**(1985), 449–457.

[66] I. E. Leonard, More on Fresnel integrals, *Amer. Math. Monthly*, **95**(1988), 431–433.

[67] T-Y Li and J. A. Yorke, Period three implies chaos, *Amer. Math. Monthly*, **82**(1975), 985–992.

[68] R. M. May, Simple mathematical models with very complicated dynamics, *Nature*, **261**(1976), 459–467.

[69] M. D. Meyerson, Every power series is a Taylor series, *Amer. Math. Monthly*, **88**(1981), 51–52.

[70] S. B. Nadler, A proof of Darboux's Theorem, *Amer. Math. Monthly*, **117**(2010), 174.

[71] I. Nemes, M. Petkovsek, H. S. Wilf, and D. Zeilberger, How to do Monthly problems with your computer, *Amer. Math. Monthly*, **104**(1997), 505–519.

[72] D. Newman, *A Problem Seminar*, Problem Books in Mathematics, Springer-Verlag, New York, 1982.

[73] I. Niven, A simple proof that π is irrational, *Bulletin of the American Mathematical Society*, **53**(1947), 509.

[74] L. Olsen, A new proof of Darboux's theorem, *Amer. Math. Monthly*, **111**(2004), 713–715.

[75] M. Petkovsek, H. Wilf, and D. Zeilberger, $A = B$, A. K. Peters, Wellesley, MA, 1996.

[76] I. Pinelis, On L'Hôpital-type rules for monotonicity, *J. Inequal. Pure Appl. Math.* **7** (2006), Article 40. Available at http://jipam.vu.edu.au/article.php?sid=657

[77] G. Pólya and G. Szegö, Problems and Theorems in Analysis (I), Springer-Verlag, New York, 1972.

[78] G. Pólya and G. Szegö, Problems and Theorems in Analysis (II), Springer-Verlag, New York, 1976.

[79] T. L. Rădulescu, V. D. Rădulescu and T. Andreescu, *Problems in Real Analysis:Advanced Calculus on the Real Axis*, Springer-Verlag, New York, 2009.

[80] K. A. Ross, *Elementary Analysis:The Theory of Calculus*, UTM, Springer-Verlag, New York, 1980.

[81] H. L. Royden, *Real Analysis*, 2nd Edition, McGraw Hill, New York, 1968.

[82] W. Rudin, *Principles of Mathematic al Analysis*, International Series in Pure and Applied Mathematics, McGraw Hill, New York, 1964.

[83] H. Samelson, On Rolle's theorem, *Amer. Math. Monthly*, **86**(1979), 486.

[84] A. N. Sharkovsky, Coexistence of cycles of a continuous map of a line into itself, *Ukr. Math. Z.*, **16**(1964), 61–71.

[85] G. Simmons, *Calculus Gems: Brief Lives and Memorable Mathematics*, McGraw-Hill, New York, 1992.

[86] N. J. A. Sloane, The On-Line Encyclopedia of Integer Sequences,
`http://www.research.att.com/\~\,njas/sequences/`

[87] R. Stanley, *Catalan Numbers*, Cambridge University Press, Cambridge, 2015.

[88] J. M. Steele, *The Cauchy-Schwarz Master Class*, Cambridge University Press, Cambridge, 2004.

[89] K. R. Stromberg, *Introduction to classical real analysis,* Wadsworth, Belmont, CA, 1981.

[90] H. Teismann, Toward a more complete list of completeness axioms, *Amer. Math. Monthly*, **120**(2013), 99–114.

[91] S. Wagon, *Mathematica in Action*, Springer-Verlag, New York, 1999.

[92] E. T. Whittaker and G. N. Watson, *A course of modern analysis*, 4th Edition, Cambridge University Press, Cambridge, 1963.

[93] R. Weinstock, Elementary evaluations of the Fresnel integrals, *Amer. Math. Monthly*, **97**(1990), 39–42.

[94] E. W. Weisstein, "Gamma Function." From MathWorld–A Wolfram Web Resource. Available at
`http://mathworld.wolfram.com/GammaFunction.html`

[95] H. Wilf, *Generatingfunctionology*, A. K. Peters, Ltd., 1994. Available at `http://www.math.upenn.edu/\~\,wilf/DownldGF.html`

[96] J. B. Wilker, Problem E3306, *Amer. Math. Monthly*, **96**(1989), 55.

[97] R. M. Young, *Excursions in Calculus*, Dolciani Mathematical Expositions, #13, The Mathematical Association of America, Washington, D. C., 1992.

[98] D. Zeilberger, The method of undetermined generalization and specialization, *Amer. Math. Monthly*, **103**(1996), 233–239.

Index

Printed in the United States
by Baker & Taylor Publisher Services